AF566651

Developments in Time Series Analysis

Maurice B. Priestley

Developments in Time Series Analysis

In honour of Maurice B. Priestley

Edited by

T. Subba Rao

Professor of Statistics
Department of Mathematics
University of Manchester Institute of Science and Technology
Manchester
UK

CHAPMAN & HALL

London · Glasgow · New York · Tokyo · Melbourne · Madras

Published by Chapman & Hall, 2–6 Boundary Row, London SE1 8HN

Chapman & Hall, 2–6 Boundary Row, London SE1 8HN, UK

Blackie Academic & Professional, Wester Cleddens Road, Bishopbriggs, Glasgow G64 2NZ, UK

Chapman & Hall Inc., 29 West 35th Street, New York NY10001, USA

Chapman & Hall Japan, Thomson Publishing Japan, Hirakawacho Nemoto Building, 6F, 1–7–11 Hirakawa-cho, Chiyoda-ku, Tokyo 102, Japan

Chapman & Hall Australia, Thomas Nelson Australia, 102 Dodds Street, South Melbourne, Victoria 3205, Australia

Chapman & Hall India, R. Seshadri, 32 Second Main Road, CIT East, Madras 600 035, India

First edition 1993

Typeset in Times by Thomson Press (India) Ltd, New Delhi
Printed in Great Britain by St Edmundsbury Press, Bury St Edmunds, Suffolk

ISBN 0 412 49260 1

A catalogue record for this book is available from the British Library

Library of Congress Cataloging-in-Publication data
Developments in time series analysis: in honour of Maurice B.
Priestley / edited by T. Subba Rao.—1st ed.
p. cm.
Includes bibliographical references and index.
ISBN 0-412-49260-1
1. Time-series analysis. I. Priestley, M. B. (Maurice Bertram)
II. Subba Rao, T.
QA280.D48 1993
519.5′5—dc20 93-3344
CIP

Printed on permanent acid-free text paper, manufactured in accordance with the proposed ANSI/NISO Z 39.48-199X and ANSI Z 39.48-1984

Contents

Contributors

C. Agiakloglou
Department of Economics
University of Illinois at Urbana-Champaign
Box 111, 330 Commerce West Building
1206 South Sixth Street
Champaign
IL 61820
USA

J. Allal
Départment de Mathématique
Université Mohammed Ier
Oujda
Morocco

I. Al-Wasel
Department of Mathematics
University of Lancaster
Lancaster
LA1 4YF
UK

R. Azrak
Institut de Statistique
Campus Plaine, CP 210
Université Libre de Bruxelles
B-1050 Bruxelles
Belgium

R.J. Bhansali
Department of Statistics and Computational Mathematics
University of Liverpool
PO BOX 147
Liverpool

L69 3BX
UK

D.R. Brillinger
Department of Statistics
University of California
Berkeley
CA 94720
USA

K.C. Chanda
Department of Mathematics
Texas Tech University
Box 41042
Lubbock
TX 79409–1042
USA

B. Cheng
Institute of Mathematics and Statistics
Cornwallis Building
The University of Canterbury
Canterbury
Kent
CT2 7NF
UK

P. Diggle
Department of Mathematics
University of Lancaster
Lancaster
LA1 4YF
UK

C. Fernandes
Fundacao Padre
Leonel Franca
Pontificia
Universidade Catolica—RJ
R. Marques de S Vicente
Rio de Janeiro
RJ 22453
Brazil

J. Franke
Mathematik
Universitat Kaiserslautern

Erwin Schrodinger Straße
Post Box 3049
6750 Kaiserslautern
Germany

M.M. Gabr
Department of Mathematics
Faculty of Science
University of Alexandria
Moharam Bek
Alexandria
Egypt

C.W.J. Granger
Department of Economics
University of California, San Diego
9500 Gilman Drive
La Jolla
CA 92093–0508
USA

M. Hallin
Institut de Statistique
Campus Plaine, CP 210
Université Libre de Bruxelles
B-1050 Bruxelles
Belgium

J.K. Hammond
ISVR
University of Southampton
Southampton
SO9 5NH

E.J. Hannan
Statistics Department
Faculty of Economics and Commerce
Australian National University
Canberra 0200
Australia

P.J. Harrison
Department of Statistics
University of Warwick
Coventry
West Midlands
CV4 7AL

R.F. Harrison
Department of Automatic Control and Systems Engineering
The University of Sheffield
Mappin Street
Sheffield
S1 4DU

A.C. Harvey
Department of Statistical and Mathematical Sciences
London School of Economics and Political Science
Houghton Street
London
WC2A 2AE

G. Janacek
School of Mathematics
University of East Anglia
Norwich
NR4 7TJ

M. Kramer
Department of Mathematics
University of California, San Diego
La Jolla
92093–0112
USA

J.S. Lee
Korea Research Institute of Ships and Ocean Engineering
PO Box 1
Daeduk
Daejeon 305–606
Korea

J. Math
KLTE
Department of Psychology
H-4010 Debrecen
Hungary

G. Mélard
Institute of Statistics
Campus Plaine, CP 210
Universite Libre de Bruxelles
B-1050 Bruxelles
Belgium

J. Miller
Department of Economics
University of Illinois at Urbana-Champaign
Box 111, 330 Commerce West Building
1206 South Sixth Street
Champaign
IL 61820
USA

P. Newbold
Department of Economics
University of Illinois at Urbana-Champaign
Box 111, 330 Commerce West Building
1206 South Sixth Street
Champaign
IL 61820
USA

J.K. Ord
Department of Management Science and Information Systems
The Mary Jean and Frank Smeal College of
Business Administration
Pennsylvania State University
303 Beam Business Administration Building
University Park
PA 16802
USA

T. Ozaki
Institute of Statistical Mathematics
4–6–7 Minami Azabu
Minato-ku
Tokyo
Japan 106

E. Parzen
Department of Statistics
Texas A & M University
College Station
Texas 77843–3143
USA

J. Pemberton
European Business Management School
University of Wales, Swansea
Singleton Park
Swansea

SA2 8PP
Wales

M. Rosenblatt
Department of Mathematics
University of California, San Diego
La Jolla
CA 92093–0112
USA

T. Seligmann
Mathematik
Universitat Kaiserslautern
Erwin Schrodinger Straße
Post Box 3049
6750 Kaiserslautern
Germany

H.J. Skaug
Department of Mathematics
University of Bergen
5007 Bergen
Norway

T. Subba Rao
Department of Mathematics
UMIST
PO Box 88
Manchester
M60 1QD

Gy Terdik
KLTE
Department of Computing, Pf 58
H-4010 Debrecen
Hungary

D. Tjøstheim
Department of Mathematics
University of Bergen
5007 Bergen
Norway

H. Tong
Institute of Mathematics and Statistics
Cornwallis Building
The University of Canterbury
Canterbury

Kent
CT2 7NF
UK

Y.H. Tsao
54 Charlton Road
Shirley
Southampton
SO1 5FN

Pham Dinh Tuan
IMAG, LMC
Laboratoire de Modelisation et Calcul
BP 53X
38041 Grenoble Cedex
France

G. Tunnicliffe Wilson
Department of Mathematics
University of Lancaster
Lancaster
LA1 4YF

P.P. Veerapen
School of Law, Management and Social Studies
University of Mauritius
Reduilt
Mauritius

A.M. Walker
Department of Probability and Statistics
University of Sheffield
Sheffield

M. Yar
3 The Dell
The Coppice
Aylesbury
Bucks
HP20 1YP

P.C. Young
Centre for Research on Environmental Systems
Institute of Environmental and Biological Sciences
Lancaster University
Lancaster
LA1 4YQ

Foreword

Professor Maurice Priestley's 60th birthday is a great memorable event for the whole international community of scientists and engineers dealing with time series analysis.

Maurice started working in the field of time series analysis more than 30 years ago when the main methods of spectral analysis of disordered fluctuations had already been developed, but this field of research was quite new and included many unsolved problems which were important and necessary for practical applications. A number of theoretical and applied problems in this field were brilliantly solved by Maurice whose many papers became classical in the field (the papers in the *Journal of the Royal Statistical Society. Ser. B*, **24**, 215–233, 1962, and *Technometrics*, **4**, 551–564, 1962 are only two examples). Moreover, most of the early papers by other authors dealt with the case of continuous spectral densities only, but Maurice studied carefully the more general and difficult case of time series with mixed (discontinuous) spectra. Later he developed quite original methods of spectral analysis for wide classes of non-stationary time series (*Journal of the Royal Statistical Society, Ser. B*, **27**, 204–237, 1965), and contributed significantly to the new field of nonlinear time series modeling (*Journal of Time Series Analysis*, **1**, 47–71, 1980). The famous books by Maurice, *Spectral Analysis and Time Series Analysis*, published by Academic Press in 1982 and 1988 are reprinted in 1991 have become classics, studied by thousands of readers throughout the world. The international *Journal of Time Series Analysis*, founded by Maurice and carefully edited by him from the first issue to the present time, has become the most popular journal in the field with a permanent wide circulation.

Maurice Priestley meets his 60th birthday at the peak of his creative activities. He is the head of the UMIST School and many of his former students are now working as professors at various universities, teaching time series analysis and contributing to its further development. I am sure that Maurice Priestley will produce many new results in the future, and his role as an acknowledged leader of the international community of time series researchers will become even more important. I am happy to wish him on his 60th birthday all success and good health, and luck at least for the next 40 years.

Professor A.M. Yaglom

Preface

It is very common in many countries and cultures to celebrate a 60th birthday and to use the occasion to recall the achievements and significant contributions which an individual has made. If this individual happens to be a 'time' series analyst to whom the number 'sixty' is very significant, the occasion will be all the more significant. The person concerned is Maurice Priestley, who will be 60 in March 1993, and whose many contributions in the field of time series analysis are significant and far reaching. To commemorate this happy occasion many of his colleagues and former students felt that it would be a fitting tribute to publish a book on time series analysis reflecting his main research interests and contributions. As a colleague of Maurice for over 25 years, I agreed to act as editor for this volume. Many distinguished time series analysts from all over the world were invited to participate, and it is very gratifying to see the overwhelming response which undoubtedly reflects the high esteem in which Maurice is held.

Maurice was born in Manchester and has spent most of his working life in this city, except for a short time when he was an undergraduate and postgraduate Diploma student at the University of Cambridge. After his studies in Cambridge he took his PhD at the University of Manchester in time series analysis under the guidance of Professor M.S. Bartlett, and was undoubtedly influenced by him as can be seen from his contributions in spectral analysis and mixed spectral analysis.

Maurice is well known in the department as a keen hi-fi enthusiast and amateur radio operator. Indeed, these engineering interests are evident in many of the examples which he gives in his books. However, it is difficult to say which one of these, time series or hi-fi, influenced the other!

His main contributions to time series can be classified into four categories: (a) analysis of mixed spectra; (b) estimation of spectral density functions, with particular emphasis on the optimal choice of bandwidths and kernels; (c) spectral analysis of non-stationary processes, filtering and prediction of non-stationary processes; (d) non-linear time series analysis, and state dependent models. His two-volume book *Spectral Analysis and Time Series* and his subsequent book *Non-linear and Non-stationary Time Series Analysis* (1988), are internationally acclaimed as standard reference sources, and widely cited in the literature. Since 1980 he has acted as the editor-in-chief of the *Journal of Time Series Analysis*. In addition to these research activities Maurice has served as Chairman of the Department of Mathematics, UMIST, over several periods totalling 12 years, and has also served as Director of the Manchester-Sheffield School of Probability and Statistics over a number of sessions.

Since the days of Professor Bartlett, Machester has always been known as a centre for time series and statistics; Maurice, through his enthusiasm and dedication, continued this tradition and has made it an internationally known centre for time series research.

I hope that this book, containing papers from a wide range of time series analysts, is one way of expressing our appreciation of his contributions. May I, on behalf of all the contributors and the community of time series analysts, wish him a very happy 60th birthday, and many happy returns!

T Subba Rao
University of Manchester
Institute of Science and Technology

Books and papers by M.B. Priestley

BOOKS

Spectral Analysis and Time Series, Vol. I, Academic Press, 1981.
Spectral Analysis and Time Series, Vol. II, Academic Press, 1981.
Essays in Time Series and Allied Processes, Gani and Priestley (eds), Appl. Prob. Trust, 1986.
Non-Linear and Non-Stationary Time Series Analysis, Academic Press, 1988.

PAPERS

Analysis of stationary time series I – Computation of correlaton coefficients on a high speed computer (with P.A. Samet), *RAE Tech. Note, M.S. 27, 1956*, Royal Aircraft Establishment. Farnborough.
Analysis of stationary time series II – Estimation of power spectra, *R.A.E. Tech. Note, M.S. 29, 1956*, Royal Aircraft Establishment, Farnborough.
The spectral analysis of time series (with G.M. Jenkins), *J.R. Statist. Soc. (B)*, **19** (1957) 1–12.
Statistical analysis of stationary time series, *Paper Technology*, **2** (2), (1959).
The analysis of stationary processes with mixed spectra – I, *J.R. Statist. Soc. (B)*, **24** (1962) 215–233.
The analysis of stationary processes with mixed spectra – II, *J.R. Statist. Soc. (B)*, **24** (1962) 511–529.
Basic considerations in the estimation of power spectra, *Technometrics*, **4** (1962) 551–564.
The spectrum of a continuous process derived from a discrete process, *Biometrika*, **50** (1963) 517–520.
Analysis of two-dimensional processes with discontinuous spectra, *Biometrika*, **51** (1964) 195–217.
Estimation of the spectral density function in the presence of harmonic components, *J.R. Statist. Soc. (B)*, **26** (1964) 123–132.
Estimation of power spectra by a wave-analyser (with C.H. Gibson), *Technometrics*, **7** (1965) 553–559.
Evolutionary spectra and non-stationary processes, *J.R. Statist. Soc. (B)*, **27** (1965) 204–237.
The role of bandwidth in spectral analysis, *Applied Statistics*, **14** (1965) 33–47.
Design relations for non-stationary processes, *J.R. Statist. Soc. (B)*, **28** (1966) 228–240.
Power spectral analysis of non-stationary random processes, *J. Sound Vibrations*, **6** (1967) 86–97.

On the prediction of non-stationary processes (with N.A. Abdrabbo), *J.R. Statist. Soc. (B)*, **29** (1967) 570–585.

Filtering non-stationary signals (with N.A. Abdrabbo), *J.R. Statist. Soc. (B)*, **31**, 150–159.

A test for non-stationarity of time series (with T. Subba Rao), *J.R. Statist. Soc. (B)*, **31**, 140–149.

Estimation of transfer functions in closed-loop stochastic systems, *Automatica*, **5** (1969) 623–632.

Control systems with time dependent parameters, *Bull. Int. Stat. Inst.*, **37** (1969).

Time-dependent spectral analysis and its applications in prediction and control, *J. Sound Vibration*, **17** (1971) 139–156.

On the physical interpretation of spectra of non-stationary processes, *J. Sound Vibration*, **17** (1971) 51–54.

Fitting relationships between time series, *Bull. Int. Stat. Inst.*, **38** (1971) 1–27 (invited paper at 38th Session of the I.S.I., Washington).

Non-parametric function fitting (with M.T. Chao), *J.R. Statist. Soc. (B)*, **34** (1972) 385–392.

Identification of the structure of multivariable stochastic systems (with T. Subba Rao and H. Tong) in *Multivariate Analysis – III* (ed. P.R. Krishnaiah), Academic Press, 1973, 351–368.

Tests of significance for the eigenvalues of a spectral matrix (with T. Subba Rao and H. Tong), Technical Report No. 32, Department of Mathematics (Statistics), University of Manchester Institute of Science and Technology, June 1972.

Asymptotic distribution of eigenvalues of a sample spectral density matrix (with T. Subba Rao and H. Tong), Technical Report No. 34, Department of Mathematics (Statistics), University of Manchester Institute of Science and Technology, June 1972.

On the analysis of bivariate non-stationary processes (with H. Tong), *J.R. Statist. Soc. (B)*, **35** (1973) 179–188.

Applications of principal component analysis and factor analysis in the identification of multivariable systems, *I.E.E.E. Trans. Automatic Control*, **AC-19** (1974) 730–734.

The estimation of factor scores and Kalman filtering for discrete parameter stationary processes (with T. Subba Rao), *Int. J. Control*, **21** (1975) 971–975.

Dimensionality reductions in multivariable stochastic systems (with V. Haggan), *Int. J. Control*, **22** (1975) 763–772.

Applications of multivariate techniques in the study of multivariable stochastic systems, *Adv. in App. Prob.*, **9** (1977) 202–205 (invited paper at the 6th International Conference on Stochastic Processes, Tel Aviv, June 1976).

A global view of time series analysis, Proc. of Conf. on Applications of Time Series Analysis, University of Southampton, September 1977.

Non-linear models in time series analysis, *The Statistician*; **27** (1978) 159–176 (invited paper, Institute of Statisticians Conference on Time Series Analysis, Cambridge, England, July 1978).

A general class of non-linear time series models, *Bull. Int. Stat. Inst.*, **42** (1979).

System identification, Kalman filtering and stochastic control (invited paper, I.M.S. Special Topics meeting on Time Series Analysis, Ames, Iowa, USA, May 1978), in *Directions in Time Series* (eds. D.R. Brillinger and G.C. Tiao), I.M.S. publication pp. 188–218, 1980.

Prediction based on a general class of non-linear models, in *Colloque Professus Aléatoires at Problèmes de Prévision*, Cahiers du Centre d'Etudes de Recherche Opérationelle (Institut des Hautes Etudes de Belgique), **22** (1980) 285–308.

A study of AR and window spectral estimation (with N. Beamish), *J.R. Statist. Soc. (C)*, **30** (1981) 41–58.

State-dependent models: a general approach to non-linear time series analysis, *J. Time Series Anal.*, **1** (1980) 47–71.

Frequency domain analysis and closed-loop systems, in *Handbook of Statistics*, Vol. 3 (ed. P.R. Krishnaiah) North-Holland; Amsterdam, 1981.

On the fitting of general non-linear time series models. Proc. 4th International Time Series Meeting (ed. O.D. Anderson), North-Holland; Amsterdam, 1982, pp. 717–731.

A study of the application of state-dependent models in non-linear time series analysis (with V. Haggan and S.M. Heravi) *J. Time Series Anal.*, **5** (1984) 69–102.

The development of time series analysis, Proc. 150th Anniversary Conference of Manchester Statistical Society (July 1984).

State-dependent models in time series and non-linear systems modelling. Proc. Intl. Time Series Symposium, Beijing, China, Nov. 1984 (Ed. Chen Zho Guo and An Hong Zhi).

The development and construction of general non-linear models in time series analysis. *Annals New York Acad. Sci.*, **452** (1985) 296–304.

Identification of non-linear systems using general state-dependent models. *J. Appl. Prob.*, **23A** (1986) 257–272.

Priestley's test for harmonic components in *Encyclopedia of Statistical Sciences*, Vol. 7, (ed. Kotz and Johnson), Wiley, 1986, pp. 176–180.

New developments in time series analysis, in *New Perspectives in Theoretical and Applied Statistics* (Ed. M.L. Puri, J.P. Vilaplana, and W. Wertz) Wiley, New York, 1987, pp. 119–131.

Current developments in time series modelling. *J. Econometrics*, **37** (1988) 67–86.

Basic theory of time series modelling: structure and identification of non-linear models. Proc. 12th Nordic Conf. on Mathematical Statistics, Turku, Finland (1989).

Non-linear modelling, in *Recent Developments in Statistical Data Analysis and Inference* (ed. Y. dodge), 1989.

Identification of non-linear systems. *Bull. Int. Stat. Inst.*, **47** (1990).

Interaction between neural networks and non-linear time series analysis. Proc. Conf. on Neural Networks (ed. F. Murtagh), DOSES, Luxembourg, 1991.

Bispectral analysis of non-stationary processes (with M.M. Gabr), in *Multivariate Analysis: Future Directions* (ed. C.R. Rao), North-Holland, Amsterdam, 1992/93.

Autoregressive model fitting and windows, in *Frontiers of Statistical Modelling* (ed. H. Bozdogan), Kluwer Academic Publishers, Holland, 1992/93.

About the volume and summary of papers

The 27 papers in this volume which are exclusively in time series analysis, deal with statistical theory, methodology and applications. The emphasis is on the recent developments in the analysis of linear and nonlinear (non-Gaussian) stationary and non-stationary time series. The papers are divided into six sections, within each section the topics covered are of a similar nature.

Part one contains five papers by Granger, Newbold, Agiakloglou and Miller; Tunnicliffe Wilson, Harrison and Veerapen; and Bhansali. Part Two contains the papers of Kramer and Rosenblatt; Pham; Chanda; and Azrak and Melard. The papers of Hannan; Parzen; Walker; and Janacek appear in Part Three. Papers by Cheng and Tong; Skaug and Tjostheim; Pemberton; and Allal and Hallin are in Part Four. Part Five contains the papers of Ozaki; Terdik and Máth; and Gabr. The final Part contains the papers of Ord, Fernandes and Harvey; Franke and Seligmann; Brillinger; Diggle and Al-Wasel; Hammond, Harrison, Tsao and Lee; Young; and Subba Rao and Yar. I summarize each paper as follows.

PART ONE. LINEAR TIME SERIES MODELS

Granger discusses the relation between positively related processes and cointegration. Newbold et al. discuss the limitations of using the parsimoniously parameterized short-term time series model for long-term forecasting. Tunnicliffe Wilson gives a method of obtaining the autocovariance function which is needed to calculate the likelihood function, and also provides an algorithm for covariance factorization. Harrison and Veerapen discuss sequential dynamic modelling where information on some variables may need either deletion or alteration. Bhansali provides a review of various methods of order determination of linear time series models.

PART TWO. ESTIMATION AND ASYMPTOTICS FOR TIME SERIES MODELS

Kramer and Rosenblatt study the asymptotic behaviour of the Gaussian likelihood function, when the process is non-Gaussian but satisfies a linear

ARMA model. Pham gives asymptotic expansions for the bias and covariance matrix of estimates of AR parameters. The asymptotic properties of autocovariances of various orders which increase with sample size are studied by Chanda. Azrak and Mélard consider the exact likelihood estimation for the extended ARIMA model.

PART THREE. SPECTRAL ANALYSIS OF STATIONARY TIME SERIES

Hannan considers the problem of determining the number of sinusoids in a regression in the presence of a stationary noise. Parzen examines the roles of information ideas and spectral analysis in time series. Asymptotic properties of periodogram for complex-valued time series were considered by Walker. Janacek considers the spectral approach to the identification and estimation of long-memory models.

PART FOUR. NONPARAMETRIC STATISTICAL INFERENCE IN TIME SERIES

The efficiency of kernel estimates of an unknown function (usually a conditional mean) possibly generated by a chaotic map is considered by Cheng and Tong. Skaug and Tjøstheim have computed nonparametric tests for serial independence and have studied their asymptotic properties. Pemberton examines the definition of nonlinearity based on martingale differences for the one-step linear prediction errors. Allal and Hallin obtain asymptotic normality and strong consistency of a class of signed-rank statistics (this class includes the signed-rank autocorrelation coefficients).

PART FIVE. NONLINEAR AND NON-GAUSSIAN TIME SERIES MODELS

Ozaki studies the properties of exponential AR processes. Terdik and Máth examines the conditions for efficiency of linear predictors when the process is non-Gaussian. Gabr considers the maximum likelihood estimation of bilinear models where some observations are missing.

PART SIX. TIME AND FREQUENCY ANALYSIS OF TIME SERIES–APPLICATIONS

Ord, Fernandes and Harvey derive multivariate models for multivariate time series of count data (non-negative integers) and also consider the modelling of goals scored by the English football team against the Scottish team at Hampden Park, Glasgow, Scotland. Franke and Seligmann consider the estimation of INAR(1) processes, and fit these models to epileptic seizure counts. Brillinger gives an application of Fourier analysis to seismic waves, and considers the estimation of parameters of earth models. Diggle and Al-Wasel consider spectral analysis of replicated time series with an application to data on hormone concentration in several blood samples. Hammond, Harrison, Tsao and Lee provide a review of recent developments in time and frequency analysis of time series and illustrate the methods of analysis with examples. The examples include a study of the vehicle motion over a rough terrain; acoustic signals perceived by a fixed observer. Young provides a unified approach to non-stationary and nonlinear time series based on time variable and state dependent estimation. He illustrates the methods with data simulated from Lorenz's 'strange attractor' model, and rainfall-flow data. Subba Rao and Yar consider the problem of demodulation of phase modulated signals using $P(\lambda)$ statistic.

All the papers in this volume have been refereed, and most of the refereeing has been carried out by the contributors. I must thank all the contributors, not only for agreeing to contribute a paper to this volume, but also for willingly accepting the responsibility of refereeing (in some cases several times) papers and making every effort to improve the presentation. Several other statisticians have kindly refereed papers for the volume, and I thank them all. They are Professors B Silverman, University of Bath; Richard Baille, Michigan State University East Lansing; Stefan Mittnik, State University of New York, Stony Brook; Lanh Tat Tran, Indiana University, Bloomington; Ed McKenzie, University of Strathclyde; and Xiobar Wang, University of California, Berkely.

I wish to thank Ms Nicki Dennis of Chapman & Hall for showing enormous interest in this project and to Mrs Sandra Kershaw who has cheerfully accepted the responsibility of typing some of the manuscripts and doing all the necessary paper work from the beginning to the end; without her help I could not have edited this volume.

T. Subba Rao

Part One

Linear Time Series Models

1

Positively related processes and cointegration

C.W.J. Granger

1.1 BASIC DEFINITIONS

An $I(0)$ series may be defined as one having a spectrum that is bounded above at all frequencies and away from zero at zero frequency. An $I(1)$ series is an accumulation of an $I(0)$ series starting in the finite past, so that

$$Y_t = \sum_{j=0}^{t} q_{t-j}$$

where q_t is $I(0)$. A pair of $I(1)$ series are said to be cointegrated if there is a linear combination of them that is $I(0)$. A necessary and sufficient condition for cointegration is that the two series have the representation

$$X_t = AW_t + \tilde{X}_t$$

$$Y_t = W_t + \tilde{Y}_t$$

where $\tilde{X}_t$, $\tilde{Y}_t$ are each $I(0)$ and W_t is a random walk.

In what follows it will be assumed that an $I(0)$ series q_t has zero mean and finite variance so that its optimum least-square forecast with horizon h,

$$f^q_{t,h} = E[q_{t+h} | q_{t-j}, \quad j \geqslant 0],$$

will tend to zero as h becomes large. If an $I(1)$ series Y_t is the accumulation of a zero mean $I(0)$ series, it will contain no deterministic drift but its long-term forecast $f^Y_{t,h}$ will not tend to zero. In particular if Y_t is a random walk, so that $\Delta Y_t = q_t$ is white noise, $f^Y_{t,h} = Y_t$ for all h. Because of these properties of forecasts, an $I(0)$ series can be considered 'short-memory' and an $I(1)$ series 'long-memory'. It will generally be assumed below that linear optimal forecasts are optimal. Series without spectra or having time-varying spectra as discussed in Priestley (1981), may usually be classified as short- or long-memory (in mean). A series may be short-memory in mean but have a strong heteroscedastic residual and thus also be long-memory in variance.

Discussions of cointegration and its interpretation can be found in Granger (1986) and Engle and Granger (1990). It has been found to be a useful concept for considering long-run relationships in macroeconomics, finance and elsewhere, and it may be equated with a type of equilibrium.

1.2 ASSOCIATED SERIES

A vector random variable $\boldsymbol{T}$ is said to have (positively) associated components if

$$\operatorname{cov}[f(\boldsymbol{T}); \quad g(\boldsymbol{T})] \geqslant 0 \tag{1.1}$$

for all functions f and g that are non-decreasing in each argument, and for which $Ef(\boldsymbol{T})$, $Eg(\boldsymbol{T})$ and $Ef(\boldsymbol{T})g(\boldsymbol{T})$ exist. The definition was suggested by Esary, Proschan and Walkup (1967) in connection with reliability theory. They obtained many of the important results in the area, which have been summarized by Boland and Proschan (1988).

The definition of association given here is a generalization of a sequence of alternative definitions of positive relatedness for a pair of random variables T_1, T_2

$$\operatorname{cov}(T_1, T_2) \geqslant 0$$

$$\operatorname{cov}[f(T_1), \quad g(T_2)] \geqslant 0$$

for all pairs of non-decreasing functions f, g, and

$$\operatorname{cov}[f(T_1, T_2), \quad g(T_1, T_2)] \geqslant 0$$

for all pairs of functions f, g which are non-decreasing in each argument. These definitions are in a natural order of increasing strength and generality.

A potentially important result is that if $\boldsymbol{T}$ is associated and $S_i = f_i(\boldsymbol{T})$ and if f_i are non-decreasing functions $i = 1, \ldots, k$ then

$$\operatorname{Prob}[S_i \leqslant t_i, \quad i = 1, \ldots, k] \geqslant \prod_{i=1}^{k} \operatorname{Prob}(S_i \leqslant t_i) \tag{1.2}$$

$$\operatorname{Prob}[S_i > t_i, \quad i = 1, \ldots, k] \geqslant \prod_{i=1}^{k} \operatorname{Prob}(S_i < t_i) \tag{1.3}$$

for all $t_1, \ldots, t_k$.

It was shown by Pitt (1982) that if $\boldsymbol{T}' = (T_1, \ldots, T_n)$ has a multivariate normal distribution, then $\boldsymbol{T}$ is associated provided that

$$\operatorname{cov}(T_i, T_j) \geqslant 0 \text{ for all } i, j$$

In this case, and if $S_i = f_i(T_i)$, the probabilities on the right hand side of (1.2) and (1.3) can be easily evaluated.

It may be noted that if a series X_t is a Gaussian $I(1)$ series which started long enough ago, then $X'_{t,k} = (X_t, \ldots, X_{t-k})$ will be associated. Suppose that X_t is generated by

$$\Delta X_t = C(B)\varepsilon_t$$
$$= C(1) + C^*(B)\Delta\varepsilon_t$$

so that

$$X_t = C(1)\frac{\varepsilon_t}{\Delta} + C^*(B)\varepsilon_t$$

Thus X_t has two components, one a random walk and the other $I(0)$. Then

$$\operatorname{cov}(X_t, X_s) = \operatorname{var} X_s + \operatorname{cov}[X_s, C^*(B)\varepsilon_t]$$

if $s < t$.

Provided that var ε is constant, the first term is positive and proportional to s and will necessarily dominate in size the second term, which is bounded, if s is large enough, so that $\operatorname{cov}(X_t, X_s) \geqslant 0$, $s < t$, s large. With the assumption that the process is Gaussian, it follows that

$$\boldsymbol{X}'_{t,k} \equiv (X_t, \ldots, X_{t-k})$$

is associated, if k is large.

It also follows, using similar reasoning, that if X_t, Y_t are a pair of $I(1)$ cointegrated series, with $Z_t = X_t - \alpha Y_t$, that is $I(0)$ and if $\alpha > 0$, the stacked vector

$$\begin{pmatrix} \boldsymbol{X}_{t,k} \\ \boldsymbol{Y}_{t,k} \end{pmatrix}$$

is associated. It follows, for example, that

$$\operatorname{cov}[f(X_t), \quad g(Y_t)] \geqslant 0$$

for all non-decreasing functions f, g for which this covariance exists. Further, the expressions (1.2) and (2.3) can be used to obtain bounds on some probabilities.

1.3 POSITIVE AUTOREGRESSIVE DEPENDENT SERIES

Esary, Pruschan and Walkup (1967) defined for a pair of random variables T, S a 'positive regression dependence' of T on S, if $\text{Prob}(T > t | S = s)$ is non-decreasing in s. They prove that in this case the pair T and S are associated. In the spirit of this definition, a vector $\boldsymbol{X}_t$ generated by the Markov scheme

$$\boldsymbol{X}_t = \boldsymbol{P}\boldsymbol{X}_{t-1} + \boldsymbol{\varepsilon}_t \tag{1.4}$$

will be called a 'positive autoregressive dependent series' if $\boldsymbol{P} > 0$, so that all elements of $\boldsymbol{P}$ are positive. Here $\boldsymbol{\varepsilon}_t$ is a zero mean white noise vector, so that $\operatorname{cov}(\varepsilon_{it}, \varepsilon_{jt}) = 0$, $s \neq t$, all i,j, and it is assumed that the distribution of $\boldsymbol{\varepsilon}_t$ does not depend on $\boldsymbol{X}_{t-j}$, $j \geqslant 0$.

The long-run behaviour of the system (1.4) is determined by the size of the largest eigenvalue (λ_1) of P. If $|\lambda_1| < 1$ the system is stable with each component of $\boldsymbol{X}$ being $I(0)$, if $\lambda_1 = 1$ the system contains a unit root and some components of $\boldsymbol{X}$ are $I(1)$ and if $|\lambda_1| > 1$ some components are exponentially explosive. These results follow directly by noting that the optimum h-step forecast vector of $\boldsymbol{X}_{t+h}$ is given by

$$f^x_{t,h} = \boldsymbol{P}^h \boldsymbol{X}_t \tag{1.5}$$

and that the maximum eigenvalue of $\boldsymbol{P}^h$ is λ_1^h.

A particularly useful set of results for considering the system (1.4) is the Perron–Frobenius theorem which proves that if $\boldsymbol{P}$ is an $n \times n$ positive matrix there exists an eigenvalue r with the properties:

(a) r is a simple root of the equation $\det[\boldsymbol{P} - r\boldsymbol{I}] = 0$, is real and positive;
(b) it is the maximum eigenvalue, so that

$$r > |\lambda|$$

where λ is any other eigenvalue of $\boldsymbol{P}$;
(c) if $\boldsymbol{v}'$ $\boldsymbol{w}$ are the left and right eigenvectors associated with r, so that

$$\boldsymbol{v}'\boldsymbol{P} = r\boldsymbol{v}' \quad \text{and} \quad \boldsymbol{P}\boldsymbol{w} = r\boldsymbol{w}, \tag{1.6}$$

then these vectors have components all the same sign, and so can be taken to be positive. Note $b\boldsymbol{v}'$ is a left eigenvalue for any b.

r may be called the Perron–Frobenius (or P–F) eigenvalue of $\boldsymbol{P}$ and denoted r_P. A good account of this theory is provided by Seneta (1973). It follows from the theorem that the P–F eigenvalue of $\boldsymbol{P}^k$ is $(r_P)^k$. An important consequence of the theorem is that

$$\boldsymbol{P}^h = r^h \boldsymbol{w}\boldsymbol{v}' + \boldsymbol{P}_2 \tag{1.7}$$

where $\boldsymbol{v}'\boldsymbol{w} = 1$, which is just a normalization,

$$\boldsymbol{v}'\boldsymbol{P}_2 = \boldsymbol{P}_2\boldsymbol{w} = 0$$

$$|\boldsymbol{P}_2| = 0(\alpha^h)$$

for some $0 < \alpha < r$, as given by Harris (1963).

Thus $\boldsymbol{P}^h/r^h \to \boldsymbol{w}\boldsymbol{v}'$ and so is of reduced rank. Applying this result to (1.5) gives

$$f^x_{t,h}/r^h \to \boldsymbol{w}\, W_t \tag{1.8}$$

where $W = \boldsymbol{v}'\boldsymbol{X}_t$ is a single linear combination of the components of $\boldsymbol{X}_t$, having positive coefficients. Note also that $\boldsymbol{w}$ has all positive coefficients in (1.8).

If $r = 1$, $\boldsymbol{W}_t$ must be $I(1)$ as it is seen that the forecasts in (1.8) are long-memory. Multiplying (3.5) by $\boldsymbol{v}'$ gives

$$f^{w}_{t,h} = W_t$$

from (1.7) and with $r = 1$, so that W_t is a random walk. It follows that the system (1.4) with $\boldsymbol{P} > 0$ and $r_P = 1$ has a single $I(1)$ common factor and that all pairs of components of $\boldsymbol{X}_t$ are cointegrated. Thus, a strong conclusion is reached from a rather limited set of assumptions. The system (1.4) may be thought of as a rather simple one but the $r = 1$ results continue to hold for some more general systems:

(a) if $\boldsymbol{X}_t$ is generated by (1.4) but with $\boldsymbol{P}$ a 'primitive matrix' so that $\boldsymbol{P}^k > 0$ for some positive integer k, and the P–F eigenvalue of $\boldsymbol{P}^k$ is one;

(b)
$$\boldsymbol{X}_t = \boldsymbol{P}(B)X_{t-1} + \varepsilon_t$$

where

$$\boldsymbol{P}(B) = \boldsymbol{P}(1) + \boldsymbol{P}^*(B)\Delta$$

and with $\boldsymbol{P}(1)$ a positive matrix with P–F eigenvalue $r = 1$, and;

(c)
$$\boldsymbol{X}_t = \sum_j b_j \boldsymbol{P}^j \boldsymbol{\varepsilon}_{t-j}$$

where $\boldsymbol{P} > 0$, $r_P = 1$ and all $b_j < 0$.

The size of r_P is clearly important. Bounds on its size can be easily found using the inequalities

$$\min_i R_i \leqslant r \leqslant \max_i R_i$$

where $R_i =$ the sum of the components of the ith row of P. The equality signs only occur if $R_i = R$ for all i, and then $r = R$.

The results can be extended to cases where $r \neq 1$. Define a univariate exponential $E(\alpha)$ process to be generated by

$$Y_t = \alpha Y_{t-1} + \varepsilon_t$$

where ε_t is white noise and $\alpha > 1$. It follows directly from the previous results that if $r > 1$, then $\boldsymbol{X}_t = wW_t + E(\alpha)$ processes, where W_t is $E(r)$ and $\alpha < r$. Thus every component of X_t contains an $E(r)$ component but some linear combinations of every pair of components of $\boldsymbol{X}_t$ contains at most $E(\alpha)$, $\alpha < r$ components. This may be expressed in terms of the process $\boldsymbol{Y}_t = \boldsymbol{X}_t/r^t$.

$\boldsymbol{Y}_t$ will be $I(1)$ but all pairs of components cointegrated to $I(0)$. However, if var $\boldsymbol{\varepsilon}_t$ is constant, $\boldsymbol{Y}_t$ will effectively be a deterministic series $\boldsymbol{Y}_0 r^t$, where $\boldsymbol{Y}_0$ is a constant vector.

Somewhat similar results can be obtained with $r < 1$, but seem to have a less clear interpretation. An $I(0)\boldsymbol{X}_t$ process can be converted into the $I(1)$ process $\boldsymbol{Y}_t$ where, as before, $\boldsymbol{Y}_t = \boldsymbol{X}_t/r^t$. Thus an $I(0)$ process is multiplied by an exponential trend, but severe heteroscedasticity in the resulting model makes interpretation difficult.

1.4 PAD PROCESSES AND ASSOCIATION

If $\boldsymbol{X}_t$ is generated by (1.4) with $\boldsymbol{\varepsilon}_t$ an i.i.d. zero mean Gaussian vector, then all pairs of components are cointegrated, it follows from the results of the section on cointegrated Gaussian series that $\boldsymbol{X}_t$ will be associated. Further, recalling if T, S are a pair of random variables such that Prob $(T > t | S = s)$ is non decreasing in S then T, S are associated, it follows directly from the form of (1.4) that:

(a) X_{jt}, $X_{k,t-1}$ are associated for every j, k since any increase in $X_{k,t-1}$ implies an increase in the conditional mean of X_{jt},
and;
(b) X_{jt}, X_{kt} are associated for every j, k if the matrix $\text{cov}(\boldsymbol{\varepsilon}_t)$ is positive, as every increase in X_{kt} will imply an increase in the conditional mean of X_{jt}.

These results hold whether or not $\boldsymbol{\varepsilon}_t$ is Gaussian. Undoubtably more general results are available.

1.5 CONCLUSIONS

A Markov process generated by (1.4) with $\boldsymbol{P} > 0$ may well be considered to be a reasonable model in many situations, such as when modelling a group of interest rates or exchange rates or highly related commodity prices or stock prices. A Bayesian may want to impose priors on $\boldsymbol{P}$ making it positive. The results found here indicate that in the long run a very simple structure arises from such a model and, if the dominant eigenvalue of $\boldsymbol{P}$ is $\geqslant 1$, very special forms of cointegration arise. The fact that the processes are also associated allows further of their properties to be derived.

REFERENCES

Boland, P.J. and Proschan, F. (1988) The impact of reliability theory on some branches of mathematics and statistics, in *Handbook of Statistics*, (eds P.R. Krishnaiah and C.R. Rao), Elsevier Science Publishers, pp. 159–174.

Engle, R.F. and Granger, C.W.J. (1990) *Long-Run Relationships Between Economic Variables*: Readings in Cointegration, Oxford University Press.

Esary, J.D., Proschan, F. and Walkup, D.W. (1967) Association of random variables, with applications. *Ann. Math. Stats.* **38** 1466–74.

Granger, C.W.J. (1986) Developments in the study of cointegrated economic variables. *Oxford Bulletin of Economics and Statistics*, **8**, 213–28.

Harris, T.E. (1963) *The Mathematical Theory of Branching Processes*, Springer-Verlag, Berlin.

Pitt, L.D. (1982) Positively correlated normal random variables are associated. *Annals of Probability*, **10**, 496–99.

Priestley, M.S. (1981) *Spectral Analysis and Time Series*, Academic Press, New York.

Seneta, E. (1973) *Non-negative Matrices*, George Allen and Unwin, London.

2

Long-term inference based on short-term forecasting models

P. Newbold, C. Agiakloglou and J. Miller

2.1 INTRODUCTION

One approach to short-term forecasting is to begin by contemplating a broad class of possible time series generating models, such as ARIMA models (Box and Jenkins, 1970) or structural models (Harvey, 1989). Based on the available data, a specific model is selected from the general class and its unknown parameters estimated. The fitted model is then extrapolated to generate forecasts. Given modern computing power, such an approach is relatively straightforward, as many alternative models can be estimated quickly, and a choice based on an order selection criterion, such as AIC or SBC. Although elaborations, allowing for nonlinearity or non-stationarity for example, are possible, an approach based on a sufficiently rich class of linear models will often yield adequate short-term forecasts. When there is uncertainty as to what might be the appropriate generating model, it typically emerges that competing models yield very similar forecasts, so that conclusions based on competent analyses should not differ to any great extent.

In the last few years, economists have felt it important to ask questions about longer-run properties of time series. In effect, these amount to questions about the behavior of the spectrum at zero and very low frequencies. Answers to these questions have often been developed from a short-term forecasting methodology, such as that discussed above. We will discuss the pitfalls inherent in this approach. To illustrate, consider the ARIMA class of models

$$\Phi(B)X_t = \theta(B)\varepsilon_t \tag{2.1}$$

where Φ and θ are polynomials in the back-shift operator B, and ε_t is white noise. If all the roots of $\Phi(z) = 0$ have modulus greater than one, the process is stationary, while if d of these roots are equal to one, the process is said to be integrated of order d, leading to the ARIMA class

$$\phi(B)(1-B)^d X_t = \theta(B)\varepsilon_t. \tag{2.2}$$

Let p denote the number of parameters in the autoregressive operator ϕ, and q the number of parameters in the moving average operator θ, so that (2.2) is designated an ARIMA (p, d, q) model. When interest is in short-term forecasting, the principle of parsimony is a useful element in model selection. Essentially, we seek the model with the fewest number of parameters that adequately describes the major features of the data. In this way, sampling errors in the estimators of redundant or near-redundant parameters cannot have a negative impact on forecast quality. Ledolter and Abraham (1981) elaborate on this principle.

One does not have to take the unrealistic point of view that some simple model represents the true generating process, rather than a convenient approximation to it, for the principle of parsimony to be relevant in short-term forecasting. However, the position is far more problematic if interest is in longer-run properties of the data. As we will see in later sections, the models (2.1) and (2.2) have a peculiar feature that renders them an uncertain basis for longer-run inference. Suppose that a particular model from the class (2.2) is deemed reasonable as a generating model. Then, it is always possible to add both an extra autoregressive and extra moving average term by multiplying through (2.2) by $(1-\alpha B)$. If the original model were 'true', the extra parameters would then, of course, be redundant. However, on the basis of data evidence, such redundancy can never be established with certainty. When the more elaborate model is estimated, it may have parameter estimates suggesting near-redundancy, that is, near-cancellation in the estimated autoregressive and moving average operators. This presents no great difficulty when the goal is short-term forecasting. The principle of parsimony, which is reflected in AIC, SBC, and other model selection criteria, suggests a preference for the simpler model.

The simple and more elaborate models may yield quite different conclusions about long-run behavior. In this context, once one abandons the fiction that time series are *a priori* likely to be generated by simple models, it is difficult to find any rationale for discarding more elaborate formulations. There may therefore be the possibility that an inference based on parsimonious models can be quite unreliable. This difficulty is compounded by an unfortunate property of parameter estimators. It can frequently happen that, when $\theta(B)$ of (2.2) has a large positive root, maximum likelihood estimation leads to an estimated root of exactly or very nearly one. This 'pile-up' effect is discussed by Ansley and Newbold (1980), Cryer and Ledolter (1981), and Shephard and Harvey (1990). As can be seen from the structure of (2.1) and (2.2) this necessarily leads to uncertainty about the correct amount of differencing, d. This concern is briefly noted by Clark (1989).

2.2 TESTING FOR UNIT AUTOREGRESSIVE ROOTS

In the last few years, a huge volume of theoretical and empirical work has appeared in the econometrics literature on the issue of unit autoregressive roots, that is, a unit root in the operator Φ of (2.1). The presence of such a root has consequences for the asymptotic distributions of estimators and test statistics, and possibly in strategies for modelling relationships among time series. (See, for example, Engle and Granger (1987), Park and Phillips (1988, 1989), Sims, Stock and Watson (1990), and Johansen (1991).)

The simplest practical case is of a single time series, where in (2.2) it is assumed that d is either zero or one. In the econometrics literature, this has most often been viewed as a hypothesis testing question, with $d = 1$ taken to be the null hypothesis. Naturally, as in all such settings, for moderate-sized samples, the null hypothesis will not be rejected very frequently when the true alternative is close to the null. For example, consider the competing models

$$(1 - B)X_t = \varepsilon_t \tag{2.3}$$

and

$$(1 - \phi B)(X_t - \mu) = \varepsilon_t. \tag{2.4}$$

The power of a test of (2.3) against (2.4) will of course decrease as the parameter ϕ gets closer to one. This is inevitable, though from an economic theory point of view not inconsequential according to West (1988). Of course, we would expect very high power if the true generating process were white noise, so that $\phi = 0$ in (2.4). We might view white noise as being very far from the space in which the null hypothesis of a unit root is true.

This view is, however, illusory, for consider the ARIMA(0, 1, 1) model

$$X_t - X_{t-1} = \varepsilon_t - \theta\varepsilon_{t-1}. \tag{2.5}$$

For any value of θ strictly less than one, the null hypothesis of a unit autoregressive root is true, but $\theta = 1$ implies that X_t is white noise. An extension of this argument to models more elaborate than (2.5) shows that **any** model with $d = 0$ is arbitrarily close to **some** model with $d = 1$. Two conclusions follow. First, if the true generating model has $d = 1$ and a large positive moving average root, tests may be misleading. Secondly, if one abandons any prior preference for simple models, the hypothesis-testing paradigm is irrelevant, for we can always come to a model sufficiently elaborate that the power of any test will be arbitrarily to its true size.

Although the second of these conclusions is even more destructive of the unit roots testing industry than the first, some illustration of difficulties caused even when data are generated by simple models is useful. Both Schwert (1989) and Agiakloglou and Newbold (1992) have reported simulation evidence based on (2.5) showing unsatisfactory performance of some commonly used test statistics in moderately large samples. The latter authors examine the augmented Dickey–Fuller tests [Said and Dickey (1984)], based on the approximating autoregression of order $K + 1$,

$$\Delta X_t = \alpha + \beta X_{t-1} + \sum_{j=1}^{K} \gamma_j \Delta X_{t-j} + e_t, \tag{2.6}$$

where in (2.6) K must be chosen sufficiently large to adequately approximate the effects of the moving average part of the true generating process. A test can then be based on the usual t-ratio associated with the least-squares estimate of β, compared with critical values tabulated by Fuller (1976). For processes generated by (2.5), with large positive θ, and samples of 100 observations, it was found that a very large K was necessary before empirical significance levels approximated the asymptotic values, and use of the AIC criterion failed to yield satisfactory values in this sense. However, if such large values of K are to be used, then the power of the test against simple alternatives is very low. For example with $\theta = 0.8$ in (2.5), the empirical significance level of a nominal 5% level test does not fall to 5% until $K = 10$, and with $\theta = 0.9$ this empirical significance level is still 14.3%. On the other hand, if the true generating process is white noise, with $K = 10$, the null hypothesis is rejected only 51.7% of the time by a 5% level test.

The model (2.5) is just the simplest possible model with a moving average term, and already we have seen serious difficulties in traditional unit roots tests, even when the moving average parameter is far from one. The issue can be viewed in a rather different way through examination of a real data set. A study by Nelson and Plosser (1982) has sometimes been taken as providing strong support for the hypothesis that economic time series are generated by processes with a unit autoregressive root. In fact, the authors **failed to reject** that hypothesis for all but one of their time series. They applied a variant of the augmented Dickey–Fuller test, with (2.6) augmented by a time trend, since many economic series grow over time. (This influences the asymptotic distribution of the test statistic, so that alternative critical values, also tabulated by Fuller (1976), are required.)

Since a rejection of the unit root hypothesis was achieved for one series in the Nelson and Plosser study—the logarithm of the US unemployment rate, observed annually over the period 1890–1970—we felt that further examination of this series might be useful. Although Guilkey and Schmidt (1991) have proposed alternative tests in the presence of trend, we begin by asking how the test applied by Nelson and Plosser is influenced by the choice of K in (2.6). In fact, using rather informal means, these authors chose, in our notation, $K = 3$, leading to rejection of the null hypothesis at the 5% level. Applying the same test, we found that for the values $K = 4$–10 the unit root hypothesis would have been rejected even at the 10% level only for $K = 7$. This seems to weaken the evidence against that hypothesis. We should note that $K = 3$ would have been chosen by the AIC criterion, but in view of the above discussion do not find that very reassuring. In fact, there is no good reason to include a time trend in a model for unemployment, and certainly the data do not suggest its necessity. Accordingly, we dropped that term, leaving us with the formulation (2.6). We tried all values $K = 0$–10. In every case, the null hypothesis of a unit autoregressive root is rejected at the 10% level, and in all but two cases at the 5% level. Further details of these results, which now appear to provide very strong evidence against the null hypothesis, can be found in Agiakloglou (1992). We decided also to look at these data through a conventional ARIMA analysis. Using the SPSS program, we fitted all ARIMA$(p, 1, q)$ models with $p + q \leqslant 5$. The ARIMA$(1, 1, 2)$ model was chosen by both the AIC and SBC criteria. The fitted model was

$$\begin{aligned}(1 - 0.57B)(1 - B)X_t &= (1 - 0.45B - 0.55B^2)\varepsilon_t \\ &= (1 - B)(1 + 0.55B)\varepsilon_t. \qquad (2.7)\end{aligned}$$

This, of course, suggests over-differencing, so that an analyst who had started by looking at first differences would naturally be driven to look at stationary models. In fact, fitting all ARIMA$(p, 0, q)$ models with $p + q \leqslant 5$, we found the model selected by SBC to be ARIMA$(1, 0, 1)$

$$(1 - 0.50B)(X_t - \mu) = (1 + 0.62B)\varepsilon_t, \qquad (2.8)$$

which is very close to (2.7). Does this suggest very strong evidence that $d = 0$? We have already said that we do not believe such evidence can exist. For example, the unit moving average root in the estimated model (2.7) could well be a manifestation of the 'pile-up' effect. It simply will never be reasonable to assert that we have exact cancellation rather than near-cancellation between autoregressive and moving average operators. Strong belief in (2.8) as the 'true' generating model can only follow from an *a priori* preference for simple models.

A different reservation about the conclusions of standard unit roots tests has been expressed by Perron (1989), who showed that conclusions of such tests could be altered when structural change is incorporated. Perron considers formulations of the form (2.6), augmented by both a trend term and dummy variables to allow for a deterministic change, at a given point in time. Having recomputed appropriate critical values for the Dickey–Fuller tests, he concludes that many of the Nelson and Plosser conclusions can be reversed, and that for many economic time series the unit root hypothesis can be rejected. Newbold and Agiakloglou (1992) have re-examined one of these series, logarithms of US common stock prices, observed annually over the period 1871–1970. Taking 1929 as the break-point, Perron rejects the unit root hypothesis at the 2.5% level. Newbold and Agiakloglou tried other break points, achieving the best fitting model by the usual criteria with a break in 1931. For this model, the hypothesis of a unit autoregressive root could not be rejected at the 10% level. Moreover, when two break points rather than one were permitted in the model, the evidence against the unit roots hypothesis was further weakened.

Introspection, simulation, and analysis of real data lead to a single conclusion—that testing for unit autoregressive roots is misguided. Uncertainty on this issue is surely inevitable, and the most benign effect of a formal test is to introduce spurious certainty where none is warranted. A more malign effect is that the analyst can be seriously misled through basing analysis on a false model, or on one that is an inadequate approximation for these purposes of the true generating process. We would propose an alternative line of enquiry. Since all models are necessarily wrong, and since all inference about unit autoregressive roots is necessarily problematic, it seems important to ask if, and how, it matters one way or the other. If two models yield virtually identical conclusions on a question of interest, all well and good. On the other hand, if two models that, for all practical purposes, are observationally indistinguishable, yield quite different conclusions, it is important to be aware of this uncertainty. We can offer little guidance on this question. However, one area of concern is with standard errors of regression parameter estimators. Newbold and Davies (1978) demonstrated the potential for serious problems here when the error terms are generated by an ARIMA(0, 1, 1) model with a large moving average parameter. As we have already seen, it is rather easy to be misled into believing that such processes are in fact stationary.

2.3 UNOBSERVED STOCHASTIC COMPONENTS

It is often felt useful to think of an economic time series as the sum of components such as trend, cycle, seasonal, and irregular, so that we might write

$$X_t = T_t + C_t + S_t + I_t. \tag{2.9}$$

The modern view is that it is inadequate to regard the individual components as fixed. Rather, they are generally modelled as stochastic processes. At the outset, this leads to the question of how to define such concepts as 'local linear trend', and as to whether (and if so, why) the individual components should be taken to be orthogonal. The recent literature speaks with surprisingly many voices on these questions, an issue explored in some detail by Newbold (1991). Indeed it appears that components estimation is an attempt to estimate that which remains to be satisfactorily defined. As Quah (1992) has elegantly demonstrated, permitting a range of possible components specifications leads to substantial differences in inference about the relative importance of individual components.

One approach, taken in the model-based seasonal adjustment literature (for example, Hillmer and Tiao (1982)), and also by Beveridge and Nelson (1981), is to begin by estimating an ARIMA model. The unobserved components are then estimated indirectly as a by-product of the fitted model. An alternative approach is through structural models, introduced by Harvey and Todd (1983) and discussed in considerably more detail by Harvey (1989).

Here, plausible models for the individual components are specified at the outset, and the model for the actual series X_t is merely an indirect consequence of these specifications. The great virtue of this view is its intuitive appeal. As a practical matter, for non-seasonal time series, the two approaches will very often yield quite similar forecasts for X_t since the structural models are generally special cases of relatively simple ARIMA models. However, this is not necessarily the case for seasonal time series.

We would expect applications of traditional ARIMA analysis and structural modelling to produce similar short-term forecasts very often. However, considerably more uncertainty can arise in attempting to estimate the individual unobserved components. Essentially, this is because the estimation of unobserved components depends on inference about the long-run behavior of a time series. Both Watson (1986) and Clark (1987) demonstrate that ARIMA and unobserved components models that appear to fit data about equally well can yield quite different conclusions about long-run behavior. The point arises in these authors' analyses because the components models can be viewed as constrained versions of sufficiently general ARIMA models. Although the implied constraints cannot be rejected by standard statistical tests, the two competing models can correspond to quite different low-frequency behavior.

In some circumstances, the question can turn on the number of unit autoregressive roots in the generating process, recalling our discussion of the previous section. To illustrate, Crafts, Leybourne and Mills (1989), analysing 214 years of historical data on British industrial production, consider a nonseasonal variant of (2.9)

$$X_t = T_t + C_t + I_t. \tag{2.10}$$

The three components are assumed to be orthogonal, with the cycle modelled as a second-order autoregression and the irregular term taken to be white noise. The trend is taken to be locally linear, with

$$T_t = T_{t-1} + \beta_{t-1} + \eta_{1t} \tag{2.11}$$

$$\beta_t = \beta_{t-1} + \eta_{2t} \tag{2.12}$$

where η_{1t} and η_{2t} are independent white noises. The data are analysed in logarithms, so that β_t is a time-varying growth rate. Part of the object of the analysis was to estimate this growth rate over time. The implied model for X_t is integrated of order 2, unless η_{2t} is identically zero, in which case X_t is integrated of order 1, and growth rates are constant. Newbold and Agiakloglou (1991) applied a relatively routine ARIMA analysis to these data, finding adequate fit for models that are integrated of order 1, and no evidence whatever of the need for second differencing. We conjecture that the two models will yield approximately the same short-term forecasts. However, their conclusions about an important subject of interest—the evolution over time of growth rates—are quite different. We do not conclude that one or the other model is incorrect. Indeed, the theme of this paper is that it will often be impossible to choose between competing models of this sort through the usual criteria.

It is useful here to put our central theme in a different way. Alternative models—nested or otherwise—are typically compared through such statistics as mean-squared errors, maximized likelihoods, or selection criteria dominated by these quantities and a preference for parsimony. In essence, the major element is simply an in-sample estimate of the one-step forecast error variance. The fitted models should therefore be excellent tools for short-term forecasting. There is no reason to expect that they will be in agreement, or reliable for inference, about longer-term behavior.

2.4 FRACTIONAL DIFFERENCE MODELS

The dichotomy between integrated processes of different orders can be relaxed by permitting the difference parameter d of (2.2) to be non-integer. In the resulting fractional difference, or ARFIMA (p, d, q) model, d can now be viewed as another parameter to be estimated. Some properties of fractional difference models were introduced by Granger and Joyeux (1980) and Hosking (1981). In addition to the usual conditions on ϕ and θ, stationarity requires $d < \frac{1}{2}$, and invertibility $d > -\frac{1}{2}$. Perhaps because they provide a smooth transition from ARMA to unit root models, fractional difference processes have attracted some recent interest in economics (see, for example, Diebold and Rudebusch (1989, 1991a)).

In light of our discussion of section 2.2, it seems infeasible to separate inference about d from inference about the autoregressive and moving average parts of the generating model. Attempts to do so have most often been based on an approach due to Geweke and Porter-Hudak (1983) (see also Porter-Hudak (1990)). Unfortunately, as demonstrated by Agiakloglou, Newbold and Wohar (1993), the resulting estimator of d can be seriously biased, and associated tests badly misleading. Essentially the difficulty is that, while useful information about d is available in estimates of the spectrum at low frequencies, this information can still be contaminated by information coming from the ARMA part of the generating model.

Because their autocorrelations decay only slowly with increasing lag length, stationary ARFIMA models with positive d are said to exhibit long memory'. This characterization might suggest that a relatively long data set is necessary for reliable inference. We believe that series of length typically available in economics are quite inadequate for this purpose. An interesting feature of this model is that inference can be considerably less precise when the mean of the process is unknown than when it is known. As discussed for example by Samarov and Taqqu (1988), the variances of the sample mean and of the maximum likelihood estimator of the mean are proportional to n^{2d-1}, when

n is the number of observations. Nevertheless, Dahlhaus (1989) has proved that the usual asymptotic distributional results on maximum likelihood estimators of d and the ARMA parameters continue to hold in the unknown mean case. On the other hand, Cheung and Diebold (1990) report simulation results suggesting that, in moderate sample sizes, d can be far more precisely estimated when the true mean is known.

Here we illustrate this issue through a simple hypothesis testing problem. Consider the fractional noise model

$$(1 - B)^d X_t = \varepsilon_t \tag{2.13}$$

where the null hypothesis that d is zero is to be tested against a two-sided alternative. Robinson (1991) has noted that the Lagrange multiplier statistic for this test is

$$S \propto \sum j^{-1} r_j, \tag{2.14}$$

where r_j are the sample autocorrelations

$$r_j = \sum_{t=j+1}^{n} (X_t - \bar{X})(X_{t-j} - \bar{X}) \Big/ \sum_{t=1}^{n} (X_t - \bar{X})^2. \tag{2.15}$$

In (2.15), the sample mean would be replaced by the true mean, if it were known. We will assume that the error terms in (2.13) are normally distributed, and truncate the sum in (2.14), giving the statistic

$$S = [n(n+2)]^{1/2} \sum_{j=1}^{m} j^{-1} r_j \Bigg/ \left[\sum_{j=1}^{m} j^{-2}(n-j) \right]^{1/2} \tag{2.16}$$

which has an asymptotic standard normal distribution under the null hypothesis.

It is possible to get approximate values for the power of this test through approximations to the mean and variance of the test statistic. Assume, without loss of generality, that the error term ε_t has variance one, and that the true process mean is zero. Then, for any series X_t of n observations, the covariance matrix $\boldsymbol{\Gamma}$ is known (Hosking, 1981), and we can find the Choleski factorization $\boldsymbol{\Gamma} = \boldsymbol{T}\boldsymbol{T}'$, where $\boldsymbol{T}$ is lower triangular. Now, let $\boldsymbol{\eta}$ be a vector of n independent standard normal random variables. Then, we can write the vector of observations on our time series as $\boldsymbol{X} = \boldsymbol{T}\boldsymbol{\eta}$. This device has been employed by, for example, Diebold and Rudebusch (1991b) to generate observations from (2.13). Here it is used as a purely theoretical construct.

Taking the sum to have m terms, the right hand side of (2.14) can be written

$$\sum_{j=1}^{m} j^{-1} r_j = \frac{\boldsymbol{\eta}'\boldsymbol{T}'\boldsymbol{F}\boldsymbol{Q}\boldsymbol{F}\boldsymbol{T}\boldsymbol{\eta}/2n}{\boldsymbol{\eta}'\boldsymbol{T}'\boldsymbol{F}\boldsymbol{T}\boldsymbol{\eta}/n} = \frac{\boldsymbol{\eta}'\boldsymbol{M}\boldsymbol{\eta}/2}{\boldsymbol{\eta}'\boldsymbol{K}\boldsymbol{\eta}} = \frac{A}{B}. \tag{2.17}$$

where $\boldsymbol{F} = (\boldsymbol{I} - ll'/n$, with a l vector of ones. Also $\boldsymbol{Q} = \sum_{j=1}^{m} j^{-1}\boldsymbol{C}_j$, where $\boldsymbol{C}_j$ is an

$n \times n$ matrix with ones on the jth super- and sub-diagonals, and zeros elsewhere. Thus for any combination (d, m, n), (2.17) is simply a ratio of quadratic forms in normal deviates, with known matrices. Then, as for example in Marriott and Pope (1954), we can write approximately

$$E\left(\frac{A}{B}\right) \approx \frac{E(A)}{E(B)}\left[1 - \frac{E(AB)}{E(A)E(B)} + \frac{E(B^2)}{[E(B)]^2}\right] \tag{2.18}$$

and also

$$\operatorname{Var}\left(\frac{A}{B}\right) \approx \left[\frac{E(A)}{E(B)}\right]^2\left[\frac{\operatorname{Var}(A)}{[E(A)]^2} + \frac{\operatorname{Var}(B)}{[E(B)]^2} - \frac{2\operatorname{Cov}(A,B)}{E(A)E(B)}\right]. \tag{2.19}$$

The expectations on the right-hand sides of these equations can be obtained directly from expressions in Kumar (1973). The mean and variance of the statistic S of (2.16) follow immediately once (2.18) and (2.19) have been computed. Then, taking the statistic to have an approximate normal distribution, probabilities of rejection of the null hypothesis for any significance level are routinely calculated. Of course, when the true mean of the process is known, the same exercise can be repeated, replacing $\boldsymbol{F}$ in (2.17) by the identity matrix. Table 2.1 shows some results for series of 100 observations, taking $m = 20$ in (2.16). We see that knowledge of the true mean is not irrelevant—it can have a substantial impact on the power of tests. This conclusion does not hold when d is negative, where the true mean is far more precisely estimated by the sample mean. Further results, and supporting simulation evidence, are reported in Agiakloglou (1992).

Although the results in Table 2.1 are interesting in demonstrating the additional uncertainty that can be induced by ignorance of the true mean, they are not of great practical interest. This is so since, in practice, we will never know that (2.13), or any other, is the true generating model. Instead, an analyst might first fit an ARMA or ARIMA model to data, and then test the possibility of fractional d. Agiakloglou (1992) shows that the Lagrange

Table 2.1 Probability of rejecting white noise against fractional noise ($n = 100$), for tests with significance level α against a 2-sided alternative

	d					
α	0.35		0.25		0.15	
0.01	0.89	0.84	0.80	0.64	0.49	0.26
0.05	0.92	0.89	0.85	0.75	0.63	0.42
0.10	0.93	0.91	0.88	0.80	0.69	0.50

In each cell the first entry is for the mean known case, and the second for the mean unknown.

multiplier test of ARIMA$(p, 0, q)$ against ARIMA(p, d, q) is of the form (2.14), but with the sample autocorrelations in that expression replaced by the residual autocorrelations from the fitted model under the null hypothesis. Two practical variants of this test were investigated and found to perform satisfactorily under the null hypothesis in moderate-sized samples. For series generated from the fractional noise process (2.13), the power of the test on d, evaluated by simulation, was found to fall dramatically when autoregressive and moving average terms were permitted in the model under the null hypothesis. For example, for series of 100 observations, with $d = 0.25$, and mean unknown, Table 2.1 shows power of 0.75 against white noise, for a 5% level test. When a first-order autoregression was permitted under the null, the power of one test fell to 0.19, while under an ARMA(1, 1) null the power was only 0.07.

The conclusion seems pretty clear. In moderate-sized samples it is likely to be very difficult to distinguish fractional difference models from even very simple ARMA or ARIMA models. Of course, as the number of parameters in these models is allowed to increase, that difficulty becomes a practical impossibility. This should not be surprising. In section 2.2 we stressed the difficulty of honest distinction between $d = 0$ and $d = 1$. Naturally, it would be more difficult yet to determine intermediate values.

2.5 PERSISTENCE OF SHOCKS

Let X_t be a time series assumed to be generated by a process that is integrated of order one, so that we can write

$$\phi(B)[(1 - B)X_t - \mu] = \theta(B)\varepsilon_t. \tag{2.20}$$

If $\hat{X}_t(k)$ denotes the forecast of X_{t+k} made at time t, it is straightforward to show that, at time t, forecasts are updated according to

$$\hat{X}_t(k) - \hat{X}_{t-1}(k+1) = \left[1 - \sum_{j=1}^{k} \psi_j\right]\varepsilon_t \tag{2.21}$$

where

$$(1 - \psi_1 B - \psi_2 B^2 - \ldots) = \psi(B) = \phi^{-1}(B)\theta(B). \tag{2.22}$$

In most economic applications, X_t is the logarithm of an observed series, such as gross national product. The white noise error term is viewed as an unanticipated 'shock' entering the system at time t. Thus, the impact of a 1% shock in the current period on forecasts of future values is, from (2.21), a percentage amount

$$A_k = 1 - \sum_{j=1}^{k} \psi_j. \tag{2.23}$$

As discussed by Campbell and Mankiw (1987), economists view the quantity

(2.23) for large k as being of some importance, providing a measure of the extent to which economic shocks persist into the future. The limiting value is

$$\lim_{k \to \infty} A_k = \psi(1) \qquad (2.24)$$

and this quantity is defined by Jaeger and Kunst (1990) as 'persistence'. It is, of course, a simple function of the spectrum of the first difference process at zero frequency.

A possible attraction of the quantity (2.24) is that it may provide some relaxation of the dichotomy between stationary and integrated processes. Of course, for all models (2.2) with $d=0$ $\psi(1)$ is zero, while it is infinite for all models with $d=2$. More interesting possibilities occur for processes that are integrated of order one. For example, for the ARIMA (0,1,1) model (2.5), $\psi(1)$ has a maximum value of 2 when $\theta = -1$, and approaches zero as θ approaches one. For large positive values of θ, persistence will be small, and in that sense it might be taken as unimportant that such models are difficult to distinguish from stationary processes. For a random walk generating model, persistence is one, and values of persistence estimated from data are often compared with that standard.

Although the interpretation of this persistence measure might be interesting, its reliable estimation is problematic. Campbell and Mankiw, and others, were particularly concerned with US real gross national product. They fitted ARMA models to the first differences of the logarithms of this series, using quarterly data over the period 1947–85. To illustrate the issues involved, we repeated part of their analysis using data for 1950–89. Given the large overlap, similarities in our numerical results are to be expected. Table 2.2 shows persistence estimates derived from fitted ARIMA$(p,1,q)$ models for all $p+q \leqslant 6$. All models were estimated through full maximum likelihood, using the SPSS program. The great majority of the fitted models, including these

Table 2.2 Persistence estimates for US real GNP, based on fitted ARIMA (p, 1, q) models

q/p	0	1	2	3	4	5	6
0	–	1.671[a]	1.881	1.660	1.572	1.493	1.610
1	1.299	1.822	1.711	1.509	0.051	1.559	–
2	1.589[b]	1.774	1.634	0.007	1.512	–	–
3	1.766	1.567	0.014	1.738	–	–	–
4	1.834	0.034	1.739	–	–	–	–
5	1.556	1.534	–	–	–	–	–
6	1.498	–	–	–	–	–	–

[a] Model selected by SBC.
[b] Model selected by AIC.

selected by AIC and SBC, show persistence measures much greater than one, suggesting more persistence than in a random walk. In the Campbell and Mankiw study the MA(2) model was selected by both AIC and SBC, giving a persistence estimate of 1.573. As in our Table 2.2, Campbell and Mankiw also found a few heavily parameterized models yielding very low persistence estimates. These authors tended to dismiss these models, arguing that simpler models could not be rejected against them, and citing the pile-up of moving average parameter estimates on the boundary of the invertibility region. We would agree that, if sufficient parameters are added to the model, such results are indeed inevitable. But this does not constitute grounds for dismissal. We would view these results as indicating the inevitable extreme uncertainty about the true value of persistence, unless one is prepared to impose strong *a priori* preference for simple models. It is difficult to find any honest justification for such a preference.

It would not be surprising to find that parsimonious selection from some alternative class of models yields quite different conclusions about persistence. Indeed, Clark (1987), working with unobserved components models fitted to the Campbell and Mankiw data, obtained the estimate 0.64. We see no reason to prefer one parsimonious model over another and take this result as a partial indication of the unrealiability of such an approach.

A different critique of the Campbell and Mankiw findings was offered by Cochrane (1988), whose analysis was based on the variance ratio statistic that has been used as a test for random walk (see, for example, Lo and MacKinlay (1988, 1989)). If X_t is generated by a random walk, then the variance of $(X_t - X_{t-k})$ is k times the error variance. Then, unbiased estimators of the error variance, based on $X_0, X_1, \ldots, X_n$ are provided by

$$\hat{\sigma}_k^2 = \frac{n}{k(n-k)(n-k+1)} \sum_{t=k}^{n} (X_t - X_{t-k} - k\hat{\mu})^2, \tag{2.25}$$

where $\hat{\mu}$ is the sample mean of the series of first differences. Cochrane plots $\hat{\sigma}_k^2$ against k for a series on real US gross national product, finding, for increasing k, decreasing values, suggesting less persistence than a random walk. Unfortunately, these results are not directly comparable with those of Campbell and Mankiw, as Cochrane used annual data.

To standardize, it is natural to compute the variance ratios

$$\hat{V}_k = \hat{\sigma}_k^2 / \hat{\sigma}_1^2 \tag{2.26}$$

and a plot of these quantities might be found useful. For a random walk, Lo and MacKinlay (1988) establish that these statistics have an asymptotic normal distribution, with mean one, and variance $2(2k-1)(k-1)/3kn$, which increases approximately as a linear function of k. However, as Campbell and Mankiw (1989) note, this is not correct for other generating models. Cochrane further shows that, approximately

$$\hat{V}_k \approx \left(1 + 2\sum_{j=1}^{k-1} k^{-1}(k-j)r_j\right), \tag{2.27}$$

where r_j is the jth sample autocorrelation of the first differenced series. As discussed, for example by Priestley (1982), this is proportional to the Bartlett estimator of the spectral density at zero frequency, so that its asymptotic standard error is $(4k/3n)^{1/2}\ V_k$, where V_k is the corresponding population value. It follows that a nonparametric estimator of persistence can be obtained from

$$\hat{\psi}(1) = (\hat{\sigma}_{\Delta X}/\hat{\sigma}_\varepsilon)\hat{V}_k^{\frac{1}{2}}, \tag{2.28}$$

where $\hat{\sigma}_{\Delta X}$ and $\hat{\sigma}_\varepsilon$ are estimators of the standard deviations of the first differenced series and of the white noise innovations in the generating model. Generally, this last term has been estimated through fitting a parsimonious model, though a nonparametric approach is of course also possible. (See, for example, Pukkila and Nyquist (1985)). Variants of this procedure have been implemented by Campbell and Mankiw (1989), Cogley (1990), and Jaeger and Kunst (1990). A difficulty is that the standard error of the estimator will be high for large k, while, as is clear from (2.27), substantial bias is possible for small k if the true generating process contains a large autoregressive root.

One way of viewing the variance ratio statistics, in the spirit of Cochrane (1988), is as checks on the adequacy of some specific fitted model. We applied these checks to two of our fitted models, ARIMA(1, 1, 0) and ARIMA(3, 1, 2), for the gross national product data. From Table 2.2, note that the corresponding persistence estimates are 1.671 and 0.007. Our approach employs the

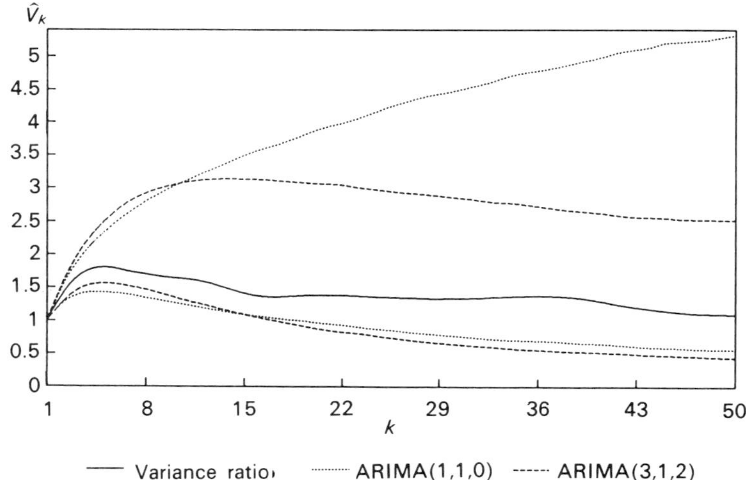

Figure 2.1. Variance ratio for log GNP with 90% bounds for an ARIMA (1, 1, 0) and an ARIMA (3, 1, 2) model.

parametric bootstrap discussed by Tsay (1992). Figure 2.1 shows values of the variance ratios calculated from the data for k up to 50, together with 90% bounds computed from 4 000 replications of a series of 160 observations generated from each of the two fitted models. Although these models imply radically different estimates of persistence, the variance ratio statistics provide no ground to question the adequacy of either. These results can be interpreted as casting serious doubts on the value of the estimator (2.28), since a very wide range of values of the variance ratio statistics are compatible with radically different persistence values. Of course, this finding arises in part because our ARIMA(3, 1, 2) model contains a very large autoregressive root, and an even larger moving average root. Unless, for some reason, the analyst is prepared to discount such processes on *a priori* grounds, it appears that reliable estimation of persistence is a practical impossibility.

2.6 SUMMARY

In this paper we have argued that parsimoniously parametrized time series models cannot be relied on as a basis for inference about the long-run behavior of a time series. In the context of ARIMA models, the difficulty arises because both the autoregressive and moving average operators in (2.1) could contain large positive roots, either of which may or may not be precisely equal to one. Unfortunately, in long-term inference, this is exactly the issue in question. Apparently valid parametric inference procedures in fact rely for their validity on *a priori* rejection of processes with near-cancelling large autoregressive and moving average roots. Unfortunately, unless sample sizes are very large indeed, this difficulty does not appear to be circumvented by an appeal to nonparametric approaches, such as the Geweke and Porter–Hudak estimator of the fractional difference parameter, or the variance ratio statistic.

Parsimoniously parametrized time series models were developed as aids to short-term forecasting, where the fiction that the analyst has discovered the 'true' model is innocuous. Such fiction, however, is far from innocuous when attempting to base inference about long-run behavior on these fitted models.

REFERENCES

Agiakloglou, C. (1992) *Experience in the Application of Unit Roots and Fractional Difference Models and Tests.* Ph.D. thesis, Department of Economics, University of Illinois, Urbana-Champaign.

Agiakloglou, C. and Newbold, P. (1992). Empirical evidence on Dickey–Fuller-type tests. *J. Time Series Anal.* **13**, 471–483.

Agiakloglou, C., Newbold, P. and Wohar, M. (1993) Bias in an estimator of the fractional difference parameter. *J. Time Series Anal.*, **14**, accepted for publication.

Ansley, C.F. and Newbold, P. (1980) Finite sample properties of estimators for autoregressive moving average models. *J. Econometrics*, **13**, 159–183.

Beveridge, S. and Nelson, C.R. (1981). A new approach to the decomposition of economic time series into permanent and transitory components with particular attention to measurement of the business cycle. *J. Monet. Econ.* **7**, 151–174.

Box, G.E.P. and Jenkins, G.M. (1970) *Time Series Analysis, Forecasting and Control*, Holden Day, San Francisco.

Campbell, J.Y. and Mankiw, N.G. (1987) Are output fluctuations transitory? *Quart. J. Econ.*, **102**, 857–880.

Campbell, J.Y. and Mankiw, N.G. (1989) International evidence on the persistence of economic fluctuations. *J. Monet. Econ.* **23**, 319–333.

Cheung, Y.W. and Diebold, F.X. (1990) On maximum likelihood estimation of the differencing parameter of fractionally integrated noise with unknown mean. Discussion paper 34, Institute for Empirical Macroeconomics, Federal Reserve Bank of Minneapolis.

Clark, P.K. (1987) The cyclical component of US economic activity. *Quart. J. Econ.* **102**, 797–814.

Clark, P.K. (1989) Trend reversion in real output and unemployment. *J. Econometrics* **40**, 15–32.

Cochrane, J.H. (1988) How big is the random walk in GNP? *J. Polit. Econ.*, **96**, 893–920.

Cogley, T. (1990) International evidence on the size of the random walk in output. *J. Polit. Econ.*, **98**, 501–518.

Crafts, N.F.R., Leybourne, S.J. and Mills, T.C. (1989) Trends and cycles in British industrial production, 1700–1913. *J. Roy. Statist. Soc., Ser. A*, **152**, 43–60.

Cryer, J.D. and Ledolter, J. (1981). Small sample properties of the maximum likelihood estimator in the first order moving average model. *Biometrika*, **68**, 691–694.

Dahlhaus R. (1989) Efficient parameter estimation for self-similar processes. *Ann. Statist.*, **17**, 1749–1766.

Diebold, F.X. and Rudebusch, G.D. (1989) Long memory and persistence in aggregate output. *J. Monet. Econ.*, **24** 189–209.

Diebold, F.X. and Rudebusch, G.D. (1991a) Is consumption too smooth? Long memory and the Deaton paradox. *Rev. Econ. Statist.*, **73**, 1–9.

Diebold, F.X. and Rudebusch, G.D. (1991b) On the power of Dickey–Fuller tests against fractional alternatives. *Econ. Letters*, **35**, 155–160.

Engle, R.F. and Granger, C.W.J. (1987) Cointegration and error correction: representation, estimation and testing. *Econometrica*, **55**, 251–276.

Fuller, W.A. (1976) *Introduction to Statistical Time Series*, Wiley, New York.

Geweke, J. and Porter-Hudak, S. (1983) The estimation and application of long memory time series models. *J. Time Series Anal.* **4**, 221–238.

Granger, C.W.J. and Joyeux, R. (1980) An introduction to long memory time series models and fractional differencing. *J. Time Series Anal.*, **1**, 15–39.

Guilkey, D.K. and Schmidt, P. (1991) An interpolation test for a unit root in the presence of drift. *Biometrika*, **78** 809–815.

Harvey, A.C. (1989). *Forecasting, Structural Time Series Models and the Kalman Filter*, University Press, Cambridge.

Harvey, A.C. and Todd, P.H.J. (1983) Forecasting economic time series with structural and Box-Jenkins models: a case study. *J. Bus. Econ. Statist*, **1**, 299–307.

Hillmer, S.C. and Tiao, G.C. (1982) An ARIMA-model-based approach to seasonal adjustment, *J. Amer. Statist. Assoc.*, **77**, 63–70.

Hosking, J.R.M. (1981) Fractional differencing. *Biometrika*, **68**, 165–176.

Jaeger, A. and Kunst, R.M. (1990) Seasonal adjustment and measuring persistence in output. *J. Appl. Econometrics*, **5**, 47–58.

Johansen, S. (1991) Estimation and hypothesis testing of cointegration vectors in Gaussian vector autoregressive models. *Econometrica*, **59**, 1551–1580.

Kumar, A. (1973) Expectation of product of quadratic forms. *Sankhya, Ser. B*, **35**, 359–362.

Ledolter, J. and Abraham, B. (1981) Parsimony and its importance in time series forecasting. *Technometrics*, **23**, 411–414.

Lo, A.W. and Mackinlay, A.C. (1988) Stock market prices do not follow random walks: evidence from a simple specification test. *Rev. Fin. Stud.*, **1**, 41–66.

Lo, A.W. and Mackinlay, A.C. (1989) The size and power of the variance ratio test in finite samples: a Monte Carlo investigation. *J. Econometrics*, **40**, 203–238.

Marriott. F.H.C. and Pope, J.A. (1954) Bias in the estimation of autocorrelations. *Biometrika*, **41**, 390–402.

Nelson, C.R. and Plosser, C.I. (1982) Trends and random walks in macroeconomic time series: some evidence and implications. *J. Monet. Econ.* **10**, 139–162

Newbold, P. (1991) Structural decomposition of time series with implications in economics, accounting, and finance research. *Rev. Quant Fin. Account.* **1**, 259–279.

Newbold, P. and Agiakloglou C. (1991) Looking for evolving growth rates and cycles in British industrial production, 1700–1913. *J. Roy. Statist. Soc., Ser. A*, **154**, 341–348.

Newbold, P. and Agiakloglou, C. (1992) US common stock prices, 1871–1970: playing with dummies. *Rev. Quant. Fin. Account*, **2**, 215–220.

Newbold, P. and Davies, N. (1978) Error mis-specification and spurious regressions. *Int. Econ. Rev.*, **19**, 513–519.

Park, J.Y. and Phillips, P.C.B. (1988) Statistical inference in regressions with integrated processes: part 1. *Econometric Theory*, **4**, 468–497.

Park, J.Y. and Phillips, P.C.B. (1989) Statistical inference in regressions with integrated processes: part 2. *Econometric Theory*, **5**, 95–131.

Perron, P. (1989) The great crash, the oil price shock, and the unit root hypothesis. *Econometrica*, **57**, 1361–1401.

Porter-Hudak, S. (1990) An application of the seasonal fractionally differenced model to the monetary aggregates. *J. Amer. Statist. Assoc.*, **85**, 338–344.

Priestley, M.B. (1982) *Spectral Analysis and Time Series*, Academic Press, London.

Pukkila, T. and Nyquist, H. (1985) On the frequency domain estimation of the innovation variance of a stationary univariate time series. *Biometrika*. **72**, 317–323.

Quah, D. (1992) The relative importance of permanent and transitory components: identification and some theoretical bounds. *Econometrica*, **60**, 107–118.

Robinson, P.M. (1991) Testing for strong serial correlation and dynamic conditional heteroskedasticity in multiple regression. *J. Econometrics*, **47**, 67–84.

Said, S.E. and Dickey, D.A. (1984) Testing for unit roots in autoregressive-moving average models of unknown order. *Biometrika*, **71**, 599–607.

Samarov, A. and Taqqu, M.S. (1988) On the efficiency of the sample mean in long memory noise. *J. Time Series Anal.* **9**, 191–200.

Schwert, G.W. (1989) Tests for unit roots: a Monte Carlo investigation. *J. Bus. Econ. Statist.*, **7**, 147–159.

Shephard, N.G. and Harvey, A.C. (1990) On the probability of estimating a deterministic component in the local level model. *J. Time Series Anal.* **11**, 339–347.

Sims, C.A., Stock, J.H. and Watson, M.W. (1990) Inference in linear time series models with some unit roots. *Econometrica*, **58**, 113–144.

Tsay, R.S. (1992) Model checking via parametric bootstraps in time series analysis. *Appl. Statist.*, **41**, 1–15.

Watson, M.W. (1986) Univariate detrending methods with stochastic trends. *J. Monet. Econ.*, **18**, 49–75.

West, K.D. (1988) On the interpretation of near random walk behavior in GNP. *Amer. Econ. Rev.*, **78**, 202–209.

3

Developments in multivariate covariance generation and factorization

G. Tunnicliffe Wilson

3.1 INTRODUCTION

Consider the multivariate autoregressive moving-average (MARMA) model with matrix coefficients for the d-dimensional vector process $\mathbf{x}_t$

$$\mathbf{\Phi}_0\mathbf{x}_t + \mathbf{\Phi}_1\mathbf{x}_{t-1} + \ldots + \mathbf{\Phi}_p\mathbf{x}_{t-p} = \mathbf{\Theta}_0\mathbf{e}_t + \mathbf{\Theta}_1\mathbf{e}_{t-1} + \ldots + \mathbf{\Theta}_q\mathbf{e}_{t-q} \quad (3.1)$$

where $\mathbf{e}_t$ is a vector white noise process having mean zero and variance matrix $\mathbf{\Sigma}$, and $\mathbf{\Phi}_0$ is the identity matrix. Conventionally $\mathbf{\Theta}_0$ is also the identity but this not required for what follows. The sign convention, though differing from that used by Box and Jenkins (1976) for univariate models, is more convenient here.

This model may conveniently be written in operator notation as

$$\mathbf{\Phi}(B)\mathbf{x}_t = \mathbf{\Theta}(B)\mathbf{e}_t \quad (3.2)$$

where B is the backward shift operator and

$$\mathbf{\Phi}(B) = \mathbf{\Phi}_0 + \mathbf{\Phi}_1 B + \cdots + \mathbf{\Phi}_p B^p,\ \mathbf{\Theta}(B) = \mathbf{\Theta}_0 + \mathbf{\Theta}_1 B + \cdots + \mathbf{\Theta}_q B^q. \quad (3.3)$$

The autocovariances of x_t are defined when the process is stationary by

$$\mathbf{\Gamma}_k = E(\mathbf{x}_t \mathbf{x}'_{t-k}), \quad (3.4)$$

and the relationship between these and the model is usefully represented by a generating function equation

$$\mathbf{\Gamma}(B) = \sum_{k=-\infty}^{\infty} \mathbf{\Gamma}_k B^k = \mathbf{\Phi}(B)^{-1}\mathbf{\Theta}(B)\mathbf{\Sigma}\mathbf{\Theta}'(B^{-1})\mathbf{\Phi}'(B^{-1})^{-1}. \quad (3.5)$$

The stationarity condition for the model is that $\mathbf{\Phi}(B)^{-1}$ has a convergent expansion for $|B| \leqslant 1$ where B is now interpreted as a complex variable. This condition must be satisfied for the autocovariances to be defined. The corresponding invertibility condition for the moving average part is again not required for what follows.

The procedure to be presented is based on expanding the generating function, the first step being to express

$$\mathbf{\Gamma}(B) = \mathbf{\Phi}(B)^{-1}\mathbf{\Lambda}(B) + \mathbf{\Lambda}'(B^{-1})\mathbf{\Phi}'(B^{-1})^{-1}, \tag{3.6}$$

where

$$\mathbf{\Lambda}(B) = \mathbf{\Lambda}_1 + \mathbf{\Lambda}_2 B^2 + \cdots + \mathbf{\Lambda}_m B^m \tag{3.7}$$

has order $m = \max(p, q)$. To solve for $\mathbf{\Lambda}(B)$ the expression is multiplied out to give

$$\mathbf{\Phi}(B)\mathbf{\Lambda}'(B^{-1}) + \mathbf{\Lambda}(B)\mathbf{\Phi}'(B^{-1}) = \mathbf{\Theta}(B)\mathbf{\Sigma}\mathbf{\Theta}'(B^{-1}) = \mathbf{\Omega}(B). \tag{3.8}$$

The right hand side of this equation is readily evaluated as

$$\mathbf{\Omega}(B) = \sum_{k=-q}^{q} \mathbf{\Omega}_k B^k \quad \text{where for} \quad k \geqslant 0 \quad \mathbf{\Omega}_k = \mathbf{\Omega}'_{-k} = \sum_{i=k}^{q} \mathbf{\Theta}_i \mathbf{\Sigma} \mathbf{\Theta}'_{i-k}. \tag{3.9}$$

Equating coefficients leads to linear equations for the unknown coefficients $\Lambda_0, \ldots, \Lambda_m$

$$\sum_{i=k}^{m} \mathbf{\Phi}_i \mathbf{\Lambda}'_{i-k} + \sum_{i=k}^{m} \mathbf{\Lambda}_i \mathbf{\Phi}'_{i-k} = \mathbf{\Omega}_k, \tag{3.10}$$

where $\mathbf{\Phi}_i$ or $\mathbf{\Omega}_k$ are taken to be zero for $i > p$ or $k > q$ in the cases $p < m$ and $q < m$ respectively.

Taken at face value and recognizing the symmetry for $k = 0$, there are $md^2 + d(d+1)/2$ independent equations and the same number of unknown elements of $\mathbf{\Lambda}_0, \ldots, \mathbf{\Lambda}_m$, because only the symmetric part of $\mathbf{\Lambda}_0$ is required, and it may be chosen to be symmetric, upper triangular or otherwise constrained to obtain a unique solution.

These equations may be solved by a standard elimination procedure which requires $O(m^3 d^6)$ arithmetic operations. The main point of this paper is to present a method of solution which to some extent exploits their special structure, achieving some gain in numerical efficiency. Having solved for the coefficients $\mathbf{\Lambda}_i$ the covariances are calculated by expanding

$$\mathbf{\Phi}(B)^{-1}\mathbf{\Lambda}(B) = \mathbf{G}_0 + \mathbf{G}_1 B + \mathbf{G}_2 B^2 + \ldots, \tag{3.11}$$

using

$$\mathbf{G}_k = -\mathbf{\Phi}_1 \mathbf{G}_{k-1} - \ldots - \mathbf{\Phi}_{\min(p,k)} \mathbf{G}_{k-\min(p,k)} + \begin{Bmatrix} \mathbf{\Lambda}_k & \text{if } k \leqslant m \\ \mathbf{0} & \text{otherwise} \end{Bmatrix} \text{for } k = 0, 1, 2 \ldots. \tag{3.12}$$

Then, $\mathbf{\Gamma}_0 = \mathbf{G}_0 + \mathbf{G}'_0$ and $\mathbf{\Gamma}_k = \mathbf{G}_k$ for $k > 0$.

An alternative set of equations is obtained by first expanding

$$\mathbf{\Phi}(B)^{-1}\mathbf{\Theta}(B) = \mathbf{\Psi}(B) = \mathbf{\Psi}_0 + \mathbf{\Psi}_1 B + \mathbf{\Psi}_2 B^2 + \ldots \tag{3.13}$$

and rewriting the generating function equation as

$$\mathbf{\Phi}(B)\mathbf{\Gamma}(B) = \mathbf{\Theta}(B)\mathbf{\Sigma}\mathbf{\Psi}'(B). \tag{3.14}$$

Equating coefficients of B^k for $k = 0, \ldots, m$ gives the same number of equations as before, with $\mathbf{\Gamma}_0, \ldots, \mathbf{\Gamma}_m$ as the unknowns. The matrix of these equations is in fact closely related to that of the previous set. These alternative equations are easily derived without recourse to generating functions and have been presented, for example, by Ansley (1980).

The prospect of improved efficiency in solving equations arising from time series models has been realized in many examples. The Durbin–Levinson algorithm for univariate series, see for example Durbin (1960), is perhaps the earliest well known example, although closely related mathematics is to be found in the earlier work of Schur.

The Durbin–Levinson algorithm solves the problem of calculating the coefficients of an autoregressive (AR) model of order p from the autocovariances to lag p in $O(p^2)$ arithmetical operations, an order less than directly solving the appropriate linear equations which have a Toeplitz structured matrix. Many methods have since been developed for similarly structured equations.

The algorithm is easily reversible to solve the covariance generation problem considered in this paper in the univariate case of a pure AR model. This is extended by Tunnicliffe Wilson (1979) to the univariate ARMA case using a form of the Euclid algorithm to solve equation (3.8) in $O(pm)$ operations. See also Mélard (1984), Demeure and Mullis (1989) for related applications and procedures.

In the multivariate case the Durbin–Levinson algorithm was generalized by Whittle (1963). It requires $O(p^2)$ matrix operations leading to $O(p^2 d^3)$ operations altogether. However, this algorithm produces both a forward and backward AR model from the covariances, and in the multivariate case the coefficients of these directional models differ. The method cannot therefore be reversed given only the model in one direction, which is of course the usual case.

The efficient univariate algorithms for covariance generation evaluate, as part of their calculations, a Schur criterion for checking the stationarity condition on the AR operator in the model. Generalizations of such criteria to matrix operators are again difficult to obtain, without having a reversed or **reflected** model. For recent work on such generalizations see Dym and Young (1990).

Another approach to covariance generation for the multivariate ARMA model is to write it in state-space form. This essentially transforms the given ARMA model to a first-order AR model of dimension md. The covariance problem is then reduced to finding the variance matrix of this transformed model. This is a classical Lyapunov problem. Some of the methods and their applications to time series models are given by Barone (1987) and Shea (1988). We shall use an implementation of the Bartels–Stewart method described by Hammarling (1982) as a benchmark for assessing the numerical efficiency of the new procedure which is described in the next section.

To end this section the spectral approach to covariance generation is considered. Defining

$$\mathbf{S}(f) = \mathbf{\Gamma}(e^{2\pi i f}) \tag{3.15}$$

the values of $\mathbf{S}(f)$ are readily evaluated at a grid of N frequencies f_j in $[0, \pi]$ and numerically inverted using the finite Fourier transform to obtain approximations to the Fourier coefficients $\mathbf{\Gamma}_k$. The number of operations required is $O(d^3 N \log N)$. The accuracy of this procedure is excellent provided N is sufficiently large. Problems arise when det $\mathbf{\Phi}(B)$ has zeros close to the unit circle. Then if a zero lies within ε of the limit circle, a value of $N \gg \varepsilon^{-1}$ is required and the numerical efficiency is reduced.

3.2 A TENSOR EUCLID ALGORITHM

In the case $p = 1$, $q = 0$ the equations (3.10) are readily rearranged to obtain the standard Lyapunov equations for $\mathbf{\Gamma}_0$

$$\mathbf{\Gamma}_0 - \mathbf{\Phi}_1 \mathbf{\Gamma}_0 \mathbf{\Phi}_1' = \mathbf{\Sigma}. \tag{3.16}$$

A direct solution using the algebra of matrices of size d is not possible because of the lack of commutativity. By vectorizing $\mathbf{\Gamma}_0$, linear equations with a matrix of size d^2 may be constructed to yield the solution. Using tensor summation notation these are

$$(\delta_{ik}\delta_{jl} - \mathbf{\Phi}_{1ik}\mathbf{\Phi}_{1jl})\mathbf{\Gamma}_{kl} = \mathbf{\Sigma}_{ij} \tag{3.17}$$

Taking symmetry into account these may in fact be reduced to $d(d+1)/2$ equations and unknowns, which require $O(d^6)$ operations to solve by standard means. By using special techniques this may be reduced to $O(d^4)$ operations and Young (1980) presents such a method besides briefly reviewing earlier ones.

In the general multivariate ARMA case, Mittnik (1993) has shown how to construct a system of $2pd^2$ equations for $\mathbf{\Gamma}_0, \ldots, \mathbf{\Gamma}_p$ in vectorized form. He shows how the matrix of these equations can be reformed so that it is block Toeplitz, with block size $2d$, so reducing the order of operations required by a factor p to $O(p^2 d^6)$.

A similar strategy is now applied to the equations (3.8). By tensorizing these equations a block Euclid algorithm generalizing the univariate procedure is obtained, with block size d^2. Again a reduction of order p is achieved in the number of required operations. Although no direct comparison has been made with Mittnik's method, the use of a smaller block size should bring some advantage.

To simplify the presentation assume that $p = q = m$, the case $p \neq q$ being considered later. Again, a repeated index summation convention is assumed as in tensor notation, though all indices are taken as subscripts. Also, z will be used in place of B to emphasize its interpretation as a complex variable in the generating functions. Introducing the Kronecker δ, (3.8) may be written

$$\Phi_{ik}(z)\delta_{jl}\Lambda_k(z^{-1}) + \Phi_{jk}(z^{-1})\delta_{il}\Lambda_{lk}(z) = \Omega_{ij}(z). \tag{3.18}$$

Note that replacing z by z^{-1} merely interchanges i and j, a symmetry which is automatically taken into account in the procedure.

It is now helpful to introduce the d^2-dimensional vector polynomials $\boldsymbol{\lambda}(z)$ and $\boldsymbol{\omega}(z)$ whose elements are appropriately ordered values of $\Lambda_{lk}(x)$ and $\Omega_{ij}(z)$, the latter being a polynomial in both z and z^{-1}. Similarly introduce the matrix polynomial $\boldsymbol{\phi}(z)$ of the same dimension and with correspondingly ordered elements given by $\phi_{ijkl}(z) = \boldsymbol{\Phi}_{ik}(z)\delta_{jl}$. It is convenient to retain i and j for labeling the rows, and k, l for labeling the columns of this matrix. The matrix $\tilde{\boldsymbol{\phi}}(z)$ with elements given by $\tilde{\phi}_{ijkl}(z) = \phi_{jikl}(z)$ is also needed to express (3.8) as

$$\boldsymbol{\phi}(z)\boldsymbol{\lambda}(z^{-1}) + \tilde{\boldsymbol{\phi}}(z^{-1})\boldsymbol{\lambda}(z) = \boldsymbol{\omega}(z). \tag{3.19}$$

The commutativity problem is now overcome by the tensor formulation, and a reduction procedure closely following the univariate case may be applied. To label the successively reduced quantities, the subscript n is attached to the terms in (3.19), starting with $n = m$.

The first step in the reduction process is to construct a matrix polynomial of reduced order by

$$\boldsymbol{\phi}_{n-1}(z) = [\boldsymbol{\phi}_n(z) - z^n\tilde{\boldsymbol{\phi}}_n(z^{-1})\boldsymbol{\alpha}_n]\boldsymbol{\beta}_n^{-1}, \tag{3.20}$$

where $\boldsymbol{\alpha}_n = \tilde{\boldsymbol{\phi}}_{n,n}$ is the coefficient of z^n in $\tilde{\boldsymbol{\phi}}_n(z)$ and postmultiplication by $\boldsymbol{\beta}_n^{-1} = (\mathbf{I} - \boldsymbol{\alpha}_n^2)^{-1}$ ensures that the constant term in $\boldsymbol{\phi}_{n-1}(z)$ is the identity, I, as it is initially for $n = m$.

The algebra of the reduction follows from using (3.20) to express

$$\boldsymbol{\phi}_n(z) = \boldsymbol{\phi}_{n-1}(z) + z^n\tilde{\boldsymbol{\phi}}_{n-1}(z^{-1})\boldsymbol{\alpha}_n$$

and

$$\tilde{\boldsymbol{\phi}}_n(z^{-1}) = \tilde{\boldsymbol{\phi}}_{n-1}(z^{-1}) + z^{-n}\boldsymbol{\phi}_{n-1}(z)\boldsymbol{\alpha}_n, \tag{3.21}$$

the second of these being identical in substance with the first, though differing in form. On substitution into (3.19) and collecting terms

$$\boldsymbol{\phi}_{n-1}(z)[\boldsymbol{\lambda}_n(z^{-1}) + \boldsymbol{\alpha}_n z^{-n}\boldsymbol{\lambda}_n(z)] + \tilde{\boldsymbol{\phi}}_{n-1}(z^{-1})[\boldsymbol{\lambda}_n(z) + \boldsymbol{\alpha}_n z^n\boldsymbol{\lambda}_n(z^{-1})] = \boldsymbol{\omega}_n(z). \tag{3.22}$$

Now express

$$\boldsymbol{\lambda}_n(z) + \boldsymbol{\alpha}_n z^n \boldsymbol{\lambda}_n(z^{-1}) = \boldsymbol{\lambda}_{n-1}(z) + \mathbf{v}_n z^n \tag{3.23}$$

which introduces the reduced order unknown polynomial $\boldsymbol{\lambda}_{n-1}(z)$ and the new coefficient $\mathbf{v}_n$. The point of this is that equating coefficients of z^n gives simply $\tilde{\mathbf{I}}\mathbf{v}_n = \boldsymbol{\omega}_{n,n}$ so that $\mathbf{v}_{n,ji} = \boldsymbol{\omega}_{n,n,ij}$, which is known. The terms involving $\mathbf{v}_n$ may now be moved to the right hand side by defining

$$\boldsymbol{\omega}_{n-1}(z) = \boldsymbol{\omega}_n(z) - z^n \tilde{\boldsymbol{\phi}}_{n-1}(z^{-1})\mathbf{v}_n - z^{-n}\tilde{\boldsymbol{\phi}}_{n-1}(z)\mathbf{v}_n \tag{3.24}$$

where by symmetry only the coefficients of $z, \ldots, z^{n-1}$ need be calculated using only the first two terms on the right. A reduced order equation of the original form is then obtained

$$\tilde{\boldsymbol{\phi}}_{n-1}(z)\boldsymbol{\lambda}_{n-1}(z^{-1}) + \tilde{\boldsymbol{\phi}}_{n-1}(z^{-1})\boldsymbol{\lambda}_{n-1}(z) = \boldsymbol{\omega}_{n-1}(z) \tag{3.25}$$

The achievement of its reduction step is equivalent to selected column operations on the original set of equations, defining new but reduced unknowns $\boldsymbol{\lambda}(z)$ and allowing the elimination of one equation and one variable. The reduction process continues until only the degree zero terms remain

$$\lambda_{0,0,ij} + \lambda_{0,0,ji} = \boldsymbol{\omega}_{0,0,i,j} \tag{3.26}$$

and a unique solution is obtained by $\boldsymbol{\lambda}_{0,0} = \boldsymbol{\omega}_{0,0}/2$ though an alternative is to make $\boldsymbol{\lambda}_{0,0}$ upper triangular.

The reduction steps are now reversed for $n = 1, \ldots, m$ to reconstruct the required solution. Setting on the right hand side of (3.23)

$$\boldsymbol{\lambda}^+_{n-1}(z) = \boldsymbol{\lambda}_{n-1}(z) + \mathbf{v}_n z^n \tag{3.27}$$

the reversed step is

$$\boldsymbol{\lambda}_n(z) = \boldsymbol{\beta}_n^{-1}[\boldsymbol{\lambda}^+_{n-1}(z) + \boldsymbol{\alpha}_n z^n \boldsymbol{\lambda}^+_{n-1}(z^{-1})] \tag{3.28}$$

In the case that $p < q$ the reduction procedure is still applied starting with $n = m$ but whilst $n > p$ the reduction step is simply to take $\boldsymbol{\phi}_{n-1}(z) = \boldsymbol{\phi}_n(z)$ and only the coefficients of $z^{n-p}, \ldots, z^{n-1}$ in $\boldsymbol{\omega}_n(z)$ need be corrected. Also the reverse step (3.28) need only be applied for $n = 1, \ldots, p$, becoming simply $\boldsymbol{\lambda}_n(z) = \boldsymbol{\lambda}^+_{n-1}(z)$ for $n > p$. The computational effort is thereby substantially reduced when $p \ll q$.

In the case that $p > q$ the reduction step starts with $n = m$, but whilst $n > q$ the quantities $\mathbf{v}_n$ are all zero so that the reduction (3.24) for $\boldsymbol{\omega}_n(z)$ is omitted.

The main computational burden is in the reduction of $\boldsymbol{\phi}_n(z)$, in particular the matrix products in (3.20). The reconstruction is relatively swift. In the scalar case Laurie (1982) dispensed with the divisions corresponding to the multiplication by $\boldsymbol{\beta}_n^{-1}$. That is one possible direction to seek for improvements in the multivariate case. Another direction is to look for structure in $\boldsymbol{\alpha}_n$, which starts for $n = p$ as the simple tensor product $\boldsymbol{\alpha}_{n,ij,kl} = \boldsymbol{\phi}_{n,n,ik}\delta_{jl}$. Multiplication by this matrix requires $O(d)$ fewer operations.

3.3 COMPARATIVE EFFICIENCY

The procedure used for comparison first requires expressing the model (3.1) in (block) state-space form, the particular choice used being

$$\begin{pmatrix} \mathbf{x}_{1,t} \\ \mathbf{x}_{2,t} \\ \vdots \\ \mathbf{x}_{m-1t} \\ \mathbf{x}_{m,t} \end{pmatrix} = \begin{pmatrix} \mathbf{0} & \mathbf{I} & \mathbf{0} & \cdots & 0 \\ \mathbf{0} & \mathbf{0} & \mathbf{I} & \cdots & 0 \\ \vdots & \vdots & \vdots & \ddots & \vdots \\ \mathbf{0} & \mathbf{0} & \mathbf{0} & \cdots & \mathbf{I} \\ -\boldsymbol{\Phi}_p & -\boldsymbol{\Phi}_{p-1} & \cdots & -\boldsymbol{\Phi}_2 & -\boldsymbol{\Phi}_1 \end{pmatrix} \begin{pmatrix} \mathbf{x}_{1,t-1} \\ \mathbf{x}_{2,t-1} \\ \vdots \\ \mathbf{x}_{m-1,t-1} \\ \mathbf{x}_{m,t-1} \end{pmatrix} + \begin{pmatrix} \boldsymbol{\Psi}_1 \\ \boldsymbol{\Psi}_2 \\ \vdots \\ \boldsymbol{\Psi}_{m-1} \\ \boldsymbol{\Psi}_m \end{pmatrix} \mathbf{e}_t \tag{3.29}$$

where $\mathbf{x}_{x,t}$ is the k-step ahead prediction of $\mathbf{x}_{t+k}$ made at time t so that

$$\mathbf{x}_{t+1} = \mathbf{x}_{1,t} + \mathbf{e}_{t+1}. \tag{3.30}$$

Summarizing (3.29) as $\mathbf{z}_t = \mathbf{T}\mathbf{z}_{l-1} + \mathbf{w}_l$ where $\mathbf{w}_t = \mathbf{K}\mathbf{e}_t$, the variance matrix of $\mathbf{w}_t$ is given by $\mathbf{W} = \mathbf{K}\boldsymbol{\Sigma}\mathbf{K}'$ and the variance matrix $\mathbf{V}$ of the stationary process $\mathbf{z}_t$ as the solution of

$$\mathbf{V} - \mathbf{T}\mathbf{V}\mathbf{T}' = \mathbf{W}. \tag{3.31}$$

The subroutine LYBAD of the SLICE library distributed by the Numerical Algorithms Group was used to solve this equation for $\mathbf{V}$ given $\mathbf{T}$ and $\mathbf{W}$ as defined above, the method being described by Hammarling (1982). Letting the first block column of $\mathbf{V}$ contain $d \times d$ matrices $\mathbf{V}_{0,0}, \ldots, \mathbf{V}_{m-1,0}$, the required covariances are given by

$$\boldsymbol{\Gamma}_k = \begin{cases} \mathbf{V}_{k,0} + \boldsymbol{\Psi}_k\boldsymbol{\Sigma} & ; \quad k = 0, \ldots, m-1 \\ -\boldsymbol{\Phi}_1\boldsymbol{\Gamma}_{k-1} - \cdots - \boldsymbol{\Phi}_m\boldsymbol{\Gamma}_{k-m} + \boldsymbol{\Theta}_m\boldsymbol{\Sigma} & ; \quad k = m \\ -\boldsymbol{\Phi}_1\boldsymbol{\Gamma}_{k-1} - \cdots - \boldsymbol{\Phi}_m\boldsymbol{\Gamma}_{k-m} & ; \quad k > m. \end{cases} \tag{3.32}$$

Whereas the new procedure proposed in the previous section takes a fixed number of arithmetic operations depending only on d, p and q, the LYBAD routine uses the Schur factorization of the transition matrix T, which though rapid is iterative. This should be remembered in comparing the two methods. The table below shows the times for the two methods for various choices of d and m, taking $p = q = m$. Although efficiencies may be gained when using the state-space method with $p \ll q$ they have not been implemented for this comparison. The times are for calculating $\boldsymbol{\Gamma}_0, \ldots, \boldsymbol{\Gamma}_m$, the calculations thereafter being the same for both methods. A modern 386 PC was used for the calculations, the smallest unit for timing being about .05 seconds.

A regression was carried out for the logarithms of the ratios of the times for the Euclid method to the times for the state-space method, the regressors being the logarithms of m and d. A good linear fit was observed giving the approximate formula for the relative times as $5.83m^{-1.656}d^{1.385}$, so that the preference would be for the Euclid method whenever $m > 2.9d^{0.84}$.

There is one numerical difficulty with the Euclid method. As mentioned earlier, it is equivalent to carrying out selected column operations on the matrix of the linear equations (3.10). This does not allow for pivoting, and in the univariate case this is not necessary because the stationarity condition ensures that the divisor $\boldsymbol{\beta}_n$ in (3.20), a scalar in that case, is always positive. In the multivariate case when $\boldsymbol{\beta}_n$ is a matrix, it might happen to be singular and cause the procedure to fail. A simple example is when

$$\boldsymbol{\Phi}(B) = \begin{pmatrix} 1 & 0 \\ 0 & 1 \end{pmatrix} + \begin{pmatrix} 0 & 1 \\ 1 & 0 \end{pmatrix} B + \begin{pmatrix} 1 & 0 \\ 0 & 0 \end{pmatrix} B^2 \tag{3.33}$$

for which the characteristic polynomial det $\boldsymbol{\Phi}(B)$ is identically one, so there is certainly no stationarity problem. In this particular case multiplying both autoregressive and moving average operators by the scalar factor $(1 + .5B)$ eliminates the numerical problem and gives the correct autocovariances. Perhaps even simpler transformations may do the same. This is one area which requires further investigation. A similar problem could well arise with the block Toeplitz approach described by Mittnik (1992) and in that case modifications given by Gover and Barnett (1985) might be extended to overcome the problem. In general the procedure appears to have a valuable advantage in the case of higher order models. Of course the search will continue for methods which combine the advantage of both the Euclid and state-space approaches.

3.4 COVARIANCE FACTORIZATION

The multivariate covariance factorization problem is to construct a moving average model with specified covariances. In the notation of the introduction to this paper it is to solve

$$\boldsymbol{\Theta}(z)\boldsymbol{\Sigma}\boldsymbol{\Theta}'(z^{-1}) = \boldsymbol{\Gamma}(z) = \sum_{k=-q}^{q} \boldsymbol{\Gamma}_k z^k \tag{3.34}$$

for $\boldsymbol{\Theta}(z) = \Sigma_{i=0}^{q} \boldsymbol{\Theta}_i z^i$ and $\boldsymbol{\Sigma}$ given $\boldsymbol{\Gamma}(z)$. It is convenient now to take $\boldsymbol{\Sigma} = \mathbf{I}$ and $\boldsymbol{\Theta}_0$ to be upper triangular to specify a unique solution, though for practical applications the transformation to the solution with $\boldsymbol{\Theta}_0 = \mathbf{I}$ is preferable.

The relevance to the foregoing sections is that an iterative solution with good convergence properties is given by calculating the sequence

$$\boldsymbol{\Theta}_{i+1}(z) = \boldsymbol{\Theta}_i(z) + \boldsymbol{\delta}_i(z) \tag{3.35}$$

where $\boldsymbol{\delta}_i(z)$ is obtained from

$$\boldsymbol{\Theta}_i(z)\boldsymbol{\delta}_i'(z^{-1}) + \boldsymbol{\delta}_i(z)\boldsymbol{\Theta}_i'(z^{-1}) = \boldsymbol{\Gamma}(z) - \boldsymbol{\Theta}_i(z)\boldsymbol{\Theta}_i'(z^{-1}) = \boldsymbol{\Delta}_i(z). \tag{3.36}$$

This is identical in form to the equations (3.8) for $\boldsymbol{\Lambda}(B)$ given in the introduction except for the convention there that $\boldsymbol{\Phi}_0 = \mathbf{I}$. In order to use the algorithm presented in section two the equation may be premultiplied by $\boldsymbol{\Theta}_{i,0}^{-1}$ and postmultiplied by its transpose so that $\boldsymbol{\Theta}_{i,0}^{-1}\boldsymbol{\Theta}_i(z)$, which does have a constant term of $\mathbf{I}$, replaces $\boldsymbol{\Theta}_i(z)$. The solution then obtained is $\boldsymbol{\Theta}_{i,0}^{-1}\boldsymbol{\delta}_i(z)$, which on premultiplication by $\boldsymbol{\Theta}_{i,0}$ gives the required solution $\boldsymbol{\delta}_i(z)$.

Details of choice of starting values and convergence properties of this iterative procedure are given by Tunnicliffe Wilson (1972) and (1979). In the scalar case Demeure and Mullis (1990) employ the Euclid algorithm in the iterations, and Johnson (1989), though concentrating also on the scalar case, makes reference of the matrix case. Because the procedure is iterative and to some extent self correcting, an approximate procedure for solving the equations (3.36) is possible. The one considered here corresponds to the spectral approach to covariance generation defined at the end of the introduction. Rewriting (3.36) as

$$\boldsymbol{\Theta}_i(z)^{-1}\boldsymbol{\delta}_i(z) + \{\boldsymbol{\Theta}_i(z^{-1})^{-1}\boldsymbol{\delta}_i(z^{-1})\}' = \boldsymbol{\Theta}_i(z)^{-1}\boldsymbol{\Gamma}(z)\boldsymbol{\Theta}_i'(z^{-1})^{-1} - \mathbf{I} \tag{3.37}$$

the solution is obtained from

$$\boldsymbol{\Theta}_i(z)^{-1}\boldsymbol{\delta}_i(z) = [\boldsymbol{\Theta}_i(z)^{-1}\boldsymbol{\Gamma}(z)\boldsymbol{\Theta}_i'(z^{-1})^{-1} - \mathbf{I}]^+, \tag{3.38}$$

where the operation $[\cdots]^+$ selects just the coefficients of positive powers of z, and makes the constant term upper triangular with its diagonal terms halved. The computations are most efficiently carried out in the frequency domain by evaluating all the expressions at a grid of points $z_j = e^{2\pi ij/N}$ on the unit circle. Then all the matrix operations are carried out pointwise for these z_j and $[\cdots]^+$ is efficiently implemented by two applications of the finite Fourier transform to each sequence element of its matrix argument. Finally,

$$\boldsymbol{\delta}_i(z) = \boldsymbol{\Theta}_i(z)[\boldsymbol{\Theta}_i(z)^{-1}\boldsymbol{\Gamma}(z)\boldsymbol{\Theta}_i'(z^{-1})^{-1} - \mathbf{I}]^+ \tag{3.39}$$

Problems arise now when det $\boldsymbol{\Gamma}(z)$ has zeros close to or on the unit circle. Then the errors which arise at each iteration due to using the finite Fourier transformation to approximate Fourier integrals, cannot be eliminated by the error correcting property. Nevertheless, even in this case, the method rapidly obtains a close approximation and then the Euclid method may be used for a few final iterations to improve the solution. The spectral method is certainly more efficient for higher dimensions d. The time taken depends on the order of the transform used rather than on the order of the moving average operator, but to compare with the examples recorded in Table 3.1, using a 256-point transform, the times were 0.11, 1.15, 3.73 and 15.88 seconds per iteration for $d = 1$, 3, 5 and 9 respectively. These times increase as $d^{2.25}$ and the dependence on the order N of the transform is very nearly linear.

Table 3.1 Times in seconds for calculation of covariances for various dimensions d and orders m: upper figures for the Euclid method, lower figures for the state-space method

	d			
m	1	3	5	9
1	0	0	.43956	13.736
	0	0	0	.10989
4	0	.27473	3.901	121.87
	0	.21978	.93407	5.659
9	.05494	.93407	14.835	460.22
	.16484	2.3626	11.978	82.473
15	.05494	2.3077	36.978	1149.99
	.60440	12.088	65.110	546.87
30	.10989	8.4615	136.04	
	4.4505	132.86	799.40	

3.5 CONCLUSION

The problems considered in this paper are still attracting interest because of their applications in multivariate time series modelling and control. Systems of high dimension and high order are being studied and in these cases improvements such as those demonstrated in this paper are valuable.

REFERENCES

Ansley, C.F. (1980) Computation of the theoretical autocovariance function for a vector ARMA process. *J. Statist. Comput. Simul.*, **12**, 15–24.

Barone, P. (1987) A method for generating independent realization of a multivariate normal stationary and invertible ARMA (p, q) process. *J. Time Series Analysis*, **8**, 125–130.

Box, G.E.P. and Jenkins, G.M. (1976) *Time Series Analysis, Forecasting and Control.* 2nd edn, Holden-Day, San Francisco.

Demeure, C.J. and Mullis, C.T. (1989) The Euclid algorithm and the fast computation of cross covariance and autocovariance sequences. *IEEE Trans. Acoust., Speech, Signal Processing*, **37**, 545–552.

Demeure, C.J. and Mullis, C.T. (1990) A Newton–Raphson method for moving-average spectral factorisation using the Euclid Algorithm. *IEEE Trans. Acoust., Speech, Signal Processing*, **38**, 1697–1709.

Durbin, J. (1960) The fitting of time series models. *Rev. Int. Inst. Stat.*, **28**, 233.

Dym, H and Young, N.J. (1990) A Schur–Cohn theorem for matrix polynomials. *Proc. Edinburgh Math. Soc.*, **33**, 337–366.

Gover, M.J.C. and Barnett, S. (1985) Inversion of Toeplitz matrices which are not strongly nonsingular. *IMA J. Numer. Anal.*, **5**, 101–110.

Hammarling, S.J. (1982) Numerical solution of the stable non-negative definite Lyapunov equation. *IMA J. Numer. Anal.*, **2**, 303–325.

Johnson, M.A. (1989) Newton schemes for polynomial spectral factorization: scalar case. *Int. J. Control*, **49**, 309–349.

Laurie, D.P. (1982) Cramèr–Wold factorisation. Algorithm AS175. *Applied Statistics*, **31**, 86–93.

Mélard, G. (1984) A fast algorithm for the exact likelihood of autoregressive moving-average models. *Applied Statistics*, **33**, 104–114.

Mittnik, S. (1993) Computing theoretical autocovariances of multivariate ARMA models using a Block-Levinson Method. *J. Royal Statist. Soc. B*, **55**, 435–440.

Shea, B.L. (1988) A note on the generation of independent realizations of a vector autoregressive moving-average process. *J. Time Series Analysis*, **9**, 403–410.

Tunnicliffe Wilson, G. (1972) The factorisation of matricial spectral densities. *SIAM J. Appl. Math.*, **23**, 420–426.

Tunnicliffe Wilson, G. (1978) A convergence theorem for spectral factorization. *J. Multivariate Analysis*, **8**, 222–232.

Tunnicliffe Wilson, G. (1979) Some efficient computational procedures for high-order ARMA models. *J. Statist. Comput. Simul.*, **8**, 301–309.

Whittle, P. (1963) On the fitting of multivariate autoregression and the approximate canonical factorisation of a spectral density matrix. *Biometrika*, **50**, 129–134.

Young, N.J. (1980) Formulae for the solution of Lyapunov matrix equations. *Int. J. Control*, **31**, 159–179.

4

Incorporating and deleting information in dynamic models

P.J. Harrison and P.P. Veerapen

4.1 INTRODUCTION

Having known Maurice for more than 30 years, it is a great pleasure to contribute to this volume in celebration of his 60th birthday and to honour his contribution to time series, as for example in his masterly work Priestley (1981).

In sequential dynamic modelling important continuing aims are forecasting, retrospective analysis, model monitoring and model maintenance. New information is combined with historic information both in order to update forecasts, and to reassess what happened in the past. This is a routine operation as long as the model monitoring statistics are satisfactory. These statistics may arise from sequences of sequential tests operating on the most recent information or from diagnostics, and may lead to a desire to eliminate a subset of the information. For example, such a subset may relate to wars in economics, strikes in commerce, suspected analytic or measurement unreliability in production, and periods of unidentified or unmeasured competitive activity in marketing. This paper is concerned with these issues in relation to dynamic models.

In section 4.2 the dynamic linear model is defined as in Harrison and Stevens (1976), and in West and Harrison (1989a). In section 4.3 basic results for incorporating and deleting information are derived in a general setting, when the observational variance is known only up to a scalar factor. A key property of this model is conditional independence as in Dawid (1979), Smith (1990) and Lauritzen *et al.* (1990). Section 4.4 derives powerful results which apply to normal models and thus also to a variety of linear estimation methods including the filtering of Kalman (1963). The particular application to dynamic linear models is then discussed. The incorporation of information is illustrated in section 4.5.1, where the full historic state distribution is given and its calculation through backward recurrence relationships seen as an immediate consequence of conditional independence. The subset of discount weighted regression dynamic models, defined in section 4.5.2, exhibits very neat procedures and provides a link with static models and least-squares procedures as featured in Ameen and Harrison (1984). Some limiting results for constant dynamic linear models are presented in section 4.5.3. The deletion of information is considered in section 4.6, where in section 4.6.1 dual recurrence relationships are derived for the elimination of a single observation. This immediately leads to a simple operation for finite truncated models where the forecasts are based on at most the last l time periods. Section 4.6.2 generalizes to the deletion of any set of past observations with a very elegant procedure for revising distributions. The resulting jack-knifed posterior state and various predictive distributions provide the basis for deriving relevant diagnostics such as those advocated by Smith and Pettit (1985), Bernardo (1985) and Johnson and Geisser (1983), and generalize their application in Harrison and West (1991). A stochastic variance model is considered in section 4.6.3 together with a method of deriving its current jack-knifed distribution. Since autoregressive integrated moving averages and linear state space models may be represented in dynamic linear model form, the paper has a direct bearing on those of de Jong (1988), Bruce and Martin (1989), and Kohn and Ansley (1989). Section 4.6.4 briefly discusses diagnostics in relation to the modelling objectives and section 4.7 concludes by outlining further applications of the results relating to subjective information and the combination of forecasts.

4.2 THE DYNAMIC LINEAR MODEL

Consider the time series $\{\mathbf{Y}_t; t = 1, 2, \ldots\}$ represented by the dynamic linear model $\{\mathbf{F}, \mathbf{G}, \mathbf{V}, \mathbf{W}\}_t$ defined as follows:

$$\mathbf{Y}_t = \mathbf{F}_t'\boldsymbol{\theta}_t + \mathbf{v}_t, \quad \mathbf{v}_t|\phi \sim N(0; \mathbf{V}_t/\phi), \tag{4.1}$$

where at time t, $\mathbf{Y}_t$ is a $k \times 1$ observation vector, $\mathbf{F}_t$ is a known $p \times k$ design matrix, $\boldsymbol{\theta}_t$ is the state $p \times 1$ vector, ϕ is a scalar and $\mathbf{v}_t$ is a $k \times 1$ observational error vector. The state evolution is represented by

$$\boldsymbol{\theta}_t = \mathbf{G}_t\boldsymbol{\theta}_{t-1} + \boldsymbol{\omega}_t, \quad \boldsymbol{\omega}_t|\phi \sim N(\mathbf{0}; \mathbf{W}_t/\phi), \tag{4.2}$$

where $\mathbf{G}_t$ is a known $p \times p$ matrix and $\boldsymbol{\omega}_t$ a $p \times 1$ evolution error vector. The error sequences $\{\mathbf{v}_t\}$ and $\{\boldsymbol{\omega}_t\}$ are, without loss in generality, assumed independent and mutually independent. The variance matrices $\mathbf{V}_t$ and $\mathbf{W}_t$ are assumed known at time t, the latter often being specified using discount factors. For each time t, let D_t denote the information available up to and including that time. If the realized value of $\mathbf{Y}_t$ is $\mathbf{y}_t$, then in the non-intervention model, which is first taken for clarity, $D_t = \{D_{t-1}, \mathbf{y}_t\}$. The model specification is completed through an initial prior distribution $(\boldsymbol{\theta}_1, \phi | D_0)$ which is assumed to be independent of the error sequences $\{\mathbf{v}_t\}$ and $\{\boldsymbol{\omega}_t\}$. The distribution $(\boldsymbol{\theta}_1 | D_0, \phi)$ is normal. The known variance model sets $\phi = 1$. In the unknown variance model $\phi | D_0$ is a gamma distribution.

The analysis is now standard, calculations following easily from the fact that, given ϕ, any collection of components of the sequences $\{\mathbf{Y}_t\}$ and $\{\boldsymbol{\theta}_t\}$ are jointly normally distributed. The sequential updating and forecasting equations, related to Kalman filter equations, will be familiar to most readers. The related retrospective filtering, or smoothing equations are of interest since they provide the basis for retrospective analysis and model assessment. Given information, $D_n = \{D_0, (\mathbf{y}_t; t = 1, \ldots, n)\}$, the state vectors will be normally distributed with

$$\boldsymbol{\theta}_t | \phi, D_n \sim N(\mathbf{a}_{n,t}; \mathbf{R}_{n,t}/\phi) \ \ (t = 1, \ldots, n). \tag{4.3}$$

It is important that $\mathbf{a}_{n,t}$ depends linearly on the observations $\{\mathbf{y}_t\}$, whilst $\mathbf{R}_{n,t}$ is independent of them. It is clear that, at any time t, given the state $(\boldsymbol{\theta}_t, \phi)$, the future is independent of the history, and that $\mathbf{Y}_t$ is independent of both. Important implications derive from the conditional independence properties of normal models developed in section 4.4, in conjunction with the results of section 4.3.

4.3 GENERAL RESULTS FOR UPDATING AND DELETION

4.3.1 General results for dynamic models

In the following theorems, the $h \times 1$ vector $\mathbf{U}$ may be any set of observations, $\mathbf{c}$ a known matrix, and $\mathbf{Z}$ any vector of observations and states, not necessarily including the states corresponding to the observational times. Further, $\mathbf{U}$ may also represent subjective information and external forecasts which are modelled as in West and Harrison (1989b, 1989a). Consequently the results are widely applicable. Throughout the paper statements such as '$\mathbf{A}$ is the regression matrix of $\mathbf{Z}$ on $\mathbf{U}$', are to be understood as being conditional upon a joint normal distribution with known variance.

Let the following be proper distributions

$$(\mathbf{U} | \mathbf{Z} = \mathbf{z}, \phi) \sim N(\mathbf{c}'\mathbf{z}; \boldsymbol{\Sigma}_{u|z}/\phi),$$
$$(\mathbf{Z} | \mathbf{U} = \mathbf{u}, \phi) \sim N(\boldsymbol{\mu}_{z|u}; \boldsymbol{\Sigma}_{z|u}/\phi),$$
$$\mathbf{Z} | \phi \sim N(\boldsymbol{\mu}_z; \boldsymbol{\Sigma}_z/\phi).$$

Write $\mathbf{A}$ as the regression matrix of $\mathbf{Z}$ on $\mathbf{U}$, and

$$\mathbf{e} = \mathbf{u} - \mathbf{c}'\boldsymbol{\mu}_z; \boldsymbol{\Sigma}_u = \boldsymbol{\Sigma}_{u|z} + \mathbf{c}'\boldsymbol{\Sigma}_z\mathbf{c}; \ \ \mathbf{A} = \mathbf{A}_{z,u} = \boldsymbol{\Sigma}_z\mathbf{c}\boldsymbol{\Sigma}_u^{-1}. \tag{4.4}$$

Standard normal results for incorporating the information $\mathbf{U} = \mathbf{u}$ are

$$\boldsymbol{\mu}_{z|u} = \boldsymbol{\mu}_z + \mathbf{A}\mathbf{e}, \tag{4.5}$$
$$\boldsymbol{\Sigma}_{z|u} = \boldsymbol{\Sigma}_z - \mathbf{A}\mathbf{c}'\boldsymbol{\Sigma}_z. \tag{4.6}$$

Dual expressions, for eliminating the information $\mathbf{U} = \mathbf{u}$, are required for calculating deletion diagnostics and discarding suspect data sets. Write

$$\mathbf{d} = \mathbf{u} - \mathbf{c}'\boldsymbol{\mu}_{z|u}; \mathbf{q} = \boldsymbol{\Sigma}_{u|z} - \mathbf{c}'\boldsymbol{\Sigma}_{z|u}\mathbf{c}; \mathbf{B} = \boldsymbol{\Sigma}_{z|u}\mathbf{c}\mathbf{q}^{-1}. \tag{4.7}$$

Theorem 4.1

Given $\mathbf{U}|\mathbf{z}$ and $\mathbf{Z}|\mathbf{u}$:

(a) the leverage of $\mathbf{U}$ on $\mathbf{Z}$, as measured by $\mathbf{A}$, is calculable as

$$\mathbf{A} = \mathbf{A}_{z,u} = \boldsymbol{\Sigma}_{z|u}\mathbf{c}\boldsymbol{\Sigma}_{u|z}^{-1}; \tag{4.8}$$

(b) the moments of the distribution of $\mathbf{Z}$ with $\mathbf{u}$ deleted are

$$\boldsymbol{\mu}_z = \boldsymbol{\mu}_{z|u} - \mathbf{B}\mathbf{d}, \tag{4.9}$$

$$\boldsymbol{\Sigma}_z = \boldsymbol{\Sigma}_{z|u} + \mathbf{B}\mathbf{c}'\boldsymbol{\Sigma}_{z|u}, \tag{4.10}$$

(c) the jack-knifed residual $\mathbf{e}$ is calculable as

$$\mathbf{e} = \boldsymbol{\Sigma}_u\boldsymbol{\Sigma}_{u|z}^{-1}\mathbf{d} = \boldsymbol{\Sigma}_{u|z}\mathbf{q}^{-1}\mathbf{d}. \tag{4.11}$$

The proof is given in the appendix to this chapter.

For the unknown variance case, with known scalars v and s let

$$vs\phi \sim \chi_v^2. \tag{4.12}$$

Given $\mathbf{U} = \mathbf{u}$, a standard result is $v_u s_u \phi | u \sim \chi_{v_u}^2$ with

$$v_u = v + h, \tag{4.13}$$

$$v_u s_u = vs + \mathbf{e}'\boldsymbol{\Sigma}_u^{-1}\mathbf{e}. \tag{4.14}$$

Theorem 4.2

Given $\phi|\mathbf{u}$ the distribution (4.12) of ϕ deleting the information $\mathbf{U} = \mathbf{u}$ may be obtained by calculating $v = v_u - h$ and

$$vs = v_u s_u - \mathbf{d}'\mathbf{q}^{-1}\mathbf{d}. \tag{4.15}$$

The proof follows from (4.11) and (4.14).

4.3.2 Dynamic linear model with unknown variance

The model of section 4.2, which is a particular form of that in section 4.3.1, is described in West and Harrison (1989a). Conditioning everything on D_{t-1}, and taking $\mathbf{U} = \mathbf{Y}_t$, $\mathbf{Z} = \boldsymbol{\theta}_t$, $\mathbf{c} = \mathbf{F}_t$, the correspondence is

$$\mathbf{Y}_t|\boldsymbol{\theta}_t, \phi \sim N(\mathbf{F}_t'\boldsymbol{\theta}_t; \mathbf{V}_t/\phi),$$

$$\boldsymbol{\theta}_t|D_t, \phi \sim N(\mathbf{a}_{t,t}; \mathbf{R}_{t,t}/\phi),$$

$$\theta_t|D_{t-1},\phi \sim N(\mathbf{a}_{t-1,t};\mathbf{R}_{t-1,t}/\phi),$$
$$v_t s_t \phi|D_t \sim \chi^2_{v_t},$$
$$\mathbf{a}_{t-1,t} = \mathbf{G}_t \mathbf{a}_{t-1,t-1},$$
$$\mathbf{R}_{t-1,t} = \mathbf{G}_t \mathbf{R}_{t-1,t-1} \mathbf{G}'_t + \mathbf{W}_t,$$
$$\mathbf{e}_t = \mathbf{y}_t - \mathbf{F}'_t \mathbf{a}_{t-1,t},$$
$$\mathrm{var}(\mathbf{Y}_t|D_{t-1},\phi) = \mathbf{Q}_t/\phi \quad \text{where } \mathbf{Q}_t = \mathbf{F}'_t \mathbf{R}_{t-1,t}\mathbf{F}_t + \mathbf{V}_t.$$

The regression matrix of θ_t on $\mathbf{Y}_t$ given D_{t-1} is written $\mathbf{A}_t$. The known variance model is equivalent to taking $\phi = 1$.

4.4 CONDITIONAL INDEPENDENCE RESULTS

In the following theorems, proofs of which are in the Appendix, let **X**, **Z**, and **U** be normal random vectors and write the regression matrices of **X** on **Z** and of **Z** on **U** as $\mathbf{A}_{x,z}$ and $\mathbf{A}_{z,u}$ so that

$$\begin{matrix}\mathbf{X}\\ \mathbf{Z}\\ \mathbf{U}\end{matrix} \sim N\left(\begin{matrix}\mu_x\\ \mu_z;\\ \mu_u\end{matrix}\begin{bmatrix}\Sigma_x & \mathbf{A}_{x,z}\Sigma_z & \mathbf{A}_{x,u}\Sigma_u\\ \cdot & \Sigma_z & \mathbf{A}_{z,u}\Sigma_u\\ \cdot & \cdot & \Sigma_u\end{bmatrix}\right).$$

Let **X** and **U** be conditionally independent given **Z**, written $\mathbf{X} \perp\!\!\!\perp \mathbf{U}|\mathbf{Z}$.

Theorem 4.3

The regression matrix $\mathbf{A}_{x,u}$ of **X** on **U** is the product of the regression matrices of **X** on **Z** and of **Z** on **U**, so that

$$\mathbf{A}_{x,u} = \mathbf{A}_{x,z}\mathbf{A}_{z,u}, \tag{4.16}$$
$$\mathrm{cov}(\mathbf{x},\mathbf{u}) = \mathbf{A}_{x,z}\mathbf{A}_{z,u}\Sigma_u. \tag{4.17}$$

Corollary 4.4

If $\mathbf{X}_1,\ldots,\mathbf{X}_n$, are random vectors such that for all i and for all $1 < j < l < n$, $\mathbf{X}_i \perp\!\!\!\perp \mathbf{X}_{i+l}|\mathbf{X}_{i+j}$, with $\mathbf{A}_{i,i+1}$ as the regression matrix of $\mathbf{X}_i$ on $\mathbf{X}_{i+1}$, then the regression matrix of $\mathbf{X}_1$ on $\mathbf{X}_n$ is

$$\mathbf{A}_{1,n} = \prod_{i=1}^{n-1} \mathbf{A}_{i,i+1}. \tag{4.18}$$

Theorem 4.5

Given $\mathbf{U} = \mathbf{u}$, write the conditional distributions $\mathbf{X}|\mathbf{U} \sim N(\mu_{x|u};\Sigma_{z|u})$ and $\mathbf{Z}|\mathbf{U} \sim N(\mu_{z|u};\Sigma_{z|u})$, then:

(a) $\mathbf{X}$ and $\mathbf{Z}$ are jointly normal, with

$$\boldsymbol{\mu}_{x|u} = \boldsymbol{\mu}_x + \mathbf{A}_{x,z}(\boldsymbol{\mu}_{z|u} - \boldsymbol{\mu}_z), \tag{4.19}$$

$$\boldsymbol{\Sigma}_{x|u} = \boldsymbol{\Sigma}_x + \mathbf{A}_{x,z}(\boldsymbol{\Sigma}_{z|u} - \boldsymbol{\Sigma}_z)\mathbf{A}'_{x,z}; \tag{4.20}$$

(b) given $\mathbf{u}$, the regression matrix $\mathbf{A}_{(x,z)|u}$ of $\mathbf{X}$ on $\mathbf{Z}$ remains as $\mathbf{A}_{x,z}$, but the regression matrix $\mathbf{A}_{(z,x)|u}$ changes so that

$$\text{cov}(\mathbf{x}, \mathbf{z}|\mathbf{u}) = \mathbf{A}_{x,z}\boldsymbol{\Sigma}_{z|u}, \tag{4.21}$$

$$\mathbf{A}_{(z,x)|u} = \boldsymbol{\Sigma}_{z|u}\mathbf{A}'_{x,z}\boldsymbol{\Sigma}^{-1}_{x|u}; \tag{4.22}$$

(c) the regression matrix of $\mathbf{X}$ on $\mathbf{U}$ may be written

$$\mathbf{A}_{x,u} = \mathbf{A}_{x,z}\mathbf{A}_{z,u} = \boldsymbol{\Sigma}_{x|u}\mathbf{A}'_{(z,x)|u}\boldsymbol{\Sigma}^{-1}_{z|u}\mathbf{A}_{z,u}. \tag{4.23}$$

4.5 INCORPORATING INFORMATION IN DYNAMIC MODELS

4.5.1 Incorporating information

The key use of the theorems in sections 4.4 and 4.5 is that the information about $\mathbf{X}$ provided by $\mathbf{U}$ may be derived simply from the posterior distribution of $\mathbf{Z}|\mathbf{U}$ together with the prior regression matrix of $\mathbf{X}$ on $\mathbf{Z}$. Consequently they are of major importance in dynamic modelling and provide a general foundation for incorporating information whether it be observational, subjective, or obtained from other forecasting systems. In learning and updating with known variance $\phi = 1$, special cases are the normal Kalman filter state updating equations, obtained when $\mathbf{Z} = \mathbf{U} = \mathbf{Y}_n$ with $\mathbf{X} = \boldsymbol{\theta}_n$ and the retrospective backward recurrence relations obtained when, for all $t < n$, $\mathbf{X} = \boldsymbol{\theta}_t$, $\mathbf{Z} = \boldsymbol{\theta}_{t+1}$ and $\mathbf{U} = \mathbf{Y}_n$. The following theorem provides a simple derivation of the full state distribution, whether or not the observation variances are known up to the scalar ϕ.

Theorem 4.6

For model section 4.3.2, given D_n, for all n and $t < t + j \leqslant n$:

(a) the regression matrices $\mathbf{A}_{t,t+j}$ of $\boldsymbol{\theta}_t$ on $\boldsymbol{\theta}_{t+j}$ remain constant and

$$\mathbf{A}_{t,t+1} = \mathbf{R}_{t,t}\mathbf{G}'_{t+1}\mathbf{R}^{-1}_{t,t+1}; \tag{4.24}$$

(b) given ϕ, the joint distribution of the historic states is normal and is defined by the marginal distributions $\boldsymbol{\theta}_t|D_n, \phi \sim N(\mathbf{a}_{n,t}; \mathbf{R}_{n,t}/\phi)$ and the covariances

$$\text{cov}(\boldsymbol{\theta}_t, \boldsymbol{\theta}_{t+j}|D_n, \phi) = \prod_{i=0}^{j-1}\mathbf{A}_{t+i,t+i+1}\frac{\mathbf{R}_{n,t+j}}{\phi} \tag{4.25}$$

where, starting with $\mathbf{a}_{n,n}$ and $\mathbf{R}_{n,n}$, the moments may be calculated using the backward recurrence relations

$$\mathbf{a}_{n,t} = \mathbf{a}_{n-1,t} + \mathbf{A}_{t,t+1}(\mathbf{a}_{n,t+1} - \mathbf{a}_{n-1,t+1}), \tag{4.26}$$

$$\mathbf{R}_{n,t} = \mathbf{R}_{n-1,t} + \mathbf{A}_{t,t+1}(\mathbf{R}_{n,t+1} - \mathbf{R}_{n-1,t+1})\mathbf{A}'_{t,t+1}; \tag{4.27}$$

(c) $v_n s_n \phi | D_n \sim \chi^2_{v_n}$, where

$$v_n = v_{n-1} + k, \tag{4.28}$$

$$v_n s_n = v_{n-1} s_{n-1} + \mathbf{e}'_n \mathbf{Q}_n^{-1} \mathbf{e}_n. \tag{4.29}$$

The marginal distributions of the $\boldsymbol{\theta}$s are multivariate T.

The proofs are straightforward, (4.24) and (4.25) follow from (4.2), (4.3) and (4.18); (4.26) and (4.27) from (4.19) and (4.20); and (4.28) and (4.29) from (4.12)–(4.14).

4.5.2 Discount weighted regression

Discounting is widely used in practical forecasting systems as in Ameen and Harrison (1984, 1985) and in West and Harrison (1989a). For each time t a particular discount model, called the discount weighted regression model, has an associated single discount factor $0 < \delta_t \leqslant 1$, takes $\mathbf{G}_t = \mathbf{I}$, and sets $\mathbf{W}_t = (\delta_t^{-1} - 1)\mathbf{R}_{t-1,t}$. The name is derived from the fact that with the usual reference prior, West and Harrison (1989a), the mean, $E(\boldsymbol{\theta}_n | D_n) = \mathbf{a}_{n,n}$, corresponds exactly to the discount regression estimates for $\boldsymbol{\theta}$ derived by minimizing the discounted weighted sum of squares Δ_n, which may be written in recursive form as

$$\Delta_n = (\mathbf{y}_n - \mathbf{F}'_n\boldsymbol{\theta})'\mathbf{V}_n^{-1}(\mathbf{y}_n - \mathbf{F}'_n\boldsymbol{\theta}) + \delta_n \Delta_{n-1}.$$

Note that this is equivalent to a procedure which derives $E(\boldsymbol{\theta}_n | D_n)$ using the regression static model $\{\mathbf{F}_t, \mathbf{I}, \mathbf{V}_t/\delta^{n-t}, \mathbf{0}\}$. The reason for featuring these dynamic models is their correspondence with regression and the neat recursive properties which arise since for all $0 < t < t + k \leqslant n$

$$\mathbf{A}_{t,t+1} = \delta_{t+1}\mathbf{I}. \tag{4.30}$$

An elegant form for deriving the full state distribution is

$$\mathbf{a}_{n,t} = \mathbf{a}_{n-1,t} + \mathbf{A}_n \mathbf{e}_n \prod_{j=t+1}^{n} \delta_j,$$

$$\mathbf{R}_{n,t} = \mathbf{R}_{n-1,t} - \mathbf{A}_n \mathbf{Q}_n \mathbf{A}'_n \prod_{j=t+1}^{n} \delta_j^2.$$

When for all t, $\delta_t = \delta$, the model is called an exponentially weighted regression model and the form is especially neat.

4.5.3 Some limit results

For the observable constant dynamic linear model $\{\mathbf{F}, \mathbf{G}, \mathbf{V}, \mathbf{W}\}$, which, after the initial prior D_0, is closed to intervention, the information $D_n = \{D_0, \mathbf{y}_1, \ldots, \mathbf{y}_n\}$. Based solely on the model form, if all the eigenvalues of $\mathbf{G}$ lie inside the unit circle then the random time series $\mathbf{Y}_n$ is itself stationary; otherwise, a linear function of the series is stationary. In either case, given D_n and given any non-negative integer l, as n increases define the $p \times k$ matrix $\mathbf{A}$, the $p \times p$ matrices $\mathbf{A}_\theta$ and $\mathbf{R}_l$ and the $k \times k$ variance matrix $\mathbf{Q}$ by

$$\lim_{n\to\infty} \mathbf{A}_n = \mathbf{A}, \quad \lim_{n\to\infty} \mathbf{A}_{n,n-l} = \mathbf{A}_\theta^l, \quad \lim_{n\to\infty} \mathbf{R}_{n,n-l} = \mathbf{R}_l, \quad \lim_{n\to\infty} \mathbf{Q}_n = \mathbf{Q}.$$

Then

$$\lim_{n\to\infty} (\mathbf{a}_{n,n-l} - \mathbf{a}_{n-1,n-l} - \mathbf{A}_\theta^l \mathbf{A}\mathbf{e}_n) = \mathbf{0}, \tag{4.31}$$

$$\mathbf{R}_l = \mathbf{R}_{l-1} - \mathbf{A}_\theta^l \mathbf{A}\mathbf{Q}\mathbf{A}'\mathbf{A}_\theta'^l. \tag{4.32}$$

4.6 DELETION OF INFORMATION AND DIAGNOSTICS

4.6.1 Deletion of a single observation $\mathbf{y}_\tau$

Consider the deletion of observation $\mathbf{y}_z$. Corresponding to (4.7) write

$$\mathbf{d}_\tau = \mathbf{y}_\tau - \mathbf{F}_\tau'\mathbf{a}_{n,\tau}; \quad \mathbf{q}_\tau = \mathbf{V}_\tau - \mathbf{F}_\tau'\mathbf{R}_{n,\tau}\mathbf{F}_\tau; \quad \mathbf{B}_\tau = \mathbf{R}_{n,\tau}\mathbf{F}_\tau\mathbf{q}_\tau^{-1}, \tag{4.33}$$

and $D = D_n - \mathbf{y}_\tau$. Then $(\boldsymbol{\theta}_\tau | D, \phi) \sim N(\mathbf{a}_{n,\tau}^*, \mathbf{R}_{n,\tau}^*/\phi)$, where from (4.9) and (4.10)

$$\mathbf{a}_{n,\tau}^* = \mathbf{a}_{n,\tau} - \mathbf{B}_\tau\mathbf{d}_\tau \quad \text{and} \quad \mathbf{R}_{n,\tau}^* = \mathbf{R}_{n,\tau} + \mathbf{B}_\tau\mathbf{F}_\tau'\mathbf{R}_{n,\tau}. \tag{4.34}$$

Since $\mathbf{A}_{t,t} = \mathbf{I}$ and, for $t < \tau \leqslant n$, $\mathbf{A}_{t,\tau} = \mathbf{A}_{t,t+1}\mathbf{A}_{t+1,\tau}$, the full jack-knifed state distribution may be obtained from the regression matrices $\mathbf{A}_{t,t+1}$ and the marginal distributions calculated as follows.

Theorem 4.7

$\boldsymbol{\theta}_t | D, \phi \sim N(\mathbf{a}_{n,t}^*, \mathbf{R}_{n,t}^*/\phi)$ and $vs\phi | D \sim \chi_v^2$, where

$$\mathbf{a}_{n,t}^* = \mathbf{a}_{n,t} + \mathbf{A}_{t,\tau}(\mathbf{a}_{n,\tau}^* - \mathbf{a}_{n,\tau}), \tag{4.35}$$

$$\mathbf{R}_{n,t}^* = \mathbf{R}_{n,t} + \mathbf{A}_{t,\tau}(\mathbf{R}_{n,\tau}^* - \mathbf{R}_{n,\tau})\mathbf{A}_{t,\tau}'. \tag{4.36}$$

and for $t \geqslant \tau$,

$$\mathbf{A}_{t,\tau} = \mathbf{R}_{n,t}\mathbf{A}_{\tau,t}'\mathbf{R}_{n,\tau}^{-1}. \tag{4.37}$$

$$v = v_n - k, \tag{4.38}$$

$$vs = v_n s_n - \mathbf{d}_\tau'\mathbf{q}_\tau^{-1}\mathbf{d}_\tau. \tag{4.39}$$

Corollary 4.8

(4.35) and (4.36) may be expressed sequentially, dualling the updating recurrences (4.26) and (4.27), so that for $t \leqslant \tau$

$$\mathbf{a}^*_{n,t} = \mathbf{a}_{n,t} + \mathbf{A}_{t,t+1}(\mathbf{a}^*_{n,t+1} - \mathbf{a}_{n,t+1}),$$

$$\mathbf{R}^*_{n,t} = \mathbf{R}_{n,t} + \mathbf{A}_{t,t+1}(\mathbf{R}^*_{n,t+1} - \mathbf{R}_{n,t+1})\mathbf{A}'_{t,t+1}.$$

The proofs are straightforward and generalize corresponding results in Harrison and West, (1991). (4.35) and (4.36) follow from (4.9), (4.10) and theorem 4.5; (4.37) simply uses (4.22) to express the regression matrices for $t > \tau$, which vary with n, in terms of the constant regression matrices; and (4.39) follows from symmetry: given D_n the probability distributions are independent of the order in which the observations have been processed, so, considering $\mathbf{y}_\tau$ to be processed last, (4.15) applies.

Sometimes practitioners like to discard all information which is older than a given age. The resulting finite truncation model of length l observations bases all forecasts on $D^*_n = \{\mathbf{y}_{n-l+1} \ldots \mathbf{y}_n\}$. Assuming proper distributions the expression for $t > \tau = n - l$ is directly applicable since $D^*_n = \{(D^*_{n-1} - \mathbf{y}_{n-l}), \mathbf{y}_n\}$ and the deletion of $\mathbf{y}_{n-l}$ may be performed either before or after updating on receipt of $\mathbf{y}_n$.

4.6.2 Deletion of a set of l observations

The recurrences may also be used to delete a set, $\{\mathbf{y}_{\tau_1}, \ldots, \mathbf{y}_{\tau_l}\}$, of l observations, where $\tau_i < \tau_{i+1}$.

Write $D_{n,i} = D_n - \{\mathbf{y}_{\tau_1} \ldots \mathbf{y}_{\tau_i}\}$. For example suppose the current distribution $\{\boldsymbol{\theta}_n, \phi | D_{n,l})$ is desired. This is swiftly achieved using a sequential procedure which requires calculations only at the $l+1$ time points $\tau_1 \ldots \tau_l, \tau_{l+1} = n$. First delete $\mathbf{y}_{\tau_1}$ using (4.7), (4.9) and (4.10) or (4.33) and (4.34) to derive $\mathbf{d}_{\tau_1}, \mathbf{q}_{\tau_1}$ and $(\boldsymbol{\theta}_{\tau_1} | D_{n,1}, \phi) \sim N(\mathbf{a}^*_{n,\tau_1}, \mathbf{R}^*_{n,\tau_1})$. Then use (4.35) and (4.36) to obtain $(\boldsymbol{\theta}_{\tau_2} | D_{n,1}, \phi)$. Now repeat the procedure applying equations (4.7), (4.9) and (4.10) to this distribution to additionally delete $\mathbf{y}_{\tau_2}$ and derive the corresponding $\mathbf{d}^*_{\tau_2}, \mathbf{q}^*_{\tau_2}$ and $(\boldsymbol{\theta}_{\tau_2} | D_{n,2}, \phi) \sim N(\mathbf{a}^*_{n,\tau_2}, \mathbf{R}^*_{n,\tau_2})$. Conditional independence ensures that the procedure can continue, so that (4.35) and (4.36) provide $(\boldsymbol{\theta}_{\tau_3} | D_{n,2}, \phi)$ and then (4.7), (4.9) and (4.10) give $\mathbf{d}^*_{\tau_3}, \mathbf{q}^*_{\tau_3}$ and $(\boldsymbol{\theta}_{\tau_3} | D_{n,3}, \phi)$. The procedure continues with a final application of (4.35) and (4.36) using $(\boldsymbol{\theta}_{\tau_l} | D_{n,l}, \phi)$ to give $(\boldsymbol{\theta}_n | D_{n,l}, \phi)$. If ϕ is unknown write $(\mathbf{d}^*_{\tau_1}, \mathbf{q}^*_{\tau_1}) \equiv (\mathbf{d}_{\tau_1}, \mathbf{q}_{\tau_1})$, when symmetry, (4.38), and (4.39) give

$$v_{n,l} s_{n,l} \phi | D_{n,l} \sim \chi^2_{v_{n,l}},$$

$$v_{n,l} = v_n - lk,$$

$$v_{n,l} s_{n,l} = v_n s_n - \sum_{i=1}^{l} \mathbf{d}^{*\prime}_{\tau_i} \mathbf{q}^{*-1}_{\tau_i} \mathbf{d}^*_{\tau_i}. \tag{4.40}$$

The procedure is particularly neat for discount weighted regression.

4.6.3 Stochastic variance

In the spirit of dynamic modelling the unknown scalar ϕ, written ϕ_t, is often considered to change slowly through time. The well used discount variance model West and Harrison (1989a), operates with a variance discount factor δ, usually such that $0.98 \leqslant \delta < 1$. The marginal distributions for ϕ_t may be modelled by a power steady model, as in Smith (1979), applied to $\psi_t = \log \phi_t$ in which, with

$$p(\psi_t = \psi | D_{t-1}) \propto \{p(\psi_{t-1} = \psi | D_{t-1})\}^{\delta}, \quad v_n s_n \phi_n | D_n \sim \chi^2_{v_n},$$
$$v_n = \delta v_{n-1} + k,$$
$$v_n s_n = \delta v_{n-1} s_{n-1} + \mathbf{e}_n' \mathbf{Q}_n^{-1} \mathbf{e}_n.$$

On deleting l observations as in section 4.6.2

$$v_{n,l} s_{n,l} \phi_n | D_{n,l} \sim \chi^2_{v_{n,l}},$$
$$v_{n,l} = v_n - k \sum_{i=1}^{l} \delta^{n-\tau_i},$$
$$v_{n,l} s_{n,l} = v_n s_n - \sum_{i=1}^{l} \delta^{n-\tau_i} \mathbf{d}^{*\prime}_{\tau_i} \mathbf{q}^{*-1}_{\tau_i} \mathbf{d}^{*}_{\tau_i}.$$

The proof follows from (4.40) and the equivalence of the calculations for $(\boldsymbol{\theta}_n, \phi_n | D_{n,l})$ with those from the model $\{\mathbf{F}_t, \mathbf{G}_t, \mathbf{V}_t/\delta^{|n-t|}; \mathbf{W}_t/\delta^{|n-t|}; \mathbf{R}_{0,1}/\delta^n\}$, in which all the known variances relating to time $0 \leqslant t \leqslant n$ are replaced by those in the original model divided by $\delta^{|n-t|}$.

4.6.4 Deletion diagnostics

With static models, the parameter vector $\boldsymbol{\theta}_t$ is constant for all times, or at least deterministic given its value at any particular time. It is then usual to derive diagnostics relating to the deletion of a set of points based upon the distributions $\boldsymbol{\theta} | D_n$ and $\boldsymbol{\theta} | D_{n,l}$, such as the divergence measures proposed by Smith and Pettit (1985) and Bernardo (1985). With dynamic models the appropriate diagnostics will depend upon the objective. For example, in a BATS univariate retrospective analysis it was the practice to graph $(y_t, a_{n,t})$ against t, and to analyse the smoothed residuals $e_{n,t} = y_t - a_{n,t}$ (West, Harrison and Pole (1987)). However, in the spirit of cross-validatory analysis, it seems preferable to examine the corresponding jack-knife residuals $d_{n,t} = y_t - E(Y_t | D_n - y_t)$. An illustration of this using an appropriate influence measure is given in Harrison and West (1991), where, following Smith and Pettit (1985), the influence measure is separated into that due to the outlyingness of the observation and that due to extremeness in the design space. For forecasting, say $\mathbf{Y}_{n+1} | D_n$, it is arguable that diagnostics should be based upon the appropriate predictive distributions $\mathbf{Y}_{n+1} | D_n$ and $\mathbf{Y}_{n+1} | D_{n,l}$. The use of predictive distributions, as proposed by Johnson and Geisser (1983), is particularly important in forecasting since there is little point in amending historic distributions which do not significantly affect forecasts.

4.7 CLOSING COMMENTS

The main point of this paper is the far reaching effects of conditional independence in enabling elegant ways of incorporating and deleting information in normal dynamic linear models. However the methods extend to very general use, applying to nonlinear non-normal models in which linear least-squares or linear Bayes procedures are used, as well as to dynamic generalized linear models: West and Harrison (1989).

For clarity, the incorporation of subjective information, including that arising from other forecast systems, was omitted in illustrating the main results. However, the usual way of conveying such information, as discussed in West and Harrison (1989a, b), models the information relating to the state $\boldsymbol{\theta}_t$ as conditionally independent of both past and future given $\boldsymbol{\theta}_t$. Hence the methods are relevant and easy to apply. In practical schemes it is very important to monitor subjective intervention since this may well be ill founded. Often monitoring is done by sequences of sequential probability ratio tests but it would be interesting to see practical examples of monitoring diagnostics applied to routine forecasting systems since these do not necessitate the specification of precise alternative hypotheses. Also particular diagnostics relating to functions of the parameters can illuminate the required intervention. Particular suspect information at time τ is easy to track and eliminate remembering that $\mathbf{A}_{\tau,n} = \mathbf{A}_{\tau,n-1}\mathbf{A}_{n-1,n}$.

Usually in forecasting systems the major concern in incorporating and deleting information relates to the present and the recent past and therefore the procedures given in this paper offer considerable advantages over those which involve reanalyses of the entire time series history.

APPENDIX

Proof of theorem 4.1

The proof utilizes standard results.

(a) From probability theory, with $\phi > 0$ known or unknown,

$$\boldsymbol{\Sigma}_{z|u}^{-1} = \boldsymbol{\Sigma}_z^{-1} + \mathbf{c}\boldsymbol{\Sigma}_{u|z}^{-1}\mathbf{c}', \tag{4.41}$$

$$\boldsymbol{\Sigma}_{z|u}^{-1}\boldsymbol{\mu}_{z|u} = \boldsymbol{\Sigma}_z^{-1}\boldsymbol{\mu}_z + \mathbf{c}\boldsymbol{\Sigma}_{u|z}^{-1}\mathbf{u}. \tag{4.42}$$

(4.8) follows from the conditional normal result

$$\boldsymbol{\mu}_{z|u} = \boldsymbol{\mu}_z + \mathbf{A}(\mathbf{u} - \mathbf{c}'\boldsymbol{\mu}_z).$$

(b) From (4.8) $\mathbf{A} = \mathbf{B}\mathbf{q}\boldsymbol{\Sigma}_{u|z}^{-1} = \mathbf{B} - \mathbf{B}\mathbf{c}'\mathbf{A} \Rightarrow \mathbf{B} = (\mathbf{I} - \mathbf{A}\mathbf{c}')^{-1}\mathbf{A}$. Premultiplying (4.41) by $\boldsymbol{\Sigma}_{z|u}^{-1}$ and post multiplying by $\boldsymbol{\Sigma}_z\mathbf{c}\boldsymbol{\Sigma}_{u|z}^{-1}$,

$$\boldsymbol{\Sigma}_z\mathbf{c}\boldsymbol{\Sigma}_{u|z}^{-1} = \mathbf{A} + \mathbf{A}\mathbf{c}'\boldsymbol{\Sigma}_z\mathbf{c}\boldsymbol{\Sigma}_{u|z}^{-1} \Rightarrow \boldsymbol{\Sigma}_z\mathbf{c}\boldsymbol{\Sigma}_{u|z}^{-1} = (\mathbf{I} - \mathbf{A}\mathbf{c}')^{-1}\mathbf{A} = \mathbf{B}.$$

(4.9) and (4.10) follow using (4.41) and (4.42).

(c) (4.11) follows from (4.4), (4.5), (4.7), and (4.9):

$$\mathbf{e} = \mathbf{u} - \mathbf{c}'\boldsymbol{\mu}_{z|u} + \mathbf{c}'(\boldsymbol{\mu}_{z|u} - \boldsymbol{\mu}_z) = \mathbf{d} + \mathbf{c}'\mathbf{A}\mathbf{e} \Rightarrow \mathbf{d} = (\mathbf{I} - \mathbf{c}'\mathbf{A})\mathbf{e} = \boldsymbol{\Sigma}_{u|z}\boldsymbol{\Sigma}_u^{-1}\mathbf{e},$$

$$\mathbf{d} = \mathbf{u} - \mathbf{c}'\boldsymbol{\mu}_z - \mathbf{c}'\mathbf{B}\mathbf{d} \Rightarrow \mathbf{e} = (\mathbf{I} + \mathbf{c}'\boldsymbol{\Sigma}_{z|u}\mathbf{c}\mathbf{q}^{-1})\mathbf{d} = \boldsymbol{\Sigma}_{u|z}\mathbf{q}^{-1}\mathbf{d}.$$

Proof of theorem 4.3

(4.16) and (4.17) simply arise from

$$\mathbf{X} \perp\!\!\!\perp \mathbf{U} \mid \mathbf{Z} \Leftrightarrow \mathrm{cov}(\mathbf{x}, \mathbf{u} \mid \mathbf{z}) = 0.$$

So $\mathbf{0} = \mathrm{cov}(\mathbf{x}, \mathbf{u}) - \mathrm{cov}(\mathbf{x}, \mathbf{z})\{\mathrm{var}(\mathbf{z})\}^{-1}\mathrm{cov}(\mathbf{z}, \mathbf{u}) = \mathrm{cov}(\mathbf{x}, \mathbf{u}) - \mathbf{A}_{x,z}\mathbf{A}_{z,u}\boldsymbol{\Sigma}_u$ and the regression matrix of $\mathbf{X}$ on $\mathbf{U}$ is $\mathbf{A}_{x,z}\mathbf{A}_{z,u}$.

Proof of theorem 4.5

This follows using standard conditional normal results.

(a)
$$\boldsymbol{\mu}_{z|u} = \boldsymbol{\mu}_z + \mathbf{A}_{z,u}(\mathbf{u} - \boldsymbol{\mu}_u) \Rightarrow \mathbf{A}_{z,u}(\mathbf{u} - \boldsymbol{\mu}_u) = \boldsymbol{\mu}_{z|u} - \boldsymbol{\mu}_z$$
$$\Rightarrow \boldsymbol{\mu}_{x|u} = \boldsymbol{\mu}_x + \mathbf{A}_{x,z}\mathbf{A}_{z,u}(\mathbf{u} - \boldsymbol{\mu}_u) = \boldsymbol{\mu}_x + \mathbf{A}_{x,z}(\boldsymbol{\mu}_{z|u} - \boldsymbol{\mu}_z).$$

Also
$$\boldsymbol{\Sigma}_{z|u} = \boldsymbol{\Sigma}_z - \mathbf{A}_{z,u}\boldsymbol{\Sigma}_u\mathbf{A}'_{z,u} \Rightarrow \mathbf{A}_{z,u}\boldsymbol{\Sigma}_u\mathbf{A}'_{z,u} = \boldsymbol{\Sigma}_z - \boldsymbol{\Sigma}_{z|u},$$
$$\Rightarrow \boldsymbol{\Sigma}_{x|u} = \boldsymbol{\Sigma}_x - \mathbf{A}_{x,z}\mathbf{A}_{z,u}\boldsymbol{\Sigma}_u\mathbf{A}'_{z,u}\mathbf{A}'_{x,z} = \boldsymbol{\Sigma}_x + \mathbf{A}_{x,z}(\boldsymbol{\Sigma}_{z|u} - \boldsymbol{\Sigma}_z)\mathbf{A}'_{x,z}.$$

(b) $\mathrm{cov}(\mathbf{x}, \mathbf{z} \mid \mathbf{u}) = \mathbf{A}_{x,z}\boldsymbol{\Sigma}_z - \mathbf{A}_{x,z}\mathbf{A}_{z,u}\boldsymbol{\Sigma}_u\mathbf{A}'_{z,u} = \mathbf{A}_{x,z}\boldsymbol{\Sigma}_{z|u}$,
and $\mathbf{A}_{(z,x)|u}\boldsymbol{\Sigma}_{x|u} = \{\mathrm{cov}(\mathbf{x}, \mathbf{z} \mid \mathbf{u})\}'$.

(c) (4.23) follows from theorem 4.3 and (4.22).

REFERENCES

Ameen, J.R.M. and Harrison, P.J. (1984) Discount weighted estimation. *J. Forecasting*, **3**, 285–96.

Ameen, J.R.M. and Harrison, P.J. (1985) Normal discount Bayesian models (with discussion). In *Bayesian Statistics 2*, (eds J.M. Bernado, M.H. DeGroot, D.V. Lindley and A.F.M. Smith, pp. 271–98), North-Holland, Amsterdam, and Valencia University Press.

Bernardo, J.M. (1985) Discussion of paper by A.F.M. Smith and L.I. Pettit. In *Bayesian Statistics 2*, (eds J.M. Bernado, M.H.D. DeGroot, D.V. Lindley and A.F.M Smith, pp. 492–3. North-Holland, Amsterdam, and Valencia University Press.

Bruce, A.G. and Martin, R.D. (1989) Leave *k*-out diagnostics for time series (with discussion). *J.R. Statist. Soc.*, **B51**, 363–424.

Dawid, A.P. (1979) Conditional independence in statistical theory (with discussion), *J.R. Statist. Soc.*, **B41**, 1–31.

de Jong, P. (1988) A cross-validation filter for time series models. *Biometrika*, **75**, 594–600.

Harrison, P.J. and Stevens, C.F. (1976) Bayesian forecasting (with discussion), *J.R. Statist. Soc.*, **B38**, 205–47.

Harrison, P.J. and West, M. (1991) Dynamic linear model diagnostics. *Biometrika*, **78**, 797–808.

Johnson, W. and Geisser, S. (1983). A predictive view of the detection and characterization of influential observations in regression analysis. *J. Am. Statist Assoc.*, **78**, 137–44.

Kalman, R.E. (1963) New methods in Wiener filtering theory. In *Proceedings of the First Symposium of Engineering Applications of Random Function Theory and Probability*, (eds J.L. Bodanoff and F.Kozin), Wiley, New York.

Kohn, R. and Ansley, C.F. (1989) A fast algorithm for signal extraction, influence and cross-validation in state space models. *Biometrika*, **76**, 65–79.

Lauritzen, S.L., Dawid, A.P., Larson, B.N. and Leimer, H.G. (1990) Independence properties of directed Markov fields. *Networks*, **20**, 491–505.

Priestley, M.B. (1981) *Spectral analysis and time series*, Academic Press.

Smith, A.F.M. and Petit, L.I. (1985) Outliers and influential observations in linear models (with discussion). In *Bayesian Statistics 2*, (eds J.M. Bernardo, M.H. DeGroot, D.V. Lindley and A.F.M. Smith, pp. 473–94, North-Holland, Amsterdam and Valencia University Press.

Smith, J.Q. (1979) A generalization of the Bayesian steady forecasting model. *J.R. Statist. Soc.*, **B41**, 378–87.

Smith, J.Q. (1990) Statistical principles on graphs (with discussion). In *Influence diagrams, belief nets and decision analysis*, (eds R.M. Oliver and J.Q. Smith), Wiley, New York, pp. 89–120.

Veerapen, P.P. (1992) *Recurrence Relationships and Model Monitoring for Dynamic Linear Models*, Unpublished Ph.D. thesis, University of Warwick.

West, M., Harrison, P.J. and Pole, A. (1987). BATS: Bayesian analysis of time series. *Professional Statistician*, **6**, 43–6.

West, M. and Harrison, P.J. (1989a) *Bayesian Forecasting and Dynamic Models*. Springer-Verlag, New York.

West, M. and Harrison, P.J. (1989b) Subjective intervention in formal models. *J. Forecasting*, **8**, 33–53.

5

Order selection for linear time series models: a review

R.J. Bhansali

5.1 INTRODUCTION

Consider the well known autoregressive-moving average model of order (m, h), ARMA(m, h), for a discrete-time stationary process $\{x_t\}$,

$$\sum_{j=0}^{m} a_{mh}(j)x_{t-j} = \sum_{j=0}^{h} \beta_{mh}(j)\varepsilon_{t-j}, a_{mh}(0) = \beta_{mh}(0) = 1, \tag{5.1}$$

where $\{\varepsilon_t\}$ is a sequence of uncorrelated random variables, each with mean 0 and variance σ^2, and the $a_{mh}(j)$ and $\beta_{mh}(j)$ are real coefficients such that if

$$A_{mh}(z) = \sum_{j=0}^{m} a_{mh}(j)z^j, \quad \Theta_{mh}(z) = \sum_{j=0}^{h} \beta_{mh}(j)z^j$$

denote their characteristic polynomials, then

$$A_{mh}(z) \neq 0, \ \Theta_{mh}(z) \neq 0, \quad |z| \leqslant 1,$$

and that $A_{mh}(z)$ and $\Theta_{mh}(z)$ do not have a common zero.

Early statistical work on this model was based on the assumption that the order (m, h) is known *a priori*; see, for example, the pioneering work of Whittle (1951) and Hannan (1969).

The order, however, is seldom (if ever) known *a priori* and a question of some importance is how to determine the value of (m, h) from the observed time series. It is clear that although the graphical methods of Box and Jenkins (1970) are interesting, they cannot usually stand by themselves because one needs various standards of comparison against which to measure the observed discrepancies. The hypothesis-testing procedures (e.g. Whittle, 1952) by contrast, are suitable when the models under the null and alternative hypotheses are prescribed *a priori*. This would usually not be the case, however, and a problem of simultaneous inference has to be faced, see Akaike (1978a).

When judged against this background, the approach pioneered by Akaike (1970, 1973), see also Mallows (1973) and Parzen (1969), and involving the use of an order selection criterion, provides a remarkable breakthrough on this important question. At one level, it transforms the order selection problem from one of hypothesis testing to that of estimation, but with two important reservations. First, Akaike himself does not agree with this particular interpretation of an order selection criterion. Secondly, Soderstrom (1977), and Terasvirta and Mellin (1986) have pointed out that if the model selection only involves two models, an order selection criterion admits an interpretation as a generalized likelihood ratio test. The implied criticism here, however, is unjustified because the main motivation for using an order selection criterion is that the class of competing models is usually large and then the two procedures are quite different, see Akaike (1978a) for a further discussion of this point.

One reason why Akaike does not accept the problem of ARMA order selection as that of estimating an unknown true order, (m_0, h_0), say, is that there is no fundamental reason why a time series need necessarily follow a 'true' ARMA model.

It is known, however, that under mild regularity conditions, Brillinger (1975), a stationary process possesses a moving average representation

$$x_t = \sum_{j=0}^{\infty} b(j)\varepsilon_{t-j}, \quad b(0) = 1, \tag{5.2}$$

and an autoregressive representation

$$\sum_{j=0}^{\infty} a(j)x_{t-j} = \varepsilon_t, \quad a(0) = 1, \tag{5.3}$$

in which the $a(j)$ and $b(j)$ are absolutely summable coefficients.

Indeed, an interpretation of an ARMA model is that it captures these representations 'parsimoniously' by postulating the $b(j)$ and $a(j)$ as functions of only a finite number of parameters, Hannan (1970).

An alternative nonparametric autoregressive model fitting approach to time series analysis, as advocated more recently by Parzen (1974), is based on the representation (5.3); early references are Durbin (1959) and Whittle (1952).

According to this approach, the behaviour of an observed time series could be described by an autoregressive model of order k, in which the value of k is chosen so as to obtain an adequate finite order approximation to an underlying infinite order process, rather than as an estimator of an unknown true order. Also, when using this approach, the question of how to select a suitable value of k from the data arises. Parzen (1974) has suggested a CAT criterion for this purpose. Also, Akaike argues that the properties of the order selected by his information criterion, AIC, should be judged to accord with this nonparametric approach.

Apart from the references already cited, there has been considerable development during the last twenty years or so on the question of time series model selection. The main objective of this paper is to review these developments, but restricted to only the class of univariate linear time series models. Although it is not our intention to provide a totally comprehensive survey of the vast literature of this topic, it is hoped that a view of the current state of the art may be gleaned from this review. References to the works in which the time series order selection techniques are reviewed from a variety of different perspectives include Priestley (1981), Andel (1982), Shibata (1985), De Gooijer *et al* (1985), Hannan (1987) and Hannan and Deistler (1988).

It will be convenient and useful to discuss first the question of autoregressive model selection, both from a nonparametric and a parametric point of view. The question of ARMA model selection, stressing in particular the computational aspects, is discussed in section 5.3. Further extensions of the model selection procedures, especially to processes with infinite variance and for prediction more than one step ahead, are discussed in section 5.4.

5.2 AUTOREGRESSIVE MODEL SELECTION

Suppose first that h in (5.1) is *a priori* known to equal zero and $\{x_t\}$ is an autoregressive process of order m, AR(m), where m is unknown but an upper bound L, for m is known, and that $\{\varepsilon_t\}$ is a sequence of independent normal variables, each with mean 0 and variance σ^2. We will now write the autoregressive coefficients more simply as $a_m(j)$ $(j=1,\ldots,m)$.

If $x_1,\ldots,x_T$ are observed, an approximation for the likelihood function of the autoregressive parameters is obtained by conditioning on $x_1,\ldots,x_m$, Priestley (1981),

$$L(x_{m+1},\ldots,x_T|x_1,\ldots,x_m;{}_m\boldsymbol{\vartheta})$$

$$=\prod_{t=m+1}^{T}\{2\pi\sigma^2(m)\}^{1/2}\exp\left[-\{2\sigma^2(m)\}^{-1}\left\{x_t+\sum_{u=1}^{m}a_m(u)x_{t-u}\right\}^2\right], \qquad (5.4)$$

where, with

$$\mathbf{a}(m)=[a_m(1),\ldots,a_m(m)]', \quad {}_m\boldsymbol{\vartheta}=[\mathbf{a}(m)',\sigma^2(m)]'.$$

The corresponding approximate maximum likelihood estimator,

$$m\hat{\boldsymbol{\vartheta}}=[\hat{\mathbf{a}}(m)',\hat{\sigma}^2(m),]'$$

of ${}_m\boldsymbol{\vartheta}$ is given by

$$\hat{\mathbf{a}}(m)=-\hat{\mathbf{R}}(m)^{-1}\hat{\mathbf{r}}(m), \qquad (5.5)$$

$$\hat{\sigma}^2(m)=\sum_{u=0}^{m}\hat{a}_m(u)D^{(T)}(0,u), \qquad (5.6)$$

where, with $\hat{a}_m(0)=1$, $\hat{\mathbf{R}}(m)=[D^{(T)}(u,v)](u,v=1,\ldots,m)$, $\hat{\mathbf{r}}(m)=[D^{(T)}(0,1),\ldots,$

$D^{(T)}(0, m)]$ and

$$D^{(T)}(u, v) = (T - m)^{-1} \sum_{t=m+1}^{T} x_{t-u} x_{t-v} (u, v = 0, 1, \ldots, m). \tag{5.7}$$

Akaike's (1973) information criterion for autoregression order selection is

$$\mathrm{AIC}(p) = T \log \hat{\sigma}^2(p) + 2p \quad (p = 0, 1, \ldots, L) \tag{5.8}$$

and the value of p that minimizes this criterion is the selected order.

As discussed by Bhansali (1986a) and Findley (1985a), a justification for using AIC is that if $p, q \geqslant m$, $\mathrm{AIC}(p) - \mathrm{AIC}(q)$ provides an asymptotically unbiased estimator of $S_T(p) - S_T(q)$, where

$$S_T(p) = -2TE_x E_y\{\log g(y|_p\hat{\vartheta})\},$$

and

$$g(y|_p\hat{\vartheta}) = \{2\pi\hat{\sigma}^2(p)\}^{\frac{1}{2}} \exp\left[-\{2\hat{\sigma}^2(p)\}^{\frac{1}{2}} \left\{ \sum_{j=0}^{p} \hat{a}_p(j) y_{n-j} \right\}^2 \right]$$

is an estimated predictive density of a future observation, y_n, of $\{y_t\}$, an independent process having the same stochastic structure as $\{x_t\}$, but with $y_{n-1}, \ldots, y_{n-p}$ treated as known. Note that $E_y[\log g(y)|_p\hat{\vartheta})]$ is also the first term of the Kullback-Leibler distance between $g(y|_p\hat{\vartheta})$ and the 'true' density $h(y|_m\vartheta)$ of y_n, the second term being a constant independent of p.

This interpretation does not hold, however, for a $p < m$, see Sawa (1978) and Chow (1981), since the second term to the right of (5.8), which is a 'penalty' term, is not correct now. On the other hand, the inclusion of a more precise penalty term is unlikely to matter asymptotically, since it is of a smaller order than the first term, see Akaike (1970).

The use of a Taylor expansion for evaluating $S_T(p)$ may be justified by a recent result of Bhansali and Papangelou (1991); see also Fuller and Hasza (1981).

A precursor of AIC is the final prediction error (FPE) criterion,

$$\mathrm{FPE}(p) = \hat{\sigma}^2(p)[1 - (pT)]^{-1}[1 + (p/T)], \tag{5.9}$$

of Akaike (1970), which was motivated by the success of Mallows' (1973) C_p criterion in selecting the regressor variables. For $p \geqslant m$, $\mathrm{FPE}(p)$ admits an interpretation as an estimator of the one-step mean squared error when predicting y_n, say, from a knowledge of $y_{n-1}, \ldots, y_{n-p}$ and the estimated autoregressive coefficient $\hat{\mathbf{a}}(p)$, Davisson (1965). A multiplying factor of $[1 - (p(T)]^{-1}$ has been introduced in (5.9) to correct for the 'degree of freedom' bias, Jones (1975).

As discussed by Bhansali and Downham (1977), if $O(T^{-2})$ terms are ignored, the FPE criterion may be viewed as a special case, with $\alpha = 2$, of an extended criterion,

$$\mathrm{FPE}_\alpha(p) = \hat{\sigma}^2(p)(1 + \alpha p/T), \tag{5.10}$$

in which α is a positive constant.

By similar reasoning, the AIC criterion, (5.8), may also be extended to the following criterion (Akaike, 1979),

$$\mathrm{AIC}_\alpha(p) = T\log\hat{\sigma}^2(p) + \alpha p, \tag{5.11}$$

in which α is a constant. A derivation of this criterion by a second order Taylor approximation to the Kullback–Leibler information, but with the bias and variance terms being given unequal weights has been given by Bhansali (1986a).

It is clear that for a fixed α, the AIC_α and FPE_α criteria are closely related. We have, for each p,

$$T\log\mathrm{FPE}_\alpha(p) = \mathrm{AIC}_\alpha(p) + O(T^{-2})$$

and thus on ignoring the $O(T^{-2})$ terms, they are seen to be equivalent.

Let $\hat{m}_\alpha$ denote the order selected by either of these two criteria. The distribution of $\hat{m}_\alpha$ for $\alpha = 2$ has been derived by Akaike (1970) and Shibata (1976) and for a fixed α by Bhansali and Downham (1977). We have, for a fixed $\alpha > O$,

$$\begin{aligned}\lim_{T\to\infty} Pr\{\hat{m}_\alpha = j\} &= p_\alpha(j-m)q_\alpha(L-j), \quad m \leqslant j \leqslant L, \quad j < m, \\ &= O \qquad\qquad\qquad\qquad j < m\end{aligned} \tag{5.12}$$

where

$$p_\alpha(n) = Pr\left\{\bigcap_{j=1}^{m} S_j > 0\right\}, q_\alpha(n) = Pr\left\{\bigcap_{j=1}^{n} S_j \leqslant 0\right\}, p_\alpha(0) = q_\alpha(0) = 1,$$

$S_j = (Z_1 - \alpha) + (Z_2 - \alpha) + \ldots + (Z_j - \alpha)$ and $\{Z_j\}$ is a sequence of independent chi-squared random variables, each with 1 degree of freedom.

This result implies, see the discussion to Tong (1977), that AIC and FPE do not estimate m consistently, but asymptotically the probability of selecting m correctly, i.e., $\hat{m}_\alpha = m$, increases as α increases. A third implication is that if $L \leqslant m$ remains fixed as $T \to \infty$, the choice of L has relatively little influence on the behaviour of $\hat{m}_\alpha$.

A certain amount of caution is required, however, when using these asymptotic results with a finite T. Bhansali and Downham (1977) report simulation results showing that if the dependence of x_t on x_{t-m} is relatively strong, the asymtotic result (5.12), holds with $T \geqslant 200$, but when this is not the case, increasing α also increases the frequency of selecting $\hat{m}_\alpha < m$, that is, of underfitting. Their conclusion is that since the degree of dependence of x_t on x_{t-m} is unknown, no one value of α is universally optimal for all T and all time series.

Hurvich and Tsai (1989) observe that if T is small but the ratio L/T is not negligibly small, AIC may select a highly non-parsimonious model.

As discussed in section 5.1, Akaike rejects the lack of consistency as a valid criticism of AIC by questioning the very idea of a 'true' finite parameter model for an observed time series. He conjectured instead that for one-step prediction, under the hypothesis of an infinite order autoregressive process, AIC would provide satisfactory results.

Shibata (1980) has proved this conjecture. On assuming that $\{x_t\}$ is **not** a finite autoregression and $L = K_T$, a function of T, such that $K_T \to \infty$, $K_T^2/T \to 0$, as $T \to \infty$, Shibata develops an asymptotic lower bound for the one-step mean squared error of prediction when the prediction is by the least-squares fitting of an autoregression of order $\tilde{k}_T$, $1 \leqslant \tilde{k}_T \leqslant K_T$, and $\tilde{k}_T$ is a random variable possibly dependent on the observed time series. He further shows that the order selection by AIC, or FPE, is asymptotically efficient because for both these criteria the lower bound is attained asymptotically as $T \to \infty$. The asymptotic efficiency of the AIC_α criterion is also examined by Shibata (1980), who shows that if the $a(j)$ decrease to 0 exponentially, the lower bound is attained with any fixed $\alpha > 1$, but not when the $a(j)$ decrease to 0 as a power of j, that is, for some $n > 0$ and a bounded constant M, $a(j) = Mj^{-n}$, and that in this sense the choice $\alpha = 2$ is justified. Shibata (1981) has extended these results to the question of autoregressive spectral estimation.

An explicit nonparametric approach to autoregressive order selection has been adopted, in work almost contemporary to Akaike's, by Parzen (1974), who introduced the criterion autoregressive transfer function (CAT). The penalty function used for deriving this criterion was later amended, Parzen (1977), and an alternative CAT* criterion proposed, so as to get around the problem of estimating σ^2, which occurs in the original CAT criterion. These two and the related CAT_α criterion are discussed in detail by Bhansali (1986b,c), who shows that for a finite autoregressive process, the CAT, CAT* and CAT_2 criteria are asymptotically equivalent to AIC in the sense that they have the same asymptotic distribution, and, for a fixed $\alpha > 0$, the CAT_α criterion is asymptotically equivalent to AIC_α. If, on the other hand, $\{x_t\}$ possesses an infinite autoregressive representation, the CAT, CAT_2 and CAT* criteria are asymptotically efficient in the sense of Shibata (1980, 1981), and if additionally the $a(j)$ decrease to 0 exponentially, so is the CAT_α criterion with any fixed $\alpha > 1$. Simulations accord with these asymptotic results especially if T is moderately large, when AIC, CAT and CAT* select the same order with a high frequency.

McClave (1975) has proposed a subset autoregressive model in which some of the intermediate autoregressive coefficients can be constrained, to equal zero; the motivation being to answer a frequent criticism, Abraham and

Ledolter (1983), that the autoregressive model fitting approach can fit a non-parsimonious model. Newbold and Granger (1974) consider a 'stepwise' procedure for this purpose.

The Yule–Walker method of estimating the autoregressive parameters (Priestley, 1981) should be mentioned here because the estimates are still given by (5.5) and (5.6) but $R^{(I)}(u-v)$ replacing $D^{(T)}(u,v)$, where

$$R^{(T)}(u) = T^{-1} \sum_{t=1}^{T-|u|} x_t x_{t+|u|} \quad (|u| \leqslant T-1) \tag{5.13}$$

is a 'positive definite' estimator of the covariance function, $R(u)$, of $\{x_t\}$. This procedure is based on the Levinson–Durbin algorithm and it has an advantage of computational economy. A disadvantage however is bias, Tjostheim and Paulsen (1983), which can have serious repercussions on the properties of the selected order.

An alternative estimation procedure is the Burg method (Beamish and Priestley, 1981), who report that this method may be preferred when $A_m(z)$ has a zero close the unit circle.

A further implication of the asymptotic distribution of the FPE_α and AIC_α criteria is that the order m, may be estimated consistently by letting $\alpha = \alpha(T)$, a function of T, such $\alpha(T) \to \infty$ as $T \to \infty$. The rate of growth of $\alpha(T)$ cannot be too fast, however, since for finite T, the risk of understanding m is then non-negligible—see Akaike (1970), in whose simulation study, $\alpha(T)$ is a polynomial function of T.

An answer to the question of an acceptable rate of growth of $\alpha(T)$ has been provided by Schwarz (1978) in a Bayesian framework, whose order selection criterion is of the same form as AIC_α but with $\alpha = \log T$. A 'tutorial' review of Schwarz's approach is given by Chow (1981).

It is interesting to observe here that the choice $\alpha(T) = \log T$ may also be justified from a variety of different perspectives. Akaike (1977) considers the question of selecting the degree of a polynomial regression and in an empirical Bayes framework, involving the use of Type II maximum likelihood procedure of Good (1965), derives his BIC criterion, whose penalty term, though not quite the same, is asymptotically equivalent to Schwarz's $\alpha = \log T$. Rissanen (1978) also arrives at this choice in a coding theory framework by adopting a minimum description length principle—see Hannan (1987) for an account of Rissanen's approach. Smith and Spiegelhalter (1980) provide a Bayesian interpretation of the Schwarz, AIC and AIC_α criteria in terms of Bayes factors. Further comments on the Schwarz criterion and its relationship to AIC have been given by Stone (1979) and Akaike (1978c).

For the specific question of autoregressive order selection, Kashyap (1977) uses a Bayesian approach to arrive at a criterion that is asymptotically equivalent to Schwarz's. However, a data-dependent prior distribution for autoregressive parameters has been specified, and his assumption concerning the initial values, $X_1, \ldots, X_m$, is rather non-standard.

Akaike (1979) also takes an unorthodox Bayesian approach to the more general question of the choice of α (see also Akaike, 1978b), whose basic thesis is that $\exp\{-\frac{1}{2}\text{AIC}(p)\}$, $p \geqslant 1$, may be interpreted as the likelihood of a pth order autoregression and hence it may be multiplied by a prior distribution, $\pi_0(p)$, say, for the order to arrive at a 'posterior' distribution, $\pi_1(p)$, say. Although the resulting procedure works well empirically, the argument is vague and unconvincing, especially because a strictly Bayesian approach involves integrating out the autoregressive parameters, ${}_p\boldsymbol{\vartheta}$, by specifying an appropriate prior distribution (Rissanen, 1987).

Hannan and Quinn (1979) bring a different perspective to the question of an appropriate rate of growth for $\alpha(T)$. They invoke the law of the iterated logarithm for an autoregressive process to show that any $\alpha(T) > 2\log\log T$ leads to a strongly consistent order estimate, and if $\alpha(T) < 2\log\log T$ a strongly consistent order estimate cannot be obtained. A lower bound for the value $\alpha(T)$ must take to achieve consistency is thus provided. In personal discussions, however, Hannan has sought to underplay the importance of this result and he has instead recommended the choice $\alpha(T) = \log T$.

The order selection criteria discussed so far have been derived by letting $T \to \infty$. Simulations show that if T is moderately large their performance accords with the asymptotic theory, but not if T is small. For the latter case, Hurvich and Tsai (1989) propose that the autoregressive order be selected by minimizing a 'bias-corrected' criterion,

$$\text{AIC}_c(p) = \text{AIC}(p) + 2(p+1)(p+2)/(T-p-2). \tag{5.14}$$

For a fixed p, as $T \to \infty$, $\text{AIC}_c(p) \to \text{AIC}(p)$, and thus the two criteria are asymptotically equivalent. These authors, however, provide simulation evidence to suggest that their correction is particularly useful when T is very small or when L is a moderately large fraction of T.

5.3 ARMA ORDER SELECTION

Suppose now that $\{x_t\}$ follows an ARMA model, (5.1), with $h > 0$. The order selection criteria discussed in section 5.2 may also be used in this situation and, with certain exceptions, the various theoretical properties discussed there can be generalized to the ARMA case.

Hannan (1980) shows that the Schwarz criterion BIC, say, provides a strongly consistent estimator of the ARMA order, but AIC does not. Moreover, the AIC_α criterion with $\alpha = \alpha(T)$ is strongly consistent with any $\alpha(T) > 2\log\log T$, but not if $\alpha(T) < 2\log\log T$.

Findley (1985a) gives a derivation of AIC for ARMA order selection and Hurvich, Shumway and Tsai (1990) consider the case when T is small.

Potscher (1990) considers the behaviour of BIC when an upper bound (P, Q) for the ARMA order has not been prespecified, and thus both these integers may approach ∞. Set $r_0 = \max(m_0, h_0)$ and suppose that only

ARMA(r, r), $r = 0, 1, 2, \ldots$, models are fitted by maximum likelihood. If now $\hat{r}$ denotes the value of r at which the first relative minimum of BIC occurs, then Potscher's main result is that $\hat{r}$ provides a strongly consistent estimator of r_0. This result is interesting because an ARMA(r_0, r_0) model connects directly with the Kalman filter and the state space representation of a stationary process developed in systems theory—see Akaike (1974), Caines (1987) and Hannon (1987). Potscher also examines the behaviour of $\hat{r}$ when the observed process is not an ARMA and thus the dimension of the state vector is infinite. His main result here is that as $T \to \infty$, $\hat{r} \to \infty$, almost surely.

An extension of the Box–Jenkins graphical technique to ARMA model identification is considered by Gray, Kelley and McIntire (1978), but Davies and Pettrucelli (1984), however, report several difficulties with this approach. A different approach, based on the notion of extended correlation functions, has been proposed by Tsay and Tiao (1984).

The role of hypothesis testing in this context has been considered by Potscher (1983), who however fails to give a rule for choosing a sequence of significance levels; a different procedure is considered by Poskitt and Tremayne (1981). Also, Poskitt (1986) has suggested a Bayesian approach.

A principal difficulty now in using an order selection criterion is computational. The likelihood function for an ARMA model is a highly nonlinear function of the parameters, and the maximized value of the likelihood function is required for a grid of values $(p, q)(p = 0, \ldots, P;\ q = 0, 1, \ldots, Q)$ where (P, Q) denotes a known upper bound for the order. An iterative procedure used for this purpose may be slow to converge, however, and sometimes the convergence may not be attained at all.

A second difficulty is that the maximum likelihood estimators of the moving average parameters are not guaranteed to be invertible and special checks may be necessary to ensure invertibility (Nicholls, 1972; Osborn 1976).

The computational difficulties just described may partially explain why, as noted by Newbold (1988), there have been only a few accounts in the literature of studies involving an ARMA order selection procedure; some relevant references are Ozaki (1977), Neftci (1982), Makridakis *et al.* (1982), Koehler and Murphree (1988), and Mbago (1979).

Two separate groups of procedures are available to reduce the computational burden of ARMA order selection and it is interesting to observe that both these groups of procedures are based on linear parameter estimation techniques suggested by Durbin (1959, 1960, 1961) some 30 years ago.

The first group of procedures is more specialized and apply when a *priori* knowledge that $m = 0$ is available so that the problem is that of moving average order determination; see Bhansali (1980, 1983a,b, 1987), who uses the inverse correlation function of $\{x_t\}$, Cleveland (1972), for this purpose.

The second group of procedures is named after Hannan–Rissanen (1982) and determines the ARMA order by a three-stage recursive procedure. As in Durbin (1960, 1961), the main idea is to estimate the ARMA parameters by a linear regression of x_t on

$$x_{t-1},\ldots,x_{t-p},\quad \hat{\varepsilon}_{t-1},\ldots,\hat{\varepsilon}_{t-q},$$

where

$$\hat{\varepsilon}_t = \sum_{u=0}^{k} \hat{a}_k(u)x_{t-u},\quad \hat{a}_k(0)=1,$$

provide an estimate of the innovations ε_t, and it is calculated at Stage I by fitting a kth order autoregression to the data, and where the value of k is determined by AIC.

The actual ARMA order selection is at Stage II and the final estimation of ARMA parameters by an asymptotically efficient technique at Stage III.

As discussed by Bhansali (1991), however, the original procedure of Hannan and Rissanen (1982) suffers from two drawbacks. First, the BIC criterion recommended there is not consistent, see Hannan and Rissanen (1983). Secondly, the computations at Stage II may be carried out more economically by using an algorithm of Franke (1985) instead of the Whittle (1953b) algorithm originally recommended.

Hannan and Kavalieris (1984a) and Poskitt (1987) suggest several different techniques for obtaining a consistent order estimate at Stage II, but these do not retain the elegant closed form simplicity of the original procedure, and in fact increase the computational cost of order determination.

Bhansali (1991) has recently investigated the use of an order determining criterion different from BIC at Stage II, and proposed selecting the ARMA order by minimizing the criterion

$$\mathrm{FPEC}_\delta(p,q)=\hat{\sigma}^2(p,q)\left[1+(k/T)\left\{\delta+\max_{p,q}\sum_{j=1}^{q}\hat{\beta}_{pq}(j)^2\right\}(p+q)\right]$$

in which $\delta > 0$ is an arbitrary constant, k is the autoregressive order selected at Stage I, and the $\hat{\beta}_{pq}(j)$ denote the Stage II estimates of the moving average coefficients of an ARMA model of order (p,q).

Bhansali (1991) shows that, as $T\to\infty$, the FPEC_δ criterion provides a consistent estimator of the ARMA order, (m_0,h_0), with any fixed $\delta\geqslant 1$. The simulation results given there indicate that if T is moderately large, and the moving average polynomial $\Theta(z)$, has a zero close to the unit circle, the generated model is selected with a frequency close to 100% and then the simulation results accord with the asymptotic theory. If, on the other hand, T is small or if the generated model is such that the various competing ARMA models do not differ appreciably in terms of their respective one-step mean-squared errors of prediction, the frequency of selecting the correct generated model is not close to 100%, and moreover, in this situation, a likelihood-based procedure selects the correct model with a higher frequency. The loss in predictive efficiency due to using the model selected by the FPEC_δ criterion is small, however.

As regards the choice of δ, the guidance provided by the simulation results is that the FPEC_δ criterion should initially be used with $\delta = 1$. A need for increasing the value of δ to say $\delta = 2$ is indicated when a non-parsimonious model, especially a long autoregression, is selected with $\delta = 1$, but δ need rarely exceed 3.

Liao (1991) has applied the FPEC_δ criterion together with the procedures of Hannan and Rissanen (1982) and Hannan and Kavalieris (1984a) to 32 data sets published in various textbooks on time series analysis and for which the graphical techniques were used by the author(s) concerned to identify an appropriate ARMA model. She reports that the proportion of times the FPEC_δ criterion selects the same model as that identified by the graphical techniques is close to 1. Moreover, when the respective models differ, that selected by the FPEC_δ criterion tends to be parsimonious and invariably passes the Box–Ljung (1978) diagnostic test.

Kavalieris (1991) has suggested an alternative consistent procedure for use at Stage II. Consistent estimation of the autoregressive parameters of an ARMA model is considered by Tuan (1988), his algorithm being related to that of Franke (1985); Tsay and Tiao (1984) earlier proposed a graphical technique for this purpose.

Recursive order selection for a multivariate ARMA model is considered by Hannan and Kavalieris (1984b) and the corresponding problem of dimensionality reduction for a multivariate time series by Tiao and Tsay (1989), see also Akaike (1976).

5.4 FURTHER EXTENSIONS

The innovations sequence, $\{\varepsilon_t\}$, has so far been treated as a sequence of independent Normal random variables, each with mean 0 and variance σ^2. Although some of the statistical properties of the selected orders may be studied without invoking the Gaussian assumption (Hannan, 1980), the existence of the first two moments is usually required.

Bhansali (1988) has shown that if $\{\varepsilon_t\}$ is instead assumed to be a sequence of independent identically distributed random variables with the common distribution belonging to the domain of attraction of an infinite variance symmetric stable law of index $\tau \varepsilon (0, 2)$, the AIC, and, with $\alpha > 1$, AIC_α, or equivalent, criteria provide a consistent estimator of the autoregressive order, and thus the principal criticism of AIC, that it is not consistent, does not now apply. The simulations approximately validate this result with finite T, especially if τ is not too close to 2.

Bhansali (1988) shows also that the moving average order selection procedure of Bhansali (1983a) and a procedure for discriminating between a pure autoregressive and a pure moving average model suggested by Bhansali (1987) are similarly consistent in this situation.

If it is explicitly recognized that for an observed time series there is no 'true' model, then, as discussed in section 5.2 (see also Whittle, 1963a), AIC and equivalent procedures are asymptotically optimal for one-step prediction. On the other hand, if the same model is used repeatedly in 'plug in' fashion for multistep prediction, then this optimality property does not hold and a different approach is called for.

Shibata (1980) has lately proposed an alternative 'direct' technique for multistep prediction. The underlying idea is to select and fit a different autoregressive model for each lead time, h. A direct linear least-squares regression of x_{t+h} on $x_t, x_{t-1}, \ldots, x_{t-k\pm 1}$ is recommended for estimating the autoregressive coefficients, where $k = \tilde{k}_h$, say, is a possibly different order of the autoregression fitted for each h; Shibata has recommended an order determining criterion, $S_h(k)$, for selecting the value of $\tilde{k}_h$.

It may be observed that this direct procedure also provides a nonparametric estimator of the h-step prediction error variance and which in turn may be employed for judging the predictability of an observed time series (Parzen, 1981).

Gersch and Kitagawa (1983) and Findley (1985b) have also endorsed the main idea of fitting a different time series model for each lead time h, and they recommend its use in a more general context with favourable results. Related references are Kabaila (1981), Stoica and Soderstrom (1989) Weiss (1991) and Tiao and Xu (1991).

A new procedure for model selection, based on the idea of minimizing the length of a predicted 'code' has recently been suggested by Rissanen (1986, 1987). This new procedure is to be contrasted with Rissanen's previous (1978) procedure, which uses the length of a non-predictive code as a criterion for model selection, and which, as discussed in section 5.2, is essentially equivalent to Schwarz's (1978) Bayesian criterion. Hannan *et al.* (1989) and Hemerley and Davis (1989, 1991) discuss this new procedure for the recursive estimation of finite autoregressive models and show that it estimates the autoregressive order consistently. However, Wei (1992) has reported several difficulties with this procedure in stochastic regression models.

5.5 CONCLUDING REMARKS

The discussion in section 5.2 shows that the various nonparametric model selection criteria, AIC, FPE, CAT, for example, are asymptotically equivalent to each other. Moreover, the various consistent criteria are equivalent to Schwarz's Bayesian criterion, BIC. It is possible thus to characterize the choice of a model selection criterion as essentially between these two classes of criteria. This choice, in turn, depends upon whether a parametric or a nonparametric approach is used for modelling an observed time series. The former is usually based on the fitting of an ARMA model and then a consistent criterion may be preferred. The computational difficulties discussed in section 5.3, however, indicate that a consistent recursive procedure may be used for

ARMA model selection. There are three reasons for making this recommendation. First, in this situation, AIC can select a highly nonparsimonious model. Secondly, an ARMA model may be used with a relatively small number of observations and then a parsimonious model may be preferred. Thirdly, little is currently known concerning the properties of an ARMA model when interpreted merely as a whitening filter. Indeed, the question of identifying an ARMA(∞, ∞) model only from a knowledge of its spectral density function is still unresolved.

This last difficulty notwithstanding, it should be emphasized that the use of a consistent procedure for ARMA model selection does not necessarily commit one to the belief in the existence of an underlying true ARMA structure. Rather, such a criterion is being used as a practical procedure for selecting a possibly acceptable model which may still be subjected to the various diagnostic tests, as in Box and Jenkins (1970).

By contrast, AIC or an equivalent criterion may be preferred when a nonparametric approach based on autoregressive model fitting is used. As in Shibata (1980, 1981) and Bhansali (1986c), the nonparametric criteria are asymptotically optimal for one-step prediction, and for autoregressive spectral estimation. Secondly, as discussed in section 5.4, the nonparametric criteria may be generalized in a straightforward fashion to the question of autoregressive order selection for prediction more than one step ahead, and to multivariate time series. Thirdly, the nature of the approximation provided by a finite autoregression is well understood theoretically. A difficulty, however, is that this approach may be less suitable with a relatively small series.

A flexible alternative to choosing between a parametric and a nonparametric criterion is to recognize that AIC and BIC are both special cases, with $\alpha = 2$ and $\alpha = \log T$ respectively, of the AIC_α criterion. One may, therefore, use the AIC_α criterion with several different values of $\alpha \in [2, \log T]$ and then choose a final model subjectively in the light of the properties of the various models selected. If further prior information concerning the nature of the model generating an observed time series is available then more objective methods, discussed in sections 5.2 and 5.3, may also be used for finally choosing an appropriate model.

REFERENCES

Abraham, B and Ledolter, J. (1983) *Statistical Methods for Forecasting*, Wiley, New York.

Akaike, H. (1970) Statistical predictor identification. *Ann. Inst. Statist. Math.*, **22**, 203–217.

Akaike, H. (1973) Information theory and an extension of the maximum likelihood principle, in *2nd International Symposium on Information Theory* (eds. B.N. Petrov and F. Csaki), Akaemia Kiado, Budapest, pp. 267–281.

Akaike, H. (1974) Stochastic theory of minimal realizations. *IEEE Trans. Auto Control*, **AC-19**, 667–674.

Akaike, H. (1976) Canonical correlation analysis of time series and the use of an information criterion. In *Systems Identification: Advances and Case Studies* (eds. R.K. Mehira and D.G. Lainiotis), pp. 27–96, New York: Academic Press.

Akaike, H. (1977) On entropy maximisation principle, in *Applied Statistics*, (ed. P.R. Krishnaiah), North Holland, Amsterdam, pp. 27–41.

Akaike, H. (1978a) Comments on 'On model structure testing in system identification. *Int. J. Control*, **27**, 323–324.

Akaike, H. (1978b) On the likelihood of a time series model. *Statistician*, **27**, 217–235.

Akaike, H. (1978c) A Bayesian analysis of the minimum AIC procedure. *Ann. Inst. Statist. Math.*, **30**, A, 9–14.

Akaike, H. (1979) A Bayesian extension of the minimum AIC procedure of autoregressive model fitting. *Biometrika*, **66**, 237–242.

Andel, J. (1982) Fitting models in time series analysis. *Math. Operationsfortsch Statist.*, Ser. Statistics, **13**, 121–143.

Beamish, N. and Priestley, M.B. (1981) A study of autoregressive and window spectral estimation. *Applied Statistics*, **30**, 41–58.

Bhansali, R.J. (1980) Autoregressive and window estimates of the inverse correlation function. *Biometrika*, **67**, 551–566.

Bhansali, R.J. (1983a) Estimation of the order of moving average process from autoregressive and window estimates of the inverse correlation function. *J. Time Series Anal*, **4**, 137–162.

Bhansali, R.J. (1983b) The inverse partial correlation function of a time series and its applications. *J. Multivariate Anal.*, **13**, 310–327.

Bhansali, R.J. (1986a) A derivation of the information criteria for selecting autoregressive models. *Adv. Appl. Prob.*, **18**, 360–387.

Bhansali, R.J. (1986b) The criterion autoregressive transfer function of Parzen. *J. Time Series Anal.*, **7**, 79–104.

Bhansali, R.J. (1986c) Asymptotically efficient estimation of the order by the criterion autoregressive transfer function. *Ann. Statist.*, **13**, 315–325.

Bhansali, R.J. (1987) The discrimination between autoregressive and moving average models from the estimated inverse correlations. *Metron*, **45**, 115–135.

Bhansali, R.J. (1988) Consistent order determination for processes with infinite variance. *J. Royal Statist. Soc.*, **B50**, 46–60.

Bhansali, R.J. (1991) Consistent recursive estimation of the order of an autoregressive moving average process. *Int. Statist. Review*, **59**, 81–96.

Bhansali, R.J. and Downham, D.Y. (1977) Some properties of the order of an autoregressive process selected by a generalization of Akaike's FPE criterion. *Biometrika*, **64**, 546–551.

Bhansali, R.J. and Papengelou, F. (1991) Convergence of moments of least squares estimators for the coefficients of an autoregressive process of unknown order. *Ann. Statist*, **19**, 1151–1162.

Box, G.E.P. and Jenkins, G.M. (1970) *Time Series Analysis: Forecasting and Control*, Holden Day, San Francisco.

Box, G.E.P. and Ljung, G.M. (1978) On a measure of lack of fit in time series models. *Biometrika*, **65**, 297–303.

Brillinger, D.R. (1975) *Time Series: Data Analysis and Theory*, Holt, New York.

Caines, P.E. (1987) *Linear Stochastic Systems*, Wiley, New York.

Chow, G.C. (1981) A comparison of the information and posterior probability criterion for model selection. *J. Econometrics*, **16**, 21–33.

Cleveland, W.S. (1972) The inverse autocorrelations of a time series and their applications. *Technometrics*, **14**, 277–293.

Davies, N. and Pettrucelli, J.D. (1984) On the use of the general partial autocorrelation function for order determination in ARMA(p, q), processes. *J. Amer. Statist. Assoc.*, **79**, 374–377.

Davisson, I.D. (1965) The prediction error of a stationary Gaussian time series of unknown covariance. *IEEE Trans on Info. Theory* **IT-11**, 527–532.

Durbin, J. (1959) Efficient estimation of parameters in moving average models. *Biometrika*, **46**, 306–316.

Durbin, J. (1960) The fitting of time series models. *Rev. Int. Stat. Inst.*, **28**, 233–244.

Durbin, J. (1961) Efficient fitting of linear models for continuous time series from discrete data. *Bull. Int. Statist. Inst*, **38**, 273–282.

De Gooijer, J.G., Abraham, B., Gould, A. and Robinson, L. (1985) Methods for determining the order of an autoregressive-moving average process: A survey. *Int. Statist. Review*, **53**, 301–329.

Findley, D.F. (1985a) On the unbiasedness property of AIC for exact or approximating linear stochastic time series models. *J. Time Series Anal.*, **6**, 229–252.

Findley, D.F. (1985b) Model selection for multi-step-ahead forecasting, in *Identification and System Parameter Estimation*, 7th IFAC/IFORS Symposium, Pergamon, Oxford, pp. 1039–1044.

Franke, J. (1985) A Levison–Durbin recursion for autoregressive-moving average processes. *Biometrika*, **72**, 573–581.

Fuller, W.A. and Hasza, D.P. (1981) Properties of predictors for autoregressive time series. *J. Amer. Statist. Assoc.*, **76**, 155–161.

Gersch, W. and Kitagawa, G. (1983) The prediction of a time series with trends and seasonalities. *J. Bus. and Eco. Statist.*, **1**, 253–264.

Good, I.J. (1965) *The Estimation of Probabilities*, MIT Press, Cambridge, Mass.

Gray, H.L., Kelley, G.D. and McIntire, D.D. (1978) A new approach to ARMA modelling. *Comm. Statist.*, **B7**, 1–77.

Hannan, E.J. (1969) The estimation of mixed moving average autoregressive systems. *Biometrika*, **56**, 579–593.

Hannan, E.J. (1970) *Multiple Time Series*, Wiley, New York.

Hannan, E.J. (1980) The estimation of the order of an ARMA process. *Ann. Statist.*, **8**, 1071–1081.

Hannan, E.J. (1987) Rational transfer function approximation. *Statist. Science*, **2**, 135–161.

Hannan, E.J. and Deistler, M. (1988) *The Statistical Theory of Linear Systems*, Wiley, New York.

Hannan, E.J. and Kavalieris, L. (1984a) A method for autoregressive-moving average estimation. *Biometrica*, **72**, 273–280.

Hannan, E.J. and Kavalieris, L. (1984b) Multivariate linear time series models. *Adv. Appl. Prob.*, **16**, 492–561.

Hannan, E.J., McDougall, A.J. and Poskitt, D.S. (1989) Recursive estimation of autoregressions. *J. Royal Statist. Soc.*, **B51**, 217–234.

Hannan, E.J. and Quinn, B.G. (1979) The determination of the order of an autoregression. *J. Royal Statist. Soc.*, **B41**, 190–195.

Hannan, E.J. and Rissanen, J. (1982). Recursive estimation of mixed autoregressive moving average order. *Biometrika*, **69**, 81–94; Correction (1983), **70**, 303.

Hemerley, E.M. and Davis, M.H.A. (1989) Strong consistency of the PLS criterion for order determination of autoregressive processes. *Ann. Statist.*, **17**, 941–946.

Hemerley, E.M. and Davis, M.H.A. (1991) Recursive order estimation of autoregressions without bounding the model set. *J. Royal Statist. Soc.*, **B53**, 201–210.

Hurvich, C.M. and Tsai, L. (1989) Regression and time series model selection in small samples. *Biometrika*, **76**, 297–307.

Hurvich, C.M., Shumway, R. and Tsai, Chih-Ling (1990) Improved estimators of

Kullback–Leibler information for autoregressive model selection in small samples. *Biometrika*, **77**, 709–720.

Jones, R.H. (1975). Fitting autoregressions. *J. Amer. Statist. Assoc.*, **70**, 590–592.

Kabaila, P.V. (1981) Estimation based on one step ahead prediction versus estimation based on multi-step ahead prediction. *Stochastics*, **6**, 43–45.

Kashyap, R.L. (1977) A Bayesian comparison of different classes of dynamic models using empirical data. *IEEE Trans. Auto. Control.*, **AC-22**, 715–727.

Kavalieris, L. (1991). A note on estimating autoregressive moving average order. *Biometrika*, **78**, 920–922.

Koehler, A.B. and Murphree, E.S. (1988) A comparison of the Akaike and Schwarz criteria for selecting model order. *Appl. Statist.*, **37**, 187–195.

Liao, Pyng-Mei (1991) *Recursive Order Selection for Autoregressive Moving Average Models: An Application*, MSc Thesis, Liverpool University.

Makridakis, S., Anderson, A., Carbone, R., Fildes, R., Hibon, M., Lewandowski, R., Newton, J., Parzen, E. and Winkler, R. (1982) The accuracy of extrapolation (time series) methods: results of a forecasting competition. *J. Forecast.*, **2**, 111–153.

Mallows, C.L. (1973) Some comments on C_p. *Technometrics*, **17**, 661–675.

McClave, J. (1975) Subset autoregression. *Technometrics*, **17**, 213–220.

Mbago, M. (1979) *Finite Sample Properties of the Order of Linear Time Series Models Selected by Akaike's Information Criterion and its Extensions: A Simulation Study*. Ph.D. Thesis, Liverpool University.

Newbold, P. (1988) Some recent developments in time series analysis—III. *Int. Statist. Review*, **56**, 17–29.

Newbold, P. and Granger, C.W.J. (1974) Experience with forecasting univariate time series and the combination of forecasts (with discussions). *J. Royal Statist. Assoc.*, **A137**, 131–156.

Neftci, S.N. (1982) Specification of economic time series models using Akaike's criterion. *J. Amer. Statist. Assoc.*, **77**, 537–540.

Nicholls, D.F. (1972) On Hannan's estimation of ARMA models. *Austral. J. Statist.*, **3**, 262–269.

Osborn, O.R. (1976) Maximum likelihood estimation of moving average processes. *Ann. Eco. Soc. Meas.*, **5**, 75–87.

Ozaki, T. (1977) On the order determination of ARIMA models. *Appl. Statist.* **26**, 290–301.

Parzen, E. (1969) Multiple time series modelling, in *Multivariate Analysis II*, (ed. P.R. Krishnaiah) Academic Press, New York.

Parzen, E. (1974) Some recent advances in time series modelling and prediction problems. *IEEE Trans. Auto. Control*, **AC-19**, 723–730.

Parzen, E. (1977) Multiple time series: determining the order of approximating schemes, in *Multivariate Analysis IV* (ed. P.R. Krishnaiah), Academic Press, New York, pp. 283–295.

Parzen, E. (1981) Time series model identification and prediction variance horizon, in *Applied Time Series Analysis II* (ed. D. Findley), Academic Press, New York, pp. 415–447.

Poskitt, D.S. (1986) A Bayes procedure for the identification of univariate time series models. *Ann. Statist.* **14**, 502–516.

Poskitt, D.S. (1987) A modified Hannan–Rissanen strategy for mixed autoregressive-moving average order determination. *Biometrika*, **74**, 781–790.

Poskitt, D.S. and Tremayne, A.R. (1981) An approach to testing linear time series models. *Ann. Statist.* **9**, 974–986.

Potscher, B.M. (1983) Order estimation in ARMA models by Lagrange multiplier tests. *Ann. Statist.* **11**, 872–885.

Potscher, B.M. (1990) Estimation of autoregressive moving average order given an infinite number of models and approximation of spectral densities. *J. Time Series Analysis*, **11**, 165–179.

Priestley, M.B. (1981) *Spectral Analysis and Time Series*, Academic Press, New York.

Rissanen, J. (1978) Modelling by shortest data description. *Automatica*, **14**, 465–471.

Rissanen, J. (1986) Stochastic complexity and predictive modelling. *Ann. Statist.*, **14**, 1080–1100.

Rissanen, J. (1987) Stochastic complexity (with discussion). *J. Royal Statist. Soc.*, **B49**, 223–239.

Sawa, T. (1978) Information criteria for discriminating among alternative regression models. *Econometrica*, **46**, 1273–1291.

Schwarz, G. (1978) Estimating the dimension of a model. *Ann. Statist.*, **6**, 461–464.

Shibata, R. (1976) Selection of the order of an autoregressive model by Akaike's information criterion. *Biometrika*, **63**, 117–126.

Shibata, R. (1980) Asymptotically efficient selection of the order of the model for estimating the parameters of a linear process. *Ann. Statist.*, **8**, 147–164.

Shibate, R. (1981) An optimal autoregressive spectral estimate. *Ann. Statist.*, **9**, 300–306.

Shibata, R. (1985) Various model selection techniques in time series analysis, in *Handbook of Statistics, Volume 5*, (eds E.J. Hannan, P.R. Krishnaiah and M.M. Rao), Elsevier Amsterdam, pp. 179–187.

Smith, A.F.M. and Spiegelhalter, J.D. (1980) Bayes factors and choice criteria for linear models. *J. Royal Statist. Soc.*, **B42**, 213–220.

Soderstrom, T. (1977) On model structure testing in system identification. *Int. J. Control*, **26**, 1–18.

Stoica, P. and Soderstrom, T. (1984). Uniqueness of estimated *k*-step prediction models of ARMA processes. *Systems & Control Letters*, **4**, 325–331.

Stone, M. (1979) Comments on model selection criteria of Akaike and Schwarz. *J. Royal Statist Soc.*, **B41**, 276–278.

Terasvirta, T. and Mellin, I. (1986) Model selection criteria and model selection tests in regression models. *Scand. J. Statist.*, **13**, 159–171.

Tjostheim, D. and Paulsen, J. (1983) Bias of some commonly-used time series estimates. *Biometrika*, **70**, 389–399.

Tong. H. (1977). Some comments on the Canadian Lynx data. *J. Royal Statist. Soc.*, **A40**, 432–436.

Tiao, G.C. and Tsay, R.S. (1989) Model specification in multivariate time series (with Discussion). *J. Royal Statist. Soc.*, **B51**, 157–214.

Tiao, G.C. and Xu, D. (1991) Robustness of MLE for multistep predictions: the exponential smoothing case. *Biometrica*, in press.

Tsay, R.S. and Tiao, G.C. (1984) Consistent estimates of autoregressive parameters and extended sample autocorrelation function for stationary and nonstationary ARMA models. *J. Amer. Statist. Assoc.*, **79**, 84–96.

Tuan, P.D. (1988) Estimation of autoregressive parameters and order selection for ARMA models. *J. Time Series Anal.*, **9**, 265.

Wei, C.Z. (1992) On predictive least-squares principle. *Ann. Statist.*, **20**, 1–42.

Weiss, A.A. (1991) Multi-step estimation and forecasting in dynamic models. *J. Econometrics*, **48**, 135–149.

Whittle, P. (1951) *Hypothesis Testing in Time Series Analysis*, Almqvist and Wiksell, Uppsala.

Whittle, P. (1952) Tests of fit in time series. *Biometrika*, **39**, 309–318.

Whittle, P. (1963a) *Prediction and Regulation by Linear Least-Square Methods.* London University Press.

Whittle, P. (1963b) On the fitting of multivariate autoregression and the approximate canonical factorisation of a spectral density matrix. *Biometrika*, **50**, 129–134.

Part Two

Estimation and Asymptotics for Time Series Models

6

The Gaussian log likelihood and stationary sequences

M. Kramer and M. Rosenblatt

6.1 INTRODUCTION

In a number of finite parameter problems in time series, a likelihood function is computed as if the process is Gaussian and this likelihood is maximized to obtain estimates of the unknown parameters. Our object is to get qualitative information about the global asymptotic behavior of this 'Gaussian' likelihood function. In particular we shall look at the sequence (X_t) as an ARMA stationary (p, q) process satisfying the relation

$$X_t - \phi_1 X_{t-1} - \cdots - \phi_p X_{t-p} = Z_t + \theta_1 Z_{t-1} + \cdots + \theta_q Z_{t-q},$$

where the Z_t terms are independent and identically distributed with mean 0 and variance σ^2, with finite 4th moment, and the polynomials $\phi(\zeta) = 1 - \phi_1 \zeta - \cdots - \phi_p \zeta^p$ and $\theta(\zeta) = 1 + \theta_1 \zeta + \cdots + \theta_q \zeta^q$ have no roots on the unit circle. If $\beta = (\phi_1, \ldots, \phi_p, \theta_1, \ldots, \theta_q)$, the spectral density of (X_t) can be written as

$$\frac{\sigma^2}{2\pi} g_\beta(\lambda)$$

where

$$g_\beta(\lambda) = |\theta(e^{-i\lambda})/\phi(e^{-i\lambda})|^2.$$

The covariance matrix of $(X_1, \ldots, X_n)$ can be written as $\sigma^2 G_n(\beta)$ where the (k, ℓ)th component of $G_n(\beta)$ is

$$\frac{1}{2\pi} \int_{-\pi}^{\pi} e^{i(k-\ell)\lambda} g_\beta(\lambda) d\lambda.$$

Suppose that (X_t) is an ARMA(p, q) process with underlying parameter (β_0, σ_0^2). The formal Gaussian log likelihood as a function of (β, σ^2) with observations $X_n = (X_1, \ldots, X_n)$ is

$$L_n(\beta, \sigma^2) = -\frac{n}{2}\log 2\pi - \frac{1}{2}\log|\sigma^2 G_n(\beta)| - \frac{1}{2\sigma^2} X_n' G_n^{-1}(\beta) X_n.$$

This is, of course, not the actual likelihood function unless (X_t) is a Gaussian process. If (X_t) is a non-Gaussian process that is not causal invertible, it is well known that maximizing this 'Gaussian' likelihood will not lead to consistent estimates of the unknown parameters (β_0, σ_0^2), see Rosenblatt (1985), for example. However, if the sequence is causal invertible, maximizing this likelihood will lead to asymptotically normal consistent estimates (see Brockwell and Davis (1991) which we shall subsequently refer to as BD). However, from a finite sample point of view not much has been said about the likelihood function globally.

6.2 MEAN BEHAVIOUR

We shall actually consider

$$\frac{1}{n} L_n(\beta, \sigma^2) = -\frac{1}{2}\log 2\pi - \frac{1}{2n}\log|\sigma^2 G_n(\beta)| - \frac{1}{2\sigma^2}\frac{1}{n} X_n' G_n^{-1}(\beta) X_n.$$

The first Szegö limit tells us that $\lim_{n\to\infty} 1/n \log|G_n(\beta)|$ is zero and the second Szegö limit theorem, Grenander and Szegö (1958) tells us that this is uniformly approached with an error of order $1/n$ as long as the zeros Z_j, $\tilde{Z}_k$ of the polynomials $\phi(\zeta)$, $\theta(\zeta)$ satisfy $\varepsilon^{-1} > |Z_j| > 1 + \varepsilon$, $\varepsilon^{-1} > |\tilde{Z}_k| > 1 + \varepsilon$, $|Z_j - \tilde{Z}_k| > \varepsilon$, $j = 1, \ldots, p$, $k = 1, \ldots, q$ for some fixed small $\varepsilon > 0$. Call this condition $C(\varepsilon)$. Also,

$$E\frac{1}{n} X_n' G_n^{-1}(\beta) X_n = \frac{1}{n}\operatorname{tr}(G_n(\beta_0) G_n^{-1}(\beta))\sigma_0^2$$

and one can approximate the spectral density $f_\beta(\lambda)$ by a trigonometric polynomial $P_m(\lambda)$ of degree m such that for all λ

$$|f(\lambda) - P_m(\lambda)| \leqslant A_k M m^{-k}$$

where M is an absolute bound for $f(\lambda)$ which is k times differentiable and A_k depends only on k (Zygmund (1959) vol. 1, p. 115). Given condition $C(\varepsilon)$ this can be done uniformly in β in such a way that we have theorem 6.1.

Theorem 6.1

If X_t is ARMA(p, q) and $C(\varepsilon)$ is satisfied

$$\frac{1}{n}EL_n(\beta, \sigma^2) = -\frac{1}{2}\log 2\pi - \frac{1}{2\sigma^2}\frac{\sigma_0^2}{2\pi}\int_{-\pi}^{\pi}\frac{g_{\beta_0}(\lambda)}{g_\beta(\lambda)}\,\mathrm{d}\lambda + 0(n^{-1+\delta})$$

for a given $\delta > 0$ with a uniform 0 as long as $\varepsilon^{-1} > \sigma^2 > \varepsilon > 0$.

6.3 FLUCTUATION

We will be looking at the asymptotic properties of the fluctuation of $L_n(\beta, \sigma^2)$ about its mean, i.e. $L_n(\beta, \sigma^2) - EL_n(\beta, \sigma^2) = (2\sigma^2)^{-1}(X_n' G_n^{-1} X_n - EX_n' G_n^{-1} X_n)$. It will be established that $n^{-1/2}(X_n' G_n^{-1}(\beta) X_n - EX_n' G_n^{-1} \beta X_n)$ is asymptotically normal with mean 0 and variance

$$(\eta - 3)\frac{\sigma_0^4}{4\pi^2}\left(\int_{-\pi}^{\pi}\frac{g_{\beta_0}(\lambda)}{g_\beta(\lambda)}\,\mathrm{d}\lambda\right)^2 + \frac{\sigma_0^4}{\pi}\int_{-\pi}^{\pi}(g_{\beta_0}(\lambda)/g_\beta(\lambda))^2\,\mathrm{d}\lambda.$$

Here $\eta = EZ_t^4/\sigma_0^4$.

In the multiparameter case, i.e. where we are looking at the fluctuation of the Gaussian likelihood surface jointly at a finite number of parameters $(\beta_1, \ldots, \beta_k)$, we will see that

$$n^{-1/2}(X' G_n^{-1}(\beta_i) X - EX' G_n^{-1}(\beta_i) X)_{i=1,\ldots,k} \text{ is AN}(0, \textstyle\sum)$$

(asymptotically normal with mean 0 and covariance $\sum$) where

$$\sum(i, j) = (\eta - 3)\frac{\sigma_0^4}{4\pi^2}\int_{-\pi}^{\pi}\frac{g_{\beta_0}(\lambda)}{g_{\beta_i}(\lambda)}\,\mathrm{d}\lambda\int_{-\pi}^{\pi}\frac{g_{\beta_0}(\lambda)}{g_{\beta_j}(\lambda)}\,\mathrm{d}\lambda + \frac{\sigma_0^4}{\pi}\int_{-\pi}^{\pi}\frac{g_{\beta_0}(\lambda)^2}{g_{\beta_i}(\lambda) g_{\beta_j}(\lambda)}\,\mathrm{d}\lambda.$$

To establish these results we will replace the quadratic form $X' G_n^{-1}(\beta) X$ by $X' \tilde{G}_n(\beta) X$ where

$$\tilde{G}_n(\beta)(k, \ell) = \frac{1}{2\pi}\int_{-\pi}^{\pi} \mathrm{e}^{i(k-\ell)\lambda} g_\beta^{-1}(\lambda)\,\mathrm{d}\lambda.$$

In section 6.4 it will be shown that asymptotically $n^{-1/2}(X' G_n^{-1} X - EX' G_n^{-1} X)$ and $n^{-1/2}(X' \tilde{G}_n X - EX' \tilde{G}_n X)$ have the same distribution i.e. their difference is $o_p(1)$. Section 6.5 focuses on the asymptotic distribution of $n^{-1/2} X' \tilde{G}_n (X - EX' G_n X)$ and section 6.6 is devoted to the multiparameter situation.

6.4 PART I

Theorem 6.2

The difference between

$$X_n' G^{-1}(\beta) X_n - E X_n' G_n^{-1}(\beta) X_n$$

and

$$X_n' \tilde{G}_n(\beta) X_n - E X_n' \tilde{G}_n(\beta) X_n$$

is $o_p(n^{1/2})$, i.e.

$$X_n'(G_n^{-1}(\beta) - \tilde{G}_n(\beta)) X_n - E X_n'(G_n^{-1}(\beta) - \tilde{G}_n(\beta)) X_n$$

is $o_p(n^{1/2})$.

Proof

We shall use the fact that a k-times differentiable periodic function $f(\lambda)$ such that $|f(\lambda)| \leqslant M$ can be approximated by a trigonometric polynomial $P_m(\lambda)$ of degree m such that $|f(\lambda) - P_m(\lambda)| \leqslant A_k M m^{-k}$ for all λ where A_k depends on k only (Zygmund (1959) vol. 1, p. 115).

Since for fixed β there is a ρ such that $\rho < g_\beta(\lambda) < \rho^{-1}$ it follows that for any given k and any $m \geqslant m(k, \rho)$ there will be a trigonometric polynomial $P_m^{(k)}(\lambda)$ such that $|g_\beta(\lambda) - P_m^{(k)}(\lambda)|$ and $|g_\beta^{-1}(\lambda) - P_m(k)(\lambda)^{-1}|$ are smaller than $K(k, \rho) m^{-k}$.

Using $P_m^{(k)}(\lambda)$ we decompose $G_n^{-1}(\beta) - \tilde{G}_n(\beta)$ as follows:

$$G_n^{-1} - \tilde{G}_n = (G_n^{-1}(\beta) - H_n^{-1}(\beta)) + (H_n^{-1}(\beta) - \tilde{H}_n(\beta)) + (\tilde{H}_n(\beta) - \tilde{G}_n(\beta))$$

where

$$H_n(\beta)(s, t) = \frac{1}{2\pi} \int_{-\pi}^{\pi} e^{i(s-t)\lambda} P_m^{(k)}(\lambda) d\lambda$$

and

$$\tilde{H}_n(\beta)(s, t) = \frac{1}{2\pi} \int_{-\pi}^{\pi} e^{i(s-t)\lambda} P_m^{(k)}(\lambda)^{-1} d\lambda$$

Note the hidden dependence of H_n and $\tilde{H}_n$ on m and k. Theorem 6.2 will be established by showing that

$$\mathrm{Var}(n^{-1/2} X'(G_n^{-1} - H_n^{-1}) X) \to 0, \tag{6.1}$$

$$\mathrm{Var}(n^{-1/2} X'(H_n^{-1} - \tilde{H}_n) X) \to 0, \tag{6.2}$$

$$\mathrm{Var}(n^{-1/2} X'(\tilde{H}_n - \tilde{G}_n) X) \to 0, \tag{6.3}$$

as $n \to 0$ if k is chosen large enough and m is an appropriate function of n. (As it turns out $k = 20$ and $m = n^{1/5}$ will do it.)

Proof of (6.1)
Let $D_n = G_n - H_n$ so

$$D_n(s,t) = \frac{1}{2\pi}\int_{-\pi}^{\pi} e^{i(s-t)\lambda}(g_\beta(\lambda) - P_m^{(k)}(\lambda))\,d\lambda.$$

Since $|g_\beta(\lambda) - P_m^{(k)}(\lambda)| < K(k,\rho)m^{-k}$ it is easily seen that for any $a, b \in \mathbb{R}^n$ $(a'D_n b)^2 \leqslant n^2\|a\|^2\|b\|^2 K(k,\rho)m^{-k}$. Hence

$$\begin{aligned}(X'(G_n^{-1} - H_n^{-1})X)^2 &= (X'G_n^{-1}D_nH_n^{-1}X)^2 \\ &\leqslant n^2K(k,\rho)m^{-k}\|G_n^{-1}X\|^2\|H_n^{-1}X\|^2.\end{aligned}$$

But the eigenvalues of G_n and H_n are smaller than $1/\rho$ and $2/\rho$ respectively as $g_\beta(\lambda)$ and $P_m^{(k)}(\lambda)$ are smaller than $1/\rho$ and $2/\rho$ respectively (the latter is true if m is large enough). As a consequence the eigenvalues of G_n^{-1} and H_n^{-1} are larger than $\rho/2$. Hence $(X'(G_n^{-1} - H_n^{-1})X)^2 \leqslant n^2K(k,\rho)m^{-k}(\rho^4/16)\|X\|^4$. Taking expectations we see that

$$\operatorname{Var}(n^{-1/2}X'(G_n^{-1} - H_n^{-1})X) \leqslant n^{-1}K(k,\rho)(\rho^4/16)E(\hat{\gamma}_n(0)^2)$$

where we used $m = n^{1/5}$, $k = 20$ and $\hat{\gamma}_n(0) = 1/n\sum_{t=1}^{n} X_t^2$. Since $E(\hat{\gamma}_n(0))^2$ is bounded (BD section 7.3.1.). (6.1) follows with the above selection of k and m.

Proof of (6.2)
Let $\hat{D}_n := H_n^{-1} - \tilde{H}_n$. Recall that H_n corresponds to $P_m^{(k)}(\lambda)$ while $\tilde{H}_n$ correponds to $P_m^{(k)}(\lambda)^{-1}$. It is well known that $P_m^{(k)}(\lambda)$ can be written as $A_m|a(e^{-i\lambda})|^2$ where a is a real polynomial of degree m with constant coefficient equal to 1 and $A_m > 0$. Using the same argument as found in BD (proposition 10.8.3 p. 381) we see that $\hat{D}_n$ has at most $2m^2$ nonzero components that are uniformly bounded as $n \to \infty$. So

$$\operatorname{Var}(n^{-1/2}X'\hat{D}_nX) = \frac{1}{n}\sum_{\substack{4m^4\\ \text{terms}}} d_{k\ell}d_{st}\operatorname{Cov}(X_kX_\ell, X_sX_t)$$

where the d_{ij} and $\operatorname{Cov}(X_kX_\ell, X_sX_t)$ are bounded (proof of BD, proposition 7.8.1). Hence

$$\operatorname{Var}(n^{-1/2}X'\hat{D}X) \leqslant \frac{1}{n}4m^4K = n^{-1/5}4K$$

with the choice of $m = n^{1/5}$ mentioned earlier. We conclude that $\operatorname{Var}(n^{-1/2}X'\hat{D}_nX) \to 0$.

Proof of (6.3)
Let $\tilde{D}_n = \tilde{H}_n - \tilde{G}_n$ so

$$\tilde{D}_n(s,t) = \frac{1}{2\pi}\int_{-\pi}^{\pi}(P_m^{(k)}(\lambda)^{-1} - g_\beta(\lambda)^{-1})e^{i(s-t)\lambda}\,d\lambda.$$

As in (6.1)

$$|g_\beta^{-1}(\lambda) - P_m^{(k)}(\lambda)^{-1}| < K(k,\rho)m^{-k} \text{ once } m > m(k,\rho).$$

Now

$$X'\tilde{D}_n X - EX'\tilde{D}_n X = \frac{n}{2\pi}\int_{-\pi}^{\pi} \sum_{|k|<n} (\hat{\gamma}_n(k) - E\hat{\gamma}_n(k))e^{ik\lambda}\delta(\lambda)\,d\lambda$$

where $\delta(\lambda) = P_m^{(k)}(\lambda)^{-1} - g_\beta(\lambda)^{-1}$ and

$$\hat{\gamma}_n(k) = \frac{1}{n}\sum_{t=1}^{n-|k|} X_t X_{t+|k|}.$$

If we set $\tilde{\gamma}_n(k) := \tilde{\gamma}_n(k) - E\hat{\gamma}_n(k)$ then

$$\mathrm{Var}(n^{-1/2}X'\tilde{D}_n X) = nE\left(\frac{1}{2\pi}\int_{-\pi}^{\pi}\sum_{|k|<n}\tilde{\gamma}_n(k)e^{ik\lambda}\delta(\lambda)\,d\lambda\right)^2.$$

Since the inner integrand is real and $1/(2\pi)d\lambda$ is a probability measure on $[-\pi,\pi]$ we can apply the Schwarz inequality to get

$$\begin{aligned}\mathrm{Var}(n^{-1/2}X'\tilde{D}_n X) &\leqslant nE\left(\frac{1}{2\pi}\int_{-\pi}^{\pi}\sum_{|k|,|\ell|<n}\tilde{\gamma}_n(k)\tilde{\gamma}_n(\ell)e^{i(k-\ell)\lambda}\delta(\lambda)^2\,d\lambda\right)\\ &\leqslant (K(k,\rho)m^{-k})^2 n\sum_{|k|<n}\mathrm{Var}\,\hat{\gamma}_n(k)\\ &= K(20,\rho)^2 n^{-7}\sum_{|k|<n}\mathrm{Var}\,\hat{\gamma}_n(k).\end{aligned}$$

But $\mathrm{Var}\,\hat{\gamma}_n(k)\to 0$ for all k as $n\to\infty$ (BD, proposition 7.3.1). Hence $\mathrm{Var}(n^{-1/2}X'\tilde{D}_n X)\to 0$ with $k=20$ and $m=n^{1/5}$. This proves theorem 6.2.

6.5 PART II

We now turn to the investigation of the asymptotic properties of

$$n^{-1/2}(X_n' G_n^{-1}(\beta)X_n - EX_n' G_n^{-1}(\beta)X_n).$$

Due to theorem 6.2 it suffices to look at the asymptotic distribution of $n^{-1/2}(X_n'\tilde{G}_n(\beta)X_n - EX_n'\tilde{G}_n(\beta)X_n)$. To this extent we will employ the fact that

$$(\hat{\gamma}_n(0),\ldots,\hat{\gamma}_n(m)) - E(\hat{\gamma}_n(0),\ldots,\hat{\gamma}_n(m)) \text{ is } \mathrm{AN}(0, B_m/n) \tag{6.4}$$

with

$$B_m(s,t) = (\eta-3)\gamma(s)\gamma(t) + \sum_{k=-\infty}^{\infty}\{\gamma(k)\gamma(k-s+t) + \gamma(k-s)\gamma(k+t)\}$$

where $EZ_t^4 = \eta\sigma_0^4$ as noted earlier. This is a direct consequence of BD proposition 7.3.4.

Theorem 6.3

$n^{-1/2}(X'\tilde{G}_n(\beta)X - EX'\tilde{G}_n(\beta)X)$ is asymptotically normal with mean 0 and variance

$$(\eta-3)\frac{\sigma_0^4}{4\pi^2}\left(\int_{-\pi}^{\pi}\frac{g_{\beta_0}(\lambda)}{g_\beta(\lambda)}\mathrm{d}\lambda\right)^2+\frac{\sigma_0^4}{\pi}\int_{-\pi}^{\pi}(g_{\beta_0}(\lambda)/g_\beta(\lambda))^2\mathrm{d}\lambda.$$

Proof

Let us first define

$$\tilde{g}(k):=\frac{1}{2\pi}\int_{-\pi}^{\pi}\mathrm{e}^{ik\lambda}g_\beta^{-1}(\lambda)\mathrm{d}\lambda$$

so that $\tilde{g}(s-t)=\tilde{G}_n(\beta)(s,t)$. Let

$$U_n:=n^{-1/2}(X'\tilde{G}_nX-EX'\tilde{G}_nX)=n^{1/2}\sum_{|k|<n}\tilde{g}(k)(\hat{\gamma}_n(k)-E\hat{\gamma}_n(k)).$$

If we let

$$V_{nm}:=n^{1/2}\sum_{|k|<m}\tilde{g}(k)(\hat{\gamma}_n(k)-E\hat{\gamma}_n(k))$$

then, using (6.4) we find that as $n\to\infty$, $V_{nm}\Rightarrow V_m$ weakly where $V_m\sim N(0,t_m'B_mt_m)$ with $t_m'=(\tilde{g}(0),2\tilde{g}(1),\ldots,2\tilde{g}(m))$. To prove the theorem we have to show that:

(a) V_m converges in distribution as $m\to\infty$ to a normal random variable U with mean 0 and the variance specified in the theorem;
(b) $\lim_m \overline{\lim_n} P(|U_n-V_{nm}|>\varepsilon)=0$ for all $\varepsilon>0$.

(BD proposition 6.3.9 justifies these criteria) as this would show that $U_n\Rightarrow U$ weakly.

Proof of (a)
We have to show that

$$t_m'B_mt_m\to\frac{\sigma_0^4}{\pi}\int_{-\pi}^{\pi}(g_{\beta_0}(\lambda)/g_\beta(\lambda))^2\mathrm{d}\lambda+(\eta-3)\frac{\sigma_0^4}{4\pi^2}\left(\int_{-\pi}^{\pi}\frac{g_{\beta_0}(\lambda)}{g_\beta(\lambda)}\mathrm{d}\lambda\right)^2.$$

From the definition of the matrix B_m and the fact that $\gamma(k)$ is the kth Fourier coefficient of

$$\sigma_0^2g_{\beta_0}(\lambda)=\left|\frac{\theta_0(\mathrm{e}^{-i\lambda})}{\phi_0(\mathrm{e}^{-i\lambda})}\right|^2\sigma_0^2$$

it can easily be derived that

$$B_m(p,q) = (\eta - 3)\gamma(p)\gamma(q) + \frac{\sigma_0^4}{2\pi}\int_{-\pi}^{\pi} g_{\beta_0}^2(\lambda)(\mathrm{e}^{i(p-q)\lambda} + \mathrm{e}^{i(p+q)\lambda})\mathrm{d}t$$

(use e.g. Zygmund, vol. 1, ch. II, theorem 1.12.). So

$$t'_m B_m t_m = \sum_{p,q=-m}^{m} (\eta - 3)\gamma(p)\gamma(q)\tilde{g}(p)\tilde{g}(q)$$

$$+ \frac{\sigma_0^4}{2\pi}\int_{-\pi}^{\pi} g_{\beta_0}^2(\lambda) \sum_{p,q=-m}^{m} \tilde{g}(p)\tilde{g}(q)(\mathrm{e}^{i(p-q)\lambda} + \mathrm{e}^{i(p+q)\lambda})\mathrm{d}\lambda = I_m + II_m.$$

Now

$$I_m = (\eta - 3)\left|\sum_{|p|\leqslant m} \tilde{g}(p)\gamma(p)\right|^2 \rightarrow (\eta - 3)\left|\sum_{-\infty}^{\infty} \tilde{g}(p)\gamma(p)\right|^2$$

as $m \rightarrow \infty$. Since $\tilde{g}(p)$ and $\gamma(p)$ are the pth Fourier coefficients of $g_\beta^{-1}(\lambda)$ and $g_{\beta_0}(\lambda)\sigma_0^2$ respectively, we see that

$$\lim_{m\rightarrow\infty} I_m = (\eta - 3)\left(\frac{\sigma_0^2}{2\pi}\int_{-\pi}^{\pi} g_{\beta_0}(\lambda) g_\beta^{-1}(\lambda)\mathrm{d}\lambda\right)^2,$$

$$II_m = \frac{\sigma_0^4}{2\pi}\int_{-\pi}^{\pi} g_{\beta_0}^2(\lambda) \sum_{p=-m}^{m} \tilde{g}(p)\mathrm{e}^{ip\lambda} \sum_{q=-m}^{m} \tilde{g}(q)(\mathrm{e}^{iq\lambda} + \mathrm{e}^{-iq\lambda})\mathrm{d}\lambda.$$

Since

$$\sum_{p=-m}^{m} \tilde{g}(p)\mathrm{e}^{ip\lambda} \rightarrow g_\beta^{-1}(\lambda)$$

uniformly as $m \rightarrow \infty$ it follows that

$$II_m \rightarrow \frac{\sigma_0^4}{\pi}\int_{-\pi}^{\pi}\left(\frac{g_{\beta_0}(\lambda)}{g_\beta(\lambda)}\right)^2 \mathrm{d}\lambda.$$

This establishes *(a)*.

Proof of (b)

We will prove that

$$\overline{\lim_n} \operatorname{Var}(U_n - V_{nm}) \rightarrow 0 \text{ as } m \rightarrow \infty.$$

$$\operatorname{Var}(U_n - V_{nm}) = \| U_n - V_{nm} \|_{L^2}^2 = \left\| \sum_{k=m+1}^{n} 2n^{1/2}\tilde{g}(k)(\hat{\gamma}_n(k) - E\hat{\gamma}_n(k))\right\|_{L^2}^2$$

$$\leqslant \left\{ \sum_{k=m+1}^{n} 2n^{1/2}\tilde{g}(k) \| \hat{\gamma}_n(k) - E\hat{\gamma}_n(k) \|_{L^2} \right\}^2.$$

$$= 4\left\{\sum_{k=m+1}^{n} \tilde{g}(k)(n\,\mathrm{Var}\,\hat{\gamma}_n(k))^{1/2}\right\}^2.$$

Since $\tilde{g}(k) \to 0$ exponentially fast and $n\,\mathrm{Var}\,\tilde{\gamma}_n(k) < nEX_t^4$, (b) follows.

6.6 PART III

In Part II we were looking at the asymptotic distribution of

$$f_n(\beta) := n^{-1/2}(X' G_n^{-1}(\beta) X - EX' G_n^{-1}(\beta) X)$$

at a single parameter β which corresponds to polynomials whose roots are off the unit circle. We shall now extend those results to the situation

$$F_n(\beta) = F_n(\beta_1, \ldots, \beta_k) := (f_n(\beta_1), \ldots, f_n(\beta_k)).$$

Just as in the one-parameter situation we will replace $G_n^{-1}(\beta_i)$ by $\tilde{G}_n(\beta_i)$, i.e. we are going to work with $\tilde{F}_n(\beta_1, \ldots, \beta_k) = (\tilde{f}_n(\beta_1), \ldots, \tilde{f}_n(\beta_k))$, with

$$\tilde{f}_n(\beta_i) = n^{-1/2}(X_n' \tilde{G}_n(\beta_i) X_n - EX_n' \tilde{G}_n(\beta_i) X_n)$$

It was shown in theorem 6.2 that $F_n(\beta_1, \ldots, \beta_k) - \tilde{F}_n(\beta_1, \ldots, \beta_k)$ is $o_p(1)$ componentwise. Hence the asymptotic distributions of F_n and $\tilde{F}_n$ are identical.

Theorem 6.4

The asymptotic distribution of $\tilde{F}_n$ and hence of F_n is a k-variate normal distribution with mean 0 and covariance matrix Σ where Σ was defined earlier.

Proof

We have to show that for any $c' = (c_1, \ldots, c_k)$ in $\mathbb{R}^k$ $c'\tilde{F}_n$ is $\mathrm{AN}(0, c'\Sigma c)$. Since

$$\tilde{f}_n(\beta_i) = n^{1/2} \sum_{|k|<n} (\tilde{\gamma}_n(k) - E\tilde{\gamma}_n(k))\tilde{g}_{\beta_i}(k) \text{ for all } \beta_i,$$

$$c'\tilde{F}_n(\beta) = n^{1/2} \sum_{|t|<n} (\hat{\gamma}_n(t) - E\hat{\gamma}_n(t))(c_1 \tilde{g}_{\beta_1}(t) + \cdots + c_k \tilde{g}_{\beta_k}(t)).$$

With $\tilde{c}_t := c_1 \tilde{g}_{\beta_1}(t) + \cdots + c_k \tilde{g}_{\beta_k}(t)$ we get

$$c'\tilde{F}_n(\beta) = \mathrm{n}^{1/2}(\hat{\gamma}_n(0) - E\hat{\gamma}_n(0))\tilde{c}_0 + 2n^{\frac{1}{2}} \sum_{t=1}^{n-1} (\hat{\gamma}_n(t) - E\hat{\gamma}_n(t))\tilde{c}_t.$$

Looking at the truncated version

$$V_{nm} := n^{1/2}(\hat{\gamma}(0) - E\hat{\gamma}(0))\tilde{c}_0 + 2n^{1/2} \sum_{t=1}^{m} (\hat{\gamma}(t) - E\hat{\gamma}(t))\tilde{c}_t$$

we see that as $n \to \infty$ $V_{nm} \Rightarrow V_m$ in distribution where V_m is normal with mean 0 and variance

$$(\tilde{c}_0, 2\tilde{c}_1, \ldots, 2\tilde{c}_m) B_m (\tilde{c}_0, 2\tilde{c}_1, \ldots, 2\tilde{c}_m)^T,$$

see (6.4).

To prove theorem 6.4 two points remain to be shown:

(a) $\overline{\lim_n} P(|c'\tilde{F}_n(\beta) - V_{nm}| > \varepsilon) \to 0$ as $m \to \infty$ for all $\varepsilon > 0$;

(b) $V_m \Rightarrow U$ in distribution where $U \sim N(0, c'\sum c)$ (BD proposition 6.3.9).

Proof of (a)

$$c'\tilde{F}_n(\beta) - V_{nm} = 2n^{1/2} \sum_{i=1}^{k} c_i \sum_{t=m+1}^{n-1} (\tilde{\gamma}_n(t) - E\tilde{\gamma}_n(t))\tilde{g}_{\beta_i}(t).$$

It was shown in section 6.5 that the ith summand satisfies for all $\varepsilon > 0$

$$\overline{\lim_n} P\left(\left| n^{1/2} \sum_{t=m+1}^{n-1} (\tilde{\gamma}_n(t) - E\tilde{\gamma}_n(t))\tilde{g}_{\beta_i}(t) \right| > \varepsilon \right) \to 0$$

as $m \to \infty$. So (a) follows at once.

Proof of (b)

It is enough to prove that $(\tilde{c}_0, 2\tilde{c}_1, \ldots, 2\tilde{c}_m) B_m (\tilde{c}_0, 2\tilde{c}_1, \ldots, 2\tilde{c}_m)^T \to c^T \sum c$ as $m \to \infty$. Using the definition B_m we find that

$$\begin{aligned}(\tilde{c}_0, 2\tilde{c}_1, \ldots, 2\tilde{c}_m) B_m (\tilde{c}_0, 2\tilde{c}_1, \ldots, 2\tilde{c}_m)^T &= (\eta - 3) \sum_{p,q=-m}^{m} \gamma(p)\gamma(q)\tilde{c}_p\tilde{c}_q \\ &\quad + \sum_{p,q=-m}^{m} \frac{\sigma_0^4}{2\pi} \int_{-\pi}^{\pi} g_{\beta_0}^2(\lambda)(\mathrm{e}^{i(p-q)\lambda} \\ &\quad + \mathrm{e}^{i(p+q)\lambda})\mathrm{d}\lambda\, \tilde{c}_p\tilde{c}_q = I_m + II_m.\end{aligned}$$

Looking at the first summand I_m separately and using the definition of $\tilde{c}_p$ we get

$$\begin{aligned}\lim_m I_m &= (\eta - 3)\left(\sum_{i=1}^{k} c_i \sum_{p=-\infty}^{\infty} \gamma(p)\tilde{g}_{\beta_i}(p) \right)^2 \\ &= (\eta - 3)\left(\sum_{i=1}^{k} c_i \frac{\sigma_0^2}{2\pi} \int_{-\pi}^{\pi} \frac{g_{\beta_0}(\lambda)}{g_{\beta_i}(\lambda)} \mathrm{d}\lambda \right)^2 \\ &= (\eta - 3) \frac{\sigma_0^4}{4\pi^2} c^T \left(\int_{-\pi}^{\pi} \frac{g_{\beta_0}(\lambda)}{g_{\beta_i}(\lambda)} \mathrm{d}\lambda \int_{-\pi}^{\pi} \frac{g_{\beta_0}(\lambda)}{g_{\beta_j}(\lambda)} \mathrm{d}\lambda \right) c =: I,\end{aligned}$$

$i, j = 1, \ldots, k.$

Now turn to the second summand II_m:

$$II_m = \frac{\sigma_0^4}{2\pi}\int_{-\pi}^{\pi} g_{\beta_0}^2(\lambda) \sum_{p=-m}^{m} \tilde{c}_p e^{ip\lambda} \sum_{q=-m}^{m} \tilde{c}_q(e^{iq\lambda} + e^{-iq\lambda})d\lambda.$$

Letting m go to ∞ and using the definition of $\tilde{c}_p$ we see that

$$\lim_m II_m = \frac{\sigma_0^4}{2\pi}\int_{-\pi}^{\pi} g_{\beta_0}^2(\lambda)\cdot 2\left(\sum_{i=1}^{k} c_i g_{\beta_i}^{-1}(\lambda)\right)^2 d\lambda$$

$$= \frac{\sigma_0^4}{\pi} c^T \left(\int_{-\pi}^{\pi} \frac{g_{\beta_0}^2(\lambda)}{g_{\beta_i}(\lambda) g_{\beta_j}(\lambda)} d\lambda\right) c = II.$$

We conclude that

$$(\tilde{c}_0, 2\tilde{c}_1, \ldots, 2\tilde{c}_m) B_m (\tilde{c}_0, 2\tilde{c}_1, \ldots, 2\tilde{c}_m)' \to I + II = c'\Sigma c.$$

This proves (b), so in fact $c'\tilde{F}_n$ is $\mathrm{AN}(0, c'\Sigma c)$.

Let us note that on heuristic grounds one would expect that if X_t is $\mathrm{ARMA}(p,q)$ and condition $C(\varepsilon)$ is satisfied one would have

$$\frac{\sigma^2}{n} L_n(\beta, \sigma^2) = -\frac{\sigma^2}{2} 2\pi - \frac{1}{2}\frac{\sigma^2}{2\pi}\int_{-\pi}^{\pi} \frac{g_{\beta_0}(\lambda)}{g_\beta(\lambda)} d\lambda + \hat{n}(\beta) + o(1)$$

where $\hat{n}(\beta)$ is the Gaussian process with covariance function $\Sigma(\beta, \beta')$ and $o(1)$ is independent of β.

This will be hopefully derived in a later paper.

REFERENCES

Brockwell, P. and Davis, R. (1991) *Time Series: Theory and Methods*, Springer Verlag.

Grenander, U. and Szegö, G. (1958) *Toeplitz Forms and Their Applications*, University of California Press.

Rosenblatt, M. (1985) *Stationary Sequences and Random Fields*, Birkhaüser.

Zygmund, A. (1959) *Trigonometric Series*, 2 vols, CUP, Cambridge.

7

On the asymptotic expansions for the bias and covariance matrix of autoregressive estimators

T.D. Pham

7.1 INTRODUCTION

Large sample theory for autoregressive (AR) estimators is well established. However, when the AR polynomial has roots near the unit circle, the standard normal approximation for the distribution of autoregressive (AR) estimators can be very bad, even for fairly large sample sizes. The finite sample distribution of the estimators in these situations can be quite skewed, can have significant bias and a covariance matrix far from that of the approximate value. Thus, there is a need to develop more accurate approximations to the distribution of such estimators. In the case of the first-order AR model, this problem has been studied in detail by various authors, using asymptotic expansions (Phillips, 1978) or approximations based on the Orstein–Ulenbeck process (Chan and Wei, 1987, Pham, 1992). However, due to their extreme complexity, the results do not seem to generalize easily to higher order models. For these reasons, we shall restrict ourselves to the asymptotic expansions for the bias and covariance matrix of the estimators, which have the most interesting characteristics and are commonly used as performance indices. Therefore, our expansions could serve for comparison between various AR estimators in the literature such as the least-squares (LS), forward-backward least-squares (FBLS), Burg's (1975) and Kay's (1983), sample partial autocorrelation (Dégerine, 1993), maximum likelihood (ML),..., estimators. Classical large sample theory cannot distinguish between them because they are all asymptotically unbiased and have the same asymptotic covariance matrix. Note that the bias of the LS and the Yule–Walker estimators have been obtained by Shaman and Stein (1988).

In this paper we provide an alternative derivation, which is more general in the sense that it is applicable to a wider class of estimators, including those mentioned above. We show that all of them have the same asymptotic bias. However, their performances differ in terms of asymptotic covariance matrix: the best is achieved by the ML estimator for which this matrix equals the Cramer–Rao (C–R) bound plus a term representing the 'curvature' of the model (and another term which disappears when bias correction is applied). For simplicity, we shall limit ourselves to the zero mean case. Results for the non-zero mean case can be obtained through a similar approach. For ease of reading proofs of results are relegated to the appendix.

7.2 SUFFICIENT STATISTICS AND AUTOREGRESSIVE ESTIMATORS

We here describe various autoregressive estimators proposed in the literature. The LS estimator excepted, they are functions of a sufficient statistic, defined below. Recall that the zero mean AR model of order p is defined by

$$\sum_{i=0}^{p} a_i X_{t-i} = e_t, \quad a_0 = 1, \tag{7.1}$$

where e_t, $t = \dots, -1, 0, 1, \dots$ are independent random variables with mean zero and variance σ^2. The polynomial

$$\sum_{i=0}^{p} a_i z^{p-i}$$

is assumed to be stable in the sense that its roots lie inside the unit circle, so that the model is stationary. Under the Gaussian assumption, the log likelihood function based on a sample $X_1, \dots, X_n$ from the model is

$$-\frac{1}{2}[\log \det(2\pi \mathbf{\Gamma}_n) + (X_1 \cdots X_n)\mathbf{\Gamma}_n^{-1}(X_1 \cdots X_n)'] \tag{7.2}$$

where Γ_n is the covariance matrix of $(X_1 \cdots X_n)'$, $'$ denoting the transpose. The last matrix is Toeplitz and hence its inverse can be computed by the formula (see, for example, Godolphin and Unwin, 1983)

$$\sigma^2 \mathbf{\Gamma}_n^{-1} = \begin{bmatrix} a_0 & & 0 \\ \vdots & \ddots & \\ a_{n-1} & \cdots & a_0 \end{bmatrix} \begin{bmatrix} a_0 & \cdots & a_{n-1} \\ & \ddots & \vdots \\ 0 & & a_0 \end{bmatrix} - \begin{bmatrix} a_n & & 0 \\ \vdots & \ddots & \\ a_1 & \cdots & a_n \end{bmatrix} \begin{bmatrix} a_n & \cdots & a_1 \\ & \ddots & \vdots \\ 0 & & a_1 \end{bmatrix}, \tag{7.3}$$

where by convention $a_i = 0$ for $i > p$. From this result, one obtains (Pham, 1988):

$$(X_1 \cdots X_n)\mathbf{\Gamma}_n^{-1}(X_1 \cdots X_n)' = n(a_0 \cdots a_p)\mathbf{Q}(a_0 \cdots a_p)'/\sigma^2 \tag{7.4}$$

where $\mathbf{Q}$ is the symmetric $(p+1)\times(p+1)$ matrix with general element

$$Q_{ij} = \frac{1}{n}\sum_{t=1}^{n-i-j} X_{t+i}X_{t+j}, \quad \text{if } i+j<n,$$

$$= -Q_{n-j,n-i}, \quad \text{if } i+j \geqslant n, \quad i,j=0,\ldots,p. \tag{7.5}$$

It becomes clear that the $Q_{ij}, 0 \leqslant j \leqslant i \leqslant p$, constitute a sufficient statistic (Arato, 1961). Since $\mathrm{E}\, Q_{ij} = [1-(i+j)/n]\gamma_{i-j}$ where $\gamma_k = \mathrm{E}(X_{t-k}X_t)$ denotes the autocovariance of lag k of the process, one has $\mathrm{E}(\mathbf{Q}) = \mathbf{\Gamma}_{p+1} - \mathbf{D}\mathbf{\Gamma}_{p+1} - \mathbf{\Gamma}_{p+1}\mathbf{D}$, $\mathbf{D}$ denoting the diagonal matrix with diagonal elements $0, 1/n, \ldots, p/n$.

The LS estimator is defined as the solution of

$$(1 a_1 \cdots a_p)\hat{\mathbf{\Gamma}}_{p+1} = (\sigma^2 0 \cdots 0) \tag{7.6}$$

where $\hat{\mathbf{\Gamma}}_{p+1}$ is the matrix having

$$\left(\sum_{t=p+1}^{n} X_{t-i}X_{t-j}\right)\Big/(n-p)$$

at the (i,j) place, $i,j=0,\ldots,p$. This is the only estimator which cannot be expressed in term of $\mathbf{Q}$, due to its non reversibility (reversing the order of the data sequence changes the value of the estimator). However, the FBLS estimator, which minimizes

$$\sum_{t=p+1}^{n}\left(X_t + \sum_{i=1}^{p} a_i X_{t-i}\right)^2 + \sum_{t=1}^{n-p}\left(X_t + \sum_{i=1}^{p} a_i X_{t+i}\right)^2$$

instead of the usual LS criterion, can be expressed in term of $\mathbf{Q}$. It is also the solution of (2.6) but with $\hat{\mathbf{\Gamma}}_{p+1}$ replaced by the matrix having $\frac{1}{2}(Q_{ij} + Q_{p-j,p-i})/(1-p/n)$ at the (i,j) place, $i,j=0,\ldots,p$ (Pham and Dégerine, 1990). Note that the simpler Yule–Walker estimator is also defined by (7.6) with $\hat{\mathbf{\Gamma}}_{p+1}$ being the $(p+1)\times(p+1)$ symmetric Toeplitz matrix having Q_{i0} on the ith diagonal. This estimator may suffer serious bias due to the bias of Q_{i0} as estimators of γ_i. We will not consider it, for simplicity, but our method may be adapted to its study.

Many AR estimators are recursive in order. The estimated AR coefficients for the kth order model $\hat{a}_{1,k}, \ldots, \hat{a}_{k,k}$, say, are related to the those for the $(k-1)$th order model through the Levinson–Durbin algorithm

$$\hat{a}_{i,k} = \hat{a}_{i,k-1} + \hat{a}_{k,k}\hat{a}_{k-i,k-1}, \quad i=1,\ldots,k-1. \tag{7.7}$$

Thus for each order, only the last AR coefficient (the negative of the partial autocorrelation) needs to be estimated. A recent method due to Dégerine (1993), called the sample partial autocorrelation method (SPAC), consists of taking $\hat{a}_{k,k}$ to be the last estimated AR coefficient in the FBLS method for the kth order model (this $\hat{a}_{k,k}$ is always less than one in absolute value, thus ensuring the stability of the estimated AR polynomial, a property does not share by the LS or FBLS method). A similar method has been introduced earlier by Dickinson (1978). Burg's method can be shown (Pham and Dégerine, 1980) to amount to taking $\hat{a}_{k,k} = -2B_k/(A_k + C_k)$ where

$$A_k = \sum_{i=1}^{k}\sum_{j=1}^{k} \hat{a}_{k-i,k-1} Q_{ij} \hat{a}_{k-j,k-1},$$

$$B_k = \sum_{i=0}^{k-1}\sum_{j=1}^{k} \hat{a}_{i,k-1} Q_{ij} \hat{a}_{k-j,k-1},$$

$$C_k = \sum_{i=0}^{k-1}\sum_{j=0}^{k-1} \hat{a}_{i,k-1} Q_{ij} \hat{a}_{j,k-1}, \quad (\hat{a}_{0,k-1} = 1). \tag{7.8}$$

Finally, Kay's method consists of taking $\hat{a}_{k,k}$ to be the solution of the maximization of the 'partial log likelihood' corresponding to the kth order model

$$-\frac{1}{2}\left\{ n\log(2\pi\sigma^2) + \log\left[\prod_{i=1}^{k}(1-\hat{a}_{ii}^2)^{-i}\right] + \frac{n}{\sigma^2}\sum_{i=0}^{k}\sum_{j=0}^{k}\hat{a}_{ik}Q_{ij}\hat{a}_{jk}\right\}$$

with $\hat{a}_{1,k},\ldots,\hat{a}_{k-1,k}$ given by (7.7) with $\hat{a}_{1,1},\ldots,\hat{a}_{k-1,k-1}$ being set equal to the coefficients of the previously estimated $(k-1)$th order model. This is equivalent to minimizing (with respect to $\hat{a}_{k,k}^2 \in]-1,1[$) of

$$n\log(A_k\hat{a}_{k,k}^2 + 2B_k\hat{a}_{k,k} + C_k) - k\log(1-\hat{a}_{k,k}^2) \tag{7.9}$$

where A_k, B_k, and C_k are given by (7.8).

Write the above estimators in the form $\hat{\mathbf{a}}(\mathbf{Q}) = [\hat{a}_1(\mathbf{Q})\cdots\hat{a}_p(\mathbf{Q})]$. One then obtains this important property: for any symmetric positive definite Toeplitz matrix $\mathbf{G}$:

$$[1 \ \ \hat{\mathbf{a}}(\mathbf{G}-\mathbf{DG}-\mathbf{GD})]\mathbf{G} = [* \ \ 0,\ldots,0] \tag{E.1}$$

where $*$ denotes an arbitrary number. That this is true for the FBLS and the SPAC estimator is clear, since when $\mathbf{Q} = \mathbf{G}-\mathbf{DG}-\mathbf{GD}$, the corresponding $\hat{\mathbf{\Gamma}}_{p+1}$ of (7.6) simply equals $\mathbf{G}$. The following lemma shows that (7.10) also holds for the Burg's and Kay's estimators.

Lemma 7.1

Suppose that $\mathbf{Q}$ is of the form $\mathbf{G}-\mathbf{DG}-\mathbf{GD}$ where $\mathbf{G}$ is a symmetric positive definite Toeplitz matrix, and that $\hat{a}_{1,k-1},\ldots,\hat{a}_{k-1,k-1}$ satisfy the equations

$$G_{0j} + \sum_{i=1}^{k-1}\hat{a}_{i,k-1}G_{ij} = 0, \quad j = 1,\ldots,k-1.$$

Then $\hat{a}_{1,k},\ldots,\hat{a}_{k-1,k}$, defined by (7.7), also satisfy similar equations

$$G_{0j} + \sum_{i=1}^{k}\hat{a}_{i,k}G_{ij} = 0, \quad j = 1,\ldots,k,$$

provided that $\hat{a}_{k,k} = -2B_k/(A_k + C_k)$ or $\hat{a}_{k,k}$ realizes the minimum of (7.9) in $]-1, 1[$, A_k, B_k and C_k being given by (7.8).

The ML estimator also satisfies (E1). To show this, observe that when $\mathbf{Q} = \mathbf{G} - \mathbf{DG} - \mathbf{GD}$, (2.7) reduces to

$$\left[\sum_{i=0}^{p}\sum_{j=0}^{p} a_i(n-i-j)G_{ij}a_j\right]\Big/\sigma^2.$$

But by the same computations as the one proving (7.4), and using the Toeplitz property of $\mathbf{G}$, it may be shown that this expression equals $\mathrm{tr}(\mathbf{\Gamma}_n^{-1}\mathbf{G}_n)$ where $\mathbf{G}_n$ is **any** $n \times n$ symmetric Toeplitz matrix for which the sub-matrix formed by its first $(p+1)$ rows and columns equals $\mathbf{G}$. Thus, the log likelihood equals $-\frac{1}{2}[\log\det(2\pi\mathbf{\Gamma}_n) + \mathrm{tr}(\mathbf{\Gamma}_n^{-1}\mathbf{G}_n)]$. Clearly, one may choose $\mathbf{G}_n$ such that

$$(1\hat{a}_1 \cdots \hat{a}_p 0 \cdots 0)\mathbf{G}_n = (\hat{\sigma}^2 0 \cdots 0) \tag{7.10}$$

where $\hat{a}_1, \ldots, \hat{a}_p$, $\hat{\sigma}^2$ are the solution of $(1\hat{a}_1 \cdots \hat{a}_p)\mathbf{G} = (\hat{\sigma}^2 0 \cdots 0)$. Now it is well known that $\log\det(\mathbf{\Gamma}_n) + \mathrm{tr}(\mathbf{\Gamma}_n^{-1}\mathbf{G}_n) \geqslant \log\det(\mathbf{G}_n) + n$, is equality if and only if $\mathbf{\Gamma}_n = \mathbf{G}_n$. The last equality holds if $a_i = \hat{a}_i$, $i = 1, \ldots, p$. and $\sigma^2 = \hat{\sigma}^2$, since $\mathbf{\Gamma}_n$ is uniquely determined from $a_1, \ldots, a_p$, σ^2 through an equation of the same form as (7.10).

7.3 ASYMPTOTIC BIAS OF AUTOREGRESSIVE ESTIMATORS

From the result of section 7.2, we are led to consider an estimator of $\mathbf{a} = (a_1 \cdots a_p)$ of the form $\hat{\mathbf{a}}(\mathbf{Q})$, $\mathbf{Q}$ being defined by (7.5), where $\hat{\mathbf{a}}(\cdot)$ satisfies (E1) and possibly depends on n. Some further 'regular' conditions for $\hat{\mathbf{a}}(\cdot)$ will be needed.

$\hat{\mathbf{a}}(\cdot)$ is bounded uniformly in n, for all n large enough, (E2)

There is a neighborhood of $\mathbf{\Gamma}_{p+1}$ (the true covariance matrix of $(X_1 \cdots X_p)'$) such that for sufficiently large n, $\hat{\mathbf{a}}(\cdot)$ is three times continuously differentiable in it with bounded derivatives up to third order, uniformly in n. (E3)

For convenience, we introduce the vector $\mathbf{q} = (q_0 \cdots q_{p(p+3)/2})$, related to $\mathbf{Q}$ through a given invertible linear map (a possibility is to take q_i as the elements of the matrix $\mathbf{Q}$ on and below the main diagonal, in a given order). When $\mathbf{Q}$ is expressed in terms of $\mathbf{q}$, $\hat{\mathbf{a}}(\mathbf{Q})$ becomes a function of $\mathbf{q}$ which we still denote by $\hat{\mathbf{a}}(\mathbf{q})$ to avoid new notation. Since $\mathrm{E}(\mathbf{Q}) = \mathbf{\Gamma}_{p+1} - \mathbf{D\Gamma}_{p+1} - \mathbf{\Gamma}_{p+1}\mathbf{D}$, by (E1) $\hat{a}(\bar{\mathbf{q}}) = \mathbf{a}$ where $\bar{\mathbf{q}} = \mathrm{E}(\mathbf{q})$. Thus, expanding $\hat{a}(\mathbf{q})$ around $\bar{\mathbf{q}}$, one gets:

$$\hat{\mathbf{a}}(\mathbf{q}) = \mathbf{a} + \sum_i \frac{\partial \hat{a}}{\partial q_i}(q_i - \bar{q}_i) + \frac{1}{2}\sum_{i,j}\frac{\partial^2 \hat{\mathbf{a}}}{\partial q_i \partial q_j}(q_i - \bar{q}_i)(q_j - \bar{q}_j) + \mathbf{R}(\mathbf{q}) \tag{7.11}$$

(the range of summation indexes being $0, \ldots, p(p+3)/2$ where $\mathbf{R}(\mathbf{q})$ denotes the remainder term and **all derivatives are evaluated at $\bar{\mathbf{q}}$**.

We first prove that the contribution of the remainder term $\mathbf{R}(\mathbf{q})$ to the bias is $O(n^{-3/2})$ (it is possible, under stronger conditions, to prove that it is actually $O(n^{-2})$). Let U be a neighborhood of $\bar{\mathbf{q}}$ for which (E3) holds and define T_n equal to 1 if $\mathbf{Q}\in U$ and 0 otherwise. Them, from (E3), $\|\mathbf{R}(\mathbf{q})T_n\| \leqslant K\|\mathbf{q}-\bar{\mathbf{q}}\|^3$, K being a constant and $\|\cdot\|$ denoting the euclidian norm. On the other hand $E\|\mathbf{q}-\bar{\mathbf{q}}\|^{2k}=O(n^{-k})$ provided that $E(e_t^{4k})<\infty$ (Yamoto and Kunimoto, 1984). Thus, under the assumption that $E(e_t^8)<\infty$, $E[\mathbf{R}(\mathbf{q})T_n]=O(n^{-3/2})$. Also from the inequality $P(\|\mathbf{q}-\bar{\mathbf{q}}\|>c)\leqslant E\|\mathbf{q}-\bar{\mathbf{q}}\|^4/c^4$, one gets $P(T_n=0)=O(n^{-2})$ and hence $E[\hat{\mathbf{a}}(\mathbf{q})(1-T_n)]=O(n^{-2})$ since $\hat{\mathbf{a}}(\cdot)$ is bounded by (E2). To obtain the result, one needs only to show that $E[\|\mathbf{q}-\bar{\mathbf{q}}\|(1-T_n)]=O(n^{-3/2})$ and $E[\|\mathbf{q}-\bar{\mathbf{q}}\|^2(1-T_n)]=O(n^{-2})$, but this follows easily from Schwarz's inequality.

Thus, the bias of $\hat{\mathbf{a}}(\mathbf{q})$ equals

$$\frac{1}{2}\sum_{i,j}(\partial^2\hat{\mathbf{a}}/\partial q_i\partial q_j)\mathrm{cov}(q_i,q_j)+O(n^{-3/2}).$$

Now, one may choose $\mathbf{q}$ such that most of the $\mathrm{cov}(q_i,q_j)$ are of the order n^{-2}. Indeed, take

$$q_i=[n/(n-i)]Q_{i0}\quad i=0,\ldots,p.$$

$$q_{p+k(k-1)/2+l}=Q_{kl}-[(n-k-l)/n]q_{k-l},\quad 1\leqslant l\leqslant k\leqslant p. \tag{7.12}$$

Then the last right hand side equals (assuming $k+l\leqslant n$)

$$\frac{1}{n-k+l}\left(\frac{n-k+l}{n}\sum_{t=1}^{n-k-l}X_{t+k}X_{t+l}-\frac{n-k-l}{n}\sum_{t=1}^{n-k+k}X_tX_{t-k+1}\right)=\frac{1}{n-k-l}$$

$$\times\left[\frac{2l}{n}\sum_{t=1}^{n-k-l}X_{t+k}X_{t+l}-\frac{n+k+l}{n}\left(\sum_{m=0}^{l-1}X_{k-m}X_{l-m}+\sum_{m=n+1}^{n+l}X_{m-k}X_{m-l}\right)\right]$$

and it may be checked that $\mathrm{cov}(q_i,q_j)=O(n^{-2})$ as soon as i or j is greater then p (while it is $O(n^{-1})$ if $i,j\leqslant p$). Note further that $\bar{\mathbf{q}}=(\gamma_0\cdots\gamma_p0\cdots0)'$. Thus by (E2), the bias of $\hat{\mathbf{a}}(\mathbf{q})$ equals

$$\frac{1}{2}\sum_{i=0}^{p}\sum_{j=0}^{p}(\partial^2\hat{\mathbf{a}}/\partial q_i\partial q_j)\mathrm{cov}(q_i,q_j)+O(n^{-3/2}).$$

An interesting consequence of (E1) is that the derivatives $\partial^2\hat{\mathbf{a}}/\partial q_i$ and $\partial^2\hat{\mathbf{a}}/\partial q_i\partial q_j$ at $\bar{\mathbf{q}}=(\gamma_0\cdots\gamma_p0\cdots0)'$, for $i,j=0,\ldots,p$, do not depend on the form of $\hat{\mathbf{a}}$. Indeed, when $q_{p+1}=\cdots=q_{p(p+3)/2}=0$, $\hat{\mathbf{a}}(\mathbf{q})$ is related to $q_0\cdots q_p$ by the same relations relating the AR coefficients $a_1,\ldots,a_p$ to the autocovariances $\gamma_0,\ldots,\gamma_p$, namely

$$(1\,a_1\cdots a_p)\begin{bmatrix}\gamma_0 & \cdots & \gamma_p\\ \vdots & \ddots & \vdots\\ \gamma_{-p} & \cdots & \gamma_0\end{bmatrix}=[\sigma^2 0\cdots 0]. \tag{7.13}$$

Thus the derivatives $\partial\hat{\mathbf{a}}/\partial q_i$ and $\partial^2\hat{\mathbf{a}}/\partial q_i\partial q_j, i,j=0,\ldots,p$, at $\bar{\mathbf{q}}$, are simply $\partial\mathbf{a}/\partial\gamma_i$ and $\partial^2\mathbf{a}/\partial\gamma_i\partial\gamma_j$. Hence, the asymptotic bias of $\hat{\mathbf{a}}(\mathbf{q})$ is equivalent to the expectation of

$$\frac{1}{2}\delta^2\mathbf{a}=\frac{1}{2}\sum_{i=0}^{p}\sum_{j=0}^{p}\frac{\partial^2\mathbf{a}}{\partial\gamma_i\partial\gamma_j}(q_i-\gamma_i)(q_j-\gamma_j) \tag{7.14}$$

which is **the same for all estimators**. To compute $\delta^2\mathbf{a}$, first differentiate (7.13), yielding

$$\delta\mathbf{a}=\sum_{i=0}^{p}\frac{\partial\mathbf{a}}{\partial\gamma_i}(q_i-\gamma_i)=-[(\hat{\hat{\gamma}}-\gamma)+\mathbf{a}(\hat{\hat{\mathbf{\Gamma}}}_p-\mathbf{\Gamma}_p)]\mathbf{\Gamma}_p^{-1}=-(\hat{\hat{\gamma}}+\mathbf{a}\hat{\hat{\mathbf{\Gamma}}}_p)\mathbf{\Gamma}_p^{-1} \tag{7.15}$$

where $\hat{\hat{\gamma}}=(q_1\cdots q_p)$ and $\hat{\hat{\mathbf{\Gamma}}}_p$ is the $p\times p$ Toeplitz symmetric matrix with q_i on the ith diagonal. Then, differentiating further yields

$$\frac{1}{2}\delta^2\mathbf{a}=(\hat{\hat{\gamma}}+\mathbf{a}\hat{\hat{\mathbf{\Gamma}}}_p)\mathbf{\Gamma}_p^{-1}(\hat{\hat{\mathbf{\Gamma}}}_p-\mathbf{\Gamma}_p)\mathbf{\Gamma}_p^{-1}=(\delta\mathbf{a})\mathbf{\Gamma}_p\sum_{i=0}^{p}\left(\frac{\partial}{\partial\gamma_i}\mathbf{\Gamma}_p^{-1}\right)(q_i-\gamma_i)$$

$$=(\delta\mathbf{a})\mathbf{\Gamma}_p\left[\sum_{i=1}^{p}\frac{\partial\mathbf{\Gamma}_p^{-1}}{\partial a_i}(\delta a_i)+\frac{\partial\mathbf{\Gamma}_p^{-1}}{\partial\sigma^2}(\delta\sigma^2)\right] \tag{7.16}$$

where δa_i are the components of $\delta\mathbf{a}$ and

$$\delta\sigma^2=\sum_{i=0}^{p}(\partial\sigma^2/\partial\gamma_i)(q_i-\gamma_i).$$

Now, from the classical asymptotic theory for AR estimators, $\delta\mathbf{a}$ and $\delta\sigma^2$ are asymptotically independent normal with covariance matrix $\sigma^2\mathbf{\Gamma}_p^{-1}/n$ and variance $2\sigma^4/n$. Thus, the expectation of the last right hand side of (7.16) is asymptotically equivalent to

$$n^{-1}\sum_{i=1}^{p}\mathbf{e}_i\sigma^2\partial\mathbf{\Gamma}_p^{-1}/\partial a_i,$$

where $\mathbf{e}_i$ denotes the ith row of the identity matrix. From (7.3) (with p in place of n) the derivatives $\partial\mathbf{\Gamma}_p^{-1}/\partial a_i$ can be computed **explicitly** easily, yielding the asymptotic bias for $\hat{a}_j(\mathbf{Q})$:

$$\frac{1}{n}\left(\sum_{k=\max(1,j-[p/2])}^{[j/2]}a_{j-2k}-\sum_{k=\max\{0,[(p+1)/2]-j\}}^{[(p-j)/2]}a_{j+2k}-ja_j\right) \tag{7.17}$$

where $[\cdot]$ denotes the integer part.

The above arguments however do not apply to the LS estimator since it is not of the form $\hat{\mathbf{a}}(\mathbf{Q})$. But from (7.6) one gets directly the following expansion for this estimator

$$\hat{\mathbf{a}}=\mathbf{a}-(\hat{\gamma}+\mathbf{a}\hat{\mathbf{\Gamma}}_p)\mathbf{\Gamma}_p^{-1}+(\hat{\gamma}+\mathbf{a}\hat{\mathbf{\Gamma}}_p)\mathbf{\Gamma}_p^{-1}(\hat{\mathbf{\Gamma}}_p-\mathbf{\Gamma}_p)\mathbf{\Gamma}_p^{-1}$$
$$-(\hat{\gamma}+\mathbf{a}\hat{\mathbf{\Gamma}}_p)\mathbf{\Gamma}_p^{-1}(\hat{\mathbf{\Gamma}}_p-\mathbf{\Gamma}_p)\mathbf{\Gamma}_p^{-1}(\hat{\mathbf{\Gamma}}_p-\mathbf{\Gamma}_p)\hat{\mathbf{\Gamma}}_p^{-1}, \tag{7.18}$$

where $\hat{\gamma}$ and $\hat{\mathbf{\Gamma}}_p$ are the p-vector with components

$$\sum_{t=1}^{n-p} X_t X_{t-i}/(n-p)$$

and the $p \times p$ matrix with general elements

$$\sum_{t=1}^{n-p} X_{t-i} X_{t-j}/(n-p).$$

By a similar argument as in Yamoto and Kumitomo (1984), the contribution of the last term of the right hand side of (7.18) to the bias of $\hat{\mathbf{a}}$ may be neglected (under appropriate assumptions). Further, since $\mathrm{E}(\hat{\gamma} + \mathbf{a}\hat{\mathbf{\Gamma}}_p) = \mathbf{0}$ and $\hat{\gamma}$, $\hat{\mathbf{\Gamma}}_p$ differ from $\hat{\hat{\gamma}}$, $\hat{\hat{\mathbf{\Gamma}}}_p$ only by a term of order n^{-1}, it may be shown that the expectation of the second term of right hand side of (7.18) is the same as that of the right hand side of (7.16), up to the order n^{-2}. Therefore, the asymptotic bias of $\hat{\mathbf{a}}$ is the same as before. Note that this bias has been derived by Shaman and Stein (1988), but not in an explicit form as (7.17).

To apply the above results to the estimators introduced in section 7.2, one still needs to check the regularity conditions (E2) and (E3). Condition (E2) follows simply from the fact that the estimated AR polynomial is stable. This is not true for the FBLS estimator but the result for this estimator can be obtained by a similar argument as in the case of the LS estimator. As for condition (E3), checking it is a rather tedious task, but can be done without particular difficulty.

7.4 ASYMPTOTIC COVARIANCE MATRIX OF AUTOREGRESSIVE ESTIMATORS

Our approach to derive the asymptotic covariance matrix is similar to the above. However expansion of the type (7.11) needs to be carried out to third order and one has to take into account the third order cumulants of the q_i(only those of $q_0, \ldots, q_p$ need to be considered since the contribution of the others may be neglected as will be made clear). Higher order cumulants between the q_i will not contribute to the asymptotic covariance matrix (which is computed up to the order n^{-2}) since the kth cumulants between the q_i are $\mathrm{O}(n^{1-k})$. We have

$$\hat{\mathbf{a}}(\mathbf{q}) = \mathbf{a} + \sum \frac{\partial \hat{\mathbf{a}}}{\partial q_i}(q_i - \bar{q}_i) + \frac{1}{2}\sum_{ij} \frac{\partial^2 \hat{\mathbf{a}}}{\partial q_i \partial q_j}(q_i - \bar{q}_i)(q_j - \bar{q}_j)$$
$$+ \frac{1}{6}\sum_{i,j,k} \frac{\partial^3 \hat{\mathbf{a}}}{\partial q_i \partial q_j \partial q_k}(q_i - \bar{q}_i)(q_j - \bar{q}_j)(q_k - \bar{q}_k) + \mathbf{R}(\mathbf{q}), \qquad (7.19)$$

where again $\mathbf{R}(\mathbf{q})$ denotes the remainder term and all derivatives are evaluated at $\bar{\mathbf{q}}$. We first show that the contribution of this term to the covariance matrix of $\hat{\mathbf{a}}(\mathbf{q})$ is $O(n^{-5/2})$. For this we have to strengthen (E3) to

> As (E3) but now $\hat{\mathbf{a}}(\cdot)$ is required to be four times differentiable with bounded derivatives up to fourth order. (E4)

As in section 7.3, we take a neighbourhood U of $\bar{\mathbf{q}}$ for which (E4) holds and define T_n to be the random variable taking value 1 if $\mathbf{Q} \in U$ and 0 otherwise. Then, by (E4), $\| \mathbf{R}(\mathbf{q}) T_n \| \leqslant K \| \mathbf{q} - \bar{\mathbf{q}} \|^4$ for some constant K. Thus, by the same argument as in section 7.3, the covariance matrices between $T_n \mathbf{R}(\mathbf{q})$ and other terms of (4.1) are $O(n^{-5/2})$, provided that $E(e_t^{12}) < \infty$. Now from $P(\| \mathbf{q} - \bar{\mathbf{q}} \| > c) \leqslant E \| \mathbf{q} - \bar{\mathbf{q}} \|^6 / c^6$, one gets $P(T_n = 0) = O(n^{-3})$, yielding $E\{[\hat{\mathbf{a}}(\mathbf{q}) - \mathbf{a}]'[[\hat{\mathbf{a}}(\mathbf{q}) - \mathbf{a}](1 - T_n)\} = O(n^{-3})$ since $\hat{\mathbf{a}}(\cdot)$ is bounded by (E2). To obtain the desired result, one observes that by Schwartz's inequality $E[\| \mathbf{q} - \bar{\mathbf{q}} \|^k (1 - T_n)] = O(n^{-(k+3)/2})$, $k = 2, \ldots, 6$.

The above arguments show that the covariance matrix of $\hat{\mathbf{a}}(\mathbf{q})$ can be written, up to a term of order $n^{-5/2}$, as

$$\sum_{i,j} \frac{\partial \hat{\mathbf{a}}'}{\partial q_i} \frac{\partial \hat{\mathbf{a}}}{\partial q_j} \operatorname{cov}(q_i, q_j) + \frac{1}{2} \sum_{i,j,k} \left[\frac{\partial \hat{\mathbf{a}}'}{\partial q_i} \frac{\partial^2 \hat{\mathbf{a}}}{\partial q_j \partial q_k} + \frac{\partial^2 \hat{\mathbf{a}}'}{\partial q_j \partial q_k} \frac{\partial \hat{\mathbf{a}}}{\partial q_i} \right] \operatorname{cum}(q_i, q_j, q_k)$$
$$+ \frac{1}{2} \sum_{i,j,k,l} \left[\frac{\partial^2 \hat{\mathbf{a}}'}{\partial q_i \partial q_j} \frac{\partial^2 \hat{\mathbf{a}}}{\partial q_k \partial q_l} + \frac{\partial \mathbf{a}'}{\partial q_i} \frac{\partial^3 \hat{\mathbf{a}}}{\partial q_j \partial q_k \partial q_l} + \frac{\partial^3 \hat{\mathbf{a}}'}{\partial q_j \partial q_k \partial q_l} \frac{\partial \hat{\mathbf{a}}}{\partial q_i} \right]$$
$$\times \operatorname{cov}(q_i, q_l) \operatorname{cov}(q_j, q_k).$$

As before, we choose $\mathbf{q}$ as given in (7.12). Then $\operatorname{cov}(q_i, q_j) = O(n^{-2})$ as soon as $i > p$ and it can also be checked that $\operatorname{cum}(q_i, q_j, q_k) = O(n^{-3})$ as soon as $i > p$. Thus, the summation indexes in the above sums, **except those for the first**, may be restricted to the range $0, \ldots, p$. Further, as is shown in section 7.3, the derivatives $\partial \hat{\mathbf{a}} / \partial q_i$, $\partial^2 \hat{\mathbf{a}} / \partial q_i \partial q_j$ and $\partial^3 \hat{\mathbf{a}} / \partial q_i \partial q_j \partial q_k$, $i, j = 0, \ldots, p$, at $\bar{\mathbf{q}} = (\gamma_0 \cdots \gamma_p 0 \cdots 0)'$, equal $\partial \mathbf{a} / \partial \gamma_i$, $\partial^2 \mathbf{a} / \partial \gamma_i \partial \gamma_j$ and $\partial^3 \mathbf{a} / \partial \gamma_i \partial \gamma_j \partial \gamma_k$. Thus, the covariance matrix of $\hat{\mathbf{a}}(\mathbf{q})$ equals

$$\sum_{i,j} \frac{\partial \hat{\mathbf{a}}'}{\partial q_i} \frac{\partial \hat{\mathbf{a}}}{\partial q_j} \operatorname{cov}(q_i, q_j) + \frac{1}{2n^2} \sum_{i,j,k,l=0}^{p} \frac{\partial^2 \mathbf{a}'}{\partial \gamma_i \partial \gamma_j} \frac{\partial^2 \mathbf{a}}{\partial \gamma_k \partial \gamma_l} g_{il} g_{jk}$$
$$+ \frac{1}{2n^2} \left\{ \sum_{i,j,k=0}^{p} \left[\frac{\partial \mathbf{a}'}{\partial \gamma_i} \frac{\partial^2 \mathbf{a}}{\partial \gamma_j \partial \gamma_k} + \frac{\partial^2 \mathbf{a}'}{\partial \gamma_j \partial \gamma_k} \frac{\partial \mathbf{a}}{\partial \gamma_i} \right] T_{ijk} \right.$$
$$\left. + \sum_{i,j,k,l=0}^{p} \left[\frac{\partial \mathbf{a}'}{\partial \gamma_i} \frac{\partial^3 \mathbf{a}}{\partial \gamma_j \partial \gamma_k \partial \gamma_l} + \frac{\partial^3 \mathbf{a}'}{\partial \gamma_j \partial \gamma_k \partial \gamma_l} \frac{\partial \mathbf{a}}{\partial \gamma_i} \right] g_{il} g_{jk} \right\} + O(n^{-5/2}), \quad (7.20)$$

where g_{ij} and T_{ijk} are the leading terms of $n \operatorname{cov}(q_i, q_j)$ and $n^2 \operatorname{cum}(q_i, q_j, q_k)$, respectively (the next term is of the order $1/n$)

The above arguments do not apply to the LS estimator. However, expanding further (3.8) yields the following representation for this estimator:

$$\hat{\mathbf{a}} = \mathbf{a} - (\hat{\gamma} + \mathbf{a}\hat{\mathbf{\Gamma}}_p)\mathbf{\Gamma}_p^{-1} + (\hat{\gamma} + \mathbf{a}\hat{\mathbf{\Gamma}}_p)\mathbf{\Gamma}_p^{-1}(\hat{\mathbf{\Gamma}}_p - \mathbf{\Gamma}_p)\mathbf{\Gamma}_p^{-1}$$
$$- (\hat{\gamma} + \mathbf{a}\hat{\mathbf{\Gamma}}_p)\mathbf{\Gamma}_p^{-1}(\hat{\mathbf{\Gamma}}_p - \mathbf{\Gamma}_p)\mathbf{\Gamma}_p^{-1}(\hat{\mathbf{\Gamma}}_p - \mathbf{\Gamma}_p)\mathbf{\Gamma}_p^{-1}$$
$$+ (\hat{\gamma} + \mathbf{a}\hat{\mathbf{\Gamma}}_p)\mathbf{\Gamma}_p^{-1}(\hat{\mathbf{\Gamma}}_p - \mathbf{\Gamma}_p)\mathbf{\Gamma}_p^{-1}(\hat{\mathbf{\Gamma}}_p - \mathbf{\Gamma}_p)\mathbf{\Gamma}_q^{-1}(\hat{\mathbf{\Gamma}}_p - \mathbf{\Gamma}_p)\hat{\mathbf{\Gamma}}_p^{-1}, \quad (7.21)$$

where $\hat{\gamma}$ and $\hat{\mathbf{\Gamma}}_p$ are defined earlier in section 7.3. Again, by a similar argument as in Yamoto and Kumitomo (1984), the contribution of the last term of this expansion to the covariance matrix of $\hat{\mathbf{a}}$ may be neglected (under appropriate assumptions). Now observe that the covariance matrix of the first terms in the right hand side of (7.21), neglecting terms of higher order than n^{-2}, is composed of the covariance matrix of $(\hat{\gamma} + \mathbf{a}\hat{\mathbf{\Gamma}}_p)\mathbf{\Gamma}_p^{-1}$ and terms involving the product of covariances and third order cumulants between the elements of $\hat{\gamma}, \hat{\mathbf{\Gamma}}_p$. By analysing the differences between $\hat{\gamma} - \hat{\hat{\gamma}}$ and $\hat{\mathbf{\Gamma}}_p - \hat{\hat{\mathbf{\Gamma}}}_p$, it may be seen that the covariances between the elements of $\hat{\gamma}, \hat{\mathbf{\Gamma}}_p$ differ from those of the corresponding elements of $\hat{\hat{\gamma}}, \hat{\hat{\mathbf{\Gamma}}}_p$ by $\mathrm{O}(n^{-2})$ and the third cumulants between the elements of $\hat{\gamma}, \hat{\mathbf{\Gamma}}_p$ differ from those of the corresponding elements of $\hat{\hat{\gamma}}$, $\hat{\hat{\mathbf{\Gamma}}}_p$ by $\mathrm{O}(n^{-3})$. Finally, using (7.15) and (7.16) and the equality

$$\frac{1}{6}\sum_{j=0}^{p}\sum_{k=0}^{p}\sum_{l=0}^{p}\frac{\partial^3 \mathbf{a}}{\partial\gamma_j\partial\gamma_k\partial\gamma_l}(q_j - \gamma_j)(q_k - \gamma_k)(q_l - \gamma_l)$$
$$= (\hat{\hat{\gamma}} + \mathbf{a}\hat{\hat{\mathbf{\Gamma}}}_p)\mathbf{\Gamma}_p^{-1}(\hat{\hat{\mathbf{\Gamma}}}_p - \mathbf{\Gamma}_p)\mathbf{\Gamma}_p^{-1}(\hat{\hat{\mathbf{\Gamma}}}_p - \mathbf{\Gamma}_p)\mathbf{\Gamma}_p^{-1},$$

the covariance matrix of $\hat{\mathbf{a}}$ is again given by the formula (7.20), except that the first term in this expression has to be replaced by the covariance matrix of $(\hat{\gamma} + \mathbf{a}\hat{\mathbf{\Gamma}}_p)\mathbf{\Gamma}_p^{-1}$.

We now show that the third term in (7.20) can be eliminated by considering the biased corrected estimator $\hat{\hat{\mathbf{a}}}(\mathbf{q}) = [\hat{\mathbf{a}}(\mathbf{q}) - \mathbf{b}/n](\mathbf{I} + \mathbf{B}/n)^{-1}$ where the (constant) vector $\mathbf{b}$ and matrix $\mathbf{B}$ are defined by

$$(\mathbf{b} + \mathbf{aB})/n = \frac{1}{2}\sum_{j=0}^{p}\sum_{k=0}^{p}(\partial^2\mathbf{a}/\partial\gamma_j\partial\gamma_k)g_{jk}/n,$$

the asymptotic bias of $\hat{\mathbf{a}}(\mathbf{q})$ (see section 7.3). For this, we shall need a relation between T_{ijk} and g_{ij}, based on the fact that distribution of $n(q_0 \cdots q_q)'$ 'belongs asymptotically' to an exponential family. Indeed, in the Gaussian case, the density of $(X_1 \cdots X_n)'$ is approximately (i.e. neglecting terms of lower order)

$$[\det(2\pi\mathbf{\Gamma}_n)]^{-1/2}\exp\left(n\sum_{k=0}^{p}\theta_k q_k\right)$$

where

$$\theta_0 = -\frac{1}{2}\sum_{i=0}^{p}a_i^2/\sigma^2, \quad \theta_k = -\sum_{i=0}^{p-k}a_i a_{i+k}/\sigma^2, \quad k = 1, \ldots, p.$$

Further,

$$\frac{1}{2}\log\det\boldsymbol{\Gamma}_n \approx \frac{n}{2}\log(\sigma^2) = -\frac{n}{4\pi}\int_{-\pi}^{\pi}\log\left[-2\sum_{k=0}^{p}\theta_k\cos(k\lambda)\right]\mathrm{d}\lambda, \quad (7.22)$$

since

$$\left[-2\sum_{k=0}^{p}\theta_k\cos(k\lambda)\right]^{-1}/(2\pi)$$

is the spectral density of the process. Thus, we may expect that the function in the right hand side of (7.22) is the 'asymptotic' cumulant generating function of $n(q_0\cdots q_p)'$, in the sense that its derivatives are asymptotically equivalent to the cumulants of nq_i. That this is true, at least for cumulants up to third order, is stated in the following lemma.

Lemma 7.2

Let $\psi(\theta_0,\ldots,\theta_p)$ be $1/n$ times the function in the right hand side of (7.22). Then $\gamma_i = \partial\psi/\partial\theta_i$ and if the fourth order cumulant of e_t is zero, $g_{ij} = \partial^2\psi/\partial\theta_i\partial\theta_j$, and if moreover the third and sixth order cumulants of e_t are zero, $T_{ijk} = \partial^3\psi/\partial\theta_i\partial\theta_j\partial\theta_k$, for all $i,j,k = 0,\ldots,p$. If only the third and fourth order cumulants of e_t are zero, one still has

$$\sum_{l=0}^{p} a_l T_{|i-l|,j,k} = \sum_{l=0}^{p} a_l \partial^3\psi/\partial\theta_{|i-l|}\partial\theta_j\partial\theta_k,$$

for $i = 1,\ldots,p, j,k = 0,\ldots,p$.

Note that the above result does not require the Gaussian assumption but only that the cumulants of order three to six of e_t are zero, since one has restricted oneself to cumulants up to order three of the process. Moreover, since by (7.15)

$$\sum_{i=0}^{p}(\partial\mathbf{a}/\partial\gamma_i)\,T_{ijk}$$

is a linear combination of

$$\sum_{l=0}^{p} a_l T_{|1-l|j,k},\ldots,\quad \sum_{l=0}^{p} a_l T_{|p-l|j,k},$$

one actually needs only that the third and fourth cumulants of e_t are zero. Under this condition:

$$\sum_{i=0}^{p}\frac{\partial\mathbf{a}}{\partial\gamma_i}T_{ijk} = \sum_{i=0}^{p}\frac{\partial\mathbf{a}}{\partial\gamma_i}\frac{\partial g_{jk}}{\partial\theta_i} = \sum_{i,l=0}^{p}\frac{\partial\mathbf{a}}{\partial\gamma_i}\frac{\partial g_{jk}}{\partial\gamma_l}\frac{\partial\gamma_l}{\partial\theta_i} = \sum_{i,l=0}^{p}\frac{\partial\mathbf{a}}{\partial\gamma_i}\frac{\partial g_{jk}}{\partial\gamma_l}g_{il}$$

$$= \lim_{n\to\infty} n\mathrm{E}\left\{(\delta\mathbf{a})\left[\sum_{l=0}^{p}\frac{\partial g_{jk}}{\partial\gamma_l}(q_l-\gamma_l)\right]\right\},$$

where $\delta\mathbf{a}$ is as in (7.15). If follows that the third term of (7.20) can be written as the sum of

$$\frac{1}{n^2}\lim_{n\to\infty} n\mathrm{E}\left\{(\delta\mathbf{a}')\left[\sum_{l=0}^{p}\frac{\partial}{\partial\gamma_l}\left(\frac{1}{2}\frac{\partial^2\mathbf{a}}{\partial\gamma_j\partial\gamma_k}g_{jk}\right)(q_r-\gamma_l)\right]\right\}$$

and its transpose. Since

$$\frac{1}{2}\sum_{j=0}^{p}\sum_{k=0}^{p}(\partial^2\mathbf{a}/\partial\gamma_j\partial\gamma_k)g_{jk}=\mathbf{b}+\mathbf{aB}$$

and the covariance matrix of $\delta\mathbf{a}$ differs from

$$\mathbf{C}=\sum_{ij}\frac{\partial\hat{\mathbf{a}}'}{\partial q_i}\frac{\partial\hat{\mathbf{a}}}{\partial q_j}\operatorname{cov}(q_i,q_j)+\frac{1}{2n^2}\sum_{i,j,k,l=0}^{p}\frac{\partial^2\mathbf{a}'}{\partial\gamma_i\partial\gamma_j}\frac{\partial^2\mathbf{a}}{\partial\gamma_k\partial\gamma_l}g_{ik}g_{jl}. \tag{7.23}$$

by $\mathrm{O}(n^{-2})$, it follows that the matrix (7.20) can be written as $(\mathbf{I}+\mathbf{B}'/n)\mathbf{C}(\mathbf{I}+\mathbf{B}/n)+\mathrm{O}(n^{-5/2})$. Thus the covariance matrix of $\hat{\hat{\mathbf{a}}}(\mathbf{q})$ is simply $\mathbf{C}+\mathrm{O}(n^{-5/2})$.

In conclusion, the asymptotic covariance matrix of the biased corrected estimator consists of two terms. The first is the covariance matrix of

$$\sum_{i=0}^{p(p+3)/2}(\partial\hat{\mathbf{a}}/\partial q_i)q_i$$

(or of $(\hat{\gamma}+\mathbf{a}\hat{\mathbf{\Gamma}}_p)\mathbf{\Gamma}_p^{-1}$ in the case of the LS estimator) The second is the leading term in the covariance matrix of the random vector $\frac{1}{2}\delta^2\mathbf{a}$, defined in (7.14). This term depends only on the model parameters and represents the 'curvature' effect of the model. By (7.16) and noting that $\mathbf{\Gamma}_p^{-1}$ is inversely proportional to σ^2,

$$\frac{1}{2}\delta^2\mathbf{a}=(\delta\mathbf{a})\mathbf{\Gamma}_p\sum_{i=1}^{p}(\partial\mathbf{\Gamma}_p^{-1}/\partial a_i)(\delta a_i)-(\delta\mathbf{a})\sigma^{-2}(\delta\sigma^2).$$

But $\delta\mathbf{a}$ and $\delta\sigma^2$ are asymptotically independently normally distributed with covariance matrix $\sigma^2\mathbf{\Gamma}_p^{-1}/n$ and variance $2\sigma^4/n$ and the fourth cumulants between their elements are $\mathrm{O}(n^2)$. Thus, from the identity $\operatorname{cov}(Z_1Z_2,Z_3Z_4)=\operatorname{cum}(Z_2,Z_2,Z_3,Z_4)+\operatorname{cov}(Z_1,Z_3)\operatorname{cov}(Z_2,Z_4)+\operatorname{cov}(Z_1,Z_4)\operatorname{cov}(Z_2,Z_3)$ for any zero mean random variables Z_1,Z_2,Z_3,Z_4, the covariance matrix of $\frac{1}{2}\delta^2\mathbf{a}$ is equivalent to

$$\frac{\sigma^4}{n^2}\sum_{i=1}^{p}\sum_{j=1}^{p}\left[\frac{\partial\mathbf{\Gamma}_p^{-1}}{\partial a_i}\mathbf{\Gamma}_p\frac{\partial\mathbf{\Gamma}_p^{-1}}{\partial a_j}(\mathbf{\Gamma}_p^{-1})_{ij}+\frac{\partial\mathbf{\Gamma}_p^{-1}}{\partial a_i}\mathbf{e}_j'\mathbf{e}_i\frac{\partial\mathbf{\Gamma}_p^{-1}}{\partial a_j}\right]+2\frac{\sigma^2}{n^2}\mathbf{\Gamma}_p^{-1}, \tag{7.24}$$

where $(\mathbf{\Gamma}_p^{-1})_{ij}$ is the general element of $\mathbf{\Gamma}_p^{-1}$ and $\mathbf{e}_i$ is the ith row of the identity matrix. Using (7.3), the above expression can be expressed in term of the autocovariances $\gamma_0,\ldots,\gamma_{p-1}$ and polynomials in $\mathbf{a}$ of degree six at most. But it does not seem possible to explicitly evaluate this expression for general p.

For the first term in (7.23), one has the following result.

Lemma 7.3

Assume that the e_t have zero third and fourth cumulants. Then the covariance matrix of

$$\sum_{i=0}^{p(p+3)/2} (\partial \hat{\mathbf{a}}/\partial q_i) q_i$$

or of $(\hat{\gamma} + \mathbf{a}\hat{\mathbf{\Gamma}}_p)\mathbf{\Gamma}_p^{-1}$ is bounded below by the **C–R** bound, computed under Gaussian assumption. Further, equality is attained if $\hat{\mathbf{a}}(\mathbf{q})$ is the ML estimator (computed under the Gaussian assumption) of $\mathbf{a}$.

The above result shows that ML estimator achieves the smallest asymptotic covariance matrix among the class of all estimators satisfying (E1)–E(4) and the LS estimator. But the explicit computation of this matrix requires the knowledge of the covariance matrix of

$$\sum_{i=0}^{p(p+3)/2} (\partial \hat{\mathbf{a}}/\partial q_i) q_i,$$

which is generally very complex. However, for the ML estimator, the last matrix is the C–R bound which can be computed quickly by the method of Pham (1989). Also, for the case $p = 1$, the asymptotic variance of the LS, FBLS and ML estimators (in this simple case other estimators described in section 7.3 coincide with the FBLS estimator) have been given explicitly in Pham (1992).

7.5 APPLICATION TO THE PARTIAL AUTOCORRELATION ESTIMATORS

As an application of the above results, consider the last autoregressive coefficient a_p. By formula (7.17) the bias of $\hat{a}_p(\mathbf{Q})$ is $-(p+1)a_p/n$ if p is odd and $[1-(p+1)a_p]/n$ if p is even. Thus the bias corrested estimator for a_p is $\hat{\hat{a}} = \hat{a}_p/[1-(p+1)/n]$ if p is odd, $= (\hat{a}_p - n^{-1})/[1-(p+1)/n]$ otherwise. The 'curvature' term in the asymptotic variance of this estimator is the element in the lower right corner of the matrix (7.24). To compute it, we need to compute the last row of $\sigma^2 \partial \mathbf{\Gamma}_p^{-1}/\partial a_i$. From (7.3), with p in place of n, the last row of $\mathbf{\Gamma}_p^{-1}$ is $[(a_{p-1} \cdots a_0) - a_p(a_1 \cdots a_p)]/\sigma^2$, hence the last row of $\sigma^2 \partial \mathbf{\Gamma}_p^{-1}/\partial a_i$ equals $\mathbf{d}_i - a_p \mathbf{e}_i$ where $\mathbf{d}_i = \mathbf{e}_{p-i}$ for $i = 1, \ldots, p-1$, $-(a_1 \cdots a_p)$ for $i = p$, $\mathbf{e}_i$ denoting as before the ith row of the identity matrix. Thus, the element in the lower right corner of the matrix (7.24) is

$$\frac{1}{n^2}\left\{ \sum_{i=1}^{p} \sum_{j=1}^{p} [\mathbf{d}_i \mathbf{\Gamma}_p \mathbf{d}_j' (\mathbf{\Gamma}_p^{-1})_{ij} - 2a_p \mathbf{d}_i \mathbf{\Gamma}_p \mathbf{e}_j' (\mathbf{\Gamma}_p^{-1})_{ij} + a_p^2 \mathbf{e}_i \mathbf{\Gamma}_p \mathbf{e}_j' (\mathbf{\Gamma}_p^{-1})_{ij} \right.$$
$$\left. + \mathbf{d}_i \mathbf{e}_j' \mathbf{e}_i \mathbf{d}_j' - 2a_p \mathbf{d}_i \mathbf{e}_j' \mathbf{e}_i \mathbf{e}_j' + a_p^2 \mathbf{e}_i \mathbf{e}_j' \mathbf{e}_i \mathbf{e}_j'] + 2(1 - a_p^2) \right\}.$$

It can be seen from the definition of $\mathbf{d}_i$ and the equation $(1\ a_1 \cdots a_p)\mathbf{\Gamma}_p =$

$(\sigma^2 0 \cdots 0)$ that $\mathbf{d}_i \boldsymbol{\Gamma}_p \mathbf{d}'_j = \gamma_{i-j}$ for i or $j < p$ and $\mathbf{d}_p \boldsymbol{\Gamma}_p \mathbf{d}'_p = \gamma_0 - \sigma^2$. Since γ_{i-j} is the general element of $\boldsymbol{\Gamma}_p$, one obtains

$$\sum_{i=1}^{p} \sum_{j=1}^{p} \mathbf{d}_i \boldsymbol{\Gamma}_p \mathbf{d}'_j (\boldsymbol{\Gamma}_p^{-1})_{ij} = p - (1 - a_p^2).$$

Clearly, one also has

$$\sum_{i=1}^{p} \sum_{j=1}^{p} \mathbf{d}_i \mathbf{e}'_j \mathbf{e}_i \mathbf{d}'_j = p - 1 + a_p^2.$$

Further, it is easy to see that

$$\boldsymbol{\Gamma}_p \sum_{j=1}^{p} \mathbf{e}'_j (\boldsymbol{\Gamma}_p^{-1})_{ij} = \mathbf{e}'_i$$

and hence

$$\sum_{i=1}^{p} \sum_{j=1}^{p} \mathbf{d}_i \boldsymbol{\Gamma}_p \mathbf{e}'_j (\boldsymbol{\Gamma}_p^{-1})_{ij} = \sum_{i=1}^{p} \mathbf{d}_i \mathbf{e}'_i$$

and

$$\sum_{i=1}^{p} \sum_{j=1}^{p} \mathbf{e}_i \boldsymbol{\Gamma}_p \mathbf{e}'_j (\boldsymbol{\Gamma}_p^{-1})_{ij} = p.$$

Finally it is clear that

$$\sum_{i=1}^{p} \sum_{j=1}^{p} \mathbf{d}_i \mathbf{e}'_j \mathbf{e}_i \mathbf{e}'_j = \sum_{i=1}^{p} \mathbf{d}_i \mathbf{e}'_i$$

and

$$\sum_{i=1}^{p} \sum_{j=1}^{p} \mathbf{e}_i \mathbf{e}'_j \mathbf{e}_i \mathbf{e}'_j = p.$$

Thus, the element in the lower right corner of the matrix (7.24) equals

$$2p(1 + a_p^2) - 4a_p \sum_{i=1}^{p} \mathbf{d}_i \mathbf{e}'_i.$$

Since

$$\sum_{i=1}^{p} \mathbf{d}_i \mathbf{e}'_i = -a_p$$

if p is odd, $= 1 - a_p$ otherwise, this element is $2[p + (p+2)a_p^2]/n^2$ if p is odd and $2[p + (p+2)a_p^2 - 2a_p]/n^2$ if p is even. Note that in the simplest case where $p = 1$, the matrix (7.24) reduces $(6a^2 + 2)/n^2$, leading to the asymptotic variance of the estimators mentioned in section 7.4.

The above result is useful to derive the asymptotic variance of the estimated partial autocorrelation. We know that the partial autocorrelation of lag p may be estimated by the negative of the estimated last AR coefficient $\hat{a}_{p,p}$ of a model of order p (we have added a second index p in $\hat{a}_{p,p}$ to dencte the order). Suppose that the process obeys an AR model of order $p^* < p$, then one may use the above result which yields that the variance of $\hat{a}_{p,p}/[1-(p+1)/n]$ equals

$$\operatorname{var}\left[\sum_{i=0}^{p(p+3)/2}\left(\frac{\partial \hat{a}_{p,p}}{\partial q_i}\right)q_i\right]+\frac{2p}{n^2}.$$

If the estimator is obtained by the ML method, then the first term equals the C–R bound, which has been proved by Pham (1989) to be equal to $1/(n-p)$. In this case, the asymptotic variance of $\hat{a}_{p,p}$ is, up to the order n^{-2},

$$\left(\frac{1}{n-p}+\frac{2p}{n^2}\right)\left(\frac{p+1}{n}\right)^2=\frac{1-2/n}{n-p}+\mathrm{O}(n^{-2}).$$

Thus, the asymptotic variance of the ML estimator for the partial autocorrelation of lag p **greater** than the true order of the underlying AR process, is $(1-2/n)/(n-p)$. This is to be compared with the customary value $1/n$ of classical first order asymptotic theory. The above result has been derived for the partial autocorrelation estimated by the ML method which is costly in computations. The lemma below shows that it still holds for the Burg's, Kay's and SPAC estimators, which are much easier to compute.

Lemma 7.4

Let $\hat{a}_{p,p}$ denote the estimated last coefficient for a p-order AR model by the methods of Burg, Kay or SPAC. Then, if the true order of the model is less than p:

$$\sum_{i=0}^{p(p+3)/2}\left(\frac{\partial \hat{a}_{p,p}}{\partial q_i}\right)_{q-\bar{q}}q_i=\frac{1}{(n-p)\sigma^2}\sum_{t=p+1}^{n}e_t\left[X_{t-p}+\sum_{j=1}^{p^*}a_jX_{t-p+j}\right]$$

and has variance $1/(n-p)$.

APPENDIX

Proof of lemma 7.1

From the particular form of **Q**, we have, putting $\hat{a}_{0,k-1}=1$,

$$A_k=\sum_{j=1}^{k}\left\{(n-j)\left[\sum_{i=1}^{k}\hat{a}_{k-i,k-1}G_{ij}\right]\hat{a}_{k-j,k-1}\right\}$$
$$-\sum_{i=1}^{k}\left\{i\hat{a}_{k-i,k-1}\left[\sum_{j=1}^{k}G_{ij}\hat{a}_{k-j,k-1}\right]\right\}$$

From the equation of the lemma and using the fact that $G_{ij}=G_{ji}=G_{k-i,k-j}$, we get

$$A_k = (n-2k)\sum_{i=0}^{k-1} \hat{a}_{i,k-1} G_{i0}.$$

In the same way,

$$B_k = (n-k)\sum_{i=0}^{k-1} \hat{a}_{i,k-1} G_{ik},$$

$$C_k = n\sum_{i=0}^{k-1} \hat{a}_{i,k-1} G_{i0}.$$

Thus

$$2B_k/(A_k + C_k) = \left(\sum_{i=0}^{k-1} \hat{a}_{i,k-1} G_{ik}\right)\Big/\left(\sum_{i=0}^{k-1} \hat{a}_{i,k-1} G_{i0}\right) = -\beta_k, \text{ say.}$$

Now, from (7.7), for $j = 1,\ldots,k$:

$$G_{i0} + \sum_{i=1}^{k} \hat{a}_{i,k} G_{ij} = \sum_{i=0}^{k-1} \hat{a}_{i,k-1} G_{ij} + \hat{a}_{i,k}\sum_{i=0}^{k-1} \hat{a}_{i,k-1} G_{i,k-j},$$

which equals zero for $j < k$, by the equations of the lemma. For $j = k$, the above right hand side is zero if $\hat{a}_{i,k} = \beta_k$, yielding the first result of the lemma.

On the other hand, since $A_k/(A_k + C_k) = \frac{1}{2}(n-2k)/(n-k)$, $C_k/(A_k + C_k) = \frac{1}{2}n/(n-k)$, the right hand side of (7.9) reduces to

$$n\log[(A_k + C_k)/2] + n\log\left(\frac{n-2k}{n-k}\hat{a}_{kk}^2 - 2\beta_k\hat{a}_{kk} + \frac{n}{n-k}\right) - k\log(1-\hat{a}_{kk}^2).$$

Setting the derivative, with respect to $\hat{a}_{kk}$, of this expression to zero yields the equation

$$n\left(\frac{n-2k}{n-k}\hat{a}_{kk} - \beta_k\right)(1-\hat{a}_{kk}^2) + k\hat{a}_{kk}\left(\frac{n-2k}{n-k}\hat{a}_{kk}^2 - 2\beta_k\hat{a}_{kk} + \frac{n}{n-k}\right) = 0.$$

Simple computation shows that the left hand side of the above equation can be written as $(\hat{a}_{kk} - \beta_k)[n-(n-2k)\hat{a}_{kk}]$. Hence this equation admits the unique solution β_k in $[-1,1]$, completing the proof of the lemma.

Proof of lemma 7.2

Since

$$\left[-2\sum_{k=0}^{p} \theta_k\cos(k\lambda)\right]^{-1}/(2\pi)$$

is the spectral density $f(\lambda)$ of the process, one gets immediately

$$\partial\psi/\partial\theta_j = \int_{-\pi}^{\pi} \cos(j\lambda)f(\lambda)\mathrm{d}\lambda = \gamma_j.$$

Also, it is well known that, if the fourth order cumulant function of the process vanishes, $n\,\mathrm{cov}(q_j, q_k)$ tends, as $n \to \infty$, to

$$2\pi \int_{-\pi}^{\pi} [\mathrm{e}^{i(j+k)\lambda} + \mathrm{e}^{i(j-k)\lambda}] f^2(\lambda)\mathrm{d}\lambda = 4\pi \int_{-\pi}^{\pi} \cos(j\lambda)\cos(k\lambda) f^2(\lambda)\mathrm{d}\lambda$$

(see for example Priestley, 1981). This yields the second result of the lemma.

Now, from the Leonov and Shiryayev's (1959) rule, one has, for zero mean random variables $Z_1, \ldots, Z_6$, with zero third, fourth and sixth joint cumulants:

$$\begin{aligned}\mathrm{cum}(Z_1 Z_2, Z_3 Z_4, Z_5 Z_6) = {} & c_{13}c_{25}c_{46} + c_{13}c_{26}c_{45} + c_{14}c_{25}c_{36} \\ & + c_{14}c_{26}c_{35} + c_{15}c_{23}c_{46} + c_{15}c_{24}c_{36} \\ & + c_{16}c_{23}c_{45} + c_{16}c_{24}c_{35}\end{aligned}$$

where $c_{ij} = \mathrm{cov}(Z_i, Z_j)$. Thus $(j, k, l \geqslant 0)$

$$\begin{aligned}&(n-j)(n-k)(n-l)\mathrm{cum}(q_j, q_k, q_l) \\ &= \mathrm{cum}\left(\sum_{t=1}^{n-j} X_t X_{t+j}, \sum_{u=1}^{n-k} X_u X_{u+k}, \sum_{v=1}^{n-1} X_v X_{v+l}\right) \\ &= \sum_{t=1}^{n-j}\sum_{u=1}^{n-k}\sum_{v=1}^{n-l} (\gamma_{u-t}\gamma_{v-t-j}\gamma_{v-u+l-k} + \gamma_{u-t}\gamma_{v-t+l-j}\gamma_{v-u-k} \\ &\quad + \gamma_{u-t+k}\gamma_{v-t-j}\gamma_{v-u+l} + \gamma_{u-t+k}\gamma_{v-t+l-j}\gamma_{v-u} + \gamma_{v-t}\gamma_{u-t-j}\gamma_{v-u+l-k} \\ &\quad + \gamma_{v-t}\gamma_{u-t+k+j}\gamma_{v-u+l} + \gamma_{v-t+l}\gamma_{u-t-j}\gamma_{v-u-k} + \gamma_{v-t+l}\gamma_{u-t+k-j}\gamma_{v-u}).\end{aligned}$$

It follows from the Lebesque dominated convergence theorem, using

$$\sum_{j=-\infty}^{\infty} |\gamma_j| < \infty$$

that $n^2\,\mathrm{cum}(q_j, q_k, q_l)$ converges, as $n \to \infty$, to

$$\begin{aligned}&\sum_{u=-\infty}^{\infty}\sum_{v=-\infty}^{\infty} (\gamma_{u-k}\gamma_{v-j}\gamma_{v-u+l} + \gamma_{u+j}\gamma_{v+l}\gamma_{v-u-k} + \gamma_{u+k}\gamma_{v-j}\gamma_{v-u+l} \\ &+ \gamma_{u+k}\gamma_{v+l}\gamma_{v-u+j} + \gamma_{v+k}\gamma_{u-j}\gamma_{v-u+l} + \gamma_{v-k}\gamma_{u-j}\gamma_{v-u+l} + \gamma_{v+l}\gamma_{u-j}\gamma_{v-u-k} \\ &+ \gamma_{v+l}\gamma_{u-j}\gamma_{v-u+k}). \\ &= 8\pi^2 \int_{-\pi}^{\pi} [\mathrm{e}^{i(j+k+l)\lambda} + \mathrm{e}^{i(j-k+l)\lambda} + \mathrm{e}^{i(j+k-l)\lambda} + \mathrm{e}^{i(j-k-l)\lambda}] f^3(\lambda)\mathrm{d}\lambda \\ &= 8(2\pi)^2 \int_{-\pi}^{\pi} \cos(j\lambda)\cos(k\lambda)\cos(l\lambda) f^3(\lambda)\mathrm{d}\lambda.\end{aligned}$$

This yields the third result of the lemma.

Finally, if the sixth cumulant of the e_t is non zero, then in the above computation one must add the term

$$\sum_{u=-\infty}^{\infty} \sum_{v=-\infty}^{\infty} \operatorname{cum}(X_0, X_j, X_u, X_{u+k}, X_v, X_{v+l}).$$

Note that the last sum does not change when sign of j is changed. Thus

$$\sum_{l=0}^{p} a_l T_{|i-l|,j,k} = \sum_{l=0}^{p} a_l \partial^3 \psi/\partial\theta_{|i-l|}\partial\theta_j\partial\theta_k$$

$$+ \sum_{u=-\infty}^{\infty} \sum_{v=-\infty}^{\infty} \operatorname{cum}(X_0, \mathrm{e}_i, X_u, X_{u+j}, X_v, \mathrm{e}_{v+k}).$$

But the last vanishes since X_t is of the form

$$\sum_{m=0}^{\infty} \alpha_m \mathrm{e}_{t-m}$$

and the e_t are independent. This completes the proof.

Proof of lemma 7.3

Let $L(\theta, \mathbf{q})$ be the log likelihood function of the model, computed under the Gaussian assumption by (7.2) and (7.4), where $\theta = (\sigma^2 a_1 \cdots a_p) = (\theta_0 \cdots \theta_p)$ denotes the parameter vector. Let $\dot{L}_i(\theta, \mathbf{q}) = \partial L(\theta, \mathbf{q})/\partial\theta_i$ be the score functions and $\dot{\mathbf{L}} = (\dot{L}_0 \cdots \dot{L}_p)$. The Fisher's information matrix is $\mathbf{J} = \mathrm{E}_\theta[\dot{\mathbf{L}}'\dot{\mathbf{L}}]$ and the C–R bound for $(a_1 \cdots a_p)$ is the sub-matrix of $\mathbf{J}^{-1}$ former by its last p rows and columns.

We note the important fact, which will be useful later, that $\dot{L}_i$ is **linear** in $\mathbf{q}$. As a score function, $\dot{L}_i$ also satisfies $\mathrm{E}_\theta(g\dot{L}_i) = \partial \mathrm{E}_\theta(g)/\partial\theta_i$ for any statistic g. This can be seen by differentiating, with respect to θ_i, the relation $\mathrm{E}_\theta(g) = \int g f_\theta$ where f_θ denotes the probability density function of the observations. This result holds only under the Gaussian assumption since $\dot{L}_i$ is computed under this assumption. However, when g is a quadratic form in $X_1, \ldots, X_n$, $g\dot{L}_i$ is a polynomial of fourth in $X_1, \ldots, X_n$ without terms of odd degree, hence its expectations depends only on the covariance and fourth cumulant functions of the X_t process and not on its actual distribution. Since $\mathrm{E}_\theta(g)$ and hence $\partial \mathrm{E}_\theta(g)/\partial\theta_i$ also depends only on the covariance function of the process, the above results holds as soon as the X_t process has a zero fourth cumulant function as in the case of a Gaussian process (which is satisfied if the e_t have zero fourth cumulant). Thus, taking $g = q_j$, one gets $\mathrm{E}_\theta(q_j\dot{L}_i) = \partial\gamma_j/\partial\theta_i$ if $j \leqslant p$, $= 0$ otherwise. Since $(\partial\hat{\mathbf{a}}/\partial q_j)(\bar{\mathbf{q}}) = \partial\mathbf{a}/\partial\gamma_j, j = 0, \ldots, p$ it follows that

$$\mathrm{E}_\theta\left[\sum_{j=0}^{p(p+3)/2} (\partial\hat{\mathbf{a}}/\partial q_j) q_j \dot{L}_i\right] = \sum_{i=0}^{p} (\partial\mathbf{a}/\partial\gamma_j)(\partial\gamma_j/\partial\theta_i) = \partial\mathbf{a}/\partial\theta_i.$$

Hence the covariance matrix of the pair of vectors

$$\sum_{j=0}^{p(p+3)/2} (\partial\hat{\mathbf{a}}/\partial q_j)q_j \quad \text{and } \dot{\mathbf{L}}$$

equals

$$\begin{bmatrix} \tilde{\mathbf{C}} & (\partial\mathbf{a}/\partial\theta)' \\ \partial\mathbf{a}/\partial\theta & \mathbf{J} \end{bmatrix} \tag{7.25}$$

where $\tilde{\mathbf{C}}$ is the covariance matrix of

$$\sum_{j=0}^{p(p+3)/2} (\partial\hat{\mathbf{a}}/\partial q_j)q_j$$

and $\partial\mathbf{a}/\partial\theta$ is the matrix with rows being $\partial\mathbf{a}/\partial\theta_i, i=0,\ldots,p$. Similarly, since $\mathrm{E}_\theta(\hat{\gamma}) = (\gamma_1 \cdots \gamma_p) = \gamma$, $\mathrm{E}_\theta(\hat{\mathbf{\Gamma}}_p) = \mathbf{\Gamma}_p$, one gets $\mathrm{E}_\theta[\hat{\gamma} + \mathbf{a}\hat{\mathbf{\Gamma}}_p)\mathbf{\Gamma}_p^{-1}\dot{L}_i] = (\partial\gamma/\partial\theta_i + \mathbf{a}\partial\mathbf{\Gamma}_p/\partial\theta_i)\mathbf{\Gamma}_p^{-1} = -\partial\mathbf{a}/\partial\theta_i$. The last equality follows from $\partial\gamma/\partial\theta_i + \mathbf{a}\partial\mathbf{\Gamma}_p/\partial\theta_i = -(\partial\mathbf{a}/\partial\theta_i)\mathbf{\Gamma}_p$, which in turn is a consequence of $\gamma + \mathbf{a}\mathbf{\Gamma}_p = \mathbf{0}$. Thus the covariance matrix of the pair $-(\hat{\gamma} + \mathbf{a}\hat{\mathbf{\Gamma}}_p)\mathbf{\Gamma}_p^{-1}$, $\dot{\mathbf{L}}$ is again given by (7.25) with $\tilde{\mathbf{C}}$ being the covariance matrix of $(\hat{\gamma} + \mathbf{a}\hat{\mathbf{\Gamma}}_p)\mathbf{\Gamma}_p^{-1}$. From the positivity of the matrix (7.25), one gets $\tilde{\mathbf{C}} \geqslant (\partial\mathbf{a}/\partial\theta)'\mathbf{J}^{-1}(\partial\mathbf{a}/\partial\theta)$. But by construction $\partial\mathbf{a}/\partial\theta_i$ is the null vector for $i=0$ and equals the ith row of the identity matrix for $i=1,\ldots,p$. Thus the matrix in the right hand side of the last ineqality is the sub-matrix of $\mathbf{J}^{-1}$ formed by its last p rows and columns, yielding the first result of the lemma.

Suppose now that $\hat{\mathbf{a}}(\mathbf{q})$ is the ML estimator of $\mathbf{a}$. Put $\hat{\theta}(\mathbf{q}) = [\hat{\sigma}^2(\mathbf{q})\ \hat{\mathbf{a}}(\mathbf{q})]$ where $\hat{\sigma}^2(\mathbf{q}) = [1\ \hat{\mathbf{a}}(\mathbf{q})]\mathbf{Q}[1\ \hat{\mathbf{a}}(\mathbf{q})]'$ is the ML estimator of σ^2. Note that $\hat{\mathbf{a}}(\bar{\mathbf{q}}) = \mathbf{a}$ and hence $\hat{\sigma}^2(\bar{\mathbf{q}}) = \sigma^2$, $\hat{\theta}(\bar{\mathbf{q}}) = \theta$. Now, by definition, $\dot{L}_i[\hat{\theta}(\mathbf{q}), \mathbf{q}] = 0$ identically for all i. Taking the derivative of this equality with respect to q_k at the point $\bar{\mathbf{q}}$, one gets

$$\frac{\partial\dot{L}_i}{\partial q_k}(\theta,\bar{\mathbf{q}}) + \sum_{j=0}^{p} \ddot{L}_{ij}(\theta,\bar{\mathbf{q}})\frac{\partial\hat{\theta}_j}{\partial q_k} = 0,$$

where $\ddot{L}_{ij} = \partial\dot{L}_i/\partial\theta_j$. Remember that $\dot{L}_i(\theta,\mathbf{q})$ is a linear function of $\mathbf{q}$ and so is $\ddot{L}_{ij}(\theta,\mathbf{q})$. Thus $\ddot{L}_{ij}(\theta,\bar{\mathbf{q}}) = \mathrm{E}_\theta[\ddot{L}_{ij}(\theta,\mathbf{q})] = -J_{ij}$, the general element of the information matrix $-\mathbf{J}$. Also

$$\begin{aligned} \dot{L}_i(\theta,\mathbf{q}) - \dot{L}_i(\theta,\bar{\mathbf{q}}) &= \sum_{k=0}^{p(p+3)/2} \left[\frac{\partial\dot{L}_i}{\partial q_k}(\theta,\bar{\mathbf{q}})\right](q_k - \bar{q}_k) \\ &= \sum_{k=0}^{p(p+3)/2} \sum_{j=0}^{p} J_{ij}\frac{\partial\hat{\theta}_j}{\partial q_k}(q_k - \bar{q}_k). \end{aligned}$$

The above equations show that the random vector

$$\sum_{k=0}^{p(p+3)/2} (\partial\hat{\theta}/\partial q_k)(q_k - \bar{q}_k)$$

equals $[\dot{\mathbf{L}}(\theta,\mathbf{q}) - \dot{\mathbf{L}}(\theta,\bar{\mathbf{q}})\mathbf{J}^{-1}$ and hence admits the covariance matrix $\mathbf{J}^{-1}$. The last result of the lemma follows.

Proof of lemma 7.4

Recall that in Burg's method, the partial autocorrelation of lag p is estimated by $-\hat{a}_{p,p} = 2B_p/(A_p + C_p)$ where A_p, B_p and C_p are given by (7.8) with the $\hat{a}_{i,p-1}$ corresponding to previously estimated coefficients of the $(p-1)$th order model. As has been shown in section 7.2, $\hat{a}_{i,p-1}(\bar{\mathbf{q}}) = a_i$, $i = 1,\ldots,p-1$ ($a_i = 0$ if $p^* < i < p$ by convention) and since $(1\ a_1 \cdots a_{p-1} 0)\mathbf{\Gamma}_{p+1} = (\sigma^2\ 0 \cdots 0)$ one gets $A_p(\bar{\mathbf{q}}) = (1 - 2p/n)\sigma^2$, $B_p(\bar{\mathbf{q}}) = 0$, $C_p(\bar{\mathbf{q}}) = \sigma^2$. Hence

$$-\sum_{i=0}^{p(p+3)/2} \left(\frac{\partial \hat{a}_{p,p}}{\partial q_i}\right)_{\mathbf{q}=\bar{\mathbf{q}}} (q_i - \bar{q}_i) = \frac{n}{(n-p)\sigma^2} \sum_{i=0}^{p-1} \sum_{j=1}^{p} a_{i,p-1} Q_{ij} a_{p-j,p-1}. \tag{7.26}$$

On the other hand, Kay's method, described in section 7.2, estimates $\hat{a}_{p,p}$ by solving the cubic equation

$$n(A_p\hat{a}_{p,p} + 2B_p)(1 - \hat{a}_{p,p}^2) + p(A_p\hat{a}_{p,p}^2 + 2B_p\hat{a}_{p,p} + C_p)\hat{a}_{p,p} = 0$$

for $\hat{a}_{p,p}$ in $]-1,1[$ (there is always a unique solution in $]-1,1[$, see Pham, 1988). The A_p, B_p and C_p are as before but the $\hat{a}_{i,p-1}$ differ since they are the Kay's estimated coefficients of the $(p-1)$th order model. However, one still has $\hat{a}_{i,p-1}(\bar{\mathbf{q}}) = a_i$, $i = 1,\ldots,p-1$, $A_p(\bar{\mathbf{q}}) = (1 - 2p/n)\sigma^2$, $B_p(\bar{\mathbf{q}}) = 0$, $C_p(\bar{\mathbf{q}}) = \sigma^2$, and $\hat{a}_{i,p}(\bar{\mathbf{q}}) = a_p = 0$. Thus, by differentiating the above equation, then setting $\mathbf{q} = \bar{\mathbf{q}}$, one obtains

$$(n-2p)\sigma^2\left(\frac{\partial \hat{a}_{p,p}}{\partial q_i}\right)_{\mathbf{q}=\bar{\mathbf{q}}} + 2n\left(\frac{\partial B_p}{\partial q_i}\right)_{\mathbf{q}=\bar{\mathbf{q}}} - p\sigma^2\left(\frac{\partial \hat{a}_{p,p}}{\partial q_i}\right)_{\mathbf{q}=\bar{\mathbf{q}}} = 0,$$

which again yields (7.26).

Further, in the SPAC method, $\hat{a}_{p,p}$ is obtained by solving the equation $(1\ \hat{a}_{1,p} \cdots \hat{a}_{p,p})\hat{\mathbf{\Gamma}}_{p+1} = (\hat{\sigma}_p^2\ 0 \cdots 0)$ where $\hat{\mathbf{\Gamma}}_{p+1}$ is the matrix with general element

$$\frac{1}{2}\left[\sum_{t=p+1}^{n} X_{t-i}X_{i-j} + \sum_{t=1}^{n-p} X_{t+i}X_{t-t}\right]\Big/(n-p) = \tfrac{1}{2}(Q_{ij} + Q_{p-j,p-i})/(1 - p/n),$$

$i,j = 0,\ldots,p$. One may interpret $-\hat{a}_{1,p},\ldots,-\hat{a}_{p,p},\hat{\sigma}_p^2$ as the regression coefficients and residual variance, respectively, of the regression of X_t on $X_{t-1},\ldots,X_{t-p}$ using $\mathbf{\Gamma}_{p+1}$ as their covariance matrix. Thus $-\hat{a}_{p,p}$ may be obtained by first regressing X_{t-1} on $X_{t-p+1},\ldots,X_{t-1}$, and then regressing X_t on this residual. The residual from the first regression

$$X_{t-p} + \sum_{i=1}^{p-1} \hat{a}_{i,p-1}X_{t-p+i},$$

and the corresponding residual variance $\hat{\sigma}_{p-1}^2$ are obtained by solving

$(1\ \hat{a}_{1,p-1} \cdots \hat{a}_{p-1,p-1}\ 0)\hat{\mathbf{\Gamma}}_{p+1} = (\hat{\sigma}^2_{p-1} 0 \cdots 0\ *\ )$, *denoting an arbitrary number (note that the matrix $\mathbf{\Gamma}_{p+1}$ is centro-symmetric, in the sense that it is unchanged when reversing the order of rows and columns). Thus,

$$-\hat{a}_{p,p} = (1\ 0 \cdots 0)\hat{\mathbf{\Gamma}}_{p+1}(0\ \hat{a}_{p-1,p-1} \cdots \hat{a}_{1,p-1}\ 1)'/\hat{\sigma}^2_{p-1}.$$

The numerator of the above right hand side equals B_p, defined in (7.8) with the $\hat{a}_{i,p-1}$ computed as above. Again, $\hat{a}_{i,p-1}(\bar{\mathbf{q}}) = a_i, i = 1, \ldots, p-1$, and hence $B_p(\bar{\mathbf{q}}) = 0$, $\hat{\sigma}^2_{p-1}(\bar{\mathbf{q}}) = \sigma^2$ and (7.26) is still verified.

Finally, by direct computation of the right-hand-side of (7.26) one gets the result of the lemma.

REFERENCES

Arato, M. (1961) On the sufficient statistics for stationary Gaussian process *Theory Proc. Appl.*, **6** (1), 199–201.

Burg, J.P. (1975) *Maximum Entropy Spectral Analysis*, PhD Dissertation, Standford University, California.

Chan, N.H. and Wei, C.Z. (1987) Asymptotic inference for nearly non stationary AR(1) process. *Ann. Statist.* **15** (3) 1031–1063.

Dégerine, S. (1993). Sample partial autocorrelation function. *IEEE Trans. Signal Processing.* **41** (1) 403–407.

Dickinson, B.W. (1978) Estimation of partial correlation matrix using Cholesky decomposition. *IEEE Trans. Automat. Contr.*, **24** (2) 302–305.

Goldolphin, E.J. and Unwin, J.M. (1983) Evaluation of the covariance matrix for the maximum likelihood estimator of a Gaussian autoregressive-moving average process. *Biometrika*, **70** (1), 279–294.

Kay, S.M. (1983) Recursive maximum likelihood estimation of autoregressive process. *IEEE Trans. Acoust. Speech Signal Processing*, **31** (6), 1980–419.

Leonov, V.P. and Shiryayev, A.N. (1959) On a method of calculating semi-invariants. *Theory Proc. Appl.*, **4**, 319–329.

Phillips, P.C.B. (1978) Edgeworth and saddle point approximation in the first order noncircular autoregression. *Biometrika*, **65** (1), 91–98.

Pham, D.T. (1988) Maximum likelihood estimation of autoregressive model by relaxation on the reflection coefficients. *IEEE Trans. Acoust Speech Signal Processing*, **36** (8), 1363–1367.

Pham, D.T. (1989) Cramer–Rao bounds for AR parameter and reflection coefficient estimators. *IEEE Trans. Acoust. Speech Signal Processing*, **37** (5), 769–772.

Pham, D.T. (1992) Approximate distribution of parameter estimators for the first order autoregressive model, *J. Times Series Analysis*, in press.

Pham D.T. and Dégerine, S. (1990). Efficient computation of autoregressive estimates through a sufficient statistic. *IEEE Trans. Signal Processing*, **38** (1), 175–177.

Priestley, M.B. (1981) *Spectral Analysis and Times Series*, Academic Press, London p. 327.

Shaman, P. and Stein, R.A. (1988) The bias of autoregressive coefficient estimators. *J. Ann. Statist Assoc.*, **83** (403), 842–848.

Yamoto, T. and Kunimoto, N. (1984). Asymptotic bias for the least squares estimator for multiple autoregressive models. *Ann. Inst. Statist. Math.*, **36A**, 419–430.

8

Asymptotic properties of serial covariances of orders which increase with sample size

K.C. Chanda

8.1 INTRODUCTION

Let $\{X_t; t \in \mathbf{Z}\}$ be a strictly stationary process with $E(X_1) = 0$ and $E(X_1 X_{1+|v|}) = \gamma_v$, $v \in \mathbf{Z}$, and let $X_1, \ldots, X_n$ be a sample from $\{X_t\}$. The serial covariance $\hat{\gamma}_v$ of order v is given by

$$\hat{\gamma}_v = n^{-1} \sum_{t=1}^{n-|v|} X_t X_{t+|v|}.$$

It is well known that under mild regularity conditions (see for example Anderson, 1970) on $\{X_t\}$, the distribution of $n^{1/2}(\gamma_{v_j} - \gamma_{v_j})$ $(1 \leqslant j \leqslant l)$, where l, $v_1, \ldots, v_l$ are fixed positive integers, is asymptotically (as $n \to \infty$) normal with mean zero and a finite covariance matrix. A stronger result is due to Brillinger (1969) who has established that under a set of stringent regularity conditions which include among others the existence of all moments of $\{X_t\}$ that $\{n^{1/2}(\hat{\gamma}_v - \gamma_v),\ v = 0, \pm 1, \ldots\}$ converges weakly in a certain topology to a zero mean Gaussian process $\{Y_v, v = 0, \pm 1, \ldots\}$ with a well defined autocovariance function. The second order properties of $\hat{\gamma}_v$ are such that one may be curious to find out if asymptotic normality for $\hat{\gamma}$ holds even when $v \to \infty$ as $n \to \infty$ but at a rate slower than n. Many interesting situations where large sample properties of serial covariances whose orders increase with sample size play useful roles have been discussed in Chanda (1992). For example if $X_t = U_t + \varepsilon_t$ where $\{U_t\}$ is an AR(1) $(U_t = \phi U_{t-1} + \eta_t\, t \in \mathbf{Z})$ independent of $\{\varepsilon_t\}$ then under certain mild regularity conditions on $\{\varepsilon_t\}$, the estimator $\tilde{\phi} = \hat{\gamma}_{k+1}/\hat{\gamma}_k$ where $k \sim g \log n / 2 \log |\phi|$ and g is any number $\in (-1, 0)$, has the property that $\mathscr{L}(n^{(1+g)/2}(\tilde{\phi} - \phi)) \to \mathscr{N}(0, 1 - \phi^2)$ as $n \to \infty$.

In the next section we shall discuss the asymptotic normality of serial covariances $\hat{\gamma}_v$ where $v \to \infty$ but $v/n \to 0$ as $n \to \infty$ and state regularity conditions under which this property holds.

8.2 ASYMPTOTIC DISTRIBUTIONS OF SERIAL COVARIANCES OF ORDERS WHICH INCREASE WITH SAMPLE SIZE

The following lemma is a slight modification of the theorem in Berk (1973) and will be useful in proving Theorem 8.2 which states the main result of this article.

Lemma 8.1

Let $\{Y_{t,n}; 1 \leqslant t \leqslant k_n, n \geqslant 1\}$ be a triangular array of random variables (r.v) such that for every n, $\{Y_{t,n}; 1 \leqslant t \leqslant k_n\}$ is a strictly stationary m_n-dependent sequence with, $E(Y_{t,n}) = 0$, $E(Y_{t,n}^2) \leqslant M$ where M is a generic symbol which denotes a finite positive constant,

$$\operatorname{var}\left(\sum_{t=a+1}^{b} Y_{t,n}\right) \leqslant (b-a)M$$

for any a, b, $1 \leqslant a < b < \infty$, and as $n \to \infty$, $k_n \to \infty$, $m_n/k_n \to 0$ and

$$N_n^{-1} \operatorname{var} \sum_{t=1}^{N_n} Y_{t,n} \to \sigma^2 \; (0 < \sigma < \infty)$$

whenever $N_n \to \infty$ as $n \to \infty$. Define

$$S_n = \sum_{t=1}^{k_n} Y_{t,n}.$$

Then

$$\mathscr{L}(S_n/k_n^{1/2}) \to \mathscr{N}(0, \sigma^2). \tag{8.1}$$

Proof

For each n, choose $p_n > 2m_n$ such that $p_n \to \infty$, $p_n/k_n \to 0$ and $m_n/p_n \to 0$ as $n \to \infty$. Define t_n, r_n by $k_n = p_n t_n + r_n$ so that $t_n = [k_n/p_n]$ and $0 \leqslant r_n < p_n$. Now write

$$U_{j,n} = \sum_{t=(j-1)p_n+1}^{jp_n - m_n} Y_{t,n} \tag{8.2}$$

$$V_{j,n} = \sum_{t=jp_n-m_n+1}^{jp_n} Y_{t,n}, \quad (1 \leqslant j \leqslant t_n),$$

$$R_n = \sum_{t=p_n t_n+1}^{k_n} Y_{t,n}.$$

Now note that $k_n/p_n = t_n + r_n/p_n \leqslant t_n + 1$ which implies that $t_n \to \infty$ as $n \to \infty$. Since $\{Y_{tn}\}$ is m_n-dependent and $p_n > 2m_n$, $\{V_{jn};\ 1 \leqslant j \leqslant t_n\}$ and $\{U_{jn};\ 1 \leqslant j \leqslant t_n\}$ are two independent and identically distributed (i.i.d) sequences of r.v.s. Also

$$S_n/k_n^{1/2} = \sum_{j=1}^{t_n} U_{j,n}/k_n^{1/2} + \sum_{j=1}^{t_n} V_{j,n}/k_n^{1/2} + R_n/k_n^{1/2}.$$

Further, by hypothesis,

$$\begin{aligned}\operatorname{var}\left(\sum_{1}^{t_n} V_{j,n}/k_n^{1/2}\right) &= (t_n/k_n)\operatorname{var} \sum_{t=p_n-m_n+1}^{p_n} Y_{t,n}\\ &\leqslant M t_n m_n/k_n\\ &\leqslant M m_n/p_n \to 0.\end{aligned} \tag{8.3}$$

as $n \to \infty$, and

$$\begin{aligned}\operatorname{var}(R_n/k_n^{1/2}) &\leqslant M r_n/k_n\\ &\leqslant M/t_n \to 0,\end{aligned} \tag{8.4}$$

as $n \to \infty$. Again it is easy to see from the conditions in lemma 8.1 that

$$\operatorname{var}\left(\sum_{j=1}^{t_n} U_{j,n}/k_n^{1/2}\right) \to \sigma^2, \tag{8.5}$$

as $n \to \infty$. Since $\{U_{j,n};\ 1 \leqslant j \leqslant t_n\}$ is an i.i.d. sequence

$$\mathscr{L}\left(\sum_{j=1}^{t_n} U_{j,n}/\sigma_n t_n^{1/2}\right) \to \mathscr{N}(0, 1), \tag{8.6}$$

as $n \to \infty$ where $\sigma_n^2 = \operatorname{var}(U_{1,n}) \to \infty$ as $n \to \infty$. The result (2.1) will now follow easily from (2.3)–(2.6).

We now prove the following.

Theorem 8.2

Let $\{X_t; t \in \mathbf{Z}\}$ be a linear process defined by

$$X_t = \sum_{j=-\infty}^{\infty} \beta_j \eta_{t-j}$$

where $\{\eta_t\}$ is an i.i.d. sequence of r.vs. with $E(\eta_1) = 0$ $E(\eta_1^2) = \sigma^2 (0 < \sigma < \infty)$, $E(\eta_1^4) = \kappa_4 + 3\sigma^4 < \infty$ and

$$\sum_{j=-\infty}^{\infty} |j\beta_j| < \infty.$$

Let $k = k_n \to \infty$, but $k/n \to 0$, as $n \to \infty$. Then for any fixed $p \geqslant 1$, $n^{1/2}(\hat{\gamma}_k - \gamma_k)$, $\ldots, n^{1/2}(\hat{\gamma}_{k+p} - \gamma_{k+p})$ have, asymptotically, (as $n \to \infty$), a $(p+1)$ variate normal distribution with mean vector zero and covariance matrix $\boldsymbol{\Lambda} = [\lambda_{ij}]$ where

$$\lambda_{ij} = \sum_{s=-\infty}^{\infty} \gamma_s \gamma_{s+i-j}. \tag{8.7}$$

Proof

Let $\alpha_0, \ldots, \alpha_p$ be arbitrary real numbers and consider the quantity

$$T_n = n^{1/2} \sum_{u=0}^{p} \alpha_u (\hat{\gamma}_{k+u} - \gamma_{k+u}).$$

Note that we can write

$$T_n = \sum_{t=1}^{N} Y_{t,k}/n^{1/2} + R_n/n^{1/2},$$

where

$$Y_{t,k} = \sum_{u=0}^{p} \alpha_u \tilde{Y}_{t,ku}, \; \tilde{Y}_{t,k+u} = X_t X_{t+k+u} - \gamma_{k+u}, \; N = n - k - p$$

and

$$R_n = \sum_{t=N+1}^{N+p} \sum_{u=0}^{N+p-t} \alpha_u \tilde{Y}_{t,k+u} - \sum_{u=0}^{p} (k+u)\alpha_u \gamma_{k+u}.$$

It is easy to see that

$$\sum_{q=-\infty}^{\infty} |q\gamma_q| \leqslant M \sum_{j=-\infty}^{\infty} |\beta_j| \sum_{j=-\infty}^{\infty} |j\beta_j| < \infty$$

which implies that the second part of R_n tends to zero as $n \to \infty$. The absolute value of the first part of R_n has a finite expectation and $N/n \to 1$ as $n \to \infty$. Therefore, the asymptotic distribution of T_n is the same as that of

$$N^{-1/2} \sum_{t=1}^{N} Y_{t,k}.$$

Now

$$\operatorname{var}\left(N^{-1/2} \sum_{t=1}^{N} Y_{t,k} \right) = \sum_{|s| \leq N} (1 - |s|/N) \gamma_s^{(k)}(Y), \tag{8.8}$$

where for $s \geqslant 0$, $\gamma_{-s}^{(k)}(Y) = \gamma_s^{(k)}(Y)$ and

$$\gamma_s^{(k)}(Y) = E(Y_{t,k} Y_{t+s,k}) = \sum_{u,v=0}^{p} \alpha_u \alpha_v (\gamma_s \gamma_{s+v-u} + \gamma_{k+s+v} \gamma_{k-s+u}$$

$$+ \kappa_4 \sum_{j=-\infty}^{\infty} \beta_j \beta_{j+s} \beta_{j+k+u} \beta_{j+k+s+v}) \tag{8.9}$$

Let

$$\tilde{S}_q = \sum_{i=-\infty}^{\infty} |\gamma_i \gamma_{q-i}|.$$

Then

$$\tilde{S}_{-q} = \tilde{S}_q, q \geq 0, \left| \sum_{s=-\infty}^{\infty} \gamma_s \gamma_{s+v-u} \right| \leq \tilde{S}_{v-u},$$

$$\left| \sum_{s=-\infty}^{\infty} \gamma_{k+s+v} \gamma_{k-s+u} \right| = \left| \sum_{w=-\infty}^{\infty} \gamma_w \gamma_{2k+u+v-2w} \right| \leq \tilde{S}_{2k+u+v} \text{ and}$$

$$\left| \sum_{s=-\infty}^{\infty} \sum_{j=-\infty}^{\infty} \beta_j \beta_{j+s} \beta_{j+k+u} \beta_{j+k+s+v} \right| \leq \sum_{j=-\infty}^{\infty} \left| \beta_j \beta_{j+k+u} \right| \sum_{j=-\infty}^{\infty} \left| \beta_j \beta_{j+k+v} \right|.$$

Again

$$\sum_{q=-\infty}^{\infty} \sum_{j=-\infty}^{\infty} |\beta_j \beta_{j+q}| = \left(\sum_{j=-\infty}^{\infty} |\beta_j| \right)^2 < \infty$$

which implies that

$$\sum_{j=-\infty}^{\infty} |\beta_j \beta_{j+q}| \to 0 \text{ as } q \to \infty.$$

Also

$$\sum_{q=-\infty}^{\infty} |\tilde{S}_q| \leq \left(\sum_{j=-\infty}^{\infty} |\gamma_j| \right)^2 \leq \left(\sum_{j=-\infty}^{\infty} |\beta_j| \right)^4 < \infty$$

implies that $\tilde{S}_q \to 0$ as $q \to \infty$. Since $k \to \infty$ as $n \to \infty$ it follows immediately that

$$\operatorname{var}(T_n) \to \sum \alpha_u \alpha_v \lambda_{u,v}, \tag{8.10}$$

as $n \to \infty$, where

$$\lambda_{u,v} = \sum_{s=-\infty}^{\infty} \gamma_s \gamma_{s+v-u}.$$

Now fix $m \geq 1$, and write

$$W_{t,k} = \sum_{u=0}^{p} \alpha_u X_{t,m} X_{t+k+u,m}, \; V_{t,k} = \sum_{u=0}^{p} \alpha_u (X_{t,m} X^*_{t+k+u,m} - \gamma_{m,k+u})$$

and

$$Z_{t,k} = \sum_{u=0}^{p} \alpha_u (X^*_{t,m} X_{t+k+u} - \gamma^*_{m,k+u})$$

where

$$X_{t,m} = \sum_{|j| \le m} \beta_j \eta_{t-j}, X^*_{t,m} = X_t - X_{t,m},$$

$$\gamma_{m,u} = \sigma^2 \sum_{j=\max(m-u+1,-m)}^{m} \beta_j \beta_{j+u} = E(X_{t,m} X^*_{t+u,m})$$

and

$$\gamma^*_{m,u} = \sigma^2_\eta \sum_{|j|>m} \beta_j \beta_{j+u} \text{ for } u \geqslant 0.$$

From now on we shall assume that n is so large that $k > 2m$, so that $\gamma_{m,k+u} + \gamma^*_{m,k+u} = \gamma_{k+u}$ for $u \geqslant 0$ and $Y_{t,k} = W_{t,k} + V_{t,k} + Z_{t,k}$. Now note that $\{W_{t,k}\}$ is m_n-dependent and strictly stationary with $m_n = k + p + 2m$. Also

$$\operatorname{var}\left(N^{-1/2} \sum_{t=1}^{N} W_{t,k}\right) = \sum_{|s| \leqslant N} (1 - |s|/N) \gamma_s^{(k)}(W), \tag{8.11}$$

where for $s \geqslant 0$, $\gamma_{-s}^{(k)}(W) = \gamma_s^{(k)}(W)$,

$$\begin{aligned}
\gamma_s^{(k)}(W) &= E(W_{t,k} W_{t+s,k}) \\
&= \sum_{u,v=0}^{p} \alpha_u \alpha_v \left(\tau_{m,s} \tau_{m,s+v-u} + \kappa_4 \sum_A \beta_j \beta_{j+s} \beta_{j+k+u} \beta_{j+k+s+v} \right) \\
&= \sum_{u,v=0}^{p} \alpha_u \alpha_v \tau_{m,s} \tau_{m,s+v-u},
\end{aligned} \tag{8.12}$$

$$\tau_{m,u} = \sigma^2 \sum_{j=-m}^{m-|u|} \beta_j \beta_{j+|u|}, \ -\infty < u < \infty,$$

and $\tau_{m,s} = 0$ whenever $|u| > 2m$. This follows because $s \geqslant 0$ and $k > 2m$ imply that $A = \{j{:}|j| \leqslant m, |j+s| \leqslant m, |j+k+u) \leqslant m, |j+k+s+u| \leqslant m\}$ is an empty set. Again for any a, b, $1 \leqslant a < b < \infty$,

$$\operatorname{var}\left(\sum_{t=a+1}^{b} W_{t,k}\right) \leqslant (b-a) \sum_{u,v=0}^{p} |\alpha_u \alpha_v \tau_{m,s} \tau_{m,s+v-u}| \leqslant M(b-a)$$

where the summation is over $0 \leqslant u, v \leqslant p$ and $|s| \leqslant 2m$, $|s+v-u| \leqslant 2m$. Also, $m_n/N = (k+p+2m)/(n-k-p) \to 0$ as $n \to \infty$. Therefore, by Lemma 8.1 and fact that $N/n \to 1$ as $n \to \infty$ we have

$$\mathscr{L}\left(n^{-1/2} \sum_{t=1}^{N} W_{t,k}\right) \to \mathscr{N}(0, \tau_m^2), \tag{8.13}$$

as $n \to \infty$, where

$$\tau_m^2 = \sum_{u,v=0}^{p} \alpha_u \alpha_v \tau_{m,s} \tau_{m,s+v-u}$$

and the summation is over $0 \leqslant u, v \leqslant p$, $|s| \leqslant 2m$ and $|s+v-u| \leqslant 2m$. Note that as $m \to \infty$ $\tau_m^2 \to$

$$\sum_{s=-\infty}^{\infty} \sum_{u,v=0}^{p} \alpha_u \alpha_v \gamma_s \gamma_{s+v-u} = \sum_{u,v=0}^{p} \alpha_u \alpha_v \lambda_{u,v}.$$

Let us now write

$$V_n = N^{-1/2} \sum_{t=1}^{N} V_{t,k}.$$

Then it is easy to see that $E(V_n) = 0$ and

$$E(V_n^2) = \sum_{|s| \leqslant N} (1 - |s|/N) \gamma_s^{(k)}(V), \tag{8.14}$$

where for $s \geqslant 0$, $\gamma_{-s}^{(k)}(V) = \gamma_s^{(k)}(V)$,

$$\begin{aligned} \gamma_s^{(k)}(V) &= E(V_{t,k} V_{t+s,k}) \\ &= \sum_{u,v=0}^{p} \alpha_u \alpha_v \Bigg(\tau_{m,s} \pi_{m,s+v-u} + \gamma_{m,k+s+v} \theta_{m,k-s+u} \\ &\quad + \kappa_4 \sum_{-m}^{m-s} \beta_j \beta_{j+s} \beta_{j+k+u} \beta_{j+k+s+v} \Bigg) \qquad (8.15) \\ &= \sum \alpha_u \alpha_v \gamma_{m,k+s+v} \theta_{m,k-s+u}, \quad \text{if } s > 2m, \end{aligned}$$

for $u \geqslant 0$

$$\pi_{m,-u} = \pi_{m,u} = \sigma^2 \sum_{Q} \beta_j \beta_{j+|u|},$$

$Q = \{j{:}j < -m - |u|,\ m - |u| < j < -m,\ j > m\}$, if $|u| \geqslant 2m + 2$ and $= \{j{:}j < -m - |u|, j > m\}$, if $|u| < 2m + 2$ and $\theta_{m,u} = \gamma_{m,u}$ if $u \geqslant 0$ and $=$

$$\sigma^2 \sum_{j=-m-|u|}^{\min(-m-1, m-|u|)} \beta_j \beta_{j+|u|},$$

if $u < 0$. Routine analysis indicates that $|\pi_{m,u}| \leqslant MM_m^2$ for $u \in (-\infty, \infty)$ and

$$\begin{aligned} &\left| \sum_{s=0}^{N} \tau_{m,s} \pi_{m,s+v-u} \right| \\ &= \left| \sum_{s=0}^{2m} \tau_{m,s} \pi_{m,s+v-u} \right| \qquad (8.16) \\ &\leqslant MM_m^2 \leqslant MM_m \end{aligned}$$

where

$$M_m = \sum_{|j|>m} |\beta_j| \to 0 \text{ as } m \to \infty.$$

Also for $u \in (-\infty, \infty)$,

$$\begin{aligned}
|\theta_{m,u}| &\leqslant (E(X_{t,m}^2)E(X_{t+u,m}^{*2}))^{1/2} \\
&\leqslant M\left(\sum_{|j|>m} \beta_j^2\right)^{1/2} \leqslant MM_m \text{ and hence}
\end{aligned}$$

$$\left|\sum_{s=0}^{N} \gamma_{m,k+s+v}\theta_{m,k-s+u}\right| \tag{8.17}$$

$$\begin{aligned}
&\leqslant MM_m \sum_{l=k+v}^{k+v+N} |\gamma_{m,l}| \\
&\leqslant MM_m \sum_{j=k+v-m}^{\infty} |\beta_j| \to 0
\end{aligned}$$

as $n \to \infty$. And, finally

$$\left|\sum_{j=0}^{2m} \sum_{j=-m}^{m-s} \beta_j\beta_{j+s}\beta_{j+k+u}\beta_{j+k+s+v}\right|$$

$$\begin{aligned}
&\leqslant \sum_{j=-m}^{m} \sum_{s=0}^{m-j} |\beta_j||\beta_{j+s}||\beta_{j+k+u}||\beta_{j+k+s+v}| \\
&= \sum_{j=-m}^{m} |\beta_j||\beta_{j+k+u}| \sum_{l=j}^{m} |\beta_l||\beta_{l+k+v}| \\
&\leqslant \sum_{k+u-m}^{k+u+m} |\beta_j| \sum_{k+v-m}^{k+v+m} |\beta_j| \to 0
\end{aligned} \tag{8.18}$$

as $n \to \infty$. We now use the relations (8.14)–(8.18) and conclude that

$$\lim_{n\to 0} E(V_n^2) \leqslant MM_m. \tag{8.19}$$

Similarly we can establish that

$$\lim_{n\to 0} E(Z_n^2) \leqslant MM_m. \tag{8.20}$$

If, now, we combine (8.12), (8.19), (8.20) and use corollary 7.7.1 in Anderson (1970) then we have

$$\mathscr{L}(T_n) \to \mathscr{N}\left(0, \sum_{u,v=0}^{p} \alpha_u\alpha_v\lambda_{u,v}\right). \tag{8.21}$$

Since (8.21) holds for any real $\alpha_0, \ldots, \alpha_p$ the result (8.7) follows immediately.

Corollary 8.3

Let the conditions of theorem 8.2 hold and let l be a fixed positive integer, let $v_j (1 \leqslant j \leqslant l)$ be such that $|v_j \rightarrow v_{j'}|$ are fixed positive integers for $1 \leqslant j, j' \leqslant l$. $v_j \rightarrow \infty$ but $v_j/n \rightarrow 0$ as $n \rightarrow \infty$. Then $n^{1/2}(\hat{\gamma}_{v_j} - \gamma_{v_j})(1 \leqslant j \leqslant l)$ have, asymptotically, (as n $\rightarrow \infty$), an l variate normal distribution with means zero and covariance matrix $\mathbf{\Lambda} = [\lambda_{ij}]$ where

$$\lambda_{ij} = \sum_{s=-\infty}^{\infty} \gamma_s \gamma_{s+v_i-v_j}. \tag{8.22}$$

Proof

The result (8.22) follows easily from (8.7).

REFERENCES

Anderson, T.W. (1970) *The Statistical Analysis of Time Series.* John Wiley & Sons, Inc., New York.

Berk, K.N. (1973) A central limit theorem for *m*-dependent random variables with unbounded *m*. *Ann. Probab.*, **1**, 352–354.

Brillinger, D.R. (1969) Asymptotic properties of spectral estimates of second order. *Biometrika*, **56**, 375–390.

Chanda, K.C. (1992) Asymptotic properties of estimators for autoregressive models with errors in variables, submitted for publication.

9

Exact maximum likelihood estimation for extended ARIMA models

R. Azrak and G. Mélard

9.1 INTRODUCTION

Several extensions of ARIMA models have been considered in recent years, including:

(a) the use of time-dependent coefficients in the autoregressive and moving average polynomials (Quenouille, 1957; Whittle, 1965; Abdrabbo and Priestley, 1967; Miller, 1968 and 1969; Subba Rao, 1970; Mélard and Kiehm, 1981; Tyssedal and Tjøstheim, 1982; Grillenzoni, 1990);
(b) various types of interventions, including the usual Box and Tiao (1975) formulation and the innovational interventions (Fox, 1972) but also interventions acting on the scale (Mélard, 1981a; Tsay, 1988);
(c) additive (level) or multiplicative (scale) trend (Mélard, 1977);
(d) built-in deterministic seasonal components on the variable (Abraham and Box, 1978) or on the innovation (Mélard, 1981b);
(e) variable transformations (Box and Cox, 1964).

The purpose of that model is to encompass several deterministic variations with respect to time in the framework of the usual stochastic ARIMA models. Other extensions not explicitly considered in this paper are ARMA models with GARCH errors (Bollerslev, 1986), threshold AR models (Tong, 1983), bilinear models (Subba Rao, 1981), and fractional differencing ARIMA models (Granger and Joyeux, 1980). It should be noted, however, that some of these extensions can be handled using the same approach. For instance, threshold ARMA models (Mélard and Roy, 1988) can be seen as time-dependent ARMA models. Other approaches for time-dependent models include spectral density estimation (Priestley 1981, 1988), recursive estimation

(Ljung and Söderström, 1983; Young, 1984) and models with random coefficients (Nicholls and Quinn, 1982; Bougerol, 1993).

Motivations for the extended ARIMA model which is used here have already been discussed elsewhere (Mélard, 1982a, 1985a). An illustration has already been provided (Mélard, 1985b). The estimation procedure was however limited to the conditional least squares approach, generalizing the approach of Box and Jenkins (1976). In this paper, an algorithm for the evaluation of the exact likelihood function is described, in the case where the innovation process is Gaussian.

9.2 THE MODEL

The following notations will be used:

- $\{z_t; t \in Z\}$ is the stochastic process which generates the time series $\{z_t; t = 1, \ldots, n\}$;
- $\{w_t; t \in Z\}$ is a second order stochastic process derived from $\{z_t; t \in Z\}$—it is supposed to be Gaussian, but its mean is not necessarily constant;
- $\{b; t \in Z)$ is the generalized innovation process—the innovations b_t are assumed to be normally distributed independent random variables, but do not necessarily constitute a stationary stochastic process with a zero mean;
- $\{a_t; t \in Z\}$ is a Gaussian white noise process in the strict sense with mean zero and variance σ^2;
- $y_t^I, y_t^D, y_t^W, y_t^M, \mu_t$, and μ'_t are arbitrary functions of time;
- f_t, g_t, y_t^S and y_t^V are strictly positive functions of time;
- m_t and m'_t are periodic functions of time;
- ϕ_{ti} and θ_{tj} are either constants or functions of time;
- $C_\lambda(.)$ is an instantaneous transformation which depends on an unknown parameter set λ;
- θ_0 is a constant;
- p, q, d and D are positive integers, and s is a strictly positive integer;
- ∇ is the regular difference operator;
- ∇_s is the seasonal difference operator with periodicity s;
- B is the backshift operator such that $B \bullet_t = \bullet_{t-1}$—it is assumed that the operator acts only on the right, e.g. $f_t B g_t = f_t g_{t-1}$.

All the functions of time included in the model have a specified analytical expression depending on a finite number of unknown parameters.

Using these notations, the extended ARIMA model for a time series is defined by the equation

$$\left(1 - \sum_{i=1}^{p} \phi_{ti} B^i\right) \frac{\nabla^d \nabla_s^D \{C_\lambda(z_t - y_t^I)/f_t\} - (y_t^D + \mu_t + m_t)}{y_t^V}$$

$$= \theta_0 + y_t^W + \left(1 - \sum_{j=1}^{q} \theta_{ij} B^j\right)[g_t\{y_t^S a_t + (y_t^M + \mu_t' + m_t')\}]. \tag{9.1}$$

This is slightly different from the model considered by Mélard (1982a, 1985b). It can be subdivided into three submodels (SM). There are:

- the variable SM

$$\frac{\nabla^d \nabla_s^D \{C_\lambda(z_t - y_t^I)/f_t\} - (y_t^D + \mu_t + m_t)}{y_t^V} = w_t; \tag{9.2}$$

- the extended ARMA SM

$$w_t - \sum_{i=1}^{p} \phi_{ti} w_{t-i} = \theta_0 + y_t^W + b_t - \sum_{j=1}^{q} \theta_{tj} b_{t-j}; \tag{9.3}$$

- the innovation SM

$$b_t = g_t\{y_t^S a_t + (y_t^M + \mu_t' + m_t')\}. \tag{9.4}$$

Note that solving (9.4) for a_t gives

$$a_t = \frac{(b_t/g_t) - (y_t^M + \mu_t' + m_t')}{y_t^S}. \tag{9.5}$$

The motivations behind the extended ARIMA model are the following. Using the data z_t as a starting point, the variable SM consists of the following scheme.

- The function y_t^I is subtracted; it is the intervention, in the sense of Box and Tiao (1975), or the intervention on the variable, in the sense of Tsay (1988); the y_t^I can be equal to zero at times where there is no intervention effect.
- An instantaneous nonlinear transformation $C_\lambda(\cdot)$ is applied (see Box and Cox, 1964); in this paper it will be illustrated by the power family but this is not restrictive.
- A division by the deterministic function f_t is performed; that function represents a multiplicative trend (Mélard, 1977).
- The regular differences and the seasonal differences are applied in order to obtain a second order (possibly non-stationary) stochastic process.
- The function y_t^D is subtracted; it is the intervention on the data after differences. For a series which requires differencing, a level change can be represented by a pulse on y_t^D (Mélard, 1981a).
- The deterministic trend μ_t is subtracted; it is either a constant, or a function of time which represents an additive trend.
- The deterministic seasonal component m_t is subtracted; it is a periodic function of time, with a sum equal to zero on an interval of length equal to the period (Abraham and Box, 1978).

- A division by y_t^V is performed; it is an intervention function acting on the standard deviation of the variable (Hipel and McLeod, 1980).

The extended ARMA SM consists of the following stages.

- A deterministic function y_t^W is taken into account; it represents a special type of intervention which is added to the constant θ_0 used by Box and Jenkins (1976) in their ARMA model.
- A time-invariant ARMA or an evolving ARMA filter (Priestley, 1981) is applied. Note that if at least one coefficient depends on time, or if y_t^S and g_t are not identically equal to 1, then the ARMA filter is evolving.

Finally, the innovation SM consists of the following steps.

- A division of the residuals by the deterministic function g_t is performed; it is a trend on the standard deviation of the innovations; the purpose of this operation is to stabilize the innovation variance (Herbst, 1963; Mélard 1977).
- y_t^M is subtracted, which is the intervention function acting on the innovation mean (Fox, 1972; Mélard, 1981a; Tsay, 1988).
- μ_t' is subtracted, which represents the trend of the innovation mean (Mélard (1981b).
- m_t' is subtracted, which is a seasonal component of the innovations.
- A division by y_t^S is performed; y_t^S is an intervention function acting on the standard deviation of the innovations in order to represent, for example, a troubled time interval during which the scatter is higher (Mélard, 1981a; Wichern *et al.*, 1976).

Parameters to be estimated are usually included in the functions $y_t^I, y_t^D, y_t^W, y_t^M, \mu_t, \mu_t', f_t, g_t, y_t^S, y_t^V, m_t, m_t', \phi_{ti}, \phi_{tj}$, and $C_\lambda(\cdot)$; ϕ_i, θ_j, and σ^2 themselves are parameters.

The usual ARIMA model in the sense of Box and Jenkins is a special case when

$$y_t^I = y_t^D = y_t^W = y_t^M = \mu_t = \mu_t' = m_t = m_t' = 0,$$

$$f_t = g_t = y_t^S = y_t^V = 1,$$

$$\phi_{ti} = \phi_i (i = 1, \ldots, p),$$

$$\theta_{tj} = \theta_j (j = 1, \ldots, q),$$

for all t, and $C_\lambda(\cdot)$ is the identity transformation.

The order in which the various elements are introduced is justified by several considerations:

(a) not all extensions are expected to be used together, as a matter of fact, most of the time, only one or the other component will be used;
(b) the data are first adjusted for the intervention y_t^I, transformed, and divided by the multiplicative trend f_t;

(c) when the series is homogeneous, differencing is used to stabilize the level changes; at that point, the additive components (trend μ_t, seasonal m_t and the intervention y_t^D) are removed;
(d) a multiplicative intervention component y_t^V is then taken into account, since local variations on the scale can be better perceived when the series is stable;
(e) comparison of (9.2) and (9.5) reveals a strong analogy between the variable and the innovation submodels, where $\nabla^d \nabla_s^D \{C_\lambda(z_t - y_t^I)/f_t\}$ is replaced by b_t/g_t.

In this paper, pseudo-maximum likelihood parameter estimation is considered instead of the conditional least-squares method in Mélard (1982a). The exact likelihood function is computed as if the innovation process were Gaussian. The process $\{z_t; t \in Z\}$ which generates the time series is not Gaussian in general. It may not even be a second-order process. On the contrary, the derived process $\{w_t; t \in Z\}$ is a Gaussian process, although it is non-stationary, since the mean is not constant and the covariance between w_t and w_s does not depend only on $t - s$.

In order to give a better understanding of the problem, it is useful to consider a simpler form of the model

$$\left(1 - \sum_{i=1}^{p} \phi_{ti} B^i\right) F_t(z_t) = \xi_t + \left(1 - \sum_{j=1}^{q} \theta_{tj} B^j\right) \{\gamma_t a_t + \alpha_t\}, \tag{9.6}$$

where

$$F_t(z_t) = \frac{\nabla^d \nabla_s^D \{C_\lambda(z_t - y_t^I)/f_t\} - (y_t^D + \mu_t + m_t)}{y_t^V} \tag{9.7}$$

$$\xi_t = \theta_0 + y_t^W \tag{9.8}$$

$$\gamma_t = y_t^S g_t \tag{9.9}$$

$$\alpha_t = g_t(y_t^M + \mu_t' + m_t'). \tag{9.10}$$

The three submodels can then be written:

- the variable SM

$$F_t(z_t) = w_t; \tag{9.11}$$

- the extended ARMA SM

$$\left(1 - \sum_{i=1}^{p} \phi_{ti} B^i\right) w_t = \xi_t + \left(1 - \sum_{j=1}^{q} \theta_{tj} B^j\right) b_t; \tag{9.12}$$

- the innovation SM

$$b_t = \gamma_t a_t + \alpha_t, \tag{9.13}$$

where $\gamma_t > 0$ for all t, and $\{a_t; t \in Z\}$ is a Gaussian white noise process with mean zero and variance σ^2.

The non linear transformation F_t and the deterministic sequences ϕ_{ti}, θ_{tj}, ξ_t, γ_t, and α_t depend on a finite number of parameters so that the model can be specified by a parameter vector of finite dimension denoted by $\boldsymbol{\beta}$ and the variance σ^2 of the white noise. Let $\delta = d + Ds$ be the number of observations lost by differencing. These observations are stored in a vector $\boldsymbol{z}_0 = (z_1, z_2, \ldots, z_\delta)^{\mathrm{T}}$, where ^{T} denotes transposition. The unconditional likelihood function which will be computed is nevertheless conditional on the $\delta = d + Ds$ first observations $\boldsymbol{z}_0$.

In the sequel of the paper, we express the likelihood function of $\boldsymbol{z} = (z_{\delta 1}, \ldots, z_n)^{\mathrm{T}}$ conditional on $\boldsymbol{z}_0$ by using the density of $\boldsymbol{w} = (w_{\delta+1}, \ldots, w_n)^{\mathrm{T}}$. This implies some additional assumptions on F_t (section 9.3) and an adequate treatment of a Jacobian (section 9.6). The process $\{w_t; t \in Z\}$ satisfies a time dependent ARMA model (9.12) where the b_t are independent normal random variables with mean α_t and variance $\gamma_t^2 \sigma^2$. Hence the problem reduces to finding the joint density of $n - \delta$ consecutive values of a time dependent ARMA process (section 9.5). For this, it is necessary to centre the process and thus to determine the mean of w_t, for all t (section 9.4).

9.3 THE VARIABLE SUBMODEL

Parameter estimation by the maximum likelihood method requires the computation of the exact likelihood function $L(\boldsymbol{\beta}, \sigma^2; \boldsymbol{z}/\boldsymbol{z}_0)$ which is the density of $\boldsymbol{z} = (z_{\delta+1}, \ldots, z_n)^{\mathrm{T}}$ conditional on $\boldsymbol{z}_0$. Using (9.2), it is equal to the density of $\boldsymbol{w} = (w_{\delta+1}, \ldots, w_n)^{\mathrm{T}}$, $f(\boldsymbol{w}; \boldsymbol{\beta}, \sigma^2)$ multiplied by the Jacobian of the transformation. Since w_t depends only on z_s for $s \leqslant t$, the Jacobian matrix is triangular, and the diagonal elements are equal to $(\partial w_t)/(\partial z_t)$. Hence the Jacobian is

$$J(\boldsymbol{\beta}) = \prod_{t=\delta+1}^{n} \frac{\partial w_t}{\partial z_t} = \prod_{t=\delta+1}^{n} \frac{\partial F_t(z_t)}{\partial z_t} = \prod_{t=\delta+1}^{n} \frac{1}{f_t y_t^V} \left\{ \frac{\partial C^\lambda(z_t - y_t^I)}{\partial z_t} \right\}. \tag{9.14}$$

The function f_t will be restricted by the condition

$$\left(\sum_{t=\delta-1}^{n} f_t \right)^{1/(n-\delta)} = 1 \tag{9.15}$$

which means that its geometric mean over the interval from $\delta + 1$ to n is equal to 1. Similarly, the function y_t^V is subject to the constraint

$$\left(\prod_{t=\delta+1}^{n} y_t^V \right)^{1/(n-\delta)} = 1. \tag{9.16}$$

In the case of the power transformation

$$C_\lambda(z_t) = \begin{cases} \dfrac{z_t^\lambda - 1}{\lambda} & \lambda \neq 0 \\ \log z_t & \lambda = 0, \end{cases} \tag{9.17}$$

we have

$$\frac{\partial C_\lambda(z_t - y_t^I)}{\partial z_t} = (z_t - y_t^I)^{\lambda - 1}. \tag{9.18}$$

Let G be the geometric mean of $z_t - y_t^I$, for $t = \delta + 1, \ldots, n$. Hence

$$L(\boldsymbol{\beta}, \sigma^2; z/z_0) = J(\boldsymbol{\beta}) f(\boldsymbol{w}; \boldsymbol{\beta}, \delta^2) = G^{(n-\delta)(\lambda-1)} f(\boldsymbol{w}; \boldsymbol{\beta}, \sigma^2). \tag{9.19}$$

9.4 THE INNOVATION SUBMODEL

Since $\{w_t; t \in Z\}$ is a (non-stationary) Gaussian process, the distribution of $\boldsymbol{w}$ is multivariate normal with a mean vector denoted by $\boldsymbol{M}^w = E(\boldsymbol{w}) = (M_{\delta+1}^w, \ldots, M_n^w)^{\mathrm{T}}$, and a variance-covariance matrix V. Its density has the form

$$f(\boldsymbol{w}; \boldsymbol{\beta}, \sigma^2) = (2\pi)^{-(n-\delta)/2} (\det V)^{-1/2} \exp\left\{ -\frac{1}{2} (\boldsymbol{w} - \boldsymbol{M}^w)^{\mathrm{T}} V^{-1} (\boldsymbol{w} - \boldsymbol{M}^w) \right\}. \tag{9.20}$$

In this section, we consider the computation of $\boldsymbol{M}^w$ which relies mainly on the innovation submodel. Since $E(a_t) = 0$, we have from (9.13) and (9.12)

$$M_t^b = E\{b_t\} = \alpha_t, \tag{9.21}$$

$$M_t^w = E\{w_t\} = \sum_{i=1}^{p} \phi_{ti} M_{t-i}^w + M_t^b - \sum_{j=1}^{q} \theta_{tj} M_{t-j}^b + \xi_t. \tag{9.22}$$

To determine the initial M_t^w in order to initialize the recurrence (9.22), the following assumptions are made:

$$\phi_{ti} = \phi_{\delta+1,i} = \phi_i (i = 1, \ldots, p) \text{ for all } t \leqslant \delta;$$

$$\theta_{tj} = \theta_{\delta+1,j} = \theta_j (j = 1, \ldots, q) \text{ for all } t \leqslant \delta; \tag{H1}$$

$$\phi(B) = 1 - \sum_{i=1}^{p} \phi_i B^i \text{ has all its zeroes outside of the unit circle;} \tag{H2}$$

$$\theta(B) = 1 - \sum_{j=1}^{q} \theta_j B^j \text{ has all its zeroes outside of the unit circle;} \tag{H3}$$

$$y_t^w = y_t^M = 0 \text{ for } t \leqslant \delta; \tag{H4}$$

$$g_t = g_{\delta+1}, \ \mu_t = \mu_{\delta+1}, \text{ and } \mu_t' = \mu_{\delta+1}' \text{ for } t \leqslant \delta. \tag{H5}$$

By (H1) and (H4) (9.22) implies

$$\phi(B) M_t^w = \theta(B) M_t^b + \theta_0 \tag{9.23}$$

for $t \leqslant \delta$. In addition, (H5) and (H2) guarantee that the process $\tilde{w}_t = w_t - M_t^w$ has a stationary behaviour at least for $t \leqslant \delta$. By (H3), the process is a (possibly non-stationary) second-order invertible process in the sense of Hallin (1980).

Let us denote

$$\psi(B) = \phi^{-1}(B)\theta(B) = \sum_{h=0}^{\infty} \psi_h B^h. \tag{9.24}$$

Hence

$$M_t^w = \psi(B)M_t^b + \phi^{-1}(B)\theta_0. \tag{9.25}$$

By (H4) and (H5):

$$M_t^b = (\mu'_{\delta+1} + m'_t)g_{\delta+1} \tag{9.26}$$

for $t \leqslant \delta$. But m'_t is a periodic function of t, with period s, which implies that, for $t \leqslant \delta$, M_t^b is also a periodic function with period s, which can be computed for all t.

Noting that the coefficients

$$\psi_h = \sum_{i=0}^{\infty} \psi_{is+h} \tag{9.27}$$

$h = 0, 1, \ldots, s-1$, can be computed in a finite number of steps, we write

$$\psi(B)M_t^b = \sum_{h=0}^{s-1} \psi_h M_{t-h}^b, \tag{9.28}$$

and since

$$\phi^{-1}(B)\theta_0 = \frac{\theta_0}{\phi(1)}, \tag{9.29}$$

we obtain

$$M_t^w = \sum_{h=0}^{s-1} \psi_h M_{t-h}^b + \frac{\theta_0}{\phi(1)} \tag{9.30}$$

for $t = \delta + 1 - p, \ldots, \delta$.

At that point, the recurrence relation (9.22) can be used for $t = \delta + 1, \ldots, n$.

9.5 THE EXTENDED ARMA SUBMODEL

There remains to evaluate det V and the quadratic form $(\boldsymbol{w} - \boldsymbol{M}^w)^{\mathrm{T}} V^{-1}(\boldsymbol{w} - \boldsymbol{M}^w)$. In the case of a time-invariant model (e.g. an ARMA model), Schweppe (1965) has shown that the computation can be expressed in terms of the sample normalized process $\{\tilde{a}_t\}$, which are simply the innovations of a zero-mean process $\{\tilde{w}_t\}$ evaluated in terms of the observations, and normalized in such a way that their variance is σ^2. In the original presentation the $\tilde{a}_t$ were obtained by means of the Kalman filter. Ansley (1979) has developed an algorithm based on a Cholesky factorization for band matrices which is computationally nearly equivalent. For stationary processes, fast Kalman algorithms (also called Chandrasekar-type, Morf *et al.*, 1974) have been described (Pearlman, 1980) and implemented (Mélard, 1984).

If the model is not strictly time-invariant, the arguments of Schweppe can be generalized. The Kalman filter algorithm remains valid (Ansley and Kohn, 1983) provided that some assumptions, like (H1)–(H3) above, are used (Mélard, 1985a). The Cholesky factorization approach can also be adapted (Mélard, 1982b).

In our problem $\tilde{w}_t = w_t - M_t^w$, $t > \delta$, where the M_t^w have been computed in the previous section. Two cases are considered.

(a) Time-dependent process. If the assumptions (H1)–(H5) are fulfilled, $\{\tilde{w}_t; t > \delta\}$ is a zero-mean time-dependent (or evolving) ARMA process with a stationary behaviour in the past (before data are available). Note that these assumptions provide a natural way to extend the model in the past but alternative assumptions can be considered. For example, if the coefficients ϕ_{ti} and θ_{tj} are periodic functions of time, other assumptions will appear more natural (Li and Hui, 1988).
(b) Time-invariant process. If, beside (H1)–(H5), the assumptions

$$f_t = g_t = y_t^S = y_t^V = 1 \text{ for all } t \tag{H6}$$

$$\phi_{ti} = \phi_i (i = 1, \ldots, p) \quad \text{and} \quad \theta_{tj} = \theta_j (j = 1, \ldots, q) \text{ for all } t \tag{H7}$$

are fulfilled, then the $\{\tilde{w}_t; t > \delta\}$ consist in a zero-mean (time-invariant) ARMA process. This case is well known and will not be discussed.

In the time-dependent case, everything is based on the Wold–Cramér decomposition (Cramér, 1961; Rissanen and Barbosa, 1969) or innovation representation of the process for $t \geqslant \delta$. Since the algorithmic problems have been described in the references, we shall focus on the assumptions and the main ideas by working out an example.

Let us consider the mixed first-order autoregressive first-order moving average time-dependent process, TDARMA (1, 1), defined by the equation:

$$\tilde{w}_t = \phi_t \tilde{w}_{t-1} + \tilde{b}_t - \theta_t \tilde{b}_{t-1} \tag{9.31}$$

where the $\tilde{b}_t$ are the innovations, with mean 0 and variance $\sigma_t^2 > 0$, say. Let us assume that $\delta = 0$ to simplify matters and, in accordance with (H1)–(H3) and (H5), $\phi_t = \phi_0$, $\theta_t = \theta_0$, and $\sigma_t^2 = \sigma_0^2$ for $t \leqslant 0$, with $|\phi_0| < 1$, and $|\theta_0| < 1$.

In the time-invariant case, these equalities hold for all t and the autocovariances of the process can be expressed in terms of $\sigma_t^2 = \sigma^2$ and the coefficients of the ARMA model. For example the fast algorithm of Wilson (1979) can be used.

In the time-dependent case, the autocovariances of the stationary process defined by

$$\tilde{w}_t = \phi_0 \tilde{w}_{t-1} + \tilde{b}_t - \theta_0 \tilde{b}_{t-1} \tag{9.32}$$

are computed in the same way. They provide starting values for recurrences with respect to s and t, from which the covariances $\gamma_{ts} = \text{cov}(\hat{w}_t, \hat{w}_s)$ are derived (for more details, see Mélard, 1982b).

Note that the Wold–Cramér decomposition of a time-dependent ARMA process has no direct relation with the rational function $(1-\theta_t z)/(1-\phi_t z)$ (Mélard (1985a); see also Mélard and Herteleer-De Schutter (1989) for consequences in spectral analysis of non-stationary processes, and Hallin (1986) for related problems).

Because of the autoregresive part of the process, the Wold–Cramér decomposition of $\tilde{w}_t$ makes use of b_s for $s = 1, \ldots, t$. To simplify the decomposition, let us consider $\tilde{\boldsymbol{w}}^*$ defined by

$$\begin{cases} \tilde{w}_1^* = \tilde{w}_1 & \text{for } t = 1 \\ \tilde{w}_t^* = \tilde{w}_t - \phi_t \tilde{w}_{t-1} & \text{for } t \geqslant 2. \end{cases} \tag{9.33}$$

The process $\{\tilde{w}_t^*, t > 0\}$ is non-stationary but note that:

(a) it is q-dependent, with $q = 1$, in the sense that $\gamma_{ts}^* = \text{cov}(\tilde{w}_t^*, \tilde{w}_s^*) = 0$ for $|t-s| > 1$;
(b) $\gamma_{ts}^* = \text{cov}(\tilde{w}_t^*, \tilde{w}_s^*)$, $|t-s| = 0$ or 1, can be deduced from the γ_{ts};
(c) the densities $f(\tilde{\boldsymbol{w}}; \boldsymbol{\beta}, \sigma^2)$ and $f(\tilde{\boldsymbol{w}}^*; \boldsymbol{\beta}, \sigma^2)$ are equal because the Jacobian of the transformation from $\tilde{\boldsymbol{w}}$ to $\tilde{\boldsymbol{w}}^*$ is unity.

There remains to evaluate $\tilde{\boldsymbol{w}}^{*\mathrm{T}} V^{*-1} \tilde{\boldsymbol{w}}^*$ and det V^*, where V^* is composed of the γ_{ts}^*. But, since the process is 1-dependent, we can write its Wold-Cramér decomposition as

$$\begin{cases} \tilde{w}_1^* = h_1 \tilde{b}_1 & \text{for } t = 1 \\ \tilde{w}_t^* = h_t \tilde{b}_t - \psi_t h_{r-1} \tilde{b}_{t-1} & \text{for } t = 2, \ldots, n, \end{cases} \tag{9.34}$$

where $h_t \tilde{b}_t$ is the innovation at time t, and $\text{var}(\tilde{b}_t) = \sigma^2$.

Neglecting computational efficiency considerations, we use the Gram–Schmidt orthogonalization procedure on (9.34). Starting with $h_1^2 = \gamma_{1,1}^*/\sigma^2$ and $\tilde{b}_1 = \tilde{w}_1^*/h_1$, we have

$$\begin{cases} \psi_t = -\dfrac{\gamma_{t-1,t}^*}{h_{t-1}^2 \sigma^2} \\[2ex] h_t^2 = \dfrac{\gamma_{t,t}^*}{\sigma^2} - \psi_t^2 h_{t-1}^2 \\[2ex] \tilde{b}_t = \dfrac{\tilde{w}_t^* + \psi_t h_{t-1} \tilde{b}_{t-1}}{h_t} \end{cases} \tag{9.35}$$

for $t > 1$.

We now leave the example and consider the general case where more efficient algorithms are given by Ansley and Kohn (1983) using the Kalman filter, and Mélard (1982b), using the Cholesky factorization (incidentally the Kalman filter algorithm can be used in the vector time-dependent ARMA case, see Mélard, 1985a). These algorithms require a number of operations (multiplication and divisions) of order $O(n(p+q^2+4q))$, which is about less than twice the number of operations for equivalent algorithms in the time-invariant case.

9.6 MAXIMUM LIKELIHOOD ESTIMATION

Let us now consider the vector of sample normalized innovations $\tilde{\boldsymbol{b}} = (\tilde{b}_{\delta+1},\ldots,\tilde{b}_n)^{\mathrm{T}}$ which are independent normal random variables. Hence, taking care of the Jacobian of the transformation from the $\tilde{w}_t^*$ to the $\tilde{b}_t$, the likelihood function (9.19) is expressed by

$$L(\boldsymbol{\beta},\sigma^2;\tilde{\boldsymbol{b}}) = G^{(n-\delta)(\lambda-1)}(2\pi)^{-(n-\delta)/2}\sigma^{-(n-\delta)}\left(\prod_{t=\delta+1}^{n} h_t^{-1}\right) \times \exp\left\{-\frac{1}{2\sigma^2}\sum_{t=\delta+1}^{n}\tilde{b}_t^2(\boldsymbol{\beta})\right\}, \tag{9.36}$$

where $\tilde{b}_t(\boldsymbol{\beta})$ is the value of $\tilde{b}_t$ derived from the $\tilde{w}_t = w_t - M_t^w$, via the $\tilde{w}_t^*$, in terms of the parameter $\boldsymbol{\beta}$.

There is no analytic expression of the likelihood function. Maximizing it numerically with respect $\boldsymbol{\beta}$ and σ is difficult. Since G and h_t depend on $\boldsymbol{\beta}$, the problem is simplified by letting

$$\tilde{b}_{*t}(\boldsymbol{\beta}) = \tilde{b}_t(\boldsymbol{\beta})G^{-(\lambda-1)}\left(\prod_{t=\delta+1}^{n} h_t\right)^{1/(n-\delta)}, \tag{9.37}$$

and

$$\sigma_* = \sigma G^{-(\lambda-1)}\left(\prod_{t=\delta+1}^{n} h_t\right)^{1/(n-\delta)}. \tag{9.38}$$

It is equivalent to maximize

$$(2\pi)^{-(n-\delta)/2}\sigma_*^{-(n-\delta)}\exp\left\{-\frac{1}{2\sigma_*^2}\sum_{t=\delta+1}^{n}\tilde{b}_{*t}^2(\boldsymbol{\beta})\right\}, \tag{9.39}$$

or to minimize

$$\frac{n-\delta}{2}\log(\sigma_*^2) + \frac{1}{2\sigma_*^2}\sum_{t=\delta+1}^{n}\tilde{b}_{*t}^2(\boldsymbol{\beta}). \tag{9.40}$$

It is easy to solve the likelihood equation for σ_*^2

$$-\frac{n-\delta}{2\sigma_*^2}+\frac{1}{2\sigma_*^4}\sum_{t=\delta+1}^{n}\tilde{b}_{*t}^2(\boldsymbol{\beta})=0, \tag{9.41}$$

whose solution is

$$\hat{\sigma}_*^2=\frac{1}{n-\delta}\sum_{t=\delta+1}^{n}\tilde{b}_{*t}^2(\boldsymbol{\beta}). \tag{9.42}$$

Substituting in (9.40) leads to minimizing

$$\frac{n-\delta}{2}\log\left(\frac{1}{n-\delta}\sum_{t=\delta+1}^{n}\tilde{b}_{*t}^2(\boldsymbol{\beta})\right) \tag{9.43}$$

or the sum of squares

$$\left(\prod_{t=\delta+1}^{n}h_t^2\right)^{1/(n-\delta)}G^{-2(\lambda-1)}\sum_{t=\delta+1}^{n}\tilde{b}_t^2(\boldsymbol{\beta}). \tag{9.44}$$

Very often, the procedure to minimize the sum of squares is the Marquardt's method (1963) but other procedures can be used. Let $\hat{\boldsymbol{\beta}}$ be the maximum likelihood estimator for $\boldsymbol{\beta}$. As a consequence, the maximum likelihood estimator of σ^2 is

$$\frac{1}{(n-\delta)}\sum_{t=\delta+1}^{n}\tilde{b}_t^2(\hat{\boldsymbol{\beta}}). \tag{9.45}$$

REFERENCES

Abdrabbo, N.A. and Priestley, M.B. (1967) On the prediction of non-stationary processes. *J. Roy. Statist. Soc. Ser. B*, **29**, 570–585.

Abraham, B. and Box, G.E.P. (1978) Deterministic and forecast-adaptative time-dependent models. *J. Roy. Statist. Soc. Ser C Appl. Statist.*, **27**, 120–130.

Ansley, C.F. (1979) An algorithm for the exact likelihood of a mixed autoregressive-moving average process. *Biometrika*, **66**, 59–65.

Ansley, C.F. and Kohn, R. (1983) Exact likelihood of vector autoregressive-moving average process with missing or aggregated data. *Biometrika*, **70**, 275–278.

Bollerslev, T. (1986) Generalized autoregressive conditional heteroskedasticity. *J. Econometrics*, **31**, 307–327.

Bougerol, P. (1993) Kalman filtering with random coefficients and contractions. *Siam J. Contr. Optim.*, in preparation.

Box, G.E.P. and Cox, D.R. (1964) An analysis of transformations. *J. Roy. Statist. Soc. Ser. B*, **26**, 211–243.

Box, G.E.P. and Jenkins, G.M. (1976) *Time Series Analysis, Forecasting and Control*, Holden-Day, San Francisco (revised edition).

Box, G.E.P. and Tiao, G.C. (1975) Intervention analysis with applications to economic and environmental problems. *J. Amer. Statist. Assoc.*, **70** 70–79.

Cramér, H. (1961) On some classes of non-stationary stochastic processes, in *Proceedings of the Fourth Berkeley Symposium on Mathematical Statistics and*

Probability, University of California Press, Berkeley and Los Angeles, Vol. 2, pp. 57–78.

Fox, A.J. (1972) Outliers in time series. *J. Roy. Statist. Soc. Ser. B*, **34**, 350–363.

Granger, C.W.J. and Joyeux, R. (1980) An introduction to long-memory time series models and fractional differencing. *J. Time Series Anal.*, **1**, 15–29.

Grillenzoni, C. (1990) Modelling time-varying dynamical systems. *J. Amer. Statist. Assoc.*, **85**, 499–507.

Hallin, M. (1980) Invertibility and generalized invertibility of time series models. *J. Royal Statist. Soc. Ser. B*, **42**, 210–212; **43**, 103.

Hallin, M. (1986) Non-stationary q-dependent processes and time-varying moving average models: invertibility properties and the forecasting problem. *Adv. Appl. Prob.*, **18**, 170–210.

Herbst, L.J. (1963) A test for variance heterogeneity in the residuals of a Gaussian moving average. *J. Royal Statist. Soc. Ser. B*, **25**, 451–454.

Hipel, K.W. and McLeod, A.I. (1980) Perspectives in stochastic hydrology, in O.D. Anderson (ed.) *Time Series*, North-Holland, Amsterdam, pp. 73–102.

Li, W.K. and Hui, Y.V. (1988) An algorithm for the exact likelihood of periodic autoregressive moving average models. *Commun. Statist.-Simula*, **17**, 1483–1494.

Ljung , L. and Söderström, T. (1983) *Theory and Practice of Recursive Identification*, MIT Press, Cambridge MA.

Marquardt, D.W. (1963) An algorithm for least-squares estimation of non-linear parameters. *Journal of the Society of Industrial Applied Mathematics*, **11**, 431–441.

Mélard, G. (1977) Sur une classe de modèles ARIMA dépendant du temps. *Cahiers du Centre d'Etudes de Recherche Opérationnelle*, **19**, 285–295.

Mélard, G. (1981a) On an alternative model for intervention analysis, in O.D. Anderson and M.R. Perryman (eds) *Time Series Analysis*, North-Holland, Amsterdam, pp. 345–354.

Mélard, G. (1981b) On a deterministic sub-model for the innovation process in ARIMA model, *1981 Proceedings of the Business and Economic Statistics Section*, American Statistical Association, Washington D.C., pp. 329–333.

Mélard, G. (1982a) Software for time series analysis, in H. Caussinus, P. Ettinger and R. Tomassone (eds), *COMPSTAT 1982 Part 1: Proceedings in Computational Statistics*, Physica-Verlag, Vienna (Austria), pp. 336–341.

Mélard, G. (1982b) The likelihood function of a time-dependent ARMA model, in O.D. Anderson and M.R. Perryman (eds), *Applied Time Series Analysis*, North-Holland, Amsterdam, pp. 229–239.

Mélard, G. (1984) Algorithm AS197: A fast algorithm for the exact likelihood of autoregressive-moving average models. *J. Royal Statist. Soc. Ser. C. Appl. Statist.*, **33**, 104–114.

Mélard, G. (1985a) *Analyse de données chronologiques*, Coll. Séminaire de mathématiques supérieures—Séminaire scientifique OTAN (NATO Advanced Study Institute) n°89, Presses de l'Université de Montréal Montréal.

Mélard, G. (1985b) Illustration of the use of a general time series model, in O.D. Anderson (ed.) *Time Series Analysis: Theory and Practice 7, Proceedings of the (general interest) International Conference held at Toronto, Canada, 18–21 August 1983*, North-Holland, Amsterdam, pp. 53–75.

Mélard, G. and Herteleer-De Schutter, A. (1989) Contributions to the evolutionary spectral theory. *J. Time Series Anal.*, **10**, 41–63.

Mélard, G. and Kiehm, J.-L. (1981) ARIMA models with time-dependent coefficients for economic time series, in O.D. Anderson and M. R. Perryman (eds), *Time Series Analysis*, North-Holland, Amsterdam, pp. 355–363.

Mélard, G. and Roy, R. (1988) Modèles de séries chronologiques avec seuil. *Revue de Statistique Appliquée*, **36**, 5–24.

Miller, K.S. (1968) *Linear Difference Equations*, Benjamin, New York.

Miller, K.S. (1969) Nonstationary autoregressive processes. *I.E.E.E. Transactions Information Theory*, **IT-15**, 315–316.

Morf, M., Sidhu, G.S. and Kailath, T. (1974) Some new algorithms for recursive estimation on constant, linear, discrete-time systems. *I.E.E.E. Trans. Automatic Control* **AC-19**, 315–323.

Nicholls, D. F. and Quinn, B. G. (1982) *Random Coefficient Autoregressive Models: An Introduction*, Springer-Verlag, New York.

Pearlman, J.G. (1980) An algorithm for the exact likelihood of a high-order autoregressive-moving average process, *Biometrika*, **67**, 232–233.

Priestley, M.B. (1981) *Spectral Analysis and Time Series*, Vol. 2, Academic Press, New York.

Priestley, M.B. (1988) *Non-Linear and Non-Stationary Time Series Analysis*, Academic Press, New York.

Quenouille, M.H. (1957) *The Analysis of Multiple Time Series*, Griffin, London.

Rissanen, J. and Barbosa, L. (1969) Properties of infinite covariance matrices and stability of optimum predictors. *Information Sci.*, **1**, 221–236.

Schweppe, F.C. (1965) Evaluation of likelihood functions of Gaussian signals, *I.E.E.E. Transactions Information Theory*, **IT-11**, 61–70.

Subba Rao, T. (1970) The fitting of non-stationary time-series models with time dependent parameters. *J. Roy. Statist. Soc. Ser. B*, **32**, 312–322.

Subba Rao, T. (1981) On the theory of bilinear models. *J. Roy. Statist. Soc. Ser. B*, **42**, 245–292.

Tong, H. (1983) *Threshold Models in Non-linear Time Series Analysis*, Springer-Verlag New York.

Tsay, R.S. (1988) Outliers, level shifts, and variance changes in time series. *J. of Forecasting*, **7**, 1–20.

Tyssedal, J.S. and Tjøstheim, D. (1982) Autoregressive processes with a time-dependent variance. *J. Time Series Anal.*, **3**, 209–217.

Young, P.C. (1984) *Recursive Estimation and Time Series Analysis*, Springer-Verlag, Berlin.

Whittle, P. (1965) Recursive relations for predictors of non-stationary processes. *J. Roy. Statist. Soc. Ser. B*, **27**, 523–532.

Wilson, G.T. (1979) Some efficient computational procedures for high-order ARMA models. *J. Statist. Comput. Simul.*, **8**, 303–309.

Wichern, D.W., Miller, R.B. and Hsu, D.A. (1976) Changes of variance in first-order autoregressive time series models—with an application. *J. Roy. Statist. Soc. Ser. C Appl. Statist.*, **25**, 248–256.

Part Three

Spectral Analysis of Stationary Time Series

10

Determining the number of jumps in a spectrum

E.J. Hannan

10.1 INTRODUCTION

The model studied is

$$y(t) = \mu + \sum_{1}^{r} (\alpha_j \cos \lambda_j t + \beta_j \sin \lambda_j t) + u(t), 0 < \lambda_j < \pi, t = 0, \pm 1, \pm 2, \dots \tag{10.1}$$

We assume that $u(t)$ is stationary with zero mean and absolutely continuous spectrum and spectral density $f(\omega)$ that is continuous with $f(\omega) > 0$, $\omega \in [-\pi, \pi]$. Other conditions will be imposed later. True values will be indicated as λ_{0_j}, r_0 etc. when the emphasis is needed. We are mainly concerned with estimating r_0. The first term on the right in (10.1) could be identified with jumps in a spectral distribution for $y(t)$, hence the title of the paper.

If $u(t)$ is white Gaussian noise then the maximum likelihood estimates (MLE) of the λ_{0_j}, for r known, are obtained by forming the regression sum of squares, $Q(\lambda_1, \dots, \lambda_r)$, from the regression on 1, $\cos \lambda_j t$, $\sin \lambda_j t$, $j = 1, \dots, r$. The $\hat{\lambda}_j$ are chosen to maximize this. When the white Gaussian noise assumption is not satisfied there might be better methods but we continue to consider the ones just defined.

An asymptotically equivalent procedure is to form $Q_0(\lambda)$ from $y(t) - \bar{y}$, $t = 1, \dots, T$, and to choose $\hat{\lambda}_1$ to maximize that. Then finding $\hat{\alpha}_1$, $\hat{\beta}_1$ by regression we form $Q_1(\lambda)$ from $e_1(t) = y(t) - \bar{y} - \hat{\alpha}_1 \cos \hat{\lambda}_1 t - \hat{\beta}_1 \sin \hat{\lambda}_1 t$ and choose $\hat{\lambda}_2$ to maximize that, and so on. This might estimate the λ_{0_j} in some permutation but that is of no consequence. A further, asymptotically equivalent, procedure is to form

$$I_0(\lambda) = \frac{1}{T} \left| \sum_{1}^{r} [y(t) - \bar{y}] \, e^{it\lambda} \right|^2 \tag{10.2}$$

and to choose $\hat{\lambda}_1$ to maximize that. Then $I_1(\lambda)$ is formed from the $e_1(t)$ and $\hat{\lambda}_2$ obtained and so on. The $\hat{\lambda}_j$ obtained from the three procedures described will not be the same but will have the same asymptotic properties and we shall not introduce a special notation to distinguish them. However, the first procedure is preferable, especially if some of the λ_{0_j} are close together (in units of order T^{-1}), and the second is preferable to the third. We use the third in the simulations in section 10.2 where we also describe a method for locating the maximizing $\hat{\lambda}_j$. For a discussion of the methods see, especially, Bloomfield (1976) and also Hannan and Quinn (1989).

Wang (1993) writes (10.1) in terms of complex exponentials $\exp(i\mu_j t)$ with $\mu_j = \lambda_j$, $\mu_{-j} = -\lambda_j$, $j = 1, 2, \ldots, r$. Then taking $\mu = 0$ in (10.1) he forms,

$$\sigma^2(r) = \frac{1}{T}\sum_1^t \left| y(t) - \sum_{-r}^{r}{}' A_T(\hat{\mu}_j) e^{i\hat{\mu}_j t} \right|^2$$

$$A_T(\lambda) = \frac{1}{T}\sum_1^T y(t)\, e^{it\lambda}, \tag{10.3}$$

where the $\hat{\mu}_j$ are chosen as the locations of the r largest relative maxima of $I(\lambda) = |\sum y(t) e^{it\lambda}|^2/T$, $\lambda > 0$, subject to a separation condition, namely $|\hat{\mu}_j - \hat{\mu}_k| > b_T$, $j \neq k$, $b_T \to 0$, $b_T(T\log T)^{1/2} \to \infty$. $\sum'$ indicates omission of $j = 0$. He then considers

$$\phi_1'(r, c) = \log \sigma^2(r) + rc(\log T)/T. \tag{10.4}$$

Wang (1993) shows, under certain conditions, that if $\hat{r}$ is the location of the first relative minimum of $\phi_1'(r, c)$, i.e. the lowest r for which $\phi_1'(r, c) \leqslant \phi_1'(r+1, c)$, then $\hat{r} \to r_0$, almost surely (a.s.), provided

$$c > 2 \max_{\omega} f(\omega) \Big/ \left[\frac{1}{2\pi} \int_{-\pi}^{\pi} f(\omega)\, d\omega \right]. \tag{10.5}$$

He also shows that a.s. convergence does not hold if c is less than the right side of (10.5). In section 10.2 we shall replace $\phi_1'(r, c)$ by

$$\phi_1(r, c) = \log v(r)^2 + cr(\log T)/T, \tag{10.6}$$

where $v(r)^2 = T^{-1}\sum e_r(t)^2$. We shall show, in section 10.3, that the separation condition is then not necessary. The separation condition seems to be necessary because otherwise a side lobe from the effect at, say, λ_{0_j} might otherwise be mistaken for a main lobe. The regression procedure asymptotically eliminates that possibility.

There is a problem with (10.4) and (10.6) because the right side of (10.5) is unknown and difficult to estimate accurately. As we shall see in section 10.2 the right side of (10.5) arises from an asymptotic result (An, Chen and Hannan, 1983) but that very large sample sizes may be needed before that is fully relevant. Thus (10.4) or (10.6) may work badly. We therefore consider another criterion, which we base, somewhat indirectly, on the minimum description length principle of Rissanen (1989). The idea is to base a model choice on the minimization of an optimal code length for encoding the data, given the model. Among the parameters specifying the model is, of course, r. However, we must also model $f(\omega)$. We follow a method used in Hannan and Rissanen (1988) and Cameron, Hannan and Speed (1992). A likelihood is constructed for the data, transformed to the quantities

$$w(\omega_j) = \frac{1}{2T}\sum_1^T y(t)\mathrm{e}^{-it\omega_j}, \quad \omega_j = \frac{2\pi j}{T}.$$

Then $f(\omega)$ is taken as constant over bands of width $2\pi m/T$. This leads, via some approximations and after maximizing the likelihood, to the criterion

$$\phi_2(r, m, 5) = \log \hat{\sigma}_r^2(m) + 5r(\log T)/T + (\log m)/2m \tag{10.7}$$

$$\log \hat{\sigma}_r^2(m) = \frac{2M}{T}\sum_{k=1}^{m} \log\left[\frac{1}{m}\sum\nolimits_k I_{r-1}(\omega_j)\right]$$
$$+ \frac{2p+\delta_T}{T}\log\left\{\frac{2}{2p+\delta_T}\left[\sum_{mM+1}^{mM+p} I_{r-1}(\omega_j) + \tfrac{1}{2}\delta_T I_{r-1}(\pi)\right]\right\}.$$

Here $M = \lfloor (T-1)/2m \rfloor$, $p = \lfloor (T-1)/2 \rfloor - mM$, $\delta_T = 0$ for T odd and 1 for T even. The sum $\sum_k$ is over $(k-1)m < j \leqslant km$. The quantity $\hat{\sigma}_r^2(m)$ is an estimate of the linear prediction variance of $u(t)$ in (10.1), from the $e_r(t)$. (See above (10.2).) The last terms in the formula for $\log \hat{\sigma}_r^2(m)$ arise only when T is even and when m does not divide $(T-1)/2$. The quantity $T/2 \log \sigma_r^2(m)$ is essentially minus the log likelihood, at its maximum, and this measures the cost of encoding the data given the parameter values. ('Essentially' here means eliminating some constants and terms independent of r). The term $5r(\log T)/T$ is $2/T$ by an approximation to the cost of encoding the parameter estimates $\hat{\alpha}_j, \hat{\beta}_j, \hat{\lambda}_j$, which are needed to decode the optimally encoded data using those values. Since $\hat{\alpha}_j, \hat{\beta}_j$ are estimates with accuracy $O(T^{-1/2})$ and $\hat{\lambda}_j$ has accuracy $O(T^{-3/2})$ this cost is approximately $2 \times 1/2 \log T + (3/2)\log T$ and $2/T$ by this is the second term. The last term measures the cost of encoding the, approximately, M estimates of the values of $f(\omega)$ over the M bands. Each has accuracy $O(m^{-1/2})$ which gives $M/2 \log m$ and $(T/2)M/2 \log m$ is approximately $(\log m)/2m$. This last calculation is very rough since, even for T large, m might be quite small. Previous experience with BIC type formulae, see Hannan and Deistler (1988) for example, leads one to suspect that (10.7) may lead to m being chosen large, i.e. M small.

We might choose $\hat{r}, \hat{m}$ to minimize (10.7), i.e. we choose $\hat{r}(m)$ as the first relative minimum of (10.7) for given m and then choose m to minimize these minima, subject to an upper bound for m. Though we have not been able to complete a satisfactory proof of asymptotic properties of $\hat{r}(m)$ it seems plausible (section 10.8) that $\hat{r}(m) \to r_0$, a.s., at least for m not too small.

Having in mind the difficulty, and unreliability, of the asymptotic theory we proceed to investigate the procedures through simulations in the next section.

10.2 SIMULATIONS

We consider first the case where $2\pi f(\omega) \equiv \sigma^2$. The $u(t)$ are simulated as pseudo-Gaussian random variables. In this situation one expects that (10.6) will perform better than (10.7) since v_r^2 is a better estimate of σ^2, $r \geqslant r_0$. When m is large the difference may be small. We take $r_0 = 2$, $\lambda_{01} = 1$, $\lambda_{02} = 2$ and, putting $A^2 = (\alpha^2 + \beta^2)$ then $A_{01} = 2^{1/2}$, $A_{02} = 2^{-1/2}$. There are 100 replications for each of $T = 128$, 512, 1024. The method used to locate $\hat{\lambda}_j$, $j = 1, 2$ is to find the $\omega_k = 2\pi k/T$ maximizing $I_j(\omega_k)$, $j = 0, 1$. Then the algorithm of Quinn and Fernandes (1992) was used to locate $\hat{\lambda}_j$. This algorithm is very quick and simple to use and appeared always to fiind a $\hat{\lambda}_j$ very close to the MLE, provided the ω_k maximizing $I_j(\omega_k)$ is close to λ_{0_j}. Asymptotically this will be $O(T^{-1})$ from λ_{0_j} but with coloured noise this may not happen as we show below.

Table 10.1 confirms the asymptotic theory discussed below. The performance of $\phi_1(r, c)$ depends on c with overestimation with c too small ($\leqslant 2$) and underestimation with c too large. At $T = 128$ both methods performed poorly but especially (10.6). For $m > 4$ there is little to favour (10.6) over (10.7) and indeed (10.7) could, perhaps, be regarded as performing better.

Consider next the case where

$$u(t) + .81u(t-2) = \varepsilon(t), \tag{10.8}$$

where the $\varepsilon(t)$ were pseudo-Gaussian with variance 1. Now $2\pi f(\omega)$ has a single peak at $\omega = \pi/2$ of height 27.7. The variance of $u(t)$ is 2.91 so the number on the right of (10.5) is 19.03. Thus (10.6) will overestimate for $c < 19.03$. However, the right side of (10.5) is based on a result in An, Chen and Hannan (1983) for a suitable, rather general, class of stationary processes, namely that

$$\lim_{T\to\infty} \max_{\omega} \frac{I(\omega)}{2\pi f(\omega)\log T} = 1, \text{ a.s.} \tag{10.9}$$

In the present case a relevant number thus is max $2\pi f(\omega)\log T = 27.7 \log T$ which is 134, 173, 192 for $T = 128$, 512, 1024, respectively. A more accurate figure is obtained from Turkman and Walker (1984), for Gaussian $u(t)$, namely

$$P\left\{\max_{\omega} I(\omega)/[2\pi f(\omega)] - \log T - \frac{1}{2}\log\log T + \frac{1}{2}\log 3/\pi \leqslant x\right\} \to \exp(e^{-x}). \tag{10.10}$$

(Note that in Turkman and Walker (1984) their periodogram is $2I(\omega)$.) For

Table 10.1 Criteria performance for white noise error, $r_0 = 2$

$\hat{r}$	T	$\phi_1(r, c)$ $c=2$	8	14	20	$\phi_2(r, m, 5)$ $m=4$	8	16	32
0	128	0	0	23	95	0	0	0	0
	512	0	0	0	0	0	0	0	0
	1024	0	0	0	0	0	0	0	0
1	128	1	71	77	5	56	23	19	3
	512	0	0	5	64	77	7	0	0
	1024	0	0	0	0	13	0	0	0
2	128	36	29	0	0	39	57	63	61
	512	47	100	95	36	23	93	100	100
	1024	37	100	100	100	87	100	100	97
$\geqslant 3$	128	63	0	0	0	5	10	18	36
	512	53	0	0	0	0	0	0	0
	1024	63	0	0	0	0	0	0	3

$x = 2.25$ the right side of (10.10) is about 0.9. Thus, for max $I(\omega)$.

$$\max 2\pi f(\omega)\left(\log T - \frac{1}{2}\log\log T - \frac{1}{2}\log\frac{3}{\pi} + 4.7\right) \tag{10.11}$$

should be exceeded certainly no more than once in 100 times, for T large. For $T = 128, 512, 1024$ then (10.11) is 219, 261, 297. The nearest value of $2\pi k/T$ to $\lambda_{02} = 2$ at $T = 512$, 1024 is at $k = 163$, 326 and then $2\pi k/T$ is very near to 2, being 2.00031 in both cases. (For comparison π/T is 0.006, 0.003.) As the discussion in section 10.4 shows the contribution to $I_1(\omega)$, from the signal at $\lambda_{02} = 2$, should then be close to $A_2^2 T/4 = 16, 64, 128$. At $\lambda_{01} = 1$ the quantity $A_{01}^2 T/4$ is 64, 256, 572, at the three T values.

The effect of the choice of a false peak in $I_r(\omega_k)$ is as follows, taking $r = 1$ for illustration. Assume that λ_{01} has been accurately estimated, so that periodic component has been, effectively, removed. If $I_1(\omega_k)$ is maximized not near $\lambda_{02} = 2$ but rather at or near $\pi/2$ then using the location of this maximum as an input to the Quinn and Fernandes (1992) algorithm (or almost any other algorithm for that matter) results in an estimate $\hat{\lambda}_2$ near $\pi/2$. Then the reduction in $\log \sigma^2(2)$ as compared to $\log \sigma^2(1)$ will be small, since then $\hat{\lambda}_2$ may be rather meaningless and not corresponding to a periodic component, and this reduction will be outweighed by the increase $5 \log T/T$ in the penalty term and the criterion may increase from $r = 1$ to $r = 2$ so that $\hat{r} = 1$ may be arrived at. To judge how likely it is that $I_1(\omega_k)$ will be maximized near $\pi/2$ rather than at $\lambda_{01} = 2$ we may compare (10.11) with $A_{02}^2 T/4 = 16, 64, 128$ for the three T values. The three corresponding values, 219, 261, 297 for (10.11), at $x = 2.25$, suggests that a value near $\pi/2$ will maximize

$I_1(\omega_k)$ in a large proportion of cases. This does not happen as Table 10.2 shows. It is evident that $\phi_1(r,c)$ has performed rather badly. Of course since the number on the right side of (10.5) is 19.03 then only for $c=20$ is the necessary condition of the theorem in Wang (1992) satisfied. However, for smaller c, underestimation is the problem not over estimation! The problem is the low signal to noise ratio. The fall in $\log v(2)$ compared to $\log v(1)$ should be about $\log\{1+A_{02}^2/(2v(2)^2)\} \simeq A_{02}^2/(2v(2)^2) \rightarrow 0.086$. For this to be bigger than $c \log T/T$ we need $c \leqslant 2.27$, 7.05, 12.70 for the three T values. Unfortunately such small c values tend to result in overestimation as is evident from Table 10.2. At $c=20>19.03$ we get underestimation since $0.086<20 \log T/T=0.76$, 0.24, 0.14 for the three T values. The SNR is too small for good results.

However $\phi_2(r,m,s)$ has not performed as badly as the analysis above would suggest, as was pointed out earlier. This is for the following reason. The results in (10.9), (10.11) come from regarding $I_u(\omega)/2\pi f(\omega)$ as $I_\varepsilon(\omega)$ where $\varepsilon(t)$ is the innovation sequence for $u(t)$. The maximum of $I_\varepsilon(\omega)/\{\sigma^2 \log T\}$, σ^2 being the prediction variance, will converge to unity but the location of that maximum is equally likely to be anywhere in $(0,\pi)$. It is improbable that it will be in the small range about $\pi/2$ where $f(\omega)$ is very large. Of course $2\pi f(\omega) I_\varepsilon(\omega) \approx I_u(\omega)$ will tend to have its maximum value near to $\pi/2$ because of the influence of the $2\pi f(\omega)$ factor. However max $I_u(\omega)$ may be a good deal less than $\log T$, unless T is really very large. The more peaked $f(\omega)$ is, the more this will be so. In fact at $T=1024$ there are 85 correct values for $\phi_2(r,m,s)$ for $m=16$ and 32. There are 7 values for $\hat{r}=0$ at $T=512$. These sets of 15 and 7 estimates give wrong values, failing to pick up the component at

Table 10.2 Criteria performance for coloured noise error, $r_0=2$

		$\phi_1(r,c)$				$\phi_2(r,m,5)$			
$\hat{r}$	c	$c=2$	8	14	20	$m=4$	8	16	32
0	128	0	54	97	100	64	47	61	8
	512	0	1	6	20	7	7	7	7
	1024	0	0	0	0	1	0	0	0
1	128	0	27	3	0	32	42	29	3
	512	0	29	78	80	90	69	56	52
	1024	0	0	78	100	86	36	15	15
2	128	0	14	0	0	4	9	9	5
	512	0	27	15	0	3	24	37	38
	1024	0	17	18	0	13	64	85	85
$\geqslant 3$	128	100	5	0	0	0	2	1	84
	512	100	43	1	0	0	0	0	3
	1024	100	83	4	0	0	0	0	0

$\lambda_{02} = 2$ in the first case and $\lambda_{01} = 1$ in the second, solely due to a wrong initial value near $\pi/2$, obtained from $I_1(\omega_k)$ or $I_0(\omega_k)$ for the Quinn and Fernandes (1992) procedure.

Two further points need to be discussed, one being the choice of m from (10.7) and the other, which we discuss first, being the effect of the location of a λ_{0_j} relative to the grid of values $2\pi k/T$. This is illustrated by a separate simulation at $r_0 = 1$, $A_1 = 2^{1/2}$, $\lambda_{01} = 1$, and $u(t)$ white pseudo-Guassian noise of variance 1. At $T = 512$ the nearest $2\pi k/512$ to $\lambda_{01} = 1$ is at $k = 81$ and in fact λ_{01} is almost exactly half way between the ω_k values at $k = 81$, 82. At $T = 1024$, then, λ_{01} is almost exactly at ω_k for $k = 163$. Even though the sample size is smaller the method (10.7) performs slightly better at $T = 512$ than at 1024. Indeed the correct value, $\hat{r} = 1$, was obtained at $m = 3, 4, 8$, respectively, in 100, 99 and 100 cases, whereas at 1024 the numbers of correct values were 76, 94 and 100. At first sight this is surprising but we shall show in section 10.3 that it is not unexpected. The effect does not show up at $m = 4$ in Table 10.1 (so that at $T = 1024$ we get a few estimates $\hat{r} = 0$) but at $m = 4$ the difference in performance between $T = 512$ and $T = 1024$ was smaller in the other simulation also. It seems to be true that there is a slight favourable effect when the true λ value is halfway between two grid points as compared to being at one of them. The effect depends on m and disappears with $T \to \infty$.

In Table 10.3 the choice of m is exhibited for $T = 512$, 1024. The tables show the bivariate observed frequency distribution of $\hat{r}$, m values. The results are for the coloured noise case (10.8). The set of simulations is different from the ones in earlier tables but agrees closely with these as the value 84, for $r = 2$, shows, at $T = 1024$. The nine cases $\hat{r} = 0$ for $T = 512$ also agree with the other simulation. It is difficult to draw a conclusion though $m = 32$ performs better than $m = 4$ at $T = 1024$. The formula chooses $m = 32$ most of the time. At $T = 512$ the picture is complicated. One is left with an impression that it is better to choose a reasonably large value of m and not choose that value via (10.7). Nevertheless, the r_0 value is chosen in 34% of

Table 10.3 Frequency distribution of (m, r) optimizing values for (10.7)

	$T=512$						$T=1024$				
	m						m				
$\hat{r}$	4	8	16	32		$\hat{r}$	4	8	16	32	
0	9	0	0	0	9	0	0	0	0	0	0
1	45	6	0	2	53	1	14	0	0	2	16
2	0	0	8	76	34	2	1	0	0	83	84
$\geqslant 3$	0	0	0	4	4	$\geqslant 3$	0	0	0	0	0
	54	6	8	32	100		15	0	0	85	100

cases at $T = 512$ which compares well with 38% of cases at $m = 32$, the best case, in Table 10.2. Again, 84% compares well with 85% at $m = 16$ and at $m = 32$ in Table 10.2 for $T = 1024$.

It is evident that the ratios $A_{0j}^2/(\max 2\pi f(\omega)$ cannot be too small if the methods are to work well, for anything but very large T. In accordance with established conventions for other situations one might put $-10\log_{10}\{A_{0_j}^2/(2\max 2\pi f(\omega))\}$, in dB, for the (logarithmic) signal to noise ratio. At $j = 2$, when (10.8) holds, this is -20.0, which is a low value by most standards. See Boashash (1992) and Table 10.1 for some related, but by no means equivalent, calculations.

10.3 DISCUSSION

The results presented in the previous section were, perhaps, unkind to the methods since, as the simulation shows, the quantities $\text{SNR} = 10\log_{10}[A_j^2/\{4\pi \max f(\omega)\}]$, which are appropriate signal to noise ratios, need to be reasonably large, in relation to T, and for $j = 2$ this was -20.0, which is a very low value. For the SNR of the simulation $T = 128$ is clearly much too small. The problem of distinguishing between a peak in the noise spectrum and a sinusoidal component is always going to be difficult. One thing that might reduce the problem in practice is knowledge of a reasonably narrow range within which the frequency might lie. This would occur with a signal transmitted at a reasonably accurately known frequency but subject to Doppler shift.

The more useful method seems to be that based on a criterion $\log \hat{\sigma}_r^2(m) + 5r(\log T)/T$, for m chosen not too small. Of course for white noise the criterion $\log \sigma^2(r) + cr(\log T)/T$ would be better, possibly with c chosen as 5 (see the columns for $c = 2$ and 8 in Table 10.1). However, as Table 10.2 shows, if the noise is not white this method may grossly overestimate the number of sinusoidal components (see the column for $c = 2$ in Table 10.2).

Further research is needed but the answer provided by the ϕ_2 criterion, or something like it, seems to be near to what is needed.

10.4 THEORETICAL INVESTIGATION

The theory presented below is asymptotic and for this a basic result is (10.9), which is proved under rather general conditions in An, Chen and Hannan (1983). Others needed are

$$T^{3/2}|\hat{\lambda}_j - \lambda_{0_j}|/(\log\log T)^{1/2} = O(1), \text{ a.s.} \tag{10.12}$$

$$T^{1/2}|\hat{\alpha}_j - \alpha_{0j}|/(\log\log T)^{1/2};\ T^{1/2}|\hat{\beta}_j - \beta_{0j}|/(\log\log T)^{1/2} = O(1), \text{a.s.} \tag{10.13}$$

The result (10.12) is proved in Hannan and Mackisack (1986) and (10.13) is a fairly standard result once (10.12) is established. We go on to establish that

$\hat{r} \to r_0$, a.s. for (10.6). Take $r_0 = 2$, for example, and $A_1 > A_2$. Then (10.12), (10.13) will hold for $\hat{\lambda}_1, \hat{\alpha}_1, \hat{\beta}_1$ since λ_{01}, λ_{02}, are fixed and different so that the component at frequency λ_{02} will have negligible influence, asymptotically, on $I_0(\omega)$ near λ_{01}. We take $\mu = 0$ for simplicity. Dropping the 0 subscript for convenience, we put

$$\begin{aligned} s_1(t) &= (\alpha_1 - \hat{\alpha}_1)\cos\hat{\lambda}_1 t + (\beta_1 - \hat{\beta}_1)\sin\hat{\lambda}_1 t + \alpha_1(\cos\lambda_1 t - \cos\hat{\lambda}_1 t) \\ &\quad + \beta_1(\sin\lambda_1 t - \sin\hat{\lambda}_1 t), \\ y_1(t) &= u(t) + \alpha_2\cos\lambda_2 t + \beta_2\sin\lambda_2 t. \end{aligned}$$

Then, uniformly in ω,

$$\frac{1}{T^2}\left|\sum_1^T (y(t) - \hat{\alpha}_1\cos\hat{\lambda}_1 t - \hat{\beta}_1\sin\hat{\lambda}_1 t)\,\mathrm{e}^{it\omega}\right|^2 \tag{10.14}$$

$$\begin{aligned} &= \frac{1}{T^2}\left|\sum_1^T y_1(t)\,\mathrm{e}^{it\omega}\right|^2 + \mathrm{O}[(\log\log T)/T] \\ &\quad + 2\mathscr{R}\left[\frac{1}{T}\sum_1^T y_1(t)\,\mathrm{e}^{it\omega}\frac{1}{T}\sum_1^T s_1(t)\,\mathrm{e}^{-it\omega}\right], \text{ a.s.,} \end{aligned}$$

where the second term is of the indicated order because of (10.12), (10.13). The last term is dominated by

$$2\left|\frac{1}{T}\sum_1^T y_1(t)\,\mathrm{e}^{it\omega}\right|\mathrm{O}\{[(\log\log T)/T]^{1/2}\} = \mathrm{O}\{[(\log\log T)/T]^{1/2}\}. \tag{10.15}$$

However if ω is very near to λ_2 this term will be $\mathrm{O}\{[(\log\log T)/T^3]^{3/2}\}$ since, for example, then

$$\frac{1}{T}\sum_1^T \cos\hat{\lambda}_1 t\,\mathrm{e}^{-it\omega} = \mathrm{O}\left(\frac{1}{T}\right).$$

However, (10.15) ensures that the maximizing ω value for $I_1(\omega)$ will be $\mathrm{O}(T^{-1})$ from λ_2 (see Hannan, 1973, for example). Thus $I_1(\omega)$ may be replaced by the first term on the right in (10.14), to $\mathrm{O}[(\log\log T)/T]$. This effect is small compared to that of

$$2\mathscr{R}\left[\frac{1}{T}2u(t)\,\mathrm{e}^{it\omega}\frac{1}{T}2(\alpha_2\cos\lambda_2 t + \beta_2\sin\lambda_2 t)\,\mathrm{e}^{-it\omega}\right]$$

for ω within $\mathrm{O}(T^{-1})$ of λ_2 so that the accuracy with which $\hat{\lambda}_2$ estimates λ_2 is the same as if the estimate was obtained from the first term on the right in (10.14). Thus as r goes from 0 to 1 and then 1 to 2 the criterion (10.6) will fall, for T larger than some a.s. finite values. As r goes from 2 to 3 we must consider

$$\frac{1}{T}\left|\sum_1^T \hat{u}(t)\,\mathrm{e}^{it\omega}\right|^2 \tag{10.16}$$

with

$$\hat{u}(t) = y(t) - \hat{A}_1 \cos(\hat{\lambda}_1 t + \hat{\phi}_1) - \hat{A}_2 \cos(\hat{\lambda}_2 t + \hat{\phi}_2),$$

and again this can be replaced by

$$\frac{1}{T}\left|\sum_1^T u(t)\,\mathrm{e}^{it\omega}\right|^2$$

to $O[(\log\log T)^{1/2}]$, a.s. This will be maximized at a frequency $\hat{\omega}$ and, from (10.9), this maximum will be bounded a.s., by

$$\log T \max_\omega 2\pi f(\omega)[1 + O(1)].$$

Thus

$$\log v_2^2 - \log v_3^2 = \log v_2^2/v_3^2 = \log\left[1 + \frac{2}{v_3^2}\frac{\log T}{T}\max_\omega 2\pi f(\omega)\right]$$

$$= \left\{2\log T \max_\omega 2\pi f(\omega) \Big/ \left[T\int f(\omega)\,\mathrm{d}\omega\right]\right\}[1 + O(1)], \text{ a.s.,}$$

as in the proof by Wang (1992). Indeed the regression sum of squares for the component at $\hat{\omega}$ is, asymptotically, twice $I(\hat{\omega})$ and it is this regression sum of squares that is the difference between v_2^2 and v_3^2. Thus (10.6) will give a strongly consistent estimate of r_0, for c satisfying (10.5).

It is plausible that $\hat{r}$ from (10.7) will be strongly consistent but a precise proof is difficult. Consider the case where T is large, $r_0 = 1$ and $\pi_1 = 2\pi j_0/T$. Then the contribution of the signal to $\log \hat{\sigma}_0^2(m)$ comes only from the $\sum_{k_0}$ containing j_0 and, taking T odd and m to divide $N = (T-1)/2$, $M = N/m$,

$$\log \hat{\sigma}_0^2(m) = M^{-1}\sum_{k \neq k_0}\log\frac{1}{m}\sum_k I_u(\omega_j) + M^{-1}\log\frac{1}{m}\sum_{j\neq j_0} k_0 I_u(\omega_j)$$

$$+ \frac{1}{m}\frac{A^2 T}{4}\left[1 + O\left(\frac{\log T}{T}\right)\right]$$

$$= M^{-1}\sum_k \log\left[\frac{1}{m}\sum_k I_u(\omega_j)\right] + M^{-1}\log\frac{A^2 T}{4m} + O\left(\frac{\log\log T}{T}\right).$$

At $r = 1$ we will get

$$\log \hat{\sigma}_1^2(m) = M^{-1}\sum_k \log\left[\frac{1}{m}\sum_k I_u(\omega_j)\right] + O\left(\frac{\log\log T}{T}\right)$$

so that (10.7) will fall as r goes from zero to 1 provided

$$M^{-1}\log\frac{A^2 T}{m} = \frac{2M}{T}\log T(1 + O(1)) > 5(\log T)/T$$

which requires $m \geqslant 3$. As r goes from 1 to 2 the criterion should rise because of the penalty term.

However all of this is no more than a plausibility argument since for only occasional T will λ_1 be even very near to some $2\pi j_0/T$. To conclude the plausibility argument we examine the expectation of $\log \hat{\sigma}_r^2(m)$ for $r_0 = 1$ when $u(t)$ is white Gaussian noise as λ_1 varies for fixed T. We shall use λ in place of λ_1 since $r_0 = 1$. We again take T odd and put $N = (T-1)/2$ and $N = mM$. The contribution to $T^{-1/2}\sum y(t)\,\mathrm{e}^{it\omega_j}$ from the signal term is, now taking t to run from 0 to $T-1$,

$$c_j = A\,\mathrm{e}^{i\phi}(2T)^{-1/2}\sum_0^{T-1}\mathrm{e}^{i(\lambda-\omega_j)t} + A\,\mathrm{e}^{-i\phi}(2T^{-1/2})\sum_0^{T-1}\mathrm{e}^{-i(\lambda+\omega_j)t}$$

$$= (2T)^{-1/2}A\,\mathrm{e}^{-i\phi}\sum_0^{T-1}\mathrm{e}^{i(\lambda-\omega_j)t} + \mathrm{O}(T^{-1/2}).$$

This has squared modulus

$$\frac{A^2}{4}\frac{\sin^2 T(\lambda-\omega_j)/2}{T\sin^2(\lambda-\omega_j)/2} + \mathrm{O}(1). \tag{10.17}$$

Then O(1) term is of that order only near $\hat{\lambda}$ and for $(\lambda-\omega_j) \gg 1$ it is $\mathrm{O}(T^{-1})$. Of course

$$\sum_1^N |c_j|^2 = \frac{A^2 T}{4} + \mathrm{O}(1).$$

We now take $u(t)$ to be Gaussian white noise with variance σ^2 and, putting $\tau_j^2 = |c_j|^2/\sigma^2$, obtain

$$E\left[\frac{1}{M}\sum_1^M \log\frac{1}{M}\textstyle\sum_k I_0(\omega_j)\right]$$

$$= \sum_{n=0}^{\infty}\left\{\frac{2^n}{\pi^{1/2}}\frac{\Gamma(n+1/2)}{(2\pi)!}[\log 2 - \log m + \Psi(n+m)]\right\}$$

$$\times \frac{1}{M}\sum_1^M\left(\textstyle\sum_k \tau_j^2\right)^n \mathrm{e}^{-1/2\sum_k\tau_j^2}. \tag{10.18}$$

Here $\Psi(x) = d\log\Gamma(x)/dx$ and we have used the expansion on page 113, formula (7), of Anderson (1958) to evaluate the expectation, since $\sigma^{-2}\sum_k I_0(\omega_j)$ is non central chi square. If $\lambda = \omega_{j_0}$ then $\tau_{j.}^2 = 0$, $j \neq j_0$ and is $A^2T/4\sigma^2$ at $j = j_0$. Thus the only contribution to (10.8) is from the value of k for which the band includes j_0 and then it is

$$\frac{1}{M}(A^2T/4\sigma^2)^{1/2}\exp(-1/2A^2T/4\sigma^2)$$

The term in square brackets is always positive and decreases fast as n increases so that only small values of n need to be considered. The effect of moving

λ away from j_0 is evidently to increase the right side of (10.8). This is because it is the exponential term which will dominate as T increases so that the expectation will be increased by spreading the sum, $A^2/4\sigma^2$, over many bands, as will happen as λ moves away from j_0. This effect will decrease as m increases though the effect will be less marked if λ is near a j_0 value which is a first or last value in $\sum_k$. This accords with the experience discussed in section 10.2. For example for $m = 3$ and $\lambda = 1$ then, at $T = 1024$, λ is nearly exactly $2\pi(163)/1024$, whereas at $T = 512$ then λ is halfway between the values for $j = 81$ and 82 and the effect will be spread over many bands. Thus for $m = 3$ the method based on (10.7) finds $r_0 = 1$ correctly in 100 cases out of 100 at $T = 512$ but only in 76 out of 100 at $T = 1024$.

REFERENCES

An, H-Z., Chen, Z-G. and Hannan, E.J. (1983) The maximum of the periodogram. *J. Multivariate Anal.*, **13**, 383–400.

Anderson, T.W. (1958) *An Introduction to Multivariate Analysis*, Wiley, New York.

Boashash, B. (1992) Estimating and interpreting the instantaneous frequency of a signal—part 2. *Proc. IEEE*, **80**, 540–567.

Bloomfield, P. (1976) *Fourier Analysis of Time Series—An Introduction.* Wiley, New York.

Cameron, M.A., Hannan, E.J. and Speed, T.P. (1992) Estimating spectra and prediction variance. Submitted for publication.

Hannan, E.J. (1973) The estimation of frequency. J. App. Prob., **10**, 510–519.

Hannan, E.J. and Deistler, M. (1988) *The Statistical Theory of Linear Systems*, Wiley, New York.

Hannan, E.J. and Mackisack, M. (1986) A law of the iterated logarithm for an estimate of frequency. *Stoch. Proc. and Appns.*, **22**, 103–109.

Hannan, E.J. and Quinn, B.G. (1989) The resolution of closely adjacent spectral lines. *J. Time Series Anal.*, **10**, 13–31.

Hannan, E.J. and Rissanen, J. (1988) The width of a spectral window, in *A Celebration of Applied Statistics*, (ed. J.M. Gani), Applied Probability Trust, Sheffield, pp. 301–307.

Quinn, B.G. and Fernandes J.M. (1992) A fast efficient technique for the estimation of frequency. *Biometrika*, **28**, 489–498.

Rissanen, J. (1989) *Stochastic Complexity in Statistical Enquiry*, World Scientific, Singapore.

Turkman, K.F. and Walker, A.M. (1984) On the asymptotic maxima of trigonometric polynomials with random coefficients. *Adv. Appl. Prob.*, **16**, 819–842.

Wang, X. (1993) An AIC type estimator for the number of cosinusoids. *J. Time Series Anal.*, submitted for publication.

11

Stationary time series analysis using information and spectral analysis

E. Parzen

11.1 INTRODUCTION

This paper aims to present two emerging ideas about the practice of statistics and time series analysis: (a) to 'stand on the shoulders of giants' to see further how one should develop a framework which unifies diverse methods; (b) information ideas are central to a unified framework since they clarify and extend methods by providing many levels of relationship between time series analysis, classical statistical methods for independent samples, and signal processing problems called inverse problems with positivity constraints.

This paper discusses some roles of information ideas and spectral analysis in time series analysis. It extends spectral estimation by exponential models and goodness of fit tests by components.

A major problem of statistical theory is how to develop technology transfer from esoteric methods to exoteric methods. We define exoteric methods as belonging to an outer or less initiate circle; exoteric statistical methods are those that have reached the status of a **consumer product**, where the consumers are applied researchers. Esoteric methods are known mainly to experts who are researching the theory and are often alleged to be an **intellectual game.**

More methods need to reach the status of consumer products (applicable methods) because computing power enables us to apply several methods to a real problem and reduces the personal investment required to learn how to apply a new method. It should now be possible to implement the growing consensus that problem solving by comparison of several methods leads to conclusions which have increased confidence.

Statisticians who work in time series analysis find their work is appreciated by many researchers in the many fields in which time series analysis is applied and developed. However, they may feel undervalued by the majority of statisticians (to whom time series analysis seems to be separate from the main stream of statistical methods). I feel that time series methods provide many of the right foundations for the successful unification of statistical methods; therefore in retrospect I feel fortunate to have studied time series analysis intensively before beginning in 1977 my work on nonparametric data modeling, unification of statistical methods, and change analysis. Another benefit that I have derived from working in time series analysis has been the friendship of Maurice Priestley and his wife Nancy since we first met in 1958. As I express my esteem for Maurice Priestley and honor his 60th birthday, let me commend Priestley (1981) as the best book to read to learn about time series analysis in both the time and frequency domains.

11.2 ENTROPY, CROSS-ENTROPY, RENYI INFORMATION

The (Kullback–Liebler) information divergence between two probability distributions F and G is defined (Kullback, 1959) by a definition which differs from usual definitions by a factor of 2:

$$I(F;G) = (-2)\int_{-\infty}^{\infty} \log\{g(x)/f(x)\}f(x)\,\mathrm{d}x,$$

when F and G are continuous with probability density functions $f(x)$ and $g(x)$;

$$I(F;G) = (-2)\sum \log\{p_G(x)/p_F(x)\}p_F(x),$$

when F and G are discrete, with probability mass functions $p_F(x)$ and $p_G(x)$. A decomposition of information divergence is

$$I(F;G) = H(F;G) - H(F),$$

in terms of entropy $H(F)$ and cross-entropy $H(F;G)$:

$$H(F) = (-2)\int_{-\infty}^{\infty} \{\log f(x)\}f(x)\,\mathrm{d}x,$$

$$H(F;G) = (-2)\int_{-\infty}^{\infty} \{\log g(x)\}f(x)\,\mathrm{d}x.$$

Adapting the fundamental work of Renyi (1961, 1967) Renyi information of index λ is defined as follows for continuous F and G: for $\lambda \neq 0, -1$

$$IR_\lambda(F;G) = \frac{2}{\lambda(1+\lambda)}\log\int\left(\left\{\frac{g(y)}{f(y)}\right\}^{1+\lambda} - (1+\lambda)\left\{\frac{g(y)}{f(y)} - 1\right\}\right)f(y)\,\mathrm{d}y$$

$$IR_0(F;G) = 2\int\left\{\frac{g(y)}{f(y)}\log\frac{g(y)}{f(y)} - \frac{g(y)}{f(y)} + 1\right\}f(y)\,\mathrm{d}y$$

$$IR_{-1}(F;G) = -2\int\left\{\log\frac{g(y)}{f(y)} - \frac{g(y)}{f(y)} + 1\right\}f(y)\,\mathrm{d}y.$$

An analogous definition holds for discrete F and G.

This definition provides extensions to non-negative functions which are not densities, and also a non-negative integrand which can provide diagnostic measures at each value of y. The above definitions hold for multivariate F and G.

Information and entropy approaches to time series model identification are discussed in Akaike (1974), Jones (1989), Jones and Byrne (1990), Newton (1988), Parzen (1967, 1974, 1977, 1982, 1983a–c, 1986), Shore (1981), Whittle (1953a, b).

11.3 ASYMPTOTIC INFORMATION OF STATIONARY NORMAL TIME SERIES

This section discusses unification of information measures of stationary normal time series and information measures of non-negative functions which are spectral density functions.

When a time series $\{Y(t), t = 1, 2, \ldots\}$ is modeled by alternative probability measures P_1 and P_2 for the infinite sequence, we define asymptotic information divergence (or rate of information divergence)

$$\text{AsymIR}_\lambda(P_2; P_1) = \lim_{n\to\infty} (1/n) IR_\lambda(P_2^{(n)}; P_1^{(n)}),$$

where $P_i^{(n)}$ is the multivariate distribution under P_i of $Y(t)$, $t = 1, \ldots, n$.

When $Y(\cdot)$ is zero mean stationary with covariance function

$$R(v) = E[Y(t)Y(t-v)]$$

and correlation function

$$\rho(v) = R(v)/R(0),$$

information is used to measure the predictability of $Y(t)$ from past values $Y(t-1), \ldots, Y(t-m)$. Define the information about $Y(t)$ in $Y(t-1)$, $Y(t-2), \ldots$, its infinite past (see Parzen 1981 and 1983) by

$$I_\infty = I(Y \mid Y_{-1}, Y_{-2}, \ldots) = E_{Y_{-1}, Y_{-2}, \ldots} I(f_{Y|Y_{-1}, Y_{-2}, \ldots}; f_Y) = \lim_{m\to\infty} I_m$$

$$I_m = I(Y \mid Y_{-1}, \ldots, Y_{-m}) = E_{Y_{-1}, \ldots, Y_{-m}} I(f_{Y|Y_{-1}, \ldots, Y_{-m}}; f_Y).$$

An important classification of time series is by memory type: no memory, short memory, long memory according to $I_\infty = 0$, $0 < I_\infty < \infty$, $I_\infty = \infty$.

The spectral density function $f(\omega)$, $0 \leqslant \omega < 1$, is defined as the Fourier transform of the correlation function (assuming it exists):

$$f(\omega) = \sum_{v=-\infty}^{\infty} \exp(-2\pi i v\omega)\rho(v).$$

We call a time series **bounded** memory if the spectral density is bounded above and below:

$$0 < c_1 \leqslant f(\omega) \leqslant c_2 < \infty.$$

Let P_f denote the probability measure on the space of infinite sequences R_∞ corresponding to a **normal** zero mean stationary time series with spectral density function $f(\omega)$.

A result of Pinsker (1984, p. 196) can be interpreted as providing a formula for asymptotic information divergence between two zero mean stationary time series with respective rational spectral density functions $f(\omega)$ and $g(\omega)$. Write $\mathrm{AsymIR}_\lambda(f, g)$ for $\mathrm{AsymIR}_\lambda(P_f; P_g)$. Adapting Pinsker (1964) one can prove that

$$\mathrm{AsymIR}_{-1}(f, g) = \int_0^1 \{(f(\omega)/g(\omega)) - 1 - \log(f(\omega)/g(\omega))\}\, \mathrm{d}\omega.$$

Because spectral densities are even functions we can take the integral to be over $0 \leqslant \omega < 0.5$; then one obtains the following important theorem.

Theorem 11.1

Unification of information measures of Pinsker (1964), and Itakura and Saito (1970).

$$\mathrm{AsymIR}_{-1}(f, g) = IR_{-1}(f(\omega)/g(\omega))_{0,5}.$$

The validity of this information measure can be extended to non-normal asymptotically stationary time series (Ephraim *et al.*, 1988).

One can heuristically motivate Pinsker's information theoretic justification of the Itakura-Saito distortion measure by the formula for the information divergence between two univariate normal distributions with zero means and different variances.

For bounded memory time series (and $-1 < \lambda < 0$), Kazakos and Kazakos (1980) prove

$$\begin{aligned}&\mathrm{AsymIR}_\lambda(f, g)\\ &= (1/\lambda)\int_0^1 \{\log(f(\omega)/g(\omega)) - (1/(1+\lambda))\log\{1 + (1+\lambda)((f(\omega)/g(\omega)) - 1\}_+\}\, \mathrm{d}\omega.\end{aligned}$$

Kazakos and Kazakos (1980) also give formulas for asymptotic information of multiple stationary time series.

11.4 ESTIMATION OF FINITE PARAMETER SPECTRAL DENSITIES

This section formulates in terms of Renyi information the classic asymptotic maximum likelihood Whittle theory of time series parameter estimation.

For a random sample of a random variable with unknown probability density f, maximum likelihood estimators $\hat{\theta}$ of the parameters of a finite parameter model f_θ of the probability density f can be shown to be equivalent to minimizing

$$IR_{-1}(\tilde{f}, f_\theta),$$

where $\tilde{f}$ is a raw estimator of f (initially, a symbolic sample probability density formed from the sample distribution function $\tilde{F}$). A similar result, called Whittle's estimator (Whittle, 1953a), holds for estimation of spectral densities of a bounded memory zero mean stationary time series for which one assumes a finite parametric model $f_\theta(\omega)$ for the true unknown spectral density $f(\omega)$.

A raw fully nonparametric estimator of $f(\omega)$ from a time series sample $Y(t)$, $t = 1, \ldots, n$, is the sample spectral density (or periodogram)

$$\tilde{f}(\omega) = \left| \sum_{t=1}^{n} Y(t) \exp(-2\pi i \omega t) \right|^2 \div \sum_{t=1}^{n} |Y(t)|^2.$$

Note that $\tilde{f}(\omega)$ is not a consistent estimator of $f(\omega)$; nevertheless,

$$E[\tilde{f}(\omega)] \text{ converges to } f(\omega),$$

a fact which can be taken as the definition of the spectral density $f(\omega)$.

An estimator $\hat{\theta}$ which is asymptotically equivalent to the maximum likelihood estimator is obtained by minimizing

$$\text{AsymIR}_{-1}(\tilde{f}; f_\theta) = IR_{-1}(\tilde{f}, f_\theta)_{0.5} = \int_0^1 \{(\tilde{f}(\omega)/f_\theta(\omega)) - 1 - \log(\tilde{f}(\omega)/f_\theta(\omega))\}\, d\omega,$$

which can be interpreted as choosing θ to make $\tilde{f}(\omega)/f_\theta(\omega)$ as flat or constant as possible.

We usually use the representation

$$f_\theta(\omega) = \sigma^2/\gamma_\theta(\omega)$$

where $\gamma_\theta(\omega)$ is the square modulus of the transfer function of the whitening filter represented by the spectral density model f_θ, constructed so that

$$\int_0^1 \log f_\theta(\omega)\, d\omega = \log \sigma^2 = -I_\infty.$$

Minimizing $\text{AsymIR}_{-1}(\tilde{f}, f_\theta)$ is equivalent to minimizing

$$(1/\sigma^2)\int_0^1 \{\tilde{f}(\omega)\gamma_\theta(\omega)\}\,\mathrm{d}\omega + \log(\sigma^2)$$

which is equivalent to minimizing over θ

$$\sigma_\theta^2 = \int_0^1 \gamma_\theta(\omega)\tilde{f}(\omega)\,\mathrm{d}\omega$$

and setting

$$\hat{\sigma}^2 = \int_0^1 \gamma_{\hat{\theta}}(\omega)\tilde{f}(\omega)\,\mathrm{d}\omega = \sigma_{\hat{\theta}}^2.$$

The information divergence between the data and the fitted model is given by

$$IR_{-1}(\tilde{f}, f_{\hat{\theta}}) = \log\sigma_{\hat{\theta}}^2 - \log\hat{\sigma}^2 = \tilde{I}_\infty - \hat{I}_\infty,$$

defining $-\hat{I}_\infty = \log\hat{\sigma}^2$,

$$-\tilde{I}_\infty = \log\tilde{\sigma}^2 = \int_0^1 \log\tilde{f}(\omega)\,\mathrm{d}\omega.$$

This criterion (however, corrected for bias in $\tilde{I}_\infty$) arises from information approaches to model identification (Parzen, 1983a). A model fitting criterion (but not a parameter estimation criterion) is provided by the information increment

$$I(Y\,|\,\text{all past } Y;\, Y \text{ values in model } \theta)$$
$$= \int_0^1 -\log\{\tilde{f}(\omega)/f_{\hat{\theta}}(\omega)\} = IR_{-1}(\tilde{f}/f_{\hat{\theta}})_{0,-.5}.$$

One can regard it as a measure of the distance of the whitening spectral density

$$f^*(\omega) = \tilde{f}(\omega)/f_{\hat{\theta}}\omega)$$

from a constant function; note that $f^*(\omega)$ is constructed to integrate to 1. When one accepts that the optimal smoother of $f^*(\omega)$ is a constant, a 'parameter-free' nonparametric estimator of the spectral density $f(\omega)$ by a smoother of $\tilde{f}(\omega)$ is given by the parametric estimator $f_{\hat{\theta}}$. By 'parameter-free' we mean that we are free to choose the parameters to make the data (raw estimator) shape up to a smooth estimator. The parameters are not regarded as having any significance or interpretation; they are merely coefficients of a representation of $f(\omega)$.

Portmanteau statistics to test goodness of fit of a model to the time series use sums of squares of correlations of residuals; an analogous statistic is

$$IR_1(\tilde{f}/f\hat{\theta})_{0,0.5} = \log \int_0^{0.5} \{\tilde{f}(\omega)/f_{\hat{\theta}}(\omega)\}^2 \, d\omega.$$

Goodness of fit of the model to the data (as measured by how close $f^*(\omega)$ is to the spectral density of white noise) is the ultimate model identification criterion to decide between competing parametric models.

11.5 GOODNESS OF FIT BY COMPONENTS AND EXPONENTIAL MODELS

We argue that goodness of fit tests of a model should test for whiteness:

$$f^*(\omega) = \tilde{f}(\omega)/f_{\hat{\theta}}(\omega).$$

We propose an analogue of the concept of components introduced in the classical goodness of fit theory by Durbin and Knott (1972):

$$T^*(J) = 2^{-.5} \int_0^1 f^*(\omega) J(\omega) \, d\omega$$

for various score functions $J(\omega)$.

One usually forms a sequence of components with functions

$$J_0(\omega) = 1, J_1(\omega), J_2(\omega), \ldots$$

which are a complete orthonormal set of functions in $L_2[0, 1]$. Choices are: harmonics ($\cos 2\pi j\omega, j = 0, 1, 2, \ldots$); Legendre polynomials; Hermite polynomial functions of the standard normal quantile function Φ^{-1}.

Under the assumption that $f(\omega) = f(\omega; \theta)$ for some parameter vector θ, the asymptotic distribution of $T^*(J_j)$ is the same as

$$2^{-.5} \int_0^1 (1/f(\omega)) J_j(\omega) \tilde{f}(\omega) \, d\omega$$

which is asymptotically normal with

$$\text{mean} \int_0^1 J(\omega) \, d\omega = 0,$$

$$\text{variance } (2/n) \int_0^1 (1/2f^2(\omega)) |J_j(\omega)|^2 f^2(\omega) \, d\omega) = (1/n) \int_0^1 |J_j(\omega)|^2 \, d\omega = 1/n.$$

Thus, properly defined components are asymptotically independent normal $(0, 1/n)$.

A component-based quadratic test of the goodness of fit of the model, with an asymptotic chi-square distribution, is

$$S_{k,m} = \sum_{j=k}^{m} |T^*(J_j)|^2.$$

These component tests have the asymptotic optimality properties of score tests if we model the true spectral density $f(\omega)$ by an exponential model extending Bloomfield (1973).

We propose to estimate $f(\omega)$ by assuming an exponential model of order m using score functions $J_j(\omega)$, $j = 1, \ldots, m$; the choice of score functions and criteria for determining from the data an optimal order $\hat{m}$ require further research.

Note that an exponential model for the spectral density provides smooth estimators of the log spectral density and therefore of cepstral correlations (the Fourier coefficients of the log spectrum) and coefficients of the AR(∞) and MA(∞) representations of a time series required for prediction.

The exponential model of order m, denoted $f_{\theta,m}$, is defined

$$\log f_{\theta,m}(\omega) = \theta_0 + \theta_1 J_1(\omega) + \cdots + \theta_m J_m(\omega).$$

The coefficient θ_0 has the interpretation

$$\theta_0 = \int_0^1 \log f(\omega)\,\mathrm{d}\omega = \log \sigma_\infty^2$$

where σ_∞^2 is the infinite memory one step ahead prediction mean square error. The exponential model can be expressed

$$f_{\theta,m} = \sigma_\infty^2 \exp\left[\sum_{j=1}^{m} \theta_j J_j(\omega)\right].$$

Maximum likelihood estimators $\hat{\theta}_m = (\hat{\theta}_1, \ldots, \hat{\theta}_m)$ of $\theta^m = (\theta_1, \ldots, \theta_m)$ are equivalent to minimizing

$$V(\theta) = \int_0^1 \mathrm{d}\omega \tilde{f}(\omega) \exp\left(-\sum_{j=1}^{m} \theta_j J_j(\omega)\right)$$

and then estimating σ_∞^2 by $V(\hat{\theta}^m)$.

The estimated spectral density is given by

$$f_{\hat{\theta}}(\omega) = V(\hat{\theta}^m) \exp\left[\sum_{j=1}^{m} \hat{\theta}_j J_j(\omega)\right];$$

which satisfies

$$\int_0^1 f^{*m}(\omega)\,\mathrm{d}\omega = 1,$$

defining

$$f^{*m}(\omega) = \tilde{f}(\omega)/f\hat{\theta}_m(\omega)$$

The product of the Fisher score function (derivative with respect to θ_j of the optimization criterion $V(\theta)$) and $2^{-.5}$ is denoted $U_j(\theta)$; for $j = 1, \ldots, m$

$$U_j(\theta) = 2^{-.5} \int_0^1 \mathrm{d}\omega(\tilde{f}(\omega)/f_\theta(\omega))J_j(\omega)$$

A goodness of fit test of a model of order m is given by a score test of an order m sub-model against an order M 'full' model:

$$U_j(\hat{\theta}^m) = 0, \quad j = m + 1, \ldots, M.$$

An overall chi-square test uses the sum of squares of these score statistics. To compute the parameter estimators, let

$$U(\hat{\theta}^m) = (U_1(\hat{\theta}^m), \ldots, U_m(\hat{\theta}^m)).$$

An approximate Newton–Raphson iterative scheme for computing can be shown, following Bloomfield (1973), to be

$$\hat{\theta}^{m(n+1)} = \hat{\theta}^{m(n)} - .5U(\hat{\theta}^{m(n)}).$$

Note that the vector of correction terms in this iteration is the vector of score tests. Exponential models for the spectral density use the same score statistics for iterative evaluation of estimators as are used for component tests of goodness of fit.

An initial estimator of θ_j, adapting Bloomfield (1973), is

$$\theta_j^{(1)} = (1/n) \sum_{t=1}^{n} \log \tilde{f}(2\pi t/n)J_j(2\pi t/n).$$

It should be emphasized that the foregoing approach to goodness of fit and spectral density estimation needs further research about the problems of choosing score functions $J_j(\omega)$ and determining an optimal order m.

REFERENCES

Akaike, H. (1974) A new look at the statistical model identification. *IEEE Trans. Autom. Contr.*, **AC-19**, 716–723.

Bloomfield, P. (1973) An exponential model for the spectrum of a scalar time series. *Biometrika*, **60** (2), 217–226.

Durbin, J. and Knott, S. (1972) Components of Cramér-von Mises statistics I, *J. Roy. Statist. Soc. Ser. B*, **34**, 290–307.

Ephraim, Y., Hanoch, L., and Gray, R. (1988) Asymptotic minimum discrimination information measure for asymptotically weakly stationary processes, *IEEE Transactions on Information Theory*, **34** (5), 1033–1040.

Itakura, F. and Saito, S. (1970) A statistical method for estimation of speech spectral density and format frequencies, *Electron. Commun. Japan*, **53-A**, 36–43.

Jones, L.K. (1989) Approximation theoretic derivation of logarithmic entropy principles for inverse problems and unique extension of the maximum entropy method to incorporate prior knowledge, *SIAM J. Appl. Math.*, **49**, 650–661.

Jones, L.K. and Byrne, C.L. (1990) General entropy criteria for inverse problems, with applications to data compression, pattern classification, and cluster analysis, *IEEE Transactions on Information Theory*, **36** (1), 23–30.

Kazakos, D. and Papantoni-Kazakos, P. (1980) Spectral distance measures between Gaussian processes, *IEEE Trans. Automat. Contr.*, **AC-25** (5), 950–959.

Kullback, S. (1959) *Information Theory and Statistics*, Wiley, New York.

Newton, H.J. (1988) *TIMESLAB: A Time Series Analysis Laboratory*, Wadsworth, Pacific Grove, California.

Parzen, E. (1967) *Time Series Analysis Papers*, Holden-Day, San Francisco, California.

Parzen, E. (1969) Multiple time series modeling, *Multivariate Analysis—II*. (ed. P. Krishnaiah), Academic Press, New York, pp. 289–409.

Parzen, E. (1974) Some recent advances in time series modeling. *IEEE Transactions on Automatic Control*, **AC-19**, 723–730.

Parzen, E. (1977) Multiple time series: determining the order of approximating autoregressive schemes, *Multivariate Analysis—IV*, (ed. P. Krishnaiah), North Holland, Amsterdam, pp. 283–295.

Parzen, E. (1981) Time series model identification and prediction variance horizon, *Proceedings of Second Tulsa Symposium on Applied Time Series Analysis*, Academic Press, New York, pp. 425–447.

Parzen, E. (1982) Maximum entropy interpretation of autoregressive spectral densities, *Statistics and Probability Letters*, **1**, 2–6.

Parzen, E. (1983a) Time series model identification by estimating information, *Studies in Econometrics, Time Series, and Multivariate Statistics in Honor of T.W. Anderson*, (ed. S. Karlin, T. Amemiya, L. Goodman), Academic Press, New York, pp. 279–298.

Parzen, E. (1983b). Time series ARMA model identification by estimating information, *Proceedings of the 15th Annual Symposium on the Interface of Computer Science and Statistics*, North Holland, Amsterdam.

Parzen, E. (1983c) Time series model identification by estimating information, memory, and quantiles. *Questo*, **7**, 531–562.

Parzen, E. (1986) Quantile spectral analysis and long memory time Series, *Journal of Applied Probability*, **23A**, 41–55.

Pinsker, M.S. (1964) *Information and Information Stability of Random Variables and Processes*. Holden-Day, San Francisco, CA.

Priestley, M.B. (1981) *Spectral Analysis and Time Series*, Academic Press, London.

Renyi, A. (1961) On measures of entropy and information *Proc. 4th Berkeley Symp. Math. Statist. Probability, 1960*, **1**, 547–561, University of California Press, Berkeley.

Renyi, A. (1967) On some basic problems of statistics from the point of view of information theory, *Proc. 5th Berkeley Symp. on Math., Stat. and Probability*, 531–543.

Shore, J. (1981) Minimum cross-entropy spectral analysis. *IEEE Trans Acoust. Speech, Signal Processing*, **ASSP-29** (2), 230–237.

Whittle, P. (1953a) Estimating and information in stationary time series. *Ark. Math.* **2**, 423–434.

Whittle, P. (1953b) The analysis of multiple stationary time series. *J. Royl Statist. Soc. B.*, **15**, 125–139.

12

Periodogram analysis for complex-valued time series

A.M. Walker

12.1 INTRODUCTION

In the theoretical treatment of second order properties of weakly stationary processes $\{X(t), t \in T\}$ the random variables $X(t)$ are often taken to be complex-valued (see, for example, Yaglom (1962), Bartlett (1966), Priestley (1981, pp. 110–111). Results needed for applications are then obtained by setting to zero the imaginary parts of the $X(t)$'s. However, one can also have a weakly stationary bivariate real-valued process

$$\left\{\begin{pmatrix} U(t) \\ V(t) \end{pmatrix}, t \in T\right\}$$

and an associated complex-valued process $\{X(t), t \in T\}$ by letting $X(t) = U(t) + iV(t)$. Properties of $\{X(t)\}$ can clearly always be obtained from properties of the bivariate process. For example, the autocovariance function of $\{X(t)\}$, defined by

$$R_X(s) = E[\{X(t+s) - \mu_X\}\{X(t) - \mu_X\}^*]. \tag{12.1}$$

where $\mu_X = E[X(t)] = E[U(t)] + iE[V(t)]$, and* denotes the complex conjugate, is equal to

$$R_{UU}(s) + R_{VV}(s) - i\{R_{UV}(s) - R_{VU}(s)\}, \tag{12.2}$$

where

$$R(s) = \begin{pmatrix} R_{UU}(s) & R_{UV}(s), \\ R_{VU}(s) & R_{VV}(s) \end{pmatrix} = \text{cov}\left[\begin{pmatrix} U(t+s) \\ V(t+s) \end{pmatrix}, (U(t), V(t))\right] \tag{12.3}$$

is the autocovariance matrix for lag s of the bivariate process.

As regards statistical analysis based on a realisation for $t \in S$, where S is a subset of T, it clearly makes no difference whether we take the process to be the bivariate real-valued one or the complex-valued one. But properties of interest may be different; for example, information about the autocovariance structure of the bivariate process will be obtained from estimates

$$\begin{pmatrix} \hat{R}_{UU}(s), & \hat{R}_{UV}(s) \\ \hat{R}_{VU}(s), & \hat{R}_{VV}(s) \end{pmatrix}$$

of autocovariance matrices, while corresponding estimates for the complex-valued process, $\hat{R}_{UU}(s) + \hat{R}_{VV}(s) - i\{\hat{R}_{UV}(s) - \hat{R}_{VU}(s)\}$, clearly contain less information.

Suppose now that we have the usual discrete parameter situation, where $T = \{0, \pm 1, \pm 2, \ldots,\}$ and $S = \{1, 2, \ldots, n\}$. Standard methods, usually involving either autocovariance analysis (in the time domain) or spectral analysis (in the frequency domain) for real-valued bivariate processes have been available for a long time (see, for example, Priestley (1981, Chapter 9) for a particularly useful concise account). However, similar methods for complex-valued univariate processes seem to be not at all well-known despite no new distribution theory whatsoever being needed. In this note we consider some properties of the periodogram of a complex-valued process observed for $t = 1, 2, \ldots, n$, which we define as

$$I_{n,x}(\omega) = \frac{1}{2\pi n}\left|\sum_{t=1}^{n} X(t)e^{-i\omega t}\right|^2, \quad -\pi < \omega \leqslant \pi. \tag{12.4}$$

(Here we have assumed that $E[X(t)] = 0$. Otherwise $X(t)$ in (12.4) may be replaced by $X(t) - \bar{X}$, where

$$\bar{X} = \sum_{t=1}^{n} X(t)/n;$$

the effect of doing so will be asymptotically negligible, and we therefore suppose that the assumption holds in what follows.)

We suppose that the spectrum of the bivariate process is continuous with spectral density matrix

$$f(\omega) = \begin{pmatrix} f_{UU}(\omega), & f_{UV}(\omega) \\ f_{VU}(\omega), & f_{VV}(\omega) \end{pmatrix} = \frac{1}{2\pi}\sum_{s=-\infty}^{\infty} R(s)e^{-i\omega s}, \tag{12.5}$$

$R(s)$ being the matrix defined in (12.3). We shall also assume that

$$\sum_{s=-\infty}^{\infty} |R_{UU}(s)| < \infty, \quad \sum_{s=-\infty}^{\infty} |R_{VV}(s)| < \infty, \quad \sum_{s=-\infty}^{\infty} |R_{UV}(s)| < \infty, \tag{12.6}$$

which ensures that $f(\omega)$ is continuous (by uniformity of convergence of the series in (12.5)). These conditions are by no means necessary but should often be satisfied, and enable the derivation of the results to be kept very simple.

Periodograms of complex-valued processes have in fact been used in practical situations, namely in the analysis of synthetic aperture radar data carried out at the University of Sheffield, where part of the output consisted of quantities proportional to $I_{n,x}(\omega)$ for a large number of series, t in this case representing a spatial variable (see Suttie (1990)). Note that, in contrast to what happens with real-valued processes, we have in general $I_{n,x}(-\omega) \neq I_{n,x}(\omega)$, when $\omega \neq 0, \pi$.

12.2 FIRST ORDER RESULTS

We refer to the results which follow as first order because they are concerned with the expectations of quadratic functions of

$$\begin{pmatrix} U(t) \\ V(t) \end{pmatrix},$$

and $I_{n,x}(\omega)$ is a quadratic function of

$$\begin{pmatrix} U(t) \\ V(t) \end{pmatrix}.$$

Theorem 12.1

Let $\{X(t), t = 0, \pm 1, \ldots\}$ be a complex-valued weakly stationary process with zero mean, having a continuous spectrum and an autocovariance matrix for the associated bivariate process satisfying the conditions (12.6). Let

$$I_{n,x}(\omega) = A_n^2(\omega) + B_n^2(\omega), \; -\pi < \omega \leqslant \pi,$$

where

$$A_n(\omega) = (2\pi n)^{-1/2} \sum_{t=1}^{n} (U(t) \cos \omega t + V(t) \sin \omega t), \tag{12.7}$$

and

$$B_n(\omega) = (2\pi n)^{-1/2} \sum_{t=1}^{n} (-U(t) \sin \omega t + V(t) \cos \omega t). \tag{12.8}$$

Then as $n \to \infty$, provided that $\omega \neq 0, \pi$,

$$\text{var}\,(A_n(\omega)] \text{ and var}\,[B_n(\omega)] \text{ tend to } \tfrac{1}{2}\{f_{UU}(\omega) + f_{VV}(\omega) + 2\text{Im}(f_{UV}(\omega))\}, \tag{12.9}$$

$$\text{and cov}\,[A_n(\omega), B_n(\omega)] \text{ tends to zero.} \tag{12.10}$$

Corollary

$$\text{As } n \to \infty, \; E[I_{n,x}(\omega)] \to f_X(\omega), \tag{12.11}$$

the spectral density function of $\{X(t)\}$.

Proof

Write

$$C_{n,u}(\omega) = (2\pi n)^{-1/2} \sum_{t=1}^{n} U(t)\cos \omega t, \quad S_{n,u}(\omega) = (2\pi n)^{-1/2} \sum_{r=1}^{n} U(t)\sin \omega t \quad (12.12)$$

and define $C_{n,v}(\omega)$, $S_{n,v}(\omega)$ similarly (by replacing $U(t)$ by $V(t)$ in (12.12)). Denote by $D_n(\omega)$ the column vector $(C_{n,u}(\omega), S_{n,u}(\omega), C_{n,v}(\omega)\ S_{n,v}(\omega))'$, and let $\omega \neq 0, \pi$. Then from standard results for bivariate processes we have

$$\lim_{n\to\infty} E[D_n(\omega)D_n(\omega)'] = \begin{pmatrix} P & 0 & R & S \\ 0 & P & -S & R \\ R & -S & Q & 0 \\ S & R & 0 & Q \end{pmatrix} \quad (12.13)$$

where

$$P = \tfrac{1}{2}f_{UU}(\omega),\ Q = \tfrac{1}{2}f_{VV}(\omega),\ R + iS = \tfrac{1}{2}f_{UV}(\omega) \quad (12.14)$$

under wide conditions (see, for example, Hannan (1970, p. 250); the conditions (12.6) certainly suffice.

For example

$$\begin{aligned} \operatorname{cov}[C_{n,u}(\omega), S_{n,v}(\omega)] &= \frac{1}{2\pi n} \sum_{r,s=1}^{n} \operatorname{cov}[U(r),V(s)] \cos \omega r \sin \omega s \\ &= \frac{1}{4\pi n} \sum_{|v|\leqslant n-1} \sum_{s} R_{UV}(v)[\sin \omega(2s+v) - \sin \omega v] \end{aligned}$$

putting $r = s + v$,

$$= \frac{1}{4\pi n}\left[\sum_{|v|\leqslant n-1} -(n-|v|)R_{UV}(v)\sin \omega v + O(R_{UV}(v))\right]$$

since

$$\left|\sum_{v=1}^{m} \sin 2\omega v\right|, \quad \left|\sum_{v=1}^{m} \cos 2\omega v\right| \leqslant \frac{1}{\sin \omega} \text{ for all } m.$$

This tends to

$$-\frac{1}{4\pi}\sum_{v=-\infty}^{\infty} R_{UV}(v)\sin \omega v = \frac{1}{2}\operatorname{Im}\left\{\frac{1}{2\pi}\sum_{v=-\infty}^{\infty} R_{UV}(v)e^{-i\omega v}\right\} = S. \quad (12.15)$$

as n tends to ∞.

The limiting values of the other three covariances occurring in the left-hand side of (12.13) can be found in the same way, giving the expressions on the right-hand side. Expression (12.9) follows at once since

$$\text{var}[A_n(\omega)] = \text{var}[C_{n,u}(\omega)] + \text{var}[S_{n,v}(\omega)] + 2\text{cov}[C_{n,u}(\omega), S_{n,v}(\omega)]$$
$$\text{and var}[B_n(\omega)] = \text{var}[S_{n,u}(\omega)] + \text{var}[C_{n,v}(\omega)] - 2\text{cov}[C_{n,v}(\omega), S_{n,u}(\omega)]$$

both of which tend to

$$P + Q + 2S \tag{12.16}$$

as $n \to \infty$. Similarly, $\text{cov}[A_n(\omega), B_n(\omega)] = \text{cov}[C_{n,u}(\omega) + S_{n,v}(\omega), \ -S_{n,u}(\omega) - C_{n,v}(\omega)]$ tends to $R - R = 0$ as $n \to \infty$.

For the corollary we use the spectral representation of $\{X(t)\}$, namely

$$X(t) = \int_{-\pi}^{\pi} e^{i\omega t}(dZ_u(\omega) + i dZ_v(\omega)),$$

where $\{Z_u(\omega)\}, \{Z_v(\omega)\}$ are processes of orthogonal increments that $E[|dZ_u(\omega) + idZ_v(\omega)|^2] = f_X(\omega)d\omega$. For then we have

$$\begin{aligned} f_X(\omega)d\omega &= E[|dZ_u(\omega)|^2 + |dZ_v(\omega)|^2 + i(dZ_u(\omega)^* dZ_v(\omega) - dZ_u(\omega)dZ_v(\omega)^*)] \\ &= \{f_{UU}(\omega) + f_{VV}(\omega) + 2\text{Im}(f_{UV}(\omega))\}d\omega, \end{aligned}$$

so that $f_X(\omega) = 2\{P + Q + 2S\}$, and the result follows from (12.9).

The corollary can alternatively be derived directly as an immediate consequence of

$$E[I_{n,x}(\omega)] = \frac{1}{2\pi n} \sum_{r,s=1}^{n} E[X_r X_s^*] e^{-i\omega(r-s)}$$

For this is equal to

$$\frac{1}{2\pi n} \sum_{|v| \leqslant n-1} \left|1 - \frac{|v|}{n}\right| R_X(v) e^{-i\omega v}$$

which tends to

$$\frac{1}{2\pi} \sum_{v=-\infty}^{\infty} R_X(v) e^{-i\omega v} = f_X(\omega) \text{ as } n \to \infty.$$

When $\omega = 0$ or π these results hold with a slight modification. Since then $A_n(\omega) = C_{n,u}(\omega)$ and $B_n(\omega) = C_{n,v}(\omega)$, $\lim_{n\to\infty} \text{var}[A_n(\omega)] = f_{UU}(\omega)$, $\lim_{n\to\infty} \text{var}[B_n(\omega)] = f_{VV}(\omega)$, while $S(\omega) = \frac{1}{2}\text{Im}\{f_{UV}(\omega)\} = 0$, corollery 12.2 is still valid. However, when $n \to \infty$, $\text{cov}[A_n(\omega), B_n(\omega)]$ tends to $R(\omega) = \frac{1}{2}\text{Re}(f_{UV}(\omega))$ instead of zero (as is obvious, for example, by taking the bivariate process

$$\begin{pmatrix} U(t) \\ V(t) \end{pmatrix}$$

to be such that $E[\{U(t)\}^2] = E[\{V(t)\}^2] = 1$, $E[U(t)V(t)] = \rho$, $E[U(s)V(t)] = 0$, $s \neq t$).

A further important remark concerning first order results is that any property of the expectations of periodogram values has an exact analogue for real-valued processes. This is due to the fact that the behaviour of the exact expression for $2\pi E[I_{n,x}(\omega)]$, namely

$$\sum_{|v| \leqslant n-1} \left(1 - \frac{|v|}{n}\right) R_X(v) \mathrm{e}^{-i\omega v},$$

is the same whether the covariance function is that of a real-valued or a complex-valued process. This means that all results for real-valued processes regarding biases of spectral density estimates based on 'spectral windows', namely of the form

$$\int_{-\pi}^{\pi} I_{n,x}(\lambda) W_n(w - \lambda) \mathrm{d}\lambda,$$

where $W_n(\theta)$ is a suitable non-negative function with period 2π and

$$\int_{-\pi}^{\pi} W_n(\theta) \mathrm{d}\theta = 1,$$

are equally applicable here. It would seem that the spectral window estimates could not be computable exactly from the above formula and consequently have to be replaced by approximating sums consisting of weighted means of values of $I_{n,x}(\lambda)$, the most obvious choice being those for $\lambda = \frac{2\pi_j}{n}, j = -[\frac{n}{2}] \ldots, 0, \ldots [\frac{n}{2}]$.

This makes no difference to asymptotic results. Nevertheless it is perhaps worth noting that the definition of a spectral window estimate as a finite weighted sum of sample autocovariances

$$\hat{R}(v) = \sum_{t=1}^{n-v} X(t+v) X^*(t)/n \text{ for } v \geqslant 0,$$

and $\hat{R}(v) = \hat{R}^*(-v)$ for $v < 0$, does enable it to be computed from a finite set of periodogram values. For from the definition of $I_{n,x}(\lambda)$ given in (12.4) we have

$$2\pi n I_{n,x}(\lambda) = \sum_{|v| \leqslant n-1} n \hat{R}(v) \mathrm{e}^{-i\lambda v} \tag{12.17}$$

Hence, letting $\omega_r = r\pi/n$, we have

$$2\pi \sum_{r=-(n-1)}^{n} I_{n,x}(\omega_r) \mathrm{e}^{i\omega_r u} = \sum_{|v| \leqslant n-1} \hat{R}(v) \sum_{r=-(n-1)}^{n} \mathrm{e}^{i\omega_r(u-v)} \tag{12.18}$$

$$= 2n\hat{R}(u), \; u = -(n-1) \ldots, (n-1). \tag{12.19}$$

Theorem 12.2

Let $\{X(t)\}$, $I_{n,x}(\omega)$, $A_n(\omega)$, $B_n(\omega)$ be as in theorem 12.1, and $D_n(\omega)$, the column vector $(C_{n,u}(\omega), S_{n,u}(\omega), C_{n,v}(\omega), S_{n,v}(\omega))'$ whose elements are defined by (12.12). Let $-\pi < \omega_1 < \omega_2 \leqslant \pi$, and $\omega_1 + \omega_2 \neq 0$. Then

$$\lim_{n\to\infty} E\{D_n(\omega_1)[D_n(\omega_2)]'\} = 0_{4\times 4}, \tag{12.20}$$

where $0_{4\times 4}$ denotes the zero 4×4 matrix. Also, when $\omega \neq 0, \pi$,

$$\lim_{n\to\infty} E[A_n(\omega)A_n(-\omega)] = \tfrac{1}{2}\{f_{UU}(\omega) - f_{VV}(\omega)\} \tag{12.21}$$

$$\lim_{n\to\infty} E[B_n(\omega)B_n(-\omega)] = \tfrac{1}{2}\{f_{VV}(\omega) - f_{UU}(\omega)\} \tag{12.22}$$

and

$$\lim_{n\to\infty} E[A_n(\omega)B_n(-\omega)] = \mathrm{Re}\{f_{UV}(\omega)\} \tag{12.23}$$

where (as before), $f_{UU}(.)$, $f_{VV}(.)$, $f_{UV}(.)$ denote the spectral densities of $\{U(t)\}$, $\{V(t)\}$ and the cross-spectral density.

Proof

We have only to observe that the elements of $D_n(\omega_1)$, $D_n(\omega_2)$ are linear functions of $C_{n,u}(\omega_j)$, $S_{n,u}(\omega_j)$, $C_{n,v}(\omega_j)$, $S_{n,v}(\omega_j)$, $j = 1,2$ (defined at (12.12)). The result then follows from an elementary property of the process

$$\begin{Bmatrix} U(t) \\ V(t) \end{Bmatrix}$$

and this is easily verified by simple evaluation of expectations in the same way as those required to obtain the main result of theorem 12.1. For example,

$$\begin{aligned} 2\pi n \operatorname{cov}[C_{n,u}(\omega_1), S_{n,v}(\omega_2)] &= \sum_{r,s=1}^{n} \cos[U(r), V(s)] \cos\omega_1 r \sin\omega_2 s \\ &= \sum_{|v|\leqslant n-1} R_{UV}(v) \sum_s \sin\omega_2 s \cos\{\omega_1(s+v)\} \end{aligned}$$

and the inner sum is equal to

$$\frac{1}{2}\sum_s [\sin\{(\omega_1+\omega_2)s + \omega_1 v\} + \sin\{(\omega_2-\omega_1)s - \omega_1 v\}],$$

which is bounded since

$$\sum_s \cos(\omega_2 \pm \omega_1)s \text{ and } \sum_s \sin(\omega_2 \pm \omega_1)s \text{ are.}$$

The final results (12.21) to (12.23) follow at once using the definitions (12.7), (12.8) of $A_n(\omega)$, $B_n(\omega)$ and the limiting covariance matrix given by theorem 12.1. For example, as

$$A_n(\omega) = C_{n,u}(\omega) + S_{n,v}(\omega), \operatorname{cov}[A_n(\omega), A_n(-\omega)] = \operatorname{var}[C_{n,u}(\omega)] - \operatorname{var}[S_{n,v}(\omega)],$$

which tends to $\frac{1}{2}(f_{UU}(\omega) - f_{VV}(\omega))$ as $n \to \infty$.

12.3 SECOND ORDER RESULTS

These are mainly concerned with asymptotic variances and covariances of periodogram values $I_{n,x}(\omega)$. We first, as Priestley (1981, p. 405) does for the real-valued case, take $\{X(t)\}$ to be purely random, by which we mean that the associated process

$$\begin{pmatrix} U(t) \\ V(t) \end{pmatrix}$$

is one of the independent and identically distributed random variables.

Theorem 12.3

Let $\{X(t),\ t = 0, \pm 1, \ldots\}$ be a purely random complex-valued stationary process with zero mean such that $E[\{U(t)\}^4] < \infty$, $E[\{V(t)\}^4] < \infty$. Let $-\pi < \omega_1 < \omega_2 < \pi$ and $\omega_1 + \omega_2 \neq 0$. Then

$$(2\pi n)^2 \operatorname{cov}[I_{n,x}(\omega_1), I_{n,x}(\omega_2)] = na + b\phi_n(\omega_2 + \omega_1) + c\phi_n(\omega_2 - \omega_1), \quad (12.24)$$

where, denoting $E[\{U(t)\}^i\{V(t)\}^j]$ by μ_{ij},

$$a = \mu_{40} - 3\mu_{20}^2 + \mu_{04} - 3\mu_{02}^2 - 2\mu_{02}\mu_{20} - 4\mu_{11}^2 + 2\mu_{22} \quad (12.25)$$

$$b = (\mu_{20} - \mu_{02})^2 + 4\mu_{11}^2,\ c = (\mu_{20} + \mu_{02})^2, \quad (12.26)$$

and

$$\phi_n(\omega) = \frac{\sin^2 \frac{1}{2}n\omega}{\sin^2 \frac{1}{2}\omega}, \omega \neq 0. \quad (12.27)$$

Proof

We have

$$(2\pi n)^2 E[I_{n,x}(\omega_1) I_{n,x}(\omega_2)] = \sum_{r,s,t,u=1}^{n} E[X(r)X^*(s)X(t)X^*(u)]e^{-i\omega_1(r-s) - i\omega_2(t-u)}. \quad (12.28)$$

Write the sum on the right hand side of (12.28) as $S_1 + S_2 + S_3 + S_4$, where $S_1 \ldots S_4$ denote sums of terms with $r = s = t = u$, $r = s \neq t = u$, $r = t \neq s = u$, $r = u \neq s = t$, respectively. (Clearly $E[X(r)X^*(s)X(t)X^*(u)] = 0$ when the arguments (r,s,t,u) are not equal in pairs.) Now,

$$S_1 = nE[\{X(r)X^*(r)\}^2] = n(\mu_{40} + \mu_{04} + 2\mu_{22}), \tag{12.29}$$

$$S_2 = n(n-1)(E[X(r)X^*(r)])^2 = n(n-1)(\mu_{20} + \mu_{02})^2, \tag{12.30}$$

$$\begin{aligned} S_3 &= \sum_{r,s=1}^{n} E[X(r)X^*(s)X(r)X^*(s)]\mathrm{e}^{-i(\omega_1+\omega_2)(r-s)} \\ &= E[\{X(r)\}^2]E[\{X^*(s)\}^2]\left\{\sum_{r,s=1}^{n} \mathrm{e}^{-i(\omega_1+\omega_2)(r-s)} - n\right\} \qquad (12.31) \\ &= |(\mu_{20} - \mu_{02}) + 2i\mu_{11}|^2\{\phi_n(\omega_1 + \omega_2) - n\}, \end{aligned}$$

and similarly

$$\begin{aligned} S_4 &= \sum_{r,s=1}^{n} E[X(r)X^*(s)X(s)X^*(r)]\mathrm{e}^{i(\omega_2-\omega_1)(r-s)} \\ &= (\mu_{20} + \mu_{02})^2\{\phi_n(\omega_1 - \omega_2) - n\}. \qquad (12.32) \end{aligned}$$

Finally, as

$$\begin{aligned} E[I_{n,x}(\omega)] &= \sum_{r,s=1}^{n} E[X_r X_s^*]\mathrm{e}^{-i\omega(r-s)}/(2\pi n) \\ &= (\mu_{02} + \mu_{20})/2\pi, \quad \text{for all } \omega, \end{aligned}$$

we have

$$\begin{aligned} (2\pi n)^2 \operatorname{cov}[I_{n,x}(\omega_1), I_{n,x}(\omega_2)] &= S_1 - n(\mu_{20} + \mu_{02})^2 + S_3 + S_4 \\ &= n(\mu_{40} + \mu_{04} + 2\mu_{22}) - 2n(\mu_{20} + \mu_{02})^2 \\ &\quad - n\{(\mu_{20} - \mu_{02})^2 - 4\mu_{11}^2\} \\ &\quad + \{\mu_{20} - \mu_{02})^2 + 4\mu_{11}^2\}\phi_n(\omega_1 + \omega_2) \\ &\quad + (\mu_{20} + \mu_{02})^2\phi_n(\omega_1 - \omega_2) \\ &= na + b\phi_n(\omega_2 + \omega_1) + c\phi_n(\omega_2 - \omega_1) \end{aligned}$$

where a, b, c are given by (12.25) and (12.26), and this is (12.24).

Our expression for $\operatorname{cov}[I_{n,x}(\omega_1), I_{n,x}(\omega_2)]$ for a purely random complex-valued process is of precisely the same form as that for a real-valued process, given by Priestley (1981, p. 405, equation (6.1.53)), even though the latter is much simpler; for this a becomes just κ_4, the fourth cumulant of $X(t)$, and $b = c = \sigma_X^4$, where σ_X^2 denotes the variance of $X(t)$. (These follow at once from

(12.24) by putting $V(t) \equiv 0$.) Also, it is easy to see that when

$$\begin{Bmatrix} U(t) \\ V(t) \end{Bmatrix}$$

is a Gaussian process, $a = 0$.

Remarks similar to those of Priestley may therefore be made. In particular, the same important part is played by frequencies which are multiples of $2\pi/n$; as in the real-valued case, we see that $\operatorname{cov}[I_{n,x}(\omega_1), I_{n,x}(\omega_2)]$ then reduces to $a/4\pi^2 n$ when ω_1 and ω_2 are both of this form. We also observe that $\operatorname{var}[I_{n,x}(\omega)]$ can be found by letting ω_1, ω_2 tend to ω in (12.24), which gives

$$\begin{aligned} \operatorname{var}[I_{n,x}(\omega)] &= \{na + b\phi_n(2\omega) + n^2 c\}/(2\pi n)^2 \\ &= (c/2\pi)^2 + \mathrm{O}(\tfrac{1}{n}), \end{aligned}$$

the leading term being equal to $(E[I_{n,x}(\omega)])^2$.

These results can be extended to fairly general complex-valued processes subject to a suitable condition on joint fourth order cumulants. For example, we may consider a Gaussian process for which such cumulants are zero, or a linear process, which in the complex-valued case may be defined by

$$X(t) = \sum_{v=-\infty}^{\infty} g(v)\varepsilon(t-v), \tag{12.33}$$

where the 'weights' $g(v)$ are in general complex-valued and $\{\varepsilon(t)\}$ is a purely random complex-valued process for which $E[\varepsilon(t)] = 0$ and $E[|\varepsilon(t)|^4] < \infty$; we assume also that (at least)

$$\sum_{v=-\infty}^{\infty} |g(v)| < \infty.$$

It should be noted that (12.33) is a particular case of a bivariate linear process, for which

$$\begin{pmatrix} U(t) \\ V(t) \end{pmatrix} = \sum_{v=-\infty}^{\infty} \begin{pmatrix} g_{11}(v) & g_{12}(v) \\ g_{21}(v) & g_{22}(v) \end{pmatrix} \begin{pmatrix} \mathrm{Re}(\varepsilon(t-v)) \\ \mathrm{Im}(\varepsilon(t-v)) \end{pmatrix}. \tag{12.34}$$

For a linear process we can proceed in the same way as Priestley (1981, pp. 422–428) does for the real-valued case. Thus, we first have the following result.

Theorem 12.4

Let $\{X(t)\}$ be a linear process as defined by (12.33), and let

$$\sum_{-\infty}^{\infty} |u|^{1/2} g(u) < \infty.$$

Let

$$(2\pi n)^{1/2} Z_{n,x}(\omega) = \sum_{t=1}^{n} X(t) e^{-i\omega t},$$

and $Z_{n,\varepsilon}(\omega)$ be defined similarly.

Let $r_n(\omega) = Z_{n,x}(\omega) - \Gamma(\omega) Z_{n,\varepsilon}(\omega)$, where

$$\Gamma(\omega) = \sum_{u=-\infty}^{\infty} g(u) e^{-i\omega u}.$$

Then $E[|r_n(\omega)|^2] = O(\frac{1}{n})$, $E[|r_n(\omega)|^4] = O(\frac{1}{n^2})$, both uniformly in ω.

Proof

See Priestley (1981); this is his theorem 6.2.1.

Theorem 12.5

Let $\{X(t)\}$ be as in theorem 12.4, with now

$$\sum_{u=-\infty}^{\infty} |u| |g(u)| < \infty.$$

Then

$$\begin{aligned}(2\pi n)^2 \operatorname{cov}[I_{n,x}(\omega_1), I_{n,x}(\omega_2)] = & \frac{f_X(\omega_1) f_X(\omega_2)}{(\mu_{20} + \mu_{02})^2} \\ & \times (\tfrac{a}{n} + b\phi_n(\omega_1 + \omega_2) + c\phi_n(\omega_1 - \omega_2)) \\ & + \psi_n(\omega_1, \omega_2) \end{aligned} \tag{12.35}$$

where $-\pi < \omega_2 < \omega_1 < \pi$ with $\omega_1 \neq -\omega_2$, a, b, c, μ_{20} and μ_{02} being as in theorem 12.3, with $\varepsilon(t) = X(t)$ there, and the remainder term $\psi_n(\omega_1, \omega_2)$ is $O(\frac{1}{n^2})$ uniformly in ω_1, ω_2. This is essentially theorem 6.2.3 of Priestley (1981, p. 426). The statement there that the covergence of $\sum |u|^\alpha |g(u)|$ implies that $\psi_n(\omega_1, \omega_2)$ is $O(\frac{1}{n^\alpha})$ may not hold for sufficiently large α, but this is of no practical significance. It may, however, affect the application of the order of magnitude calculations in the proof of this theorem to the present case, but with the above convergence conditions these difficulties can be overcome by using a more detailed argument.

With a Gaussian process, for which $a = 0$, one can proceed directly, as in the real-valued case, using the fact that $E[X(r)X(s)^* X(t) X(u)^*] = E[X(r)X(s)^*]\, E[X(t)X(u)^*] + E[X(r)X(t)]\, E[X(s)^* X(u)^*] + E[X(r) \times X^*(u)]\, E[X(s)^* X(t)]$, which still holds. Here the most natural convergence conditions are

$$\sum_{u=-\infty}^{\infty} |u| |R_X(u)| < \infty, \text{ and } \sum_{u=-\infty}^{\infty} |u| |Q_X(u)| < \infty,$$

where $Q_X(u) = E[X(t+u)X(t)]$; they are clearly satisfied when

$$\sum_{u=-\infty}^{\infty} |u| \begin{pmatrix} |R_{UU}(u)| & |R_{UV}(u)| \\ |R_{VU}(u)| & |R_{VV}(u)| \end{pmatrix}$$

is finite. It is not hard to show that, under these, the conclusion (12.35) of theorem 12.5 still holds but the remainder term $\psi_n(\omega_1,\omega_2)$ can only be taken to be $o(\frac{1}{n})$, uniformly in ω_1,ω_2.

Finally, results for spectral windows such as those given by Priestley (1981, pp. 449–463) carry over with slight modifications because of theorem 12.5. The most important of these is the following.

Theorem 12.6

Let $\{X(t)\}$ be as in theorem 12.5, and $W_n(\omega)$ a spectral window satisfying suitable conditions, (for which see Priestley (1981, p. 450); essentially they ensure that $W_n(\omega)$ approaches a δ-function at the origin as $n \to \infty$). Let

$$\tilde{f}_X(\omega) = \int_{-\pi}^{\pi} I_n(\lambda) W_n(\omega - \lambda)\, \mathrm{d}\lambda$$

be a spectral window estimator of the spectral density of $\{X(t)\}$. Thus for $\omega \neq 0, \pi$,

$$\operatorname{var}[\hat{f}_X(\omega)] \sim \frac{2\pi}{n} f_X^2(\omega) \int_{-\pi}^{\pi} W_n^2(\lambda)\, \mathrm{d}\lambda \qquad (12.36)$$

and for $\omega_2 \neq \omega_1, -\omega_1$, $\lim_{n\to\infty} \operatorname{cov}(\hat{f}_X(\omega_1), \hat{f}_X(\omega_2)) = 0$.

Proof

See Priestley (1981, pp. 452–455).

12.4 ASYMPTOTIC DISTRIBUTIONS

In most situations these can be obtained by appealing to a central limit theorem whereby the joint distribution of

$$C_{n,u}(\omega) = (2\pi n)^{-1/2} \sum_{t=1}^{n} U(t) \cos \omega t, \quad S_{n,u}(\omega) = (2\pi n)^{-1/2} \sum U(t) \sin \omega t,$$

$$C_{n,v}(\omega) = (2\pi n)^{-1/2} \sum_{t=1}^{n} V(t) \cos \omega t, \quad S_{n,v}(\omega) = (2\pi n)^{-1/2} \sum V(t) \sin \omega t$$

or, more generally, ($C_{n,u}(\omega_j)$, $S_{n,u}(\omega_j)$, $C_{n,v}(\omega_j)$, $S_{n,v}(\omega_j)$, $j = 1 \ldots m$) tends to a (multivariate) normal distribution $N(O,\Sigma)$ where, if $\lim_{n\to\infty} E[D_n(\omega)D_n(\omega)'] =$

$\sigma(\omega)$, the 4×4 matrix given by equation (12.13), Σ is equal to the diagonal block matrix

$$\begin{bmatrix} \sigma(\omega_1) & 0 & . & 0 \\ 0 & \sigma(\omega_2) & . & 0 \\ . & . & . & . \\ 0 & 0 & . & \sigma(\omega_m) \end{bmatrix},$$

provided that for no(j, k), $\omega_j = -\omega_k$.

The main results are as follows.

Theorem 12.7

(a) Let $\omega \neq 0, \pi$, and $A_n(\omega) = C_{n,u}(\omega) + S_{n,v}(\omega)$, $B_n(\omega) = -S_{n,u}(\omega) + C_{n,v}(\omega)$, as in equations (12.7), (12.8). Then the joint distribution of $A_n(\omega)$, $B_n(\omega)$ tends to $N(0, V(\omega))$ as $n \to \infty$ where

$$V(\omega) = \begin{pmatrix} \frac{1}{2} f_X(\omega) & 0 \\ 0 & \frac{1}{2} f_X(\omega) \end{pmatrix},$$

and hence the distribution of $I_{n,x}(\omega) = A_n^2(\omega) + B_n^2(\omega)$ tends to that of $\frac{1}{2} f_X(\omega) \chi_2^2$.

(b) Let $A_n(\omega_j)$, $B(\omega_j)$ be defined as in (a) for $j = 1 \ldots m$ with $-\pi < \omega_1 \ldots < \omega_m < \pi$ and for all (j,k), $\omega_j \neq -\omega_k$. Then the joint distribution of $(A_n(\omega_1)$, $B_n(\omega_1) \ldots A_n(\omega_m)$, $B_n(\omega_m))$ tends to $N(\mathbf{O}, \mathbf{W})$, where $\mathbf{W}$ is the diagonal block matrix diag $(V(\omega_1), V(\omega_2), \ldots V(\omega_m))$. Thus in the limit the $I_{n,x}(\omega_j)$ are distributed independently as $\frac{1}{2} f_X(\omega_j) \chi_2^2, j = 1 \ldots m$.

Nothing further needs to be said regarding proof. The central limit theorem holds under wide conditions and in particular can be deduced very simply for a linear process from the classical central limit theorem for equal components satisfying the Lindeberg–Lévy condition; this follows from the complex-valued version of theorem 6.2.1 of Priestley (1981, p. 422) according to which

$$(2\pi n)^{-1/2} \sum_{t=1}^{n} X_t e^{-i\omega t} = \left[\sum_{u=-\infty}^{\infty} e^{-i\omega u} g(u) \right] (2\pi n)^{-1/2} \sum_{t=1}^{n} \varepsilon_t e^{-i\omega t} + r_n(\omega),$$

where $E\{|r_n(\omega)|^2\} = 0(\frac{1}{n})$, uniformly in ω.

12.5 CONCLUSIONS

We have seen that for spectral analysis of complex-valued stationary processes $\{X(t)\}$, $X(t) = U(t) + iV(t)$, interpreted as inferences based on the sample periodogram $I_{n,x}(\omega)$, most of the standard results for the real-valued case still apply. The only significant difference is that we now have to consider spectral densities for negative values of ω (since $f_X(-\omega)$ will in general not be equal to $f_X(\omega)$). The asymptotic covariance structure for estimates of $f_X(-\omega)$, $\omega > 0$ is the same as that for estimates of $f_X(\omega)$, $\omega > 0$, but estimates of $f_X(\omega)$ and $f_X(-\omega)$

are correlated. Consideration of the spectral representation indicates that this might have been expected. From the expression for asymptotic covariances given by equations (12.21) to (12.23) it is clear that spectral analysis alone cannot yield any information about these correlations. There is, however, one obvious situation in which they vanish, namely when $\{U(t)\}$ and $\{V(t)\}$ are uncorrelated with each other and have the same spectral density. This in fact was assumed to be the case in the work of Suttie (1990), referred to in the introduction.

We finally remark that it is possible for $\{X(t)\}$, with $X(t) = U(t) + iV(t)$, to be weakly stationary without

$$\begin{pmatrix} U(t) \\ V(t) \end{pmatrix}$$

being a weakly stationary bivariate process. For example, let $U(t) = Y(t)\cos\omega_0 t$, and $V(t) = Z(t)\sin\omega_0 t$, where $\{Y(t)\}$ and $\{Z(t)\}$ are uncorrelated completely random processes with the same variance σ^2. Clearly neither $\{U(t)\}$ or $\{V(t)\}$ are weakly stationary but $\{X(t)\}$ is since $E[X(t+s)\{X(t)\}^*] = 0$ for $s \neq 0$, and σ^2 for $s = 0$. However, from equations (12.7) and (12.8) we have

$$\begin{aligned} 2\pi n \operatorname{var}[A_n(\omega)] &= \sigma^2 \sum_{t=1}^{n} (\cos^2\omega_0 t \cos^2\omega t + \sin^2\omega_0 t \sin^2\omega t) \\ &= \frac{1}{4}\sigma^2 \sum_{t=1}^{n} ([\cos(\omega-\omega_0)t + \cos(\omega+\omega_0)t]^2 \\ &\quad + [\cos(\omega-\omega_0)t - \cos(\omega+\omega_0)t]^2) \qquad (12.37) \\ &= \frac{1}{2}\sigma^2 \sum_{t=1}^{n} (\cos^2(\omega-\omega_0)t + \cos^2(\omega+\omega_0)t) \\ &= \frac{1}{2}\sigma^2 \sum_{t=1}^{n} (1 + \tfrac{1}{2}\{\cos 2(\omega-\omega_0)t + \cos 2(\omega+\omega_0)t\}) \qquad (12.38) \end{aligned}$$

and similarly

$$2\pi n \operatorname{var}[B_n(\omega)] = \frac{1}{2}\sigma^2 \sum_{t=1}^{n} (1 - \tfrac{1}{2}\{\cos 2(\omega-\omega_0)t + \cos 2(\omega+\omega_0)t\}) \qquad (12.39)$$

Thus for any fixed $\omega \neq \pm\omega_0$ both var $A_n(\omega)$ and $B_n(\omega)$ tend to $\frac{\sigma^2}{4\pi}$ as $n \to \infty$, the differences between them and $\frac{\sigma^2}{4\pi}$ being $0(n^{-1})$, uniformly for all ω such that $|\omega \pm \omega_0| > \varepsilon$, any prescribed positive number. Moreover, the expected value of the periodogram at frequency ω, $E[I_{n,x}(\omega)] = \frac{\sigma^2}{2\pi}$ **exactly**, which is the (constant) spectral density of $\{X(t)\}$, equal to

$$\frac{1}{2\pi}\sum_{-\infty}^{\infty} R_X(s)\mathrm{e}^{-i\omega s} = \frac{R_X(0)}{2\pi}.$$

Thus the behaviour of $\{X(t)\}$ in the frequency domain is asymptotically the same as when

$$\begin{pmatrix} U(t) \\ V(t) \end{pmatrix}$$

is weakly stationary except when ω approaches $\pm\omega_0$ (note that from our assumptions we can see at once that $\text{cov}[A_n(\omega), B_n(\omega)] = 0$, exactly). Also the central limit theorem can be applied in the same way as for the real-valued case to show that $A_n(\omega_j)$, $B_n(\omega_j)$, $j = 1\ldots m$ are asymptotically distributed independently and normally with variances $\sigma^2/4\pi$. However, when ω is close to $\pm\omega_0$ the situation is different. In fact putting $\omega = \omega_0$ in (12.38), (12.39) we see that $\lim_{n\to\infty} 2\pi\,\text{var}[A_n(\omega_0)] = \frac{3}{4}\sigma^2$ and $\lim_{n\to\infty} 2\pi\,\text{var}[B_n(\omega_0)] = \frac{1}{4}\sigma^2$. $A_n(\omega_0)$, $B_n(\omega_0)$ will still be asymptotically normal but with different variances, so that the limiting distribution of $I_n(\omega_0)$ is no longer that of $\frac{\sigma^2}{2\pi}\cdot\frac{1}{2}\chi_2^2$.

Further examination of this situation would be possible but is perhaps of little value because it might well be an unusual one. Of course with the assumption of a (weakly) stationary

$$\begin{pmatrix} U(t) \\ V(t) \end{pmatrix},$$

everything could be done using existing theory for bivariate stationary processes, but it was felt that the algebraic calculations would be significantly heavier (for instance, instead of $I_{n,x}(\omega)$ one would have to consider (using an obvious notation) $I_{UU}(\omega) + I_{VV}(\omega) - i(I_{UV}(\omega) - I_{VU}(\omega))$.

REFERENCES

Bartlett, M.S. (1966) *An Introduction to Stochastic Processes with Special Reference to Methods and Applications*; 2nd edn, Cambridge University Press.

Hannan, E.J. (1970) *Multiple Time Series*, Wiley, New York.

Priestley, M.B. (1981) *Spectral Analysis and Time Series*, Academic Press.

Suttie, M.R. (1990) *An Investigation of Synthetic Aperture Radar Data*, MSc Thesis, University of Sheffield.

Yaglom, A.M. (1962) *An Introduction to the Theory of Stationary Random Functions*, Prentice-Hall.

13

A spectral approach to long memory time series

G. Janacek

13.1 INTRODUCTION

There has been a consistent interest in 'long memory' or 'strongly dependent' time series since the papers of Granger and Joyeux (1980) and Hosking (1981). This has covered both the theoretical aspects, e.g. Cox (1991), and the empirical, Carlin and Dempster (1989) or McLeod and Hipel (1978).

We follow Granger and Joyeux (1980) and characterize a long memory series by the 'fractional difference' model

$$(1 - B)^d \phi(B) X_t = \theta(B) a_t \tag{13.1}$$

where ϕ, θ are polynomials, of order p and q respectively, in the backshift operator B. The series $\{a_t\}$ is a zero mean white noise series with constant variance σ^2 while the parameter d is not necessarily an integer. The model (13.1) is thus a non-standard ARIMA model, however, for our purposes we find it most natural to think of the process in terms of the 'spectrum' $f_x(\omega)$ where

$$f_x(\omega) = |1 - e^{-i\omega}|^{-2d} f_y(\omega). \tag{13.2}$$

Here $f_y(\omega)$ is the spectrum of $\{Y_t\}$, a stationary series with the rational spectrum

$$f_y(\omega) = \frac{\sigma^2 |\theta(e^{-i\omega})|^2}{2\pi |\phi(e^{-i\omega})|^2}. \tag{13.3}$$

From Hosking (1981) we know that the $\{X_t\}$ series is stationary when $|d| < 0.5$ but even for such stationary models the spectra have large low frequency peaks suggesting long period effects. This can be confirmed by considering the covariance structure which decays slowly, indeed we can show that the autocorrelations are of the form

$$\rho_s \approx s^{2d-1}\frac{\Gamma(1-d)}{\Gamma(d)} \quad \text{as } s \to \infty.$$

The form of the spectrum for a long memory series has prompted several authors to look at the log of the spectrum

$$\log f_y(\omega) = -2d \log|1 - e^{-i\omega}| + \log f_x(\omega). \tag{13.4}$$

and a regression approach based on $\log f_y(\omega)$ has been pursued by Granger and Joyeux (1980) and by Geweke and Porter-Hudack (1983). Li and MacLeod (1986) on the other hand have used maximum likelihood in the time domain while Janacek (1982) has proposed another approach based on the Wiener decomposition of the spectrum.

We take the view that it is quite natural to approach the identification and estimation of long memory models from the point of view of the power spectrum. Indeed we feel that there are compelling reasons for so doing. What is more, since the spectral approach is so natural, we suggest that the simulation of long memory models is most easily done via the spectrum.

13.2 ESTIMATION IN THE FREQUENCY DOMAIN

In what follows we take an essentially practical approach to the identification and estimation of long memory series. As we have said we also take a frequency viewpoint of such series, a compelling reason being that we can use expressions which are closed forms rather than the extended recursions required for an equivalent time domain formulation. Since we see the spectrum as the natural tool we attempt to use one of the principal virtues of spectral estimates, their nice statistical properties.

We recall that a 'crude' estimate of $f(\omega)$ is the periodogram

$$I_N(\omega) = \frac{2}{N}\left|\sum_{t=1}^{N} X_t e^{i\omega t}\right|^2 \tag{13.5}$$

which has many shortcomings but has the attractive property that for a fixed set of values $\omega_j = (2\pi j/N)$, $j = 1, 2, \ldots, k$ the joint distribution of the $I_N(\omega_j)$ tends to that of k independent random variables, each distributed as $2\pi f(\omega_j)\chi^2$ with two degrees of freedom. At $\omega = 0$ and π the distributions are χ^2 with one degree of freedom, but we shall not require these points. For details and derivations see Priestley (1981, chapter 6). Since we have a sequence of independent random variables each with a known distribution from the exponential family, it would seem appropriate to consider the likelihood based on the joint distribution of the periodogram ordinates. This is easy to write in closed form (see Janacek and Swift (1993) or Harvey (1991)) especially if we omit the periodogram values at frequencies zero and π.

Thus we can contemplate the estimation of the parameter d and the spectrum, or equivalently the parameters of the expression (13.1) by finding the maximum of the frequency domain likelihood. This is a fairly straightforward proposition from a numerical viewpoint but rather than proceed to a full likelihood method we suggest that one can do as well by fitting a generalized linear model. We also feel that this rather simpler approach is a more natural one to the statistician.

We know that the periodogram ordinates are independent gamma variates so we have a common distribution from the exponential family, with a known scale factor. In addition the means are of the form

$$E[I_N(\omega_j)] = 4\pi f(\omega_j). \tag{13.6}$$

This is exactly the kind of problem one can tackle using a generalized linear model. What is more an efficient programme, GLIM, is widely available for estimation and analysis of generalized linear models. The reader will find a comprehensive account of generalized linear models in McCullagh and Nelder (1989).

13.3 METHODOLOGY

We aim to fit a generalized linear model to explain the variation in the periodogram ordinates, assuming at each frequency ω_j the periodogram has a gamma distribution with mean

$$E[I_N(\omega_j)] = \mu_j.$$

We have as usual a set of covariates $(x_{1,j}, x_{2,j}, x_{3,j}, \ldots, x_{k,j}) = \mathbf{x}_j^t$ say and a coefficient vector β thus $g(\mu_j) = \mathbf{x}_j^t\beta$ for some link function $g(\cdot)$. In the case of interest to us we shall use either the identity link $g(\mu) = \mathbf{x}_j^t\beta$ or the canonical link which for the gamma distribution is the reciprocal

$$g(\mu) = \frac{1}{\{\mathbf{x}_j^t\beta\}}.$$

In fact our situation is a little more complex since from (13.1) and (13.3) we have

$$\mu_j = \frac{4\pi \sum_{s=0}^{q} \beta_s \cos(s\omega_j)}{[2(1-\cos\omega_j)]^d \sum_{s=0}^{p} \alpha_s \cos(s\omega_j)} \sigma^2, \tag{13.7}$$

where the sine and cosine coefficients are convolutions of the ARMA coefficients. Writing $\phi_0 \equiv \theta_0 \equiv 1$ we have the explicit forms

$$\alpha_s = \sum_{r=0}^{p-|s|} \phi_r \phi_{r+s}, \quad \beta_s = \sum_{r=0}^{q-|s|} \theta_r \theta_{r+s}.$$

For us to fit this into the context of a generalized linear model we need to be able to specify d, filter in the time domain and then estimate the coefficients α_s and β_s using $\cos(s\omega_j)$ for $s = 1, 2, \ldots, k$ as covariates. The covariates would be fitted sequentially until no significant deviance reduction is achieved. We propose the strategy below.

(a) We first estimate d by fitting the model

$$\mu_j = \frac{4\pi}{[2(1-\cos\omega_j)]^d} \tag{13.8}$$

for a sequence of values of d. This is a reasonably straightforward exercise using the macro facilities of GLIM. We also use the low frequency terms in the periodogram since at these frequencies the 'difference filter' dominates the spectrum. Experience shows that it is best to restrict the periodogram values modelled to those corresponding to frequencies below $\pi/4$ or $\pi/5$. The choice of d is then the value which minimizes the deviance (likelihood ratio statistic).

(b) Having found an estimate of d one may then remove the difference effect and fit a classical ARMA model to the residual terms. This can also be done via GLIM since after the powers in d have been removed we have as our model of interest

$$\mu_j = 4\pi \frac{\sum_{s=0}^{q} \beta_s \cos(s\omega_j)}{\sum_{s=0}^{p} \alpha_s \cos(s\omega_j)} \sigma^2. \tag{13.9}$$

We can easily fit either an AR or an MA model as these correspond to either the denominator or the numerator of (13.9). A mixed model has to be fitted in two stages, first as an AR term using just the denominator, then as an MA term using the numerator **given** the estimated AR part. This procedure may be cycled until a sufficiently well fitting model is found. In fitting the ARMA part we do not restrict ourselves to a frequency band around zero. It may be necessary to repeat steps (a) and (b) until an overall model is found which is satisfactory. While it is possible to do all the estimation in one step using GLIM we have found the two stage process outlined above more satisfactory.

In passing we note that we can also think of the d parameter as a parameter of the link function rather in the fashion of Pregibon (1980). In this case if we have an initial guess for the difference parameter, say D, then since

$$\mu_j = \frac{4\pi}{[2(1-\cos\omega_j)]^d},$$

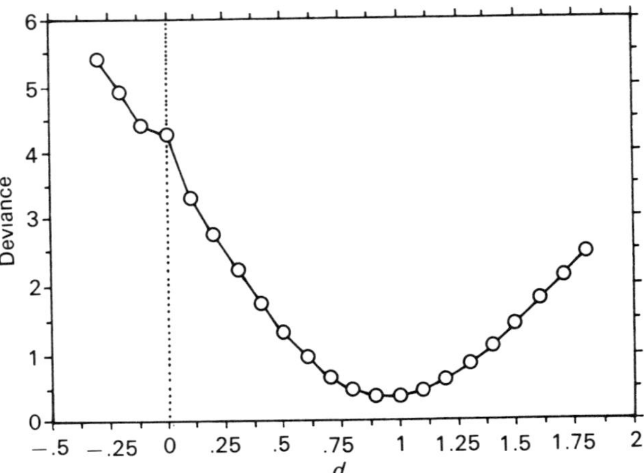

Figure 13.1. US GNP, undifferenced.

we can expand in a Taylor series to give

$$\mu_j = \frac{4\pi}{[2(1-\cos\omega_j)]^D} + (d-D)\frac{4\pi}{[2(1-\cos\omega_j)]^D}\log\left\{\frac{4\pi}{[2(1-\cos\omega_j)]}\right\}. \tag{13.10}$$

We can then modify the link function based on D by adding an extra term which can be estimated as an extra covariate.

For the moment we leave such refinements and concentrate on the initial estimation of the parameter d. We thus perform the first step of the estimation procedure once without any refinements. Figure 13.1 is a plot of the deviance of the model after fitting just a 'difference' filter, i.e.

$$\mu_j = \frac{4\pi}{[2(1-\cos\omega_j)]^d}$$

to the series consisting of the US GNP (quarterly) from the first quarter of 1947 to the last of 1966. The 80 observations are clearly nonstationary and Nelson (1973) fitted an AR(1) model to the first differences. As we can clearly see, given his data, the value of d which gives a minimum deviance is 1. This agrees with Nelson's conclusion that the series is AR(1) after differencing. That one can work with such a short series, even for such a straightforward case is most gratifying.

A rather longer series (310 observations) is series 'D' of chemical viscosity readings considered by Box and Jenkins (1970). This is modelled as a stationary series and a differenced series, the stationary model having a slightly smaller

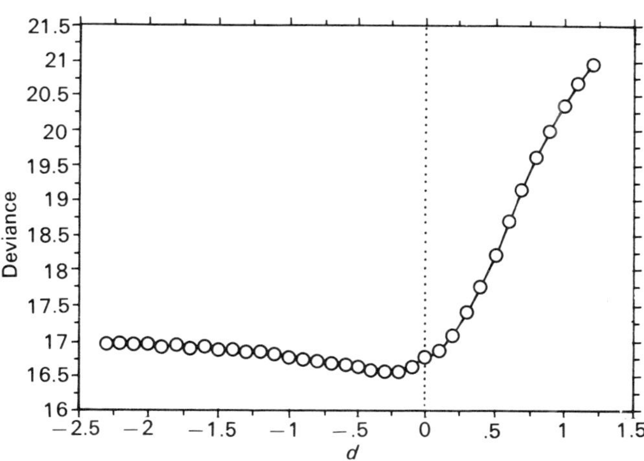

Figure 13.2. Deviance plot for series, Box and Jenkins (1970).

residual variance. Again we fit a model of the form

$$\mu_j = \frac{4\pi}{[2(1-\cos\omega_j)^d}$$

and the plot of the deviance for a set of values of the difference parameter d is given in Figure 13.2. We can see from the plot of the deviances that a small non-zero value, somewhat less than zero, for the d parameter is called for. While the value of d suggested is not positive we have some reservations about the small negative value of d which would follow from an unthinking application of our approach. Suppose for $d = -\alpha$, say, for some positive α, we have

$$\mu_j = \frac{4\pi}{[2(1-\cos\omega_j)^d} = 4\pi\{2(1-\cos\omega_j)\}\alpha$$
$$= 2^{2+\alpha}\pi\{1-\alpha\cos\omega + \alpha(\alpha-1)\cos^2\omega_j \ldots\}$$

Now, unless we have a white noise series, it is quite possible that the mean (of the periodogram) will be reasonably approximated near zero by a trigonometric polynomial. The expansion of our link function involves just such trigonometric polynomials and in consequence we would urge a refined approach to isolate the effect of the difference filter. We conclude that $d = 0$ is appropriate and also note that $d = 0$ would agree with the conclusions reached by a different approach by Janacek (1982).

As there appears to be no benchmark fractional series we fall back on an application to a stimulated series of length 256 with $d = 0.25$. The mechanism used is described in more detail in section 13.4 below. The deviance plot in

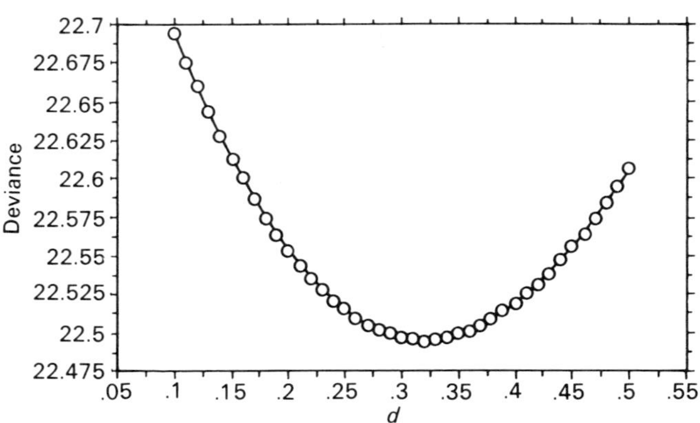

Figure 13.3. Simulated model with $d = 0.25$.

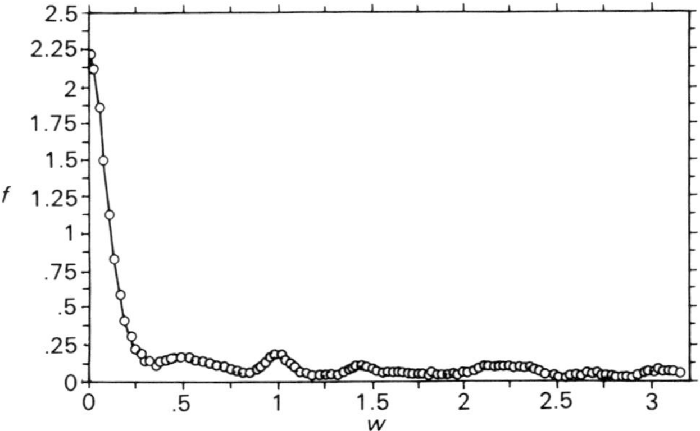

Figure 13.4. Parzen spectrum of lake mud series.

Figure 13.3 suggests that a fractional model is indeed plausible with a minimum of the order of 0.325. This is rather larger that the parameter used in the simulation but we have not used any refinements!

A similar pattern has been seen in a new data set, the thickness in mm of mud layers in Lake Saki by decades cited by Lamb (1977). The spectrum is shown in Figure 13.4 and can be see to have sharp peak at the low frequency end which is the behaviour we expect of a long memory model.

The series in fact can be modelled by a ARMA model after differencing. The deviance plot shown in Figure 13.5 is however suggestive of a d parameter of the order of 0.65. We hope to report on this series in more detail.

The referee pointed out the interesting paper by Agiaklogou, Newbold and Wohar (1993). In this the authors point out a real drawback in the use

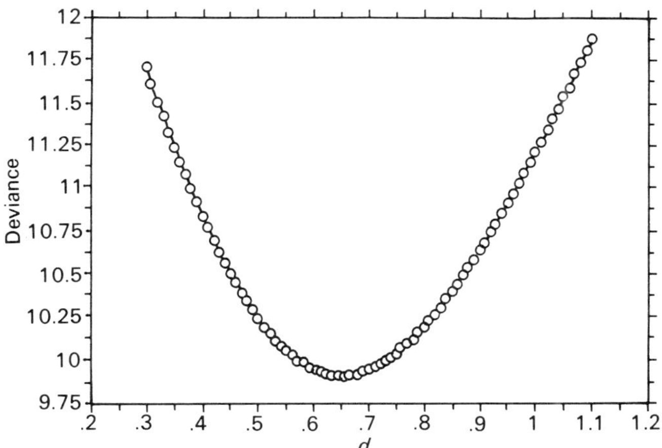

Figure 13.5. Deviance plot for lake mud series.

of the log spectrum for estimating d. Given that from (13.4)

$$\log f_y(\omega) = -2d \log|1 - e^{-i\omega}| + \log f_x(\omega)$$

if we regress $\log f_y(\omega)$ on $-2\log|1 - e^{-i\omega}|$ we must assume that the remaining term $\log f_x(\omega)$ is approximately constant, at least near the origin. If this is not true, for example when there is a large AR or MA root of appropriate sign, they demonstrate that the resulting bias may be large. It is clear that we face a similar problem when dealing with the spectrum directly as a low frequency peak or trough will distort the low frequency component of the difference filter $|1 - e^{-i\omega}|^{2d}$. Since our procedure is two stage we do estimate the remaining spectral components and any bias will eventually be corrected. It is possible however that the initial d estimate may be substantially out and the iterations will be lengthy. One solution is to take into account the larger roots of the ARMA model. To do so we modify our initial step and rather than (13.9) we use

$$\mu_j = \frac{4\pi}{[2(1 - \cos\omega_j)]}\{\beta_0 + \beta_1 \cos\omega_j\}^{-1} \tag{13.11}$$

where β_0 and β_1 are to be estimated. The aim is *not* to estimate the ARMA component but to model the terms confounded with peak caused by the difference filter. If the dominant AR root is α then

$$\mu_j = \frac{4\pi}{[2(1 - \cos\omega_j)]^d}\{1 + \alpha^2 + 2\alpha\cos\omega_j\}^{-1} \times \text{other terms}$$

If we assume that the remaining terms are approximately constant near the origin then (13.11) is a reasonable approximation. What is more, we can use the same expression for a moving average model. For a moving average

$$\mu_j = \frac{4\pi}{[2(1-\cos\omega_j)]^d}\{1+\alpha^2+2\cos\omega_j\} \times \text{other terms}$$

but

$$\frac{1}{\{1+\alpha^2+2\alpha\cos\omega_j\}} \approx \frac{1}{\{1+\alpha^2\}}\left\{1-\frac{2\alpha}{1+\alpha^2}\cos\omega_j+\cdots\right\}$$

and if we use the canonical (reciprocal) link to a first approximation we have (13.11). This gives an initial way of fitting the mean μ_j. Again the frequency range 0 to $\pi/5$ is used. If the roots are not large then the beta coefficients in the model will be small and the distortion should be slight. We can also use the flexibility of GLIM to add a further cosine term as this can sharpen the approximation and we shall be able to see if it does from the deviance changes. A further advantage over the use of the log spectrum is that we can fit the remaining terms in the model over our choice of the frequency domain, thus eliminating the range of frequencies near zero. Preliminary results show considerable promise as can be see for the deviance plots in Figure 13.6. These are for two simulated series, one generated by the fractional MA model $(1-B)^{0.4}X_t=(1-0.9\,B)a_t$ and the other by a fractional AR model $(1-B)^{0.4}(1-0.9\,B)X_t=a_t$. While the MA process is clearly satisfactory we see some bias in the AR series. We point out that the deviance is quite flat around these values and that additional refinements would follow in practice, indeed the bias is reduced by these.

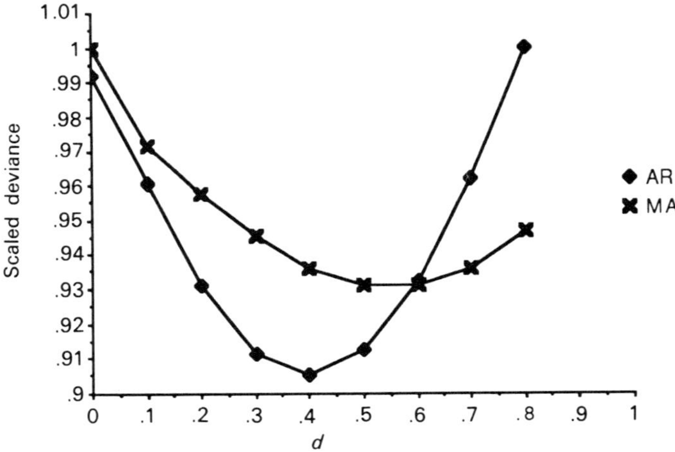

Figure 13.6. Deviances for AR1 and MA1 models with root parameter 0.9 and $d=0.4$.

13.4 SIMULATION

In any approach to estimate the d values for a fractional differenced series one is naturally lead to the idea of simulating such fractional series. It is a common activity, see for example MacLeod and Hipel (1978) or Gweke and Porter-Hudak (1983). Such simulations are usually done by using a long order approximating ARMA model or by using an explicit covariance matrix and generating a multivariate normal point using either a Cholseky decomposition or a Durbin–Levinson type recurrence.

These methods are all based on the covariance structure, whereas we believe that it is natural to consider the spectral characteristics of long memory series. In consequence we propose a method of simulation which is based directly on the spectrum of the required series. Since no parametric model is involved there are no problems with long memory series or indeed short memory ones.

We consider the harmonic series

$$X_t = \sum_{j=1}^{n} \{a_j \cos(t\omega_j) + b_j \sin(t\omega_j)\}, \tag{13.12}$$

where $\{a_j\}, \{b_j\}$ are independent sequences of independent zero mean random variables with

$$E[a_j] = E[b_j] = \sigma_j^2 \qquad j = 1, 2, \dots, n.$$

It is easily shown that the series $\{X_t\}$ has zero mean and autocovariances

$$\gamma_k = \sum_{j=1}^{n} \sigma_j^2 \cos(k\omega_j) \quad k = \dots -2, -1, 0, 1, 2, \dots. \tag{13.13}$$

In consequence the 'spectrum' is just

$$f_x(\omega) = \frac{1}{2\pi} \sum_{k=-\infty}^{\infty} \gamma_x(k) \cos(\omega k) \tag{13.14}$$

and consists of a set of lines at frequencies $\{\omega_j, j = 1, 2, \dots, n\}$, where

$$f(\omega_j) = \frac{\sigma_j^2}{2} \quad \omega_j = \pm \frac{\pi_j}{n}$$

$$0 \quad \text{otherwise}$$

For the models that interest us, the power spectra are continuous and in fact can be modelled as rational functions of trigonometric polynomials. If our spectrum of interest is of this type we can simply split the range $[0, \pi]$ into m sub-intervals and by taking $f(\omega)\delta\omega$, the spectral height at the right hand boundary times its width, to evaluate the area over the interval we can approximate this by $0.5\,\sigma_i^2$. This is the same sort of process one might follow in approximating the binomial by the normal distribution. We may thus approximate the desired spectrum $f(\omega)$ by a line spectrum and $\{X_t\}$ by a harmonic model of the form (13.12). Provided our discrete lines are close enough we shall get a reasonable approximation. As we are dealing with the spectral representation, the length of an ARMA 'equivalent' model is irrelevant, hence the attraction for long memory models.

In practice we need to generate sequences $\{a_j\}$, $\{b_j\}$ having the appropriate independence properties and with

$$\operatorname{var}(a_j) = \operatorname{var}(b_j) = \sigma_j^2$$

Now this can be easily done using standard algorithms, in fact it is probably sufficient to use uniformly distributed variates to save computational time. More time consuming is the summation for each value of t. This is not a real problem since as we shall see this is easily done via a fast fourier transform (FFT). In our computations we have used the NAG algorithm C06GFAF but the basic algebra is common to all algorithms, see Monro and Branch (1977).

We define the discrete Fourier transform of $a_0, a_1, \ldots, a_{N-1}$ to be

$$X_t = N^{-1/2} \sum_{k=0}^{N-1} a_k \exp\left(\frac{-2\pi ikt}{N}\right) t = 0, 1, \ldots, N-1, \tag{13.15}$$

with the inverse

$$a_t = N^{-1/2} \sum_{k=0}^{N-1} a_k \exp\left(\frac{2\pi ikt}{N}\right) t = 0, 1, \ldots, N-1. \tag{13.16}$$

(Note different authors use different divisors). The FFT algorithm enables one to compute sums of the form (13.5) very quickly and economically. We may easily adapt (13.5) since

$$X_t = N^{-1/2} \left\{ \sum_{k=0}^{N-1} a_k \cos\left(\frac{2\pi tk}{N}\right) - i \sum_{k=0}^{N-1} a_k \sin\left(\frac{2\pi tk}{N}\right) \right\}.$$

If we redefine $f(\omega)$ over $(0, 2\pi)$ then we can split the range into N intervals $(0, 2\pi/N, 4\pi/N, 6\pi/N, \ldots, 2\pi)$. Two applications of the FFT giving

$$X_t^{(1)} = N^{-1/2} \sum_{k=0}^{N-1} a_k \exp\left(-\frac{2\pi tk}{N}\right) \quad t = 0, 1, \ldots, N-1$$

$$X_t^{(2)} = N^{-1/2} \sum_{k=0}^{N-1} b_k \exp\left(-\frac{2\pi tk}{N}\right) \quad t = 0, 1, \ldots, N-1$$

can then be used to give $\{X_t\}$. Using FFT algorithms has two main advantages, they are numerically stable and fast!

13.5 PRACTICAL CONSIDERATIONS

To generate $X_0 \ldots X_{T-1}$ using our model (13.12) we need to choose the number of terms in the summation or the grid size for approximating the desired spectrum $f_y(\omega)$. We can get some idea of this number N as follows for a fairly smooth spectrum, say that of some ARMA model. If we consider a spectral estimate $f(\omega)$ based on the simulated series we can see (cf. Koopmans, 1973, p. 283)

$$E[\hat{f}(\omega)] = \int_{-\pi}^{\pi} W_T(\omega - \mu) E[I_N(\mu)] \mathrm{d}\mu \approx \sum_{-\infty}^{\infty} \alpha_j W_M(\omega - \omega_j),$$

so we have for the estimated spectrum a set of peaks centred at the points of discrete spectral power. These will have a bandwidth about the magnitude of the bandwidth of the smoothing window used to estimate the spectrum, say b_w. Then if the number of 'grid points' is N, the desired spectrum $f(\omega)$ will be approximated by lines of separation $g = 2\pi N^{-1}$, a distance which we would like to be small compared to b_w. Now the bandwidth of a (windowed) spectral estimate is of the form $c\pi M^{-1}$ where c is a constant depending on the form of the window and M is the truncation point. Thus for a Parzen window $c = \frac{8}{3}$ and hence if we choose $b_w g^{-1}$ to be of the order of 3 using a truncation point of $M = T/3$ gives N of the order of 2.25 T. We have tried values of $N = T$ and $N = 2T$, and our empirical results lead us to a choice of $N = 2T$.

As you might expect for a long memory model the choice of the grid size N becomes more difficult. The problem is the large peak at the zero frequency which needs to be adequately represented in the harmonic model. If we have a realization of length T then the first frequencies at which we can compute the sample periodogram are 0 and $2\pi T^{-1}$. If we are to provide some shape to the peak at zero than we need some values for the harmonic process in this range. In fact a doubling of our previous suggestion viz. $N = 4T$ seems to fit the bill as can be seen below.

13.6 RESULTS

We present here the result of simulating some series of 128 observations. We have taken two simple AR1 models

$$X_t = 0.7\,X_{t-1} + a_t \qquad X_t = -0.7\,X_{t-1} + a_t$$

and two simple fractional models

$$(1 - B)^{0.25} X_t = a_t \qquad (1 - B)^{-0.25} X_t = a_t$$

Mean sample correlations and confidence intervals (based on the sample of 100) are given in Figures 13.7 and 13.8 for the AR models and Figures 13.9 and 13.10 for the fractional models.

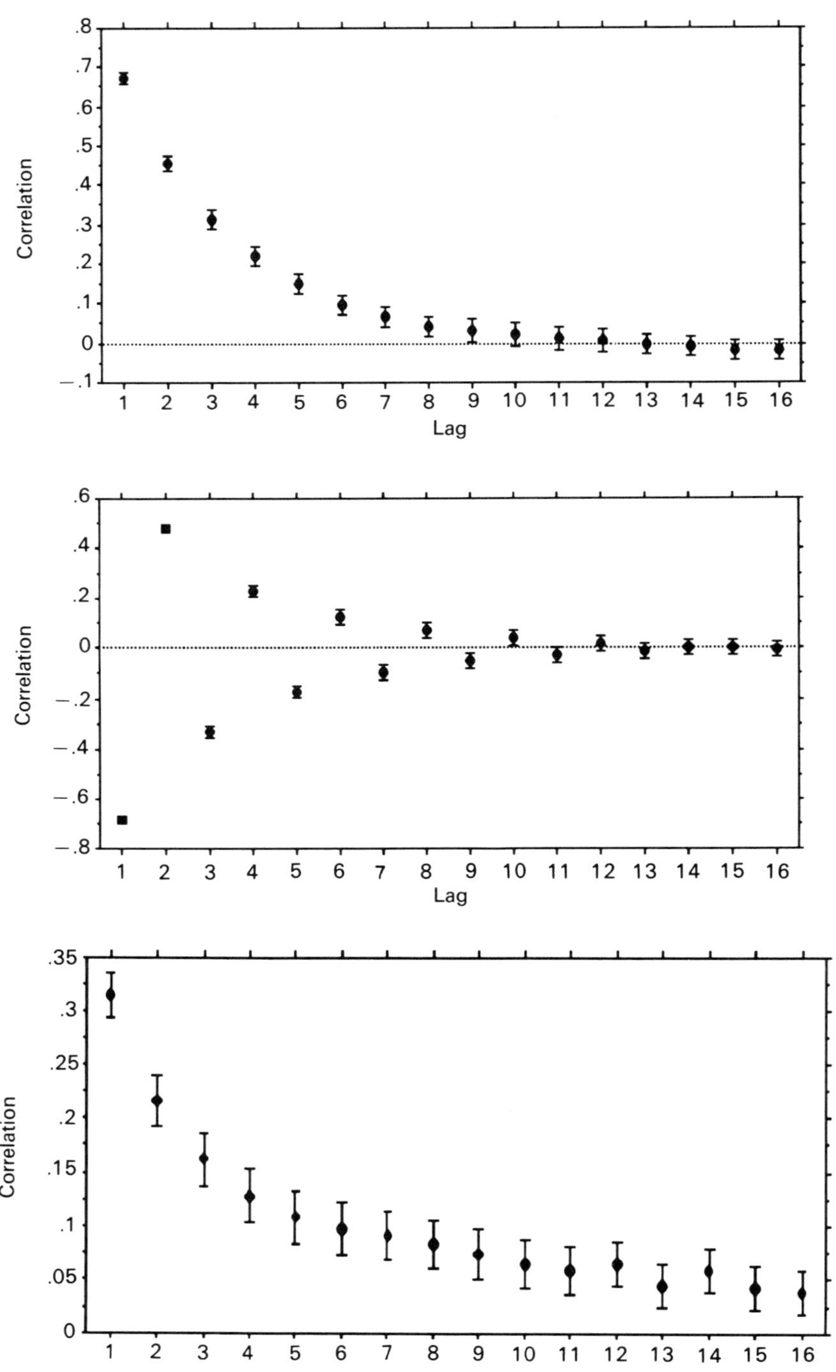

Correlation
Lag
.8
.7
.6
.5
.4
.3
.2
.1
0
−.1
1 2 3 4 5 6 7 8 9 10 11 12 13 14 15 16
.6
.4
.2
−.2
−.4
−.6
−.8
.35
.3
.25
.2
.15
.1
.05

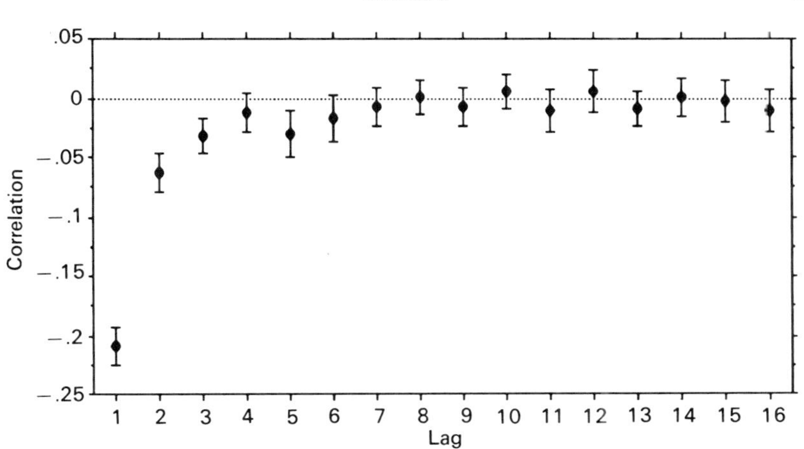

Figure 13.10. Correlations for model with $d = -0.25$.

Table 13.1 Autocorrelations when $d = 0.25$ for $T = 128$ and varying N

	N						
	$2T$		$4T$		$8T$		
Lag	Mean	S. devn	Mean	S. devn	Mean	S. devn	Actual
1	0.298	0.092	0.321	0.099	0.312	0.118	0.333
2	0.198	0.109	0.214	0.106	0.227	0.128	0.238
3	0.162	0.104	0.167	0.107	0.181	0.142	0.195
4	0.140	0.102	0.142	0.107	0.142	0.140	0.169
5	0.096	0.107	0.118	0.101	0.127	0.132	0.151
6	0.092	0.106	0.109	0.108	0.111	0.127	0.138
7	0.073	0.117	0.098	0.099	0.094	0.134	0.128
8	0.063	0.094	0.086	0.096	0.088	0.126	0.119
9	0.068	0.102	0.081	0.112	0.080	0.126	0.113
10	0.067	0.102	0.064	0.105	0.085	0.115	0.107
11	0.053	0.108	0.055	0.100	0.061	0.115	0.102
12	0.035	0.119	0.052	0.111	0.065	0.107	0.098
13	0.034	0.110	0.051	0.109	0.069	0.108	0.094
14	0.043	0.101	0.041	0.112	0.069	0.110	0.090
15	0.045	0.106	0.042	0.107	0.067	0.107	0.087
16	0.027	0.106	0.026	0.101	0.066	0.104	0.084

◄

Figure 13.7. Correlations for ARI model parameter 0.7.
Figure 13.8. Correlations for ARI model parameter −0.7.
Figure 13.9. Correlations for model with $d = 0.25$.

As can be seen there seems to be an agreement with our expectations for an AR1 process. The much more interesting case is that of a fractional model and the correlations of the two simple models

$$(1 - B)^d X_t = a_t$$

are given in Figure 13.9 and 13.10. Table 13.1 also shows the change in the estimate of d for changes in grid parameter.

We conclude that the method as outlined works reasonably well. Note, however, that we are approximating two functions and then using their Fourier transforms. There may, therefore, be the possibility of some small cyclical behaviour at frequencies between the fixed grid frequencies for small values of T. In our numerical investigations such oscillation has been small. It may be worth investigating some tapering to allow for the finite frequency range employed.

Thus as we have seen, the spectrum provides a simple closed form for the simulation of stationary time series which is reasonably efficient. In addition, by using the standard tools available, a generalized model may be fitted to the spectrum which provides information for model fitting or filtering.

REFERENCES

Agiakloglou, C., Newbold, P. and Wohar, M. (1993) Bias in an estimator of the fractional difference parameter. *J. Time Series Analysis*, in press.

Box, G. and Jenkins, G. (1970) *Time Series Analysis: Forecasting and Control*, Holden-Day, San Francisco.

Carlin, J. and Dempster, A. (1989) Sensitivity analysis of seasonal adjustments. *J. American Statist. Assn.*, **84**, 6–32.

Granger, C.W. and Joyeux, R. (1980) An introduction to long memory times series and fractional differencing. *J. Time Series Analysis*, **6**(1), 15–30.

Gweke, J. and Porter-Hudak, S. (1983) The estimation and application of long memory time series models. *J. Time Series Analysis*, **4**(4), 221–238.

Cox, D. (1991) Long-range dependence, non-linearity and time irreversibility. *J. time Series Analysis*, **12**(4), 329–336.

Harvey, A. (1989) *Forecasting Structural Time Series Models and the Kalman Filter*, Cambridge University Press, Cambridge.

Hosking, J. (1981) Fractional differencing. *Biometrika*, **68**(1), 165–176.

Janacek, G. (1982) Determining the degree of differencing for time series via the log spectrum. *J. Time Series Analysis*, **3**, 177–184.

Janacek, G. and Swift, A. (1993) *Time Series Analysis*, Ellis Horwood, Chichester.

Koopmans, L.H. (1973) *Spectral Analysis of Time Series*, Academic Press, New York.

Lamb, H. (1977) *Climate: Present, Past and Future*, Vol. 2, Methuen, London.

Li, W. and McLeod, A. (1986) Fractional Time Series Modelling. *Biometrika*, **73**, 217–221.

McLeod, A.I. and Hipel, K.W. (1978) Preservation of the rescaled range. *Water Resources Research*, **14**(3), 491–518.

Monro, P.M. and Branch, B.L. (1977) The chirp discrete fourier transform of general length. *Applied Stats.*, **26**(3), 251–361.

McCullagh, P. and Nelder, J. (1989) *Generalized Linear Models*, 2nd edn., Chapman and Hall, London.

Nelson, C.R. (1973) *Applied Time Series Analysis*, Holden-Day, San Francisco.

Pregibon, D. (1980) Goodness of link tests for generalized linear models. *Applied. Stats.*, **29**, 15–24.

Priestley, M. (1981) *Spectral Analysis and Time Series*, Academic Press, New York.

Part Four

Nonparametric Statistical Inference in Time Series

14

Nonparametric function estimation in noisy chaos

B. Cheng and H. Tong

14.1 INTRODUCTION

In 1972, Priestley and Chao discussed the use of the kernel method to estimate the unknown function/map f given observations $Z_1, Z_2, \ldots, Z_n$ corresponding to the 'covariate' x at $x_1, x_2, \ldots, x_n$ respectively in the form

$$Z = f(x) + \varepsilon$$

ε being the noise disturbance. Our paper extends the methodology to the time series context where the 'covariate' x is the 'history' of Z, with particular reference to chaotic time series.

Let $\{Z_t\}_{t \in N}$ be a time series and $F_d(Z_{t-1}, \ldots, Z_{t-d}) = \mathrm{E}[Z_t | Z_{t-1}, \ldots, Z_{t-d}]$ for $d \geqslant 1$. We assume that Z_t satisfies a nonlinear autoregressive model with order d_0, i.e.,

$$Z_t = F_{d_0}(Z_{t-1}, \ldots, Z_{t-d_0}) + \varepsilon_t \tag{14.1}$$

where $\{\varepsilon_t\}_{t \in N}$ is a stationary martingale difference with variance σ^2. We assume that d_0 if finite. Note that d_0 is related to the embedding dimension in chaos (Cheng and Tong, 1992). Indeed, model (14.1) may be motivated by a deterministic model in which ε_t is absent. Current interest in the dynamical system literature focuses on those Fs which admit an exotic attractor, namely a strange attractor, or more commonly, chaos, and on the 'inverse' problem of recovering F from noisy data as a result of observational and/or system (i.e. dynamic) noise. Model (14.1) may be considered a dynamical system with dynamic noise.

In this paper, we are interested in the determination of d_0 and the estimation of F_{d_0} simultaneously based on the observations $Z_1, \ldots, Z_n$. This is also called the problem of map reconstruction from noisy data in the chaos literature. (See, e.g., the special issue on chaos in the *Journal of the Royal Statistical*

Society (B), 1992.) Cheng and Tong (1992) have considered a consistent estimator of d_0, which is important because, if it is not chosen properly, there will be redundance or irrelevance (see, e.g., Casdagli *et al.* (1991)). However, it may be more crucial to estimate F_{d_0}. For example, as we have discussed in Cheng and Tong (1992), an estimate of F_{d_0} may enable us to understand the underlying dynamics of the 'skeleton' and to identify such exotic attractors as chaos. We shall use kernel type estimators in this paper.

The aim of the paper is to prove rigorously that under appropriate conditions pertaining to the smoothness of F and the mixing rate of the observed time series, the kernel-type reconstruction of F has some optimal large-sample properties. This is the content of our main result: theorem 14.12. Of course, for large embedding dimension, the sample may indeed have to be rather large due to the curse of dimensionality. However, we do not address the finite-sample properties in this paper, which we plan to do elsewhere.

Denote $Y_t^d = (Z_{t-1}, \ldots, Z_{t-d})^T$ for $d = 1, 2, \ldots$ and let f_d be the density function of Y_t^d and $F_d(x) = \mathrm{E}[Z_t \mid Y_t^d = x]$ for $x \in R^d$. We estimate f_d and F_d by

$$\hat{f}_d(x) = \frac{1}{nh^d} \sum_{t=1}^{n} K\left[\frac{x - Y_t^d}{h}\right] \tag{14.2}$$

and

$$\hat{F}_d(x) = \frac{1}{nh^d} \sum_{t=1}^{n} Z_t K\left[\frac{x - Y_t^d}{h}\right] \div \hat{f}_d(x) \tag{14.3}$$

where the kernel K will be defined later.

When we use $\hat{F}_d$ to approximate F_{d_0}, there are two parameters which have to be chosen, namely the bandwidth h and the order d. It is natural to consider using cross-validation criteria to choose h and d. Hart (1991) has pointed out that the ordinary cross-validation procedure could behave erratically when the data are sufficiently positively correlated. Consequently, it has been suggested that we should leave out more than just one observation. Hart and Vieu (1990), Härdle and Vieu (1990), and Hart (1991) have gone this way. We will adopt their idea and define leave-out estimators, $\hat{f}_{d,\setminus t}$ and $\hat{F}_{d,\setminus t}$, of f_d and F_d by

$$\hat{f}_{d,\setminus t}(x) = \frac{1}{n_t h^d} \sum_{|s-t| > \rho_n} K\left[\frac{Y_s^d - x}{h}\right] \tag{14.4}$$

and

$$\hat{F}_{d,\setminus t}(x) = \frac{1}{n_t h^d} \sum_{|s-t| > \rho_n} Z_s K\left[\frac{Y_s^d - x}{h}\right] \div \hat{f}_{d,\setminus t}(x) \tag{14.5}$$

where $\{\rho_n\}$ is a sequence of non-decreasing integers, called the leave-out sequence, and

$$n_t = \frac{1}{n}\#\{(t,s); |s-t| > \rho_n\}$$

In particular, when $\rho_n \equiv 0$, we have the ordinary cross-validation criteria.

We sometimes augment the argument of the estimators considered so far by the letter h so as to emphasize their dependence on the bandwidth h.

We assume that $\{Z_t\}$ is α-mixing in the sense of Rosenblatt. Specifically, defining the mixing coefficient

$$\alpha(m) = \sup_{\substack{A \in \mathcal{F}_{l+m}^{+\infty} \\ B \in \mathcal{F}_{-\infty}^{l}}} |P(A \cap B) - P(A)P(B)|, \tag{14.6}$$

where the σ-field $\mathcal{F}_l^m = \sigma(Z_l, \ldots, Z_m)$, we assume that $\alpha(m) \to 0$ as $m \to \infty$.

To keep the proofs of our results at reasonable length, we further assume that

$$\alpha(m) = 0(\beta^m) \qquad \text{(A1), (14.7)}$$

for some β such that $0 < \beta < 1$, i.e. we assume thas $\{Z_t\}$ is α-mixing with geometrically decaying coefficients. All the proofs will be collected at the end of the paper.

Let $\mathcal{K}_d$ be the family of the kernel functions on R^d such that $\forall K \in \mathcal{K}_d$,

$$K(x) = \prod_{i=1}^{d} \tilde{k}(x_i), \tag{14.8}$$

for each $x = (x_1, \ldots, x_d) \in R^d$ and $\tilde{k}$ satisfies the following assumptions.

> $\tilde{k}$ is symmetric, Lipschitz continuous, compactly supported and has an absolutely integrable Fourier transform. (A2)

$$\int_{-\infty}^{\infty} \tilde{k}(t)\,\mathrm{d}t = 1, \quad \tilde{k}(t) \geqslant 0 \quad \text{and} \quad \int_{-\infty}^{\infty} t^2\tilde{k}(t)\mathrm{d}t < \infty. \qquad \text{(A3)}$$

Let K be an element in $\mathcal{K}_d$ and assume that the sequence $\{Z_t\}$ is α-mixing. Suppose that (A1) and (A2) hold and let $t(1), \ldots, t(p)$ be p distinct positive integers, integer $d \geqslant 1$, and define

$$g(Y_{t(1)}^d, \ldots, Y_{t(p)}^d) = \prod_{i=1}^{q_1} \prod_{\substack{j=1 \\ j \neq i}}^{q_2} \left[\frac{1}{h^d} K\left[\frac{Y_{t(i)}^d - Y_{t(j)}^d}{h}\right]\right]^{\beta_{ij}} \prod_{i=1}^{p} g_i(Y_{t(i)}^d)$$

where the g_is are real-valued functions such that $|g_i| \leqslant M_i < \infty$, the $\beta_{i,j}$ are non-negative integers, $q_1 \leqslant p$ and $q_2 \leqslant p$. Let $A_1, \ldots, A_v$ be a partition of $\{t(1), \ldots, t(p)\}$. We have the following basic mixing inequality.

Proposition 14.1

There exists a finite positive constant c such that

$$\left|\int g \mathrm{d}P_{(Y_{t(1)},\ldots,Y^{d}_{t(p)})} - \int g \mathrm{d}P_{(Y_t, t\in A_1)} \times \cdots \times \mathrm{d}P_{(Y_t, t\in A_v)}\right| \leqslant C \prod_{i=1}^{p} M_i h^{-d\beta}\tilde{\alpha}(m),$$

where

$$m = \inf\{\mathrm{dist}(A_i, A_j);\ i, j = 1, \ldots, v,\ i < j\},$$

$$\mathrm{dist}(A_i, A_j) = \inf\{|u - u'|,\ u \in A_i, u' \in A_j\}$$

$$\beta = \sum_{i=1}^{q_1} \sum_{\substack{j=1 \\ j \neq i}}^{q_2} \beta_{i \cdot j},$$

and

$$\tilde{\alpha}(m) = \sup_{j \geqslant m} \alpha(j).$$

The proof is similar to the proof of proposition 1 of Hart and Vieu (1990).

The α-mixing condition is a very mild one among similar mixing conditions such as absolute regularity, ϕ-mixing and so on. If we use the absolutely regular condition, the conclusion of proposition 14.1 is a simple corollary of Yoshihara inequality (Denker and Keller (1983), Lemma 6). However, we do not know whether Yoshihara inequality is available to α-mixing sequences. In any case, for the cumulants of our kernel type estimators, the conclusion of proposition 14.1 suffices.

For the choice of the order d in our index set, it is natural to allow the candidate set to increase as the sample size increases. Thus, we define

$$D_n = \{1, 2, \ldots, L\}$$

with $L = \max\{1, c[\ln(n)]^{\delta}\}$ and $0 < \delta < 1$, where c is a positive value and $[x]$ is the largest integer less than or equal to x.

$$F_L \text{ is compactly supported in } R^L. \tag{A4}$$

Let S_L^0 be the interior open set of the compact support of f_L. Choose a closed set S_L in S_L^0. We assume the following.

$$F_L \text{ and } f_L \text{ have the first two continuous derivatives on } S_L^0 \text{ and } f_L \text{ is strictly positive on } S_L. \tag{A5}$$

Define S_d to be the projection of S_L in R^d for $1 \leqslant d \leqslant L$, and the weighting function W_d on R^d by

$$W_d(x) = \begin{cases} 1 & \text{if } x \in S_d, \\ 0 & \text{otherwise.} \end{cases} \tag{14.9}$$

We define a bandwidth interval

$$H_{n,d} = [an^{(-1/2(d+3)-\varepsilon)}, bn^{(-1/2(d+3)+\varepsilon)}] \tag{14.10}$$

where, for a small $\xi > 0$,

$$0 < \varepsilon < \frac{U_d}{(d+3)V_d} - V_d^{-1}\xi, \tag{14.11}$$

$$U_d = \min\{\tfrac{3}{4}d + 1, \tfrac{1}{2}d + 2\},$$

$$V_d = \max\{d + 8, \tfrac{1}{2}d + 4\},$$

and

$$0 < a, b < \infty.$$

The reason why we do not use the common choice for $h \sim n^{-1/(4+d)}$ is that in the rate of uniform convergence for the estimators $\hat{f}_d$ and $\hat{F}_d$, we need the condition

$$h^3 + \{(\ln n)^2/(n^{1/2}h^d)\}^{1/2} \to 0 \quad \text{as} \quad n \to \infty.$$

For the leave-out sequence $\{\rho_n\}$, we assume

$$0 \leqslant \rho_n \leqslant \rho_n^* \tag{A.6}$$

where

$$\rho_n^* = n^\tau \text{ with } 0 < \tau < U_d/(d+3) - V_d\varepsilon - \xi, \tag{14.12}$$

where U_d, V_d, ε, and ξ are as defined in the definition of $H_{n,d}$.

In addition, we assume, for $1 \leqslant d \leqslant L$, S_d is so chosen that

$$\sup_{d \in D_n} n^{2/(d+3)+4\varepsilon}(\ln n)^{2\delta} P(Y_t^d \in S_d^0 \backslash S_d) = o(1), \tag{A.7}$$

where S_d^0 is the projection of S_L^0 in R^d, ε is as defined in (14.11) and δ is as defined in the definition of D_n.

Proposition 14.2

Under (A1)–(A6), we have for any compact subset S in R^d,

$$\sup_{h \in H_{n,d}} \sup_{x \in s} |\hat{f}^*(x) - f_d(x)| = o_p(n^{-1.5/(d+3)+3\varepsilon}), \tag{14.13}$$

and

$$\sup_{h \in H_{n,d}} \sup_{x \in s} |\hat{F}^*(x) - F_d(x)| = o_p(n^{-0.5/(d+3)+3\varepsilon}), \tag{14.14}$$

where $\hat{f}^*$ is either $\hat{f}_d$ or $\hat{f}_{d,\backslash t}$ and $\hat{F}^*$ is either $\hat{F}_d$ or $\hat{F}_{d,\backslash t}$.

Proposition 2 is a direct consequence of theorems 3.3.6 and 5.3.3 of Györfi *et al.* (1989).

Theorem 14.3

Let $\hat{d}(n)$ be a data-based estimator of d. If $\{\hat{d}(\tilde{n})\}$ is a sub-sequence of $\{\hat{d}(n)\}$ such that $\hat{d}(\tilde{n}) \to d_1$ in probability, then we have

$$\sup_{h \in H_{\tilde{n},d_1}} \sup_{x \in s} |\hat{F}_{d(\tilde{n})}(x) - F_{d_1}(x)| = o_p(\tilde{n}^{-1.5/(d_1+3)+3\varepsilon}). \tag{14.15}$$

Theorem 14.3 is a direct consequence of proposition 14.2 on noting that $\hat{d}(\tilde{n}) \to_p d$ as $\tilde{n} \to +\infty$ implies that $\hat{d}(\tilde{n}) \equiv d_1$, when $\tilde{n}$ is large enough, since $\hat{d}(\tilde{n})$ and d are both positive integers.

Theorem 14.3 suggests that we may approximate F_d by $\hat{F}_d$ provided that we choose d and h properly. Cheng and Tong (1992) have given a consistent estimator of d for bounded time series.

Now, we choose d and h by minimizing the cross-validation function

$$CV(h, d) = \frac{1}{n} \sum_{t=1}^{n} [Z_t - \hat{F}_{d,\backslash t}(Y_t^d; h)]^2 W_d(Y_t^d) \tag{14.16}$$

and denote the minimizers by $\hat{h}, \hat{d}$. We then estimate $F_d(\cdot)$ by $\hat{F}_{\hat{d}}(\cdot; \hat{h})$. In the next section, we shall show that this cross-validatory estimator enjoys some asymptotic optimal properties. We measure the lack of accuracy of this estimator (i.e. the loss function) by the average squared error,

$$\text{ASE}(h, d) = n^{-1} \sum_{t=1}^{n} [\hat{F}_d(Y_t^d; h) - F_{d_0}(Y_t^{d_0})]^2 W_d(Y_t^d). \tag{14.17}$$

Shibata (1980) has defined a similar loss function for a linear autoregressive model with $d_0 = \infty$. When $d_0 < +\infty$, in the same linear setting, Akaike (1974) also defined a similar loss function based on the Kullback–Leibler measure of information.

Before discussing the asymptotic properties, it is pertinent to list various issues which could be studied with large-scale simulations. We plan to report these elsewhere.

(a) The efficiency of $\hat{F}_d$ with $d = \hat{d}$ and $h = \hat{h}$.

(b) The numerical results of ASE and the ratio

$$\frac{\text{ASE}(\hat{h}, \hat{d})}{\text{ASE}(\mathbf{h}, \mathbf{d})},$$

where $(\mathbf{h}, \mathbf{d})$ will be defined in section 14.2.

(c) The 'stability' of $\hat{F}_d$—if F_{d_0} has an attractor, how about $\hat{F}_{\hat{d}}(\cdot; \hat{h})$?

(d) Choice of leave-out sequence. Does $\rho_n \equiv 0$ work not so badly in our case?

(e) Theoretically, let

$$\frac{\partial^{j_1 + \cdots + j_d}}{\partial x_1^{j_1} \cdots \partial x_d^{j_d}} F_d(x_1 \cdots x_d)$$

be the $(j_1,\ldots,j_d)$ partial derivative of F_d, Then

$$\frac{\partial^{j_1+\cdots j_d}}{\partial x_1^{j_1}\cdots\partial x_d^{j_d}}\hat{F}_d(x_1,\cdots x_d)\to_p\frac{\partial^{j_1+\cdots+j_d}}{\partial d_1^{j_1}\cdots\partial_d^{j_d}}F_d(x_1,\ldots x_d).$$

Therefore to study the asymptotic properties of the derivatives of the F_d's, we need only study those of the derivatives of the $\hat{F}_{\hat{d}}$'s.

14.2 ASYMPTOTIC OPTIMALITY OF THE ADAPTIVE ESTIMATOR

Let

$$(\hat{h},\hat{d})=\underset{d\in D_n}{\operatorname{argmin}}\left\{\underset{h\in H_{n,d}}{\operatorname{argmin}}\ CV(h,d)\right\}, \tag{14.18}$$

and

$$(\mathbf{h},\mathbf{d})=\underset{d\in D_n}{\operatorname{argmin}}\left\{\underset{h\in H_{n,d}}{\operatorname{argmin}}\ \mathrm{ASE}(h,d)\right\}. \tag{14.19}$$

We aim to prove that

$$\frac{\mathrm{ASE}(\hat{h},\hat{d})}{\mathrm{ASE}(\mathbf{h},\mathbf{d})}\to_p 1 \text{ as } n\to\infty \tag{14.20}$$

or equivalently

$$\frac{\mathrm{ASE}(\hat{h},\hat{d})-\mathrm{ASE}(\mathbf{h},\mathbf{d})}{\mathrm{ASE}(\mathbf{h},\mathbf{d})}\to_p 0 \text{ as } n\to\infty.$$

By the definition of $\hat{h}$ and $\hat{d}$

$$CV(\mathbf{h},\mathbf{d})-CV(\hat{h},\hat{d})\geqslant 0$$

and by the definition of $\mathbf{h},\mathbf{d}$

$$\mathrm{ASE}(\hat{h},\hat{d})-\mathrm{ASE}(\mathbf{h},\mathbf{d})\geqslant 0.$$

So we only need to show

$$\frac{\mathrm{ASE}(\hat{h},\hat{d})-\mathrm{ASE}(\mathbf{h},\mathbf{d})+CV(\mathbf{h},\mathbf{d})-CV(\hat{h},\hat{d})}{\mathrm{ASE}(\mathbf{h},\mathbf{d})}=\mathrm{o}_p(1).$$

Define

$$G(h,h',d,d')=\frac{\mathrm{ASE}(h,d)-\mathrm{ASE}(h',d')+\mathrm{CV}(h',d')-\mathrm{CV}(h,d)}{\mathrm{ASE}(h',d')}.$$

Then

$$|G(\hat{h},\mathbf{h},\hat{d},\mathbf{d})|\leqslant\sup_{d,d'\in D_n}\ \sup_{h,h'\in H_{n.d}}|G(h,h',d,d')|.$$

Define

$$\overline{\mathrm{ASE}}(h,d)=n^{-1}\sum_{t=1}^{n}[\hat{F}_{d,\backslash t}(Y_d;h)-F_{d_0}(Y_t^{d_0})]^2\,W_d(Y_t^d).$$

Lemma 14.4

Under (A1)–(A6), we have

$$\sup_{d\in D_n}\sup_{h\in H_{n,d}}\frac{|\overline{\mathrm{ASE}}(h,d)-\mathrm{ASE}(h,d)|}{\mathrm{ASE}(h,d)}=o_p(1).$$

It follows from lemma 14.4 that

$$\sup_{d,d'\in D_n}\sup_{h,h'\in H_{n,d}}|G(h,h',d,d')|=o_p(1) \tag{14.21}$$

if

$$\sup_{d,d'\in D_n}\sup_{h,h'\in H_{n,d}}|\bar{G}(h,h',d,d')|=o_p(1), \tag{14.22}$$

where $\bar{G}$ is G after we have replaced ASE by $\overline{\mathrm{ASE}}$ in the denominator of G. Decompose

$$\overline{\mathrm{ASE}}(h,d)=-n^{-1}\sum_{t=1}^{n}\varepsilon_t^2\,W_d(Y_t^d)+CV(h,d)+2C_n(h,d) \tag{14.23}$$

where

$$C_n(h,d)=n^{-1}\sum_{t=1}^{n}\varepsilon_t[\hat{F}_{d,\backslash t}(Y_t^d)-F_{d_0})]\,W_d(Y_t^d). \tag{14.24}$$

Then (14.22) is true if we show that

$$\sup_{d\in D_n}\sup_{h\in H_{n,d}}\frac{|C_n(h,d)|}{\mathrm{ASE}(h,d)}=o_p(1) \tag{14.25}$$

and

$$\sup_{d,d'\in D_n}\sup_{h\in H_{n,d}}\frac{\frac{1}{n}\sum_{t=1}^{n}\varepsilon_t^2|W_d(Y_t^d)-W_{d'}(Y_t^{d'})|}{\mathrm{ASE}(h,d)}=o_p(1). \tag{14.26}$$

The above technique is standard, and has been used by, for example, Härdle and Marron (1985), Marron and Härdle (1986), and Härdle and Vieu (1990).

Now, let us denote

$$\bar{C}_n(h,d)=\frac{1}{n}\sum_{t=1}^{n}\varepsilon_t[\hat{F}_{d,\backslash t}(Y_t^d)-F_{d_0}(Y_t^{d_0})]\hat{f}_{d,\backslash t}(Y_t^d)/f_d(Y_t^d)W_d(Y_t^d)$$

and decompose it

$$\bar{C}_n(h,d) = \bar{C}_{n,1}(h,d) + \bar{C}_{n,2}(h,d),$$

where

$$\bar{C}_{n,1}(h,d) = n^{-1}\sum_{t=1}^{n} n_t^{-1} \sum_{|s-t|>\rho_n} \varepsilon_t\varepsilon_s K\left(\frac{Y_s^d - Y_t^d}{h}\right) W_d(Y_t^d) f_d^{-1}(Y_t^d) \quad (14.27)$$

and

$$\bar{C}_{n,2}(h,d) = n^{-1}\sum_{t=1}^{n} n_t^{-1} \sum_{|s-t|>\rho_n} \varepsilon_t[F_{d_0}(Y_s^{d_0}) - F_{d_0}(Y_t^{d_0})] W_d(Y_t^d) \times K\left(\frac{Y_s^d - Y_t^d}{h}\right) f_d^{-1}(Y_t^d). \quad (14.28)$$

Define $H'_{n,d}$ to be a finite subset of $H_{n,d}$ consisting of equally spaced elements and such that

$$\#H'_{n,d} = n^{\tau_d} \quad (14.29)$$

where, for an $\eta > 0$ and ε in the definition of $H_{n,d}$,

$$\tau_d > \frac{d}{2(d+3)} + 3(d+1)\varepsilon + \eta. \quad (14.30)$$

Then (14.22) follows from lemmas 14.5–14.10 stated as follows.

Lemma 14.5

Under (A1)–(A6) and if $\rho_n = \rho_n^*$, we have

$$\sup_{d\in D_n} \sup_{h\in H'_{n,d}} \frac{|\bar{C}_{n,1}(h,d)|}{\mathrm{ASE}(h,d)} = o_p(1).$$

Lemma 14.6

Under (A1)–(A6), and if $\rho_n = \rho_n^*$, we have

$$h\in H'_{n,d} \frac{|\bar{C}_{n,2}(h,d)|}{\mathrm{ASE}(h,d)} = o_p(1).$$

Lemma 14.7

Under (A1)–(A6), we have

$$\mathrm{ASE}(h,d) = \begin{cases} \mathrm{E}[F_d(Y_t^d) - F_{d_0}(Y_t^{d_0})]^2 + o_p(1) \text{ if } d < d_0 \\ \dfrac{c_1}{nh^d} + C_2 h^4 + o_p(\mathrm{ASE}(h,d)) \text{ if } d \geqslant d_0 \end{cases}$$

where the order d_0 is as given in Cheng and Tong (1992),

$$C_1 = \sigma^2 \int_{R_d} K^2(x)\mathrm{d}x$$

and

$$C_2 = \frac{1}{4}\left[\int_{R_d} x_1^2 K^2(x)\mathrm{d}x\right]^2 \int [\nabla^2 f(x)]^2 \mathrm{d}x$$

where $x = (x_1, x_2, \ldots, x_d)$ and ∇^2 denotes the Laplacian.

Lemma 14.8

Under (A1)–(A6) and if $\rho = \rho_n^*$ we have

$$\sup_{d \in D_n} \sup_{h \in H'_{n,d}} \frac{|\bar{C}_n(h, d) - C_n(h, d)|}{\mathrm{ASE}(h, d)} = \mathrm{o}_p(1).$$

Lemma 14.9

Under (A1)–(A6) and if $\rho = \rho_n^*$ we have

$$\sup_{d \in D_n} \sup_{h \in H_{n,d}} \frac{|C_n(h, d) - C_n(h^*, d)|}{\mathrm{ASE}(h, d)} = \mathrm{o}_p(1),$$

where for any $h \in H_{n,d}$, h^* is defined to be the element of $H'_{n,d}$ that is closest to h.

Lemma 14.10

Under (A1)–(A6) we have

$$\sup_{d \in D_n} \sup_{h \in H_{n,d}} \frac{|C_n(h, d) - C_n^*(h, d)|}{\mathrm{ASE}(h, d)} = \mathrm{o}_p(1),$$

where C_n^* denotes the quantity C_n which applies when $\rho_n = \rho_n^*$.

Lemma 14.11

Under (A1)–(A7) we have

$$\sup_{d, d' \in D_n} \sup_{h \in D_n} \frac{\frac{1}{n}\sum_{t=1}^{n} \varepsilon_t^2 |W_d(Y_t^d) - W_{d'}(Y_t^{d'})|}{\mathrm{ASE}(h, d)} = \mathrm{o}_p(1).$$

Finally, by lemmas 14.4–14.11, we have the main conclusion of this paper, as follows.

Theorem 14.12

Under (A1)–(A7), we have

$$\frac{\mathrm{ASE}(\hat{h},\hat{d})}{\mathrm{ASE}(\mathbf{h},\mathbf{d})} \to_p 1 \quad \text{as} \quad n \to \infty.$$

14.3 PROOFS OF THE LEMMAS

Proof of lemma 14.4

$$\overline{\mathrm{ASE}}(h,d) - \mathrm{ASE}(h,d) = -2n^{-1} \sum_{t=1}^{n} [\hat{F}_d(Y_t^d) - F_{d_0}(Y_t^{d_0})][\hat{F}_d(Y_t^d) - F_{d\setminus t}(Y_t^d)] W_d(Y_t^d) + n^{-1} \sum_{t=1}^{n} [\hat{F}_{d\setminus t}(Y_t^d) - \hat{F}_d(Y_t^d)]^2 W_d(Y_t^d). \tag{14.31}$$

By the definition of $\hat{F}$ and $\hat{F}_{d\setminus t}$, we have

$$\hat{F}_{d\setminus t}(Y_t^d) = \frac{\hat{f}_d(Y_t^d)}{\hat{f}_{d,\setminus t}(Y_t^d)}\left[\frac{n}{n_t}\hat{F}_d(Y_t^d)\right] - \frac{A_t}{\hat{f}_{d,\setminus t}(Y_t^d)}, \tag{14.32}$$

where

$$A_t = n_t^{-1} \sum_{|s-t| \leqslant \rho_n} \frac{1}{h^d} K\left(\frac{Y_s^d - Y_t^d}{h}\right) Z_s. \tag{14.33}$$

Then $\overline{\mathrm{ASE}}(h,d) - \mathrm{ASE}(h,d) = -2T_1(h,d) + T_2(h,d)$ where $-2T_1(h,d)(T_2(h,d))$ is the first (second) term of the right part of (14.31) when we use the right part of (14.32).

First we have

$$|A_t| = O_p\left(\frac{\rho_n}{nh^d}\right).$$

By proposition 14.2, and (A5) we have

$$\sup_{h,t} \frac{|A_t|}{\hat{f}_{d,\setminus t}(Y_t^d)} = O_p\left(\frac{\rho_n}{nh^d}\right).$$

Similarly we have

$$|n\hat{f}_d(Y_t^d) - n_t \hat{f}_{d,\setminus t}(Y_t^d)| = O_p\left(\frac{\rho_n}{h^d}\right).$$

So it follows that

$$\sup_{h,t} \left| \frac{\hat{f}_d(Y_t^d)n}{\hat{f}_{d,\setminus t}(Y_t^d)n_t} - 1 \right| = O_p\left(\frac{\rho_n}{nh^d}\right).$$

We have by proposition 14.2, and lemma 14.7

$$h^{-4}|T_1(h,d)| \leqslant h^{-4}\sup_{x\in S_d}|\hat{F}_d(x) - F_d(x)|\frac{\rho_n}{nh^d}$$
$$= 0(n^{-1.5/(d+3)+3\varepsilon}\rho_n^*(nh^{d+4})^{-1})$$
$$= 0(n^{-(d+4)/[2(d+3)]+(d+7)\varepsilon+\tau}),$$

where we bound $[\text{ASE}(h,d)]^{-1}$ by h^{-4} by lemma 14.7.

According to (14.11) and (14.12), we have

$$\sup_{d\in D_n}\sup_{h\in H_{n,d}}\frac{|T_1(h,d)|}{\text{ASE}(h,d)} = o_p(1).$$

Secondly,

$$h^{-4}|T_2(h,d)| = 0\left(\left[\frac{\rho_n}{nh^{d+2}}\right]^2\right)$$
$$= 0(n^{2r}),$$

where $r = \tau - (d+4)/(2d+6) + (d+2)\varepsilon$.

According to (14.11) and (14.12), we obtain

$$\sup_{d\in D_n}\sup_{h\in H_{n,d}}\frac{|T_2(h,d)|}{\text{ASE}(h,d)} = o_p(1).$$

Proof of lemma 14.5

Since

$$\sup_{d\in D_n}\sup_{h\in H'_{n,d}}\left[\frac{|\bar{C}_{n,1}(h,d)|}{\text{ASE}(h,d)}\right]^2 \leqslant \sum_{d\in D_n}\sum_{h\in H'_{n,d}}\left[\frac{|\bar{C}_{n,1}(h,d)|}{\text{ASE}(h,d)}\right]^2,$$

it suffices to show for some $b_1 > 0$ that we have, for any positive integer k,

$$\#H'_{n,d}\sup_{d\in D_n}\sup_{h\in H'_{n,d}}\text{E}\left[\frac{|\bar{c}_{n,1}(h,d)|}{\text{ASE}(h,d)}\right]^{2k} = 0(n^{-kb_1}). \tag{14.34}$$

Notice that $n_t \sim n$ as $n\to\infty$ for each t. Let us denote by

$$C^+(h,d) = n^{-2}\sum_{t=1}^{n}\sum_{s>t+\rho_n^*}U(s,t)$$

and

$$C^-(h,d) = n^{-2}\sum_{t=1}^{n}\sum_{s<t+\rho_n^*}U(s,t)$$

where

$$U(s,t) = K\left(\frac{Y_s^d - Y_t^d}{h}\right)\varepsilon_s\varepsilon_t f_d^{-1}(Y_t^d)W_d(Y_t^d). \tag{14.35}$$

We only need to prove

$$\sup_{d\in D_n}\sup_{h\in H'_{n,d}} \{\#H'_{n,d}h^{-4}\mathrm{E}[C^*(h,d)]^k\} = \mathrm{o}(n^{kb_1}) \tag{14.36}$$

where C^* is either C^+ or C^-, and we bound $\mathrm{ASE}^{-1}(h,d)$ by h^{-4} because of lemma 14.7. We show that (14.36) holds for $C^* = C^+$. The analysis of C^- is similar.

Let n_1 be the largest integer less than or equal to $n/2\rho_n^*$,

$$q' = i + (2q-1)\rho_n^* + \delta_1\rho_n^* \tag{14.37}$$

and

$$m' = j + 2(m-1)\rho_n^* + \delta_2\rho_n^*, \tag{14.38}$$

where $0 \leqslant i, j \leqslant \rho_n^*$, $1 \leqslant m$, $q \leqslant n_1$, δ_1 and $\delta_2 = 0$ or 1.

Then it is easy to check that

$$m' + \rho_n^* \leqslant q' \Leftrightarrow 1 + m \leqslant q. \tag{14.39}$$

We have

$$C^+(h,d) = n^{-2}\sum_{\delta_1\delta_2}\sum_{i,j=1}^{\rho_n^*}\sum_{q=1}^{n_1}\sum_{m=1}^{q} U(q',m'),$$

where q' and m' are defined in (14.37) and (14.38). So, in fact, we have to prove that there exists a generic constant c which is independent of d and h such that, for $d\in D_n$, $h\in H'_{n,d}$,

$$(h^{-4})^{2k}\#H'_{n,d}(n^{-1}\rho_n^*)^{4k}\sup_{i,j,\delta_1,\delta_2}\mathrm{E}\left[\sum_{q=1}^{n_1}\sum_{m=1}^{q} U(q',m')\right]^{2k} \leqslant cn^{-kb_1} \tag{14.40}$$

for some $b_1 > 0$, where we bound $\mathrm{ASE}^{-1}(h,d)$ by h^{-4} because of lemma 14.7.

Define now

$$\Delta = \{(q_1,m_1,\ldots,q_{2k},m_{2k}); 1 \leqslant m_i \leqslant q_i \leqslant n_1, \forall i = l,\ldots,2k\}$$

and decompose it into

$$\Delta = \Delta_1 \cup \Delta_2,$$

where

$$\Delta_1 = \{J = (q_1,\ldots,m_{2k})\in\Delta; \exists u\in\{q_1,\ldots,m_{2k}\} \text{ and } \forall v\in\{q_1,\ldots,m_{2k}\}\backslash\{u\}, |u-v| \geqslant 2\}$$

and

$$\Delta_2 = \Delta - \Delta_1.$$

So we have

$$\mathrm{E}\left[\sum_{q=1}^{n_1}\sum_{m=1}^{q} U(q',m')\right]^{2k} = \sum_{J\in\Delta_1} \mathrm{E}\Psi(J) + \sum_{J\in\Delta_2} \Psi(J), \tag{14.41}$$

where for $J\in\Delta$, $\Psi(J)$ is defined by

$$\Psi(J) = \prod_{i=1}^{2k} U(q',m'). \tag{14.42}$$

By the definition of Δ_1, let m_0 be the element of $\{q_1, \ldots, m_{2k}\}$ which differs from all others by at least 2. So $\forall u\in\{q_1,\ldots,m_{2k}\}\backslash\{m_0\}$, we have $|u' - m_0'| \geqslant \rho_n^*$, by proposition 14.1 and

$$\mathrm{E}[\varepsilon_s|\mathcal{F}_{-\infty}^{s-1}] = 0,$$

$$|\mathrm{E}\Psi(J)| = 0(h^{-2dk}\alpha(\rho_n^*))$$

and by $\#\Delta_1 \sim n_1^{4k} \sim n^{4k}(\rho_n^*)^{-4k}$

$$\left|\sum_{J\in\Delta_1} \mathrm{E}\Psi(J)\right| = 0(n^{4k}(\rho_n^*)^{-4k}h^{-2dk}\alpha(\rho_n^*)). \tag{14.43}$$

Let us now consider the case when $J\in\Delta_2$.

$$\Delta_2 = \{J = \{q_1,\ldots,m_{2k}\}\in\Delta : \forall u,v\in\{q_1,\ldots,m_{2k}\}, |u-v| \leqslant 1\}$$

and define

$$\Delta_2(l) = \{J\in\Delta_2; \#\{q_1,\ldots,m_{2k}\} = l\}.$$

Then

$$\Delta_2 = \bigcup_{l=1}^{4k} \Delta_2(l). \tag{14.44}$$

Since, by (A4), $\{Y_t^d\}_{t\in N}$ is bounded, for each $J\in\Delta_2(l)$, we have

$$|\mathrm{E}\Psi(J)| = 0(h^{-2dk+dl/2})$$

by integration by substitution.

Noticing that $\#\Delta_2(l) = 0(n_1^{\min\{l,2k\}})$, we have

$$\left|\sum_{J\in\Delta_2} \mathrm{E}\Psi(J)\right| = 0\left(\sum_{l=2k}^{4k} n^{2k}(\rho_n^*)^{-2k}h^{-2dk+dl/2}\right) + 0\left(\sum_{l=1}^{2k} n^l(\rho_n^*)^{-1}h^{-2dk+dl/2}\right).$$

It is easy to check

$$n\rho_n^{*-1}h^{d/2} \to +\infty$$

by (14.11) and (14.12). So we get

$$\left|\sum_{J\in\Delta_2} \mathrm{E}\Psi(J)\right| = 0(n^{2k}(\rho_n^*)^{-2k}h^{-kd}).$$

Totally we have

$$\left|\sum_{J\in\Delta} \mathrm{E}\Psi(J)\right| = 0(n^{4k}(\rho_n^*)^{-4k}h^{-2kd}\alpha(\rho_n^*)) + 0(n^{2k}(\rho_n^*)^{-2k}h^{-kd}). \quad (14.45)$$

Since $\alpha(\rho_n^*)$ is geometrically decaying, we only need to check the second term. In order to prove (14.40), we need to show

$$h^{-(8+d)}[\#'_{n,d}]^{1/k}(n^{-1}\rho_n^*)^2 = 0(n^{-b_1}) \quad (14.46)$$

for some $b_1 > 0$.

According to (14.11) and (14.12), we have

$$\begin{aligned} h^{-(8+d)}(n^{-1}\rho_n^*)^2 &= 0(n^{-2+2\tau+(8+d)/2(d+3))+(8+d)\varepsilon}) \\ &= 0(n^{-\xi}). \end{aligned}$$

In order to prove (14.46), by (14.29), we only need to choose k to be large enough such that

$$\frac{\tau_d}{k} - \xi < 0 \quad (14.47)$$

or $k > \tau_d/\xi$.

Proof of lemma 14.6

The main body of this proof is the same as in lemma 14.5 before but we need to consider cases, $1 \leqslant d < d_0$ and $d \geqslant d_0$, separately. For $1 \leqslant d < d_C$, by lemma 14.7, we only need to show

$$\#H'_{n,d}(n^{-1}\rho_n^*)^{4k} \sup_{i,j,\delta_1,\delta_2} \mathrm{E}\left[\sum_{q=1}^{n_1}\sum_{m=1}^{q} \mathrm{V}(q',m')\right]^{2k} = 0(n^{-b_1k}) \quad (14.48)$$

for some $b_1 > 0$, and for $d \geqslant d_0$. By lemma 14.7, we need to show

$$(h^{-4})^{2k}\#H'_{n,d}(n^{-1}\rho_n^*)^{4k} \sup_{i,j,\delta_1,\delta_2} \mathrm{E}\left[\sum_{q=1}^{n_1}\sum_{m=1}^{q} \mathrm{V}(q',m')^{2k}\right] = 0(n^{-b_1k}) \quad (14.49)$$

for some $b_1 > 0$, where

$$\mathrm{V}(s,t) = \frac{1}{h^d}K\left(\frac{Y_s^d - Y_t^d}{h}\right)[F_{d_0}(Y_s^{d_0}) - F_{d_0}(Y_t^{d_0})]\varepsilon_t W_d(Y_t^d) f_d^{-1}(Y_t^d)$$

and where n_1, q', and m' are defined as in lemma 14.5. The set Δ being defined as before, we decompose it in the following way.

$$\Delta = \Delta_3 \cup \left[\bigcup_{l=1}^{2k} \Delta_4(l)\right]$$

where

$$\Delta_3 = \{J = \{q_1, \ldots, m_{2k}\} \in \Delta : \exists i \in \{1, \ldots, 2k\}$$
$$\forall j \in \{1, \ldots, 2k\} \backslash \{i\}, |q_j - m_j| > 1\},$$
$$\Delta_4 = \Delta - \Delta_3,$$

and

$$\Delta_4(l) = \{J = (q_1, \ldots, m_{2k}) \in \Delta_4 : \#\{m_1, \ldots, m_{2k}\} = l\}.$$

Similarly we have

$$\left| \sum_{J \in \Delta_3} \mathrm{E}\left[\prod_{i=1}^{2k} \mathrm{V}(q_i', m_i') \right] \right| = \mathrm{o}_p(n^{4k}(\rho_n^*)^{-4k} h^{-2dk} \alpha(\rho_n^*))$$

by proposition 14.1 and $\mathrm{E}[\varepsilon_s | \mathcal{F}^s_{-\infty}] = 0$. For Δ_4, we first consider the case, $1 \leqslant d < d_0$. Since $\#\Delta_4 = 0(n_1^{2k})$, by the boundedness of K, F_{d_0}, $\{\varepsilon_t\}$, and f_d^{-1} on S_d, we have

$$\left| \sum_{J \in \Delta_4} \mathrm{E} \prod_{i=1}^{2k} \mathrm{V}(q_i', m_i') \right| = 0(n^{2k}(\rho_n^*)^{-2k} h^{-2dk}). \tag{14.50}$$

We need to show

$$[\# H_{n,d}']^{1/k} (n^{-1} \rho_n^*)^2 h^{-2d} = \mathrm{o}(n^{-b_1})$$

for some $\mathrm{b}_1 > 0$ when k is large enough. But

$$n^{-1} \rho_n^* h^{-d} = 0(n^{-\eta}),$$

where $\eta = 8\varepsilon + \xi + (d+3)^{-1}$, according to (14.11) and (14.12). So when $1 \leqslant d < d_0$, (14.49) is true if we choose k to be large enough.

Secondly we consider the case $d \geqslant d_0$. First we easily have, $\forall h \in H_{n,d}$,

$$K\left(\frac{Y_s^d - Y_t^d}{h} \right) |F_{d_0}(Y_s^{d_0}) - F_{d_0}(Y_t^{d_0})| = 0_p(h^{d+1}). \tag{14.51}$$

In the product

$$\prod_{i=1}^{2k} \mathrm{V}(m_i', q_i'),$$

we first bound all the terms

$$\frac{1}{h^d} K\left(\frac{Y_{m_i'}^d - Y_{q_i'}^d}{h} \right) [F_{d_0}(Y_{m_i'}^{d_0}) - F_{d_0}(Y_{q_i'}^{d_0})]$$

for those m_i' and q_i' such that $m_i = q_i$ and notice that when $m_i \neq q_i$

$$|m_i' - q_i'| \geqslant \rho_n^*.$$

Then apply proposition 14.1 and obtain for $J \in \Delta_4(l)$

$$\left|\mathrm{E}\left[\prod_{i=1}^{2k} \mathrm{V}(q_i', m_i')\right]\right|$$

$$= 0\left\{\mathrm{E}\left|\frac{1}{h^d} K\left(\frac{\bar{Y}_s^d - \bar{Y}_t^d}{h}\right)[F_{d_0}(\bar{Y}_s^{d_0}) - F_{d_0}(\bar{Y}_t^{d_0})]\right|\right\}^l + 0(h^{-2dk}\alpha(\rho_n^*))$$

$$= 0(h^l) + 0(h^{-2dk}\alpha(\rho_n^*))$$

where $\bar{Y}_s^d$ and $\bar{Y}_t^d$ are independent.

Since $\#\Delta_3 = 0(n_1^{4k})$ and $\#\Delta_4(l) = 0(n_1^{2k})$, we get

$$\mathrm{E}\left[\sum_{q=1}^{n_1} \sum_{m=1}^{q} \mathrm{V}(q', m')\right]^{2k}$$

$$= \sum_{J \in \Delta_3} \mathrm{E}\left[\prod_{i=1}^{2k} \mathrm{V}(m_i' q_i')\right] + \sum_{l=1}^{2k} \sum_{J \in \Delta_4(l)} \mathrm{E}\left[\prod_{i=1}^{2k} \mathrm{V}(m_i', q_i')\right]$$

$$= 0(n^{4k}(\rho_n^*)^{-4k} h^{-2dk} \alpha(\rho_n^*)) + 0\left(n^{2k}(\rho_n^*)^{-2k} \sum_{l=1}^{2k} h^l\right).$$

According to (14.11) and (14.12)

$$n^2(\rho_n^*)^{-2} h^2 \geqslant n^{\varepsilon + (d+1)\varepsilon + (d+3)^{-1}} \to +\infty.$$

So

$$\mathrm{E}\left[\sum_{q=1}^{n_1} \sum_{m=1}^{q} \mathrm{V}(m_i', q_i')\right]^{2k} = 0(n^{4k}(\rho_n^*)^{-4k} h^{-2dk} \alpha(\rho_n^*)) + 0(n^{2k}(\rho_n^*)^{-2k} h^{2k}).$$

Now the only thing we need to do is to check

$$n^{-2}(\rho_n^*)^2 h^{-6}[\#H_{n,d}']^{1/k} = 0(n^{-b_1})$$

for some $b_1 > 0$ if k is large enough. By (14.11) and (14.12)

$$n^{-1} \rho_n^* h^{-3} = 0(n^{-\xi - (1 + d/2)\varepsilon - 0.5(d+3)^{-1}}).$$

Proof of lemma 14.7

$$\mathrm{ASE}(h,d) = \frac{1}{n} \sum_{t=1}^{n} [\hat{F}_d(Y_t^d) - F_d(Y_t^d)]^2 W_d(Y_t^d)$$

$$+ \frac{2}{n} \sum_{t=1}^{n} [\hat{F}_d(Y_t^d) - F_d(Y_t^d)][F_d(Y_t^d) - F_{d_0}(Y_t^{d_0})] W_d(Y_t^d)$$

$$+ \frac{1}{n} \sum_{t=1}^{n} [F_d(Y_t^d) - F_{d_0}(Y_t^{d_0})]^2 W_d(Y_t^d),$$

where $F_d(Y_t^d) = \mathrm{E}[Z_t | Y_t^d]$.

First we have

$$\begin{aligned}&\mathrm{E}[F_d(Y_t^d)-F_{d_0}(Y_t^{d_0})]^2\,W_d(Y_t^d)\\&=\begin{cases}\mathrm{E}[F_d(Y_t^d)-F_{d_0}(Y_t^{d_0})]^2\,W_d(Y_t^d)\neq 0 & \text{if}\quad d<d_0,\\ 0 & \text{if}\quad d\geqslant d_0.\end{cases}\end{aligned}$$

So when $1\leqslant d<d_0$, by proposition 14.2, we have

$$\mathrm{ASE}(h,d)=\mathrm{E}[F_d(Y_t^d)-F_{d_0}(Y_t^{d_0})]^2\,W_d(Y_t^d)+\mathrm{o}_p(1).$$

We only need to consider the case of $d\geqslant d_0$, i.e. $F_d(Y_t^d)\equiv F_{d_0}(Y_t^{d_0})$. So

$$\mathrm{ASE}(h,d)=\frac{1}{n}\sum_{t=1}^{n}\,[\hat{F}_d(Y_t^d)-F_d(Y_t^d)]^2\,W_d(Y_t^d). \tag{14.52}$$

Define

$$\mathrm{ASE}^*(h,d)=\frac{1}{n}\sum_{t=1}^{n}\,[\hat{F}_d(Y_t^d)-F_d(Y_t^d)]^2\left[\frac{\hat{f}_d(Y_t^d)}{f_d(Y_t^d)}\right]^2 W_d(Y_t^d)$$

and denote the expectation of ASE* by

$$\mathrm{MASE}^*(h,d)=\mathrm{E}\{\mathrm{ASE}^*(h,d)\}.$$

By proposition 14.2 and (A5), we have

$$\sup_{d\in D_n}\ \sup_{h\in H_{n,d}}\frac{|\mathrm{ASE}(h,d)-\mathrm{ASE}^*(h,d)|}{\mathrm{ASE}(h,d)}=\mathrm{o}_p(1).$$

On the other hand, the general term of MASE*(h,d) has the form

$$\mathrm{E}\left\{[F_d(Y_s^d)-F_d(Y_t^d)][F_d(Y_{s'}^d)-F_d(Y_t^d)]K\left(\frac{Y_s^d-Y_t^d}{h}\right)K\left(\frac{Y_{s'}^d-Y_t^d}{h}\right)\right\}$$

or

$$\mathrm{E}\left\{\varepsilon_s\varepsilon_{s'}\,K\left(\frac{Y_s^d-Y_t^d}{h}\right)K\left(\frac{Y_{s'}^d-Y_t^d}{h}\right)\right\}.$$

So

$$\begin{aligned}&|\mathrm{MASE}^*(h,d)-\mathrm{MASE}^{*\prime}(h,d)|\\&=0\left(n^{-3}h^{-2d}\sum_{t=1}^{n}\ \sum_{s',s=1}^{n}\ \min\{\alpha(|s-t|),\alpha(|s'-t|),\alpha(|s-s'|)\}\right)\\&=0(n^{-3}h^{-2d}\sum_{s',s=1}^{n}\ \alpha(|s'-s|))\end{aligned}$$

where the MASE*′ denotes the expectation of ASE* that would apply if the variables were independent. By (A1),

$$\sup_n \sum_{s',s=1}^{n} \alpha(|s'-s|) < +\infty.$$

Therefore we have

$$\text{MASE}^*(h,d) = \text{MASE}^{*\prime}(h,d) + o((nh^d)^{-1} + h^4).$$

It is well known that MASE*′ has the following mean squared error decomposition

$$\text{MASE}^{*\prime}(h,d) = \frac{C_1}{nh^d} + C_2 h^4 + o((nh^d)^{-1} + h^4), \tag{14.53}$$

where

$$C_1 = \sigma^2 \int_{R^d} K^2(x)\mathrm{d}x$$

and

$$C_2 = \frac{1}{4}\left[\int_{R^d} x_1^2 K^2(x)\mathrm{d}x\right]^2 \int [\nabla^2 f(x)]^2 \mathrm{d}x.$$

The remaining part we need to prove is to show

$$\sup_{d\in D_n} \sup_{h\in H_{n,d}} \frac{|\text{ASE}^*(h,d) - \text{MASE}^*(h,d)|}{\text{ASE}^*(h,d)} = o_p(1).$$

The proof is long and involves computations of $2k$th order moments of quantities which have the same structure as $\bar{C}_{n,1}$ and $\bar{C}_{n,2}$ in lemmas 14.5 and 14.6 above. In addition, when $\{Z_t\}$ is absolutely regular, Cheng and Tong (1992) also proved a similar result.

Proof of lemma 14.8

We have

$$C_n(h, d) - \bar{C}_n(h, d) = D_1(h, d) + D_2(h, d),$$

where

$$D_1(h, d) = n^{-1} \sum_{t=1}^{n} \varepsilon_t [\hat{F}_{d,\setminus t}(Y_t^d) - F_{d_0}(Y_t^{d_0})]\left[\frac{f_d(Y_t^d) - \hat{f}_{d,\setminus t}(Y_t^d)}{f_d(Y_t^d)}\right]\frac{\hat{f}_{d,\setminus t}(Y_t^d)}{f_d(Y_t^d)} W_d(Y_t^d)$$

and

$$D_2(h, d) = n^{-1} \sum_{t=1}^{n} \varepsilon_t [\hat{F}_{d,\setminus t}(Y_t^d) - F_{d_0}(Y_t^d)]\left[\frac{\hat{f}_{d,\setminus t}(Y_t^d - f_d(Y_t^d)}{f_d(Y_t^d)}\right]^2 W_d(Y_t^d).$$

We have

$$|D_2(h,d)| \leqslant \sup_{x \in S_t} |\hat{f}_{d,\setminus t}(x) - f_d(x)|$$

$$\left\{ \frac{1}{n} \sum_{t=1}^{n} [\hat{f}_{d,\setminus t}(Y_t^d) - f_d(Y_t^d)]^2 W_d(Y_t^d) \right\}^{1/2} \{\mathrm{ASE}(h,d)\}^{1/2},$$

which by proposition 14.2 and lemma 14.7 is

$$0(h^{-2} n^{6\varepsilon - 3/(d+3)} \mathrm{ASE}(h,d)) = 0(n^{2\varepsilon - 2/(d+3)} \mathrm{ASE}(h,d)).$$

By (14.11) and (14.12), we obtain

$$\sup_{d \in D_n} \sup_{h \in H'_{n,d}} \frac{|D_2(h,d)|}{\mathrm{ASE}(h,d)} = o_p(1).$$

Note now that D_1 has roughly the same structure as $\bar{C}_n(h,d)$. So we can write D_1 as

$$D_1(h,d) = D_{11}(h,d) + D_{12}(h,d)$$

where, using the same notation as in lemmas 14.5 and 14.6

$$D_{11}(h,d) = n^{-1} \sum_{t=1}^{n} \sum_{|s-t| > \rho_n^*} n_t^{-1} U(s,t) \left[\frac{f_d(Y_t^d) - \hat{f}_{d,\setminus t}(Y_t^d)}{f_d^2(Y_t^d)} \right] W_d(Y_t^d),$$

and

$$D_{12}(h,d) = n^{-1} \sum_{t=1}^{n} \sum_{|s-t| > \rho_n^*} n_t^{-1} \mathrm{V}(s,t) \left[\frac{f_d(Y_t^d) - \hat{f}_{d,\setminus t}(Y_t^d)}{f_d^2(Y_t^d)} \right] W_d(Y_t^d).$$

Proceeding as in lemmas 14.5 and 14.6, we can show that

$$\sup_{d \in D_n} \sup_{h \in H_{n,d}} \frac{|D_{11}(h,d)|}{\mathrm{ASE}(h,d)} = o_p(1)$$

and

$$\sup_{d \in D_n} \sup_{h \in H_{n,d}} \frac{|D_{12}(h,d)|}{\mathrm{ASE}(h,d)} = o_p(1).$$

Proof of lemma 14.9

Since $Z_t = F_{d_0}(Y_t^{d_0}) + \varepsilon_t$,

$$\hat{F}_{d,\setminus t}(Y_t^d) = \frac{1}{n_t h^d} \sum_{|s-t| > \rho_n^*} K\left(\frac{Y_s^d - Y_t^d}{h} \right) \varepsilon_s \hat{f}_{d,\setminus t}^{-1}(Y_t^d) W_d(Y_t^d)$$

$$+ \frac{1}{n_t h^d} \sum_{|s-t| > \rho_n^*} K\left(\frac{Y_s^d - Y_t^d}{h} \right) F_{d_0}(Y_s^{d_0}) \hat{f}_{d,\setminus t}^{-1}(Y_t^d) W_d(Y_t^d),$$

$$C_n(h,d) = n^{-1} \sum_{|s-t|>\rho_n^*} n_t^{-1} h^{-d} K\left(\frac{Y_s^d - Y_t^d}{h}\right) \varepsilon_s \varepsilon_t \hat{f}_{d,\setminus t}(Y_t^d) W_d(Y_t^d)$$

$$+ n^{-1} \sum_{|s-t|>\rho_n^*} n_t^{-1} h^{-d} K\left(\frac{Y_s^d - Y_t^d}{h}\right) [F_{d_0}(Y_s^{d_0})$$

$$- F_{d_0}(Y_t^{d_0})]\, \varepsilon_t \hat{f}_{d,\setminus t}^{-1}(Y_t^d) W_d(Y_t^d).$$

We have for some c such that $0 < c < \infty$.

$$|C_n(h,d) - C_n(h^*,d)| \leqslant c\left\{\left|\frac{1}{h^d} - \frac{1}{(h^*)^d}\right| K\left(\frac{Y_s^d - Y_t^d}{h}\right)\right.$$

$$\left. + \frac{1}{(h^*)^d}\left|K\left(\frac{Y_s^d - Y_t^d}{h}\right) - K\left(\frac{Y_s^d - Y_t^d}{h^*}\right)\right|\right\}.$$

Since K is Lipschitz continuous and compactly supported, we have

$$\left|K\left(\frac{Y_s^d - Y_t^d}{h}\right) - K\left(\frac{Y_s^d - Y_t^d}{h^*}\right)\right| \leqslant c\left[h^*\left|\frac{1}{h} - \frac{1}{h^*}\right|\right].$$

So we have

$$|C_n(h,d) - C_n(h^*,d)| \leqslant c\left\{\left|\frac{1}{h^d} - \frac{1}{(h^*)^d}\right| + \frac{1}{(h^*)^{d-1}}\left|\frac{1}{h} - \frac{1}{h^*}\right|\right\}.$$

For

$$h \in H_{n,d} = [an^{-\varepsilon - 0.5(d+3)^{-1}}, bn^{\varepsilon - 0.5(d+3)^{-1}}], \quad |h - h^*| \leqslant A_n,$$

where

$$A_n \sim [\#H'_{n,d}]^{-1} n^{-1/2(d+3)+\varepsilon}.$$

$$(h^*)^{-(d-1)}\left|\frac{1}{h} - \frac{1}{h^*}\right| \leqslant B_n \quad \text{where} \quad B_n \sim n^{d/2(d+3)+(d+2)\varepsilon}[\#H'_{n,d}]^{-1}$$

and

$$\left|\frac{1}{h^d} - \frac{1}{(h^*)^d}\right| = \frac{|h - h^*|(h^{d-1} + h^{d-2}h^* + \ldots + (h^*)^{d-1})}{(h^*)^d h^d} \leqslant L_n,$$

where

$$L_n \sim n^{d/2(d+3)+3d\varepsilon}[\#H'_{n,d}]^{-1}.$$

Now $\#H'_{n,d} = n^{\tau d}$. By (14.30), we have

$$\sup_{d \in D_n} \sup_{h \in H_{n,d'}} \frac{|C_n(h,d) - C_n(h^*,d)|}{\text{ASE}(h,d)} = o_p(1).$$

Proof of lemma 14.10

Let us denote by

$$\hat{g}_{d,\setminus t}(Y_t^d) = \frac{1}{n_t h^d} \sum_{|s-t| > \rho_n} Z_s K\left(\frac{Y_s^d - Y_t^d}{h}\right)$$

and denote by $\hat{g}^*_{d,\setminus t}$, $\hat{f}^*_{d,\setminus t}$, n^*_t, the quantities $\hat{g}_{d,\setminus t}$, $\hat{f}_{d,\setminus t}$, n_t that applies when $\rho_n = \rho_n^*$, respectively.

Since $\{\varepsilon_t\}$ is a bounded sequence,

$$|C_n(h,d) - C_n^*(h,d)| \leq c \sup_{1 \leq t \leq n} |\hat{F}_{d,\setminus t}(Y_t^d) - \hat{F}^*_{d,\setminus t}(Y_t^d)|.$$

From proposition 14.2, we know that

$$\inf_{t,n} \hat{f}_{d,\setminus t}(Y_t^d) 1_{\{Y_t^d \in S_d\}} \geqslant \delta > 0.$$

So

$$|C_n(h,d) - C_n^*(h,d)| = 0(\sup_t |\hat{g}_{d,\setminus t}(Y_t^d) - \hat{g}^*_{d,\setminus t}(Y_t^d)|$$

$$+ \sup_t |\hat{f}_{d,\setminus t}(Y_t^d) - \hat{f}^*_{d,\setminus t}(Y_t^d)|)$$

we have

$$|\hat{g}_{d,\setminus t}(Y_t^d) - \hat{g}^*_{d,\setminus t}(Y_t^d)| = \frac{1}{h^d}\left|\frac{1}{n_t^*} - \frac{1}{n_t}\right| \left|\sum_{|s-t| > \rho_n^*} Z_s K\left(\frac{Y_s^d - Y_t^d}{h}\right)\right|$$

$$+ \frac{1}{h^d n_t}\left|\sum_{\rho_n < |s-t| < \rho_n^*} Z_s K\left(\frac{Y_s^d - Y_t^d}{h}\right)\right|$$

$$\leq c \frac{|n_t - n_s^*|}{n_t} \frac{1}{n_t^* h^d} \sum_{|s-t| > \rho_n^*} K\left(\frac{Y_s^d - Y_t^d}{h}\right)$$

$$+ \frac{c}{n_t h^d} \sum_{\rho_n < |s-t| < \rho_n^*} K\left(\frac{Y_s^d - Y_t^d}{h}\right).$$

The second term is

$$\frac{1}{n_t}[n_t^* \hat{f}^*_{d,\setminus t}(Y_t^d) - n_t \hat{f}_{d,\setminus t}(Y_t^d)].$$

So by proposition 14.2, we obtain

$$|\hat{g}_{d,\setminus t}(Y_t^d) - \hat{g}^*_{d,\setminus t}(Y_t^d)| = 0\left(\frac{n_t - n_t^*}{n_t}\right)$$

$$= 0\left(\frac{\rho_n^*}{n}\right)$$

since $n_t \sim n - \rho_n$ and $n_t^* \sim n - \sigma_n^*$. The same argument holds for $\hat{f}^*_{d,\setminus t}$ and $\hat{f}_{d,\setminus t}$. So by lemma 14.7,

$$\sup_{h \in H_{n,d}} \frac{|C_n(h, d) - C_n^*(h, d)|}{\text{ASE}(h, d)} = 0(h^{-4} n^{-1} \rho_n^*)$$

$$= 0(n^{4\varepsilon + \tau - (d+1)/(d+3)}).$$

By (14.11) and (14.12), we have

$$\sup_{d \in D_n} \sup_{h \in H_{n,d}} \frac{|C_n(h, d) - C_n^*(h, d)|}{\text{ASE}(h, d)} = o_p(1).$$

Proof of lemma 14.11

Without loss of generality, we only prove the case $d \geqslant d'$.

$$\frac{1}{n} \sum_{t=1}^{n} \varepsilon_t^2 |W_d(Y_t^d) - W_{d'}(Y_t^{d'})| \leq \frac{1}{n} \sum_{t=1}^{n} \varepsilon_t^2 1_{\{Y_t^d \in S_d^0 \setminus S_d\}}$$

$$+ \frac{1}{n} \sum_{t=1}^{n} \varepsilon_t^2 1_{\{Y_t^d \notin S_d^0\}} = I_1(d, d') + I_2(d)$$

where S_d^0 is the projection of S_L in R^d. By (A7), and boundedness of ε_t, we have

$$\text{E} \sup_{d,d' \in D_n} \sup_{h \in H_{n,d}} \frac{I_1(d, d')}{\text{ASE}(h, d)} \leq \sum_{d,d' \in D_n} \text{E} \sup_{h \in H_{n,d}} \frac{I_1(d, d')}{\text{ASE}(h, d)},$$

which is, by Lemma 14.7,

$$\leq c \sum_{d \in D_n} (\ln(n))^{\delta} n^{2/(d+3)+4\varepsilon} P(Y_t^d \in S_d^0 \setminus S_d) = o(1),$$

and by $P(Y_t^d \notin S_d^0) = 0$, we have

$$\text{E} \sup_{d \in D} \sup_{h \in H_{n,d}} \frac{I_2(d)}{\text{ASE}(h, d)} = 0.$$

REFERENCES

Akaike, H. (1974) A new look at the statistical model identification. *IEEE Trans Auto. Control*, **19**, 716–723.

Casdagli, M., Eubank, S., Farmer, J.D. and Gibson, J. (1991) State Space Reconstruction in the Presence of Noise. Technical Report, Los Alamos National Lab., USA.

Cheng, B. and Tong, H. (1992) On consistent non-parametric order determination and chaos. *J. Royal Statistic. Soc. (B)*, **54**, 427–474.

Denker, M. and Keller, G. (1983) On U-statistics and von Mises' statistics for weakly dependent processes. *Z. Wahr. Verw. Geb.*, **64**, 505–522.

Györfi, L., Härdle, W., Sards, P. and Vieu, P. (1989) Nonparametric curve estimation from time series. *Lecture Notes in Statistics*, **60**, Springer Verlag.

Härdle, W. and Marron, J.S. (1985) Optimal bandwidth selection in nonparametric regression function estimation. *Ann. Statist.*, **13**, 1465–1481.

Härdle, W. and Vieu, P. (1990) Kernel regression smoothing of time series. Technical Report, Université Catholique de Louvain, Belgium.

Hart, J.D. (1991) Kernel regression with time series errors. *J. Royal. Statist. Soc.*, (*B*), **53**, 173–187.

Hart, J.D. and Vieu, P. (1990) Data-driven bandwidth choice density estimation. *Ann. Statist.*, **18**, 873–890.

Marron, J.S. and Härdle, W. (1986) Random approximation to some measures of accuracy in nonparametric curve estimation. *J. Mult. Anal.*, **20**, 91–113.

Priestley, M.B. and Chao, M.T. (1972) Non-parametric function fitting. *J. Royal Statist. Soc.* (*B*), **34**, 385–392.

Shibata, R. (1980) Asymptotically efficient selection of the order of the model for estimating parameters of a linear process. *Ann. Statist.*, **8**, 147–164.

15

Nonparametric tests of serial independence

H.J. Skaug and D. Tjøstheim

15.1 INTRODUCTION

Measuring *dependence* and testing independence are important problems in all of statistics. In this paper we look at aspects of these problems in the framework of time series analysis. For a given stationary time series $\{X_t\}$ we are interested in testing independence between the X_t's.

Such tests are of interest in diagnostic fitting but also in other contexts (cf. Robinson 1992). Correlation type tests, for example the Box–Ljung test, are mostly used within the ARMA model class (see e.g. Brockwell and Davis, 1987, Ch. 9.4). However, it is known that such tests have poor power against many nonlinear alternatives. The approach in this paper is more general. We will be concerned with nonparametric tests based on estimated densities. Then the power is reduced compared to the correlation test in ARMA models, but the tests retain power in nonlinear situations.

Recently there have been some contributions in this direction. Robinson (1991) considers a test based on an entropy functional. Related work has been done by Joe (1989). Chan and Tran (1992) discuss a bootstrap test based on the absolute difference $|p_2(x, y) - p(x)p(y)|$, where p_2 and p are the bivariate and marginal densities of (X_t, X_{t-1}) in a stationary time series. Rosenblatt (1975) and Wahlen (1991) use analytic arguments on a similar expression for the related problem of testing independence between two processes each consisting of independent identically distributed (i.i.d.) random variables.

We will introduce various types of functionals in section 15.2, but we concentrate on functional measuring differences $p_2(x, y) - p(x)p(y)$. The functional can be analysed asymptotically in a relatively straightforward fashion, and it is sufficiently general to illustrate the type of problems one encounters. Both asymptotic properties and bootstrap arguments will be used in constructing the tests. The bootstrap seems crucial in moderate sample sizes. There are many open problems, some of which we plan to explore in subsequent publications. Problems currently under investigation are the use of empirical distribution functions instead of estimated density functions, functionals extending over several lags and tests of conditional independence involving more than one time series. Also, in this paper our examples are restricted to first order models, although it is fairly obvious how the analysis can be extended to higher order models.

15.2 MAIN TYPES OF FUNCTIONALS

We will focus on a comparison of bivariate and marginal densities of (X_t, X_{t-1}) for a stationary time series $\{X_t\}$. Such densities will always be assumed to exist. Doing the same for (X_t, X_{t-k}) requires mostly notational changes, but it may be essential to use $k \neq 1$ for higher order processes. Our real objective is to test the i.i.d. property, and of course examples of non-i.i.d. processes can be constructed where we have pairwise independence, but we do not think this is a very serious problem in practice.

Higher order simultaneous densities for $(X_t, X_{t-1}, \ldots, X_{t-k})$ can in principle be treated likewise, but as k increases, problems can be expected due to the curse of dimensionality in the estimation of the joint densities.

Let p_2 and p denote the joint and marginal density functions of (X_t, X_{t-1}). If we have independence, then $p_2 = p^2$, and this can be exploited to construct a number of functionals for measuring dependence and testing for independence. Some possibilities are listed below.

(a) The entropy functional (Joe 1989, Granger and Lin 1991, Robinson 1991):

$$I_1 = \int_S \log\left\{\frac{p_2(x, y)}{p(x)p(y)}\right\} p_2(x, y)\,\mathrm{d}x\,\mathrm{d}y,$$

where S is a suitably chosen set of integration so that the integral is finite.

(b) The absolute value functional (Chan and Tran 1992):

$$I_2 = \int |p_2(x, y) - p(x)p(y)|\,\mathrm{d}x\,\mathrm{d}y.$$

(c) The squared difference functional (Rosenblatt 1975, Wahlen 1991):

$$I_3 = \int \{p_2(x, y) - p(x)p(y)\}^2\,\mathrm{d}x\,\mathrm{d}y.$$

For each of these functionals we have $I_i \geqslant 0$, and $I_i = 0$ only in the independent case.

(d) The weighted difference functional:

$$I_4 = \int \{p_2(x, y) - p(x)p(y)\}\, p_2(x, y)\,\mathrm{d}x\,\mathrm{d}y.$$

This functional does not satisfy $I_4 \geqslant 0$, and it is in a sense counter-intuitive since at first glance one may believe that cancelling effects may render it useless for testing of independence. However, such a cancellation is not likely to occur in practice. Using a very rough argument, when $p_2(x, y) < p(x)p(y)$, the weight $p_2(x, y)$ is 'small', whereas when $p_2(x, y) > p(x)p(y)$, it is 'big', so that positive contributions to the integral dominate the negative ones. In the Gaussian case it is possible to prove a formal result.

Proposition 15.1

Let $\{X_t\}$ be a Gaussian stationary process. Then $I_4 \geqslant 0$. It is zero if and only if X_t and X_{t-1} are independent.

Proof

Let $\sigma^2 = \text{var}(X_t)$ and $\rho = \text{corr}(X_t, X_{t-1})$. Then it is easily shown that

$$\int p_2^2(x, y)\,\mathrm{d}x\,\mathrm{d}y - \int p^2(x)p^2(y)\,\mathrm{d}x\,\mathrm{d}y = (4\pi\sigma^2)^{-1}\{(1-\rho^2)^{-1/2} - 1\} \geqslant 0. \tag{15.1}$$

But, using the Schwarz inequality

$$\begin{aligned} I_4 &= \int p_2^2(x, y)\,\mathrm{d}x\,\mathrm{d}y - \int p(x)p(y)p_2(x, y)\,\mathrm{d}x\,\mathrm{d}y \\ &\geqslant \int p_2^2(x, y)\,\mathrm{d}x\,\mathrm{d}y - \left\{\int p_2^2(x, y)\,\mathrm{d}x\,\mathrm{d}y\right\}^{1/2}\left\{\int p^2(x)p^2(y)\,\mathrm{d}x\,\mathrm{d}y\right\}^{1/2} \end{aligned} \tag{15.2}$$

It follows from (15.1) that $I_4 \geqslant 0$. From its definition we have that $I_4 = 0$ if X_t and X_{t-1} are independent. On the other hand if $I_4 = 0$, then from (15.2),

$$\int p_2^2(x, y)\,\mathrm{d}x\,\mathrm{d}y \leqslant \int p^2(x)p^2(y)\,\mathrm{d}x\,\mathrm{d}y,$$

and it follows from (15.1) that we must have $\rho = 0$, i.e. X_t and X_{t-1} must be independent.

We do not always have $I_4 \geqslant 0$. The following counterexample shows this. Let $p_2(x, y)$ be given by

$$p_2(x, y) = \begin{cases} \alpha, & -1 < x \leqslant 0, 0 < y \leqslant 1 \\ 1 - 2\alpha, & 0 < x, y \leqslant 1 \\ \alpha, & 0 < x \leqslant 1, -1 < y \leqslant 0 \\ 0, & \text{elsewhere.} \end{cases}$$

Then it is easy to show that $I_4 = \alpha^2(4\alpha - 1)$ and I_4 is negative for $0 < \alpha < \frac{1}{4}$.

15.3 ESTIMATION AND RELATIONSHIP BETWEEN FUNCTIONALS

There are at least two ways of estimating/evaluating the functionals I_i. For both methods we need estimates $\hat{p}_2$, $\hat{p}$ for the densities involved. We have used kernel estimates, but other types of estimates are clearly possible. Chan and Tran (1992) have used histogram estimates.

The most obvious way of estimating I_i is to use numerical integration, so that e.g.

$$\hat{I}'_4 = \int \{\hat{p}_2(x, y) - \hat{p}(x)\hat{p}(y)\}\hat{p}_2(x, y)w_2(x, y)\,\mathrm{d}x\,\mathrm{d}y, \tag{15.3}$$

where $w_2(x, y)$ is a weight function so that the integral can be cut off for large values of x or y. For such values the estimators $\hat{p}_2$ and $\hat{p}$ are very unreliable due to lack of observations. However, we have chosen to work with empirical averages where, for given observations $\{X_0, X_1, \ldots, X_n\}$,

$$\hat{I}_4 = \frac{1}{n}\sum_t \{\hat{p}_2(X_t, X_{t-1}) - \hat{p}(X_t)\hat{p}(X_{t-1})\}w_2(X_t, X_{t-1}). \tag{15.4}$$

Under relatively weak assumptions, due to an ergodic theorem,

$$\hat{I}_4 \xrightarrow{p} \int \{p_2(x, y) - p(x)p(y)\}p_2(x, y)w_2(x, y)\,\mathrm{d}x\,\mathrm{d}y,$$

which, apart from the weighting factor w_2, is I_4. Introducing w_2, in (15.4) means that extreme observations can be screened off. Actually, in all our simulations for $\hat{I}_4$ we have used $w_2(x, y) \equiv 1$.

We prefer (15.4) to (15.3) mainly for computational reasons, and this is the type of estimate that will be considered henceforth for all functionals.

For I_1 corresponding to (15.4) we have

$$\hat{I}_1 = \frac{1}{n}\sum_t \log\left\{\frac{\hat{p}_2(X_t, X_{t-1})}{\hat{p}(X_t)\hat{p}(X_{t-1})}\right\}w_2(X_t, X_{t-1}), \tag{15.5}$$

where w_2 brings about the restriction to the integration set S. To avoid certain problems in the asymptotic theory of $\hat{I}_1$ when S increases with n, Robinson (1991) introduced a version of $\hat{I}_1$, given by

$$\hat{I}_{1,\gamma} = \frac{1}{n_\gamma}\sum_t c_t(\gamma)\log\left\{\frac{\hat{p}_2(X_t, X_{t-1})}{\hat{p}^2(X_{t-1})}\right\}w_2(X_t, X_{t-1}), \tag{15.6}$$

where

$$c_t(\gamma) = \begin{cases} 1 + \gamma, & t\text{ odd} \\ \gamma, & t\text{ even} \end{cases}$$

and where γ is a non-negative real number with $n_\gamma = n$ for n even and $n_\gamma = n + \gamma$ for n odd.

If we insert an extra weight factor $\hat{p}_2$ in $\hat{I}_1$, then

$$\hat{I}'_1 = \frac{1}{n}\sum_t \log\left\{\frac{\hat{p}_2(X_t, X_{t-1})}{\hat{p}(X_t)\hat{p}(X_{t-1})}\right\}\hat{p}_2(X_t, X_{t-1})w_2(X_t, X_{t-1}). \tag{15.7}$$

By series expansion of the logarithm

$$\log\left\{\frac{\hat{p}_2}{\hat{p}\hat{p}}\right\}\hat{p}_2 \sim (\hat{p}_2 - \hat{p}\hat{p})\frac{\hat{p}_2}{\hat{p}\hat{p}},$$

such that $\hat{I}'_1$ is related to $\hat{I}_4$ asymptotically, and in fact in our simulation experiments its behavior was much closer to $\hat{I}_4$ than to $\hat{I}_1$.

It is less straightforward to convert I_2 and I_3 to empirical average type estimates. Thus the estimate of I_2 should, strictly speaking, be

$$\hat{I}'_2 = \frac{1}{n}\sum_t |\hat{p}_2(X_t, X_{t-1}) - \hat{p}(X_t)\hat{p}(X_{t-1})|\hat{p}_2^{-1}(X_t, X_{t-1})w_2(X_t, X_{t-1}), \tag{15.8}$$

but to avoid division by $\hat{p}_2$ we have found it more reasonable to work with

$$\hat{I}_2 = \frac{1}{n}\sum_t |\hat{p}_2(X_t, X_{t-1}) - \hat{p}(X_t)\hat{p}(X_{t-1})|\, w_2(X_t, X_{t-1}). \tag{15.9}$$

In the case of I_3 it is observed that

$$I_3 = \int p_2^2(x, y)\,dx\,dy - 2\int p(x)p(y)p_2(x, y)\,dx\,dy + \left(\int p^2(x)\,dx\right)^2,$$

and we may write

$$\hat{I}_3 = \frac{1}{n}\sum_t \hat{p}_2(X_t, X_{t-1})w_2(X_t, X_{t-1}) - \frac{2}{n}\sum_t \hat{p}(X_t)\hat{p}(X_{t-1})w_2(X_t, X_{t-1}) + \left\{\frac{1}{n}\sum_t \hat{p}(X_t)w(X_t)\right\}^2, \tag{15.10}$$

where w is a one-dimensional weight function.

For I_3 and I_4 there is a rather direct connection between estimates obtained by numerical integration and those of (15.4) and (15.10) if kernel estimates are employed. Consider the kernel density estimators

$$\hat{p}_2(x, y) = \frac{1}{n}\sum_s k_h(X_s - x)k_h(X_{s-1} - y) \quad \text{and} \quad \hat{p}(x) = \frac{1}{n}\sum_s k_h(X_s - x), \tag{15.11}$$

where $h = h(n)$ is the bandwidth, $k_h(x) = h^{-1}k(h^{-1}x)$ with k being a kernel function satisfying standard assumptions such as $\int k(x)\,dx = 1$, $k(-x) = k(x)$, $k(x)$ bounded and $\int x^2 k(x)\,dx < \infty$. Note that the same bandwidth is used both for the two- and one-dimensional estimate. We follow Rosenblatt (1975) and Wahlen (1991) at this point. It is not quite in accordance with optimality theory of density estimation, but it simplifies our derivations. Robinson (1992) has analysed $\hat{I}_{1,\gamma}$ using different bandwidths. Using (15.11) we obtain the relation

$$\int \{\hat{p}_2(x,y)\}^2 \,\mathrm{d}x\,\mathrm{d}y$$

$$= \frac{1}{n^2}\sum_s\sum_t \int k_h(X_s - x)k_h(X_{s-1} - y)k_h(X_t - x)k_h(X_{t-1} - y)\,\mathrm{d}x\,\mathrm{d}y$$

$$= \frac{1}{n}\sum_t \frac{1}{n}\sum_s G_h(X_s - X_t)G_h(X_{s-1} - X_{t-1})$$

$$\triangleq \frac{1}{n}\sum_t \hat{p}_{2,G}(X_t, X_{t-1}),$$

where G is a kernel function obtained by convolving k with itself, i.e.

$$G(z) = \int k(x)k(z-x)\,\mathrm{d}x.$$

Similarly

$$\int \hat{p}(x)\hat{p}(y)\hat{p}_2(x,y)\,\mathrm{d}x\,\mathrm{d}y$$

$$= \frac{1}{n^3}\sum_s\sum_t\sum_u \int k_h(X_s - x)k_h(X_t - y)k_h(X_u - x)k_h(X_{u-1} - y)\,\mathrm{d}x\,\mathrm{d}y$$

$$= \frac{1}{n}\sum_t \frac{1}{n}\sum_s G_h(X_s - X_t)\frac{1}{n}\sum_u G_h(X_u - X_{t-1})$$

$$\triangleq \frac{1}{n}\sum_t \hat{p}_G(X_t)\hat{p}_G(X_{t-1}),$$

so that corresponding to $\hat{I}_4$ we have

$$\int \{\hat{p}_2(x,y) - \hat{p}(x)\hat{p}(y)\}\hat{p}_2(x,y)\,\mathrm{d}x\,\mathrm{d}y$$

$$= \frac{1}{n}\sum_t \{\hat{p}_{2,G}(X_t, X_{t-1}) - \hat{p}_G(X_t)\hat{p}_G(X_{t-1})\}.$$

The functional I_3 can be treated in exactly the same way, but not I_1 and I_2.

Note that for a Gaussian kernel with $k(x) = (2\pi)^{-1/2}\exp(-\frac{1}{2}x^2)$, G is still Gaussian with $G(x) = (2\pi)^{-1/2}2^{-1/2}\exp(-\frac{1}{2}(x/2^{1/2})^2)$. Moreover, weight-functions w can be included in the above argument. Then there are additional terms of order $\mathrm{O}(h^2)$, which can be neglected asymptotically.

15.4 SOME SIMULATION EXPERIMENTS

In section 15.5 we will take $\hat{I}_4$ as our primary object of investigation in the asymptotic analysis. This is due to its simplicity but also its good performance in the simulation experiments undertaken by us. We present some of this evidence already at this point to show the potential of the various functionals. To get a fair comparison of the tests, critical values found by simulation are used. In practice the critical values must, of course, be determined from the single realization at hand, and we return to this problem and its influence on the simulation results in sections 15.5 and 15.6.

In Figure 15.1(a)–(d) are displayed simulated power for the test functionals $\hat{I}_1$, $\hat{I}_{1,\gamma}$ ($\gamma = 0.5$), $\hat{I}'_1$, $\hat{I}_2$, $\hat{I}'_2$, $\hat{I}_3$ and $\hat{I}_4$ for four different cases, in each case using time series of length $n + 1 = 100$. We take as our null hypothesis that $\{X_t\}$ is i.i.d., and we reject this hypothesis if $\hat{I}_i \geqslant c_i$, where c_i has been determined by simulation using 8000 realizations in the null situation. The

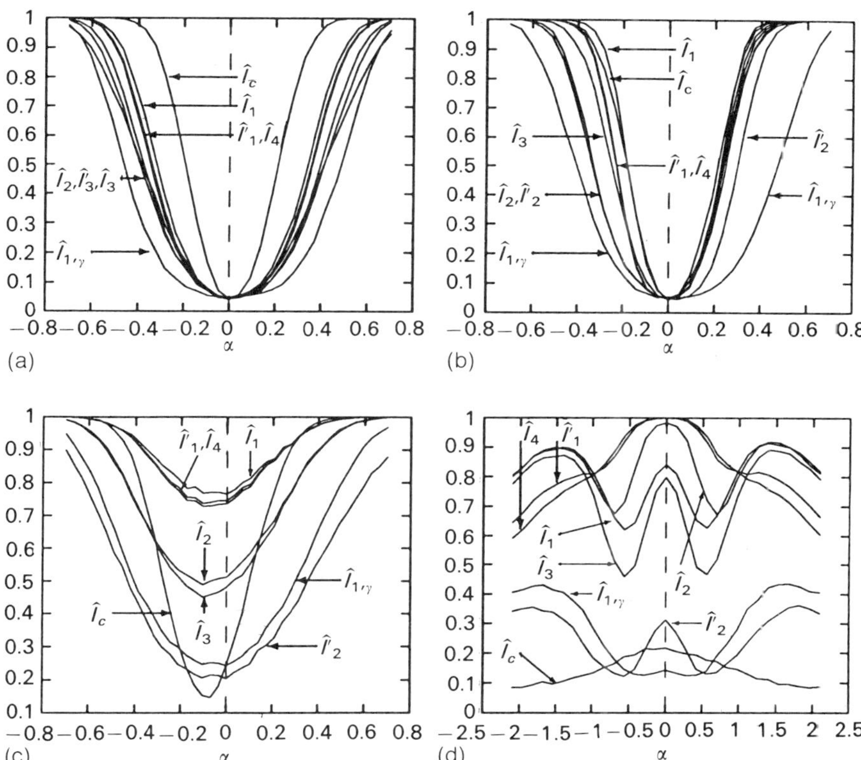

Figure 15.1. Power of tests (significance level 0.05) as function of α with $n + 1 = 100$: (a) AR (1) with $e_t \sim N(0, 1)$; (b) AR (1) with $e_t \sim \exp(1)$; (c) Model (15.15) with $e_t \sim N(0, 1)$ and $\beta = 0.4$; (d) Model (15.16) with $e_t \sim N(0, 1)$.

critical value c_i was determined so that the significance level is 0.05. Similar experiments were carried out for 0.01 and 0.1 and with similar results. In all of our simulations we have used the random number generator 'random( )' of the standard C library, and the simulations have been run on the parallel computer 'MasPar' (which has 8192 processors).

In the experiment depicted on Figure 15.1 we have also included the correlation functional

$$\hat{I}_c = \frac{\sum_t (X_t - \bar{X})(X_{t-1} - \bar{X})}{\sum_t (X_t - \bar{X})^2} \tag{15.12}$$

which is the one-lag version of the correlation type test mentioned in the introduction. Moreover, the plots are based on 8000 realizations for a number of different parameter values under the alternative hypothesis. The entropy functional is scale invariant, but to eliminate effects from different scalings for the other functionals, each process has been normalized, i.e. $\{X_t, 0 \leqslant t \leqslant n\}$ has been replaced by $\{X_t' = X_t/\hat{S}D(X)\}$, where

$$\hat{S}D(X) = \left\{\frac{1}{n}\sum_t (X_t - \bar{X})^2\right\}^{1/2}.$$

This does introduce a slight dependence between the X_t's, but it can be shown to be a higher order effect, and the simulated power functions are virtually indistinguishable from those obtained using the normalization $\{X_t'' = X_t/SD(X)\}$.

We have used the leave-one-out kernel density estimators

$$\begin{aligned} \hat{p}_2(X_t, X_{t-1}) &= \frac{1}{n-1}\sum_{s \neq t} k_h(X_s - X_t)k_h(X_{s-1} - X_{t-1}) \\ \hat{p}(X_t) &= \frac{1}{n-1}\sum_{s \neq t} k_h(X_s - X_t), \end{aligned} \tag{15.13}$$

with $k(\cdot)$ being Gaussian. The bandwidth h was taken to be $h = n^{-1/6}$. This is roughly in accordance (Silverman, 1986, p. 86) with optimality theory of bivariate density estimation in the case where $\hat{S}D(X) = 1$. The weight function $w_2(x, y)$ was taken to be $w_2(x, y) \equiv 1$ except for $\hat{I}_1$ and $\hat{I}_2'$ for which $w_2(x, y) = 1\{|x| \leqslant 2\hat{S}D(X), |y| \leqslant 2\hat{S}D(X)\}$, where $1(\cdot)$ is the indicator function. Here w_2 is data dependent, but in our next section this will be assumed not to be the case.

Since the correlation functional $\hat{I}_c$ of (15.12) is adapted to a situation where independence is measured using correlation of lag 1, it is ideally suited to the Gaussian AR(1) process

$$X_t = \alpha X_{t-1} + e_t, \quad |\alpha| < 1, \tag{15.14}$$

with $\{e_t\}$ being a sequence of i.i.d. Gaussian random variables. The simulated power for the various functionals as a function of α is portrayed in Figures 1(a). The correlation functional has uniformly better power than its nonparametric competitors. However, the power of the nonparametric functionals is quite good in view of the fact that only 100 observations are used. The functional $\hat{I}_4$ is seen to come out well among the nonparametric ones. In particular it has better power than $\hat{I}_2$, which is surprising. The somewhat inferior behavior of $\hat{I}_{1,\gamma}$ in this and in other plots could be due to our inexperience in choosing the parameter γ. Only $\gamma = 0.5$ has been tried.

The power advantage of the correlations functional may disappear if a non-Gaussian input process $\{e_t\}$ is used in (15.14). This is illustrated in Figure 15.1(b), where we have used a zero-mean exponentially distributed series $\{e_t\}$ with e_t having density function $\exp\{-(x+1)\}1(x \geqslant -1)$.

For genuinely nonlinear models the correlation functional does much worse as is illustrated for the nonlinear processes

$$X_t = (\alpha + \beta e_{t-1})X_{t-1} + e_t, \quad \alpha^2 + \beta^2 < 1, \tag{15.15}$$

in Figure 15.1(c) and

$$X_t = e_{t-1}(\alpha + e_t), \tag{15.16}$$

in Figure 15.1(d), where the input process $\{e_t\}$ in both cases has been taken to be Gaussian. In both cases the power is displayed as a function of α with $\beta = 0.4$ in (15.15). The bilinear process (15.15) is identical to that used by Chan and Tran (1992), and the functional $\hat{I}'_2$ corresponds roughly to the one they are looking at (but its power properties turn out to be inferior to those obtained for the functional in their paper). The process $\{X_t\}$ defined by (15.16) consists of uncorrelated random variables, and the correlation test could be expected to perform badly. It does so, but at least its power rises above the

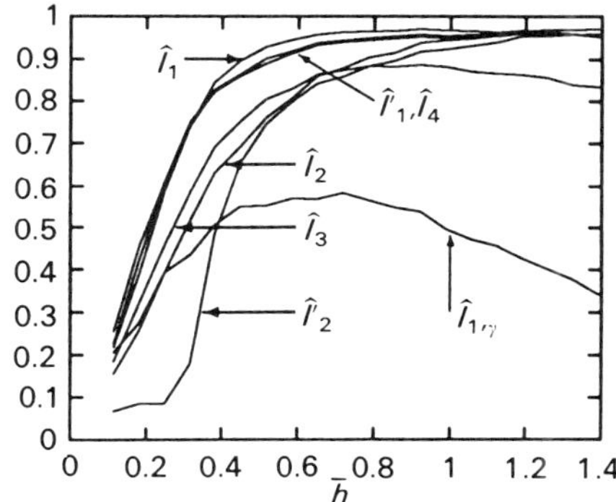

Figure 15.2. Power of the tests (significance level 0.05) as function of h for the AR (1) process with $\alpha = 0.5$, $n + 1 = 100$ and $e_t \sim N(0, 1)$.

significance level of the test. There are other interesting details of the power plots that remain unexplained, but the main message is that the nonparametric functionals have considerably better power, and that $\hat{I}_4$ does well among these. There seems to be some consistency in the internal rankings among the functionals, but clearly more experiments are needed to examine this closer.

Our results were not very sensitive to the choice of h over quite a wide range of h-values. The dependence on h for the linear Gaussian process (15.14) is illustrated in Figure 15.2. This shows the power of the tests against the particular alternative $\alpha = 0.5$ and $n + 1 = 100$ for a number of h-values. Our choice of $h = n^{-1/6} = 0.46$ is seen to be somewhat less than the optimal one for this process and this value of α, although for several of the functionals approximately the same power is attained for a wide range of h-values.

15.5 ASYMPTOTIC PROPERTIES

We limit the asymptotic analysis to $\hat{I}_4$ defined by (15.4) and the use of kernel estimates. Taylor expansion arguments are employed although the full Taylor expansion has a finite number of terms in this case and much of what we do could have been obtained by working directly with (15.4). This more general approach is chosen because it provides a framework for other functionals such as $\hat{I}_1$ and $\hat{I}_3$. The analysis of $\hat{I}_4$, although simpler, illustrates many of the difficulties that one encounters in an asymptotic analysis of the others.

The Taylor expansion for an entropy related functional has been used by Joe (1989) in an i.i.d. situation. We use a system of notation similar to his. The index '2' is omitted on bivariate expressions when no misunderstanding can arise. Thus let

$$F_n(x, y) = F_{n,2}(x, y) = \frac{1}{n} \sum_t 1(X_t \leqslant x, X_{t-1} \leqslant y)$$

be the empirical bivariate distribution function generated by $\{X_0, X_1, \ldots, X_n\}$. Then $\hat{I}_4$ may be written

$$\hat{I}_4 = \int \{\hat{p}(x, y) - \hat{p}(x)\hat{p}(y)\}\omega(x, y)\mathrm{d}F_n(x, y).$$

Moreover, let

$$F(x, y) = P(X_t \leqslant x, X_{t-1} \leqslant y) = \int_{-\infty}^{x} \int_{-\infty}^{y} p(u, v)\,\mathrm{d}u\,\mathrm{d}v$$

and

$$U_n(x, y) = n^{1/2}\{F_n(x, y) - F(x, y)\},$$

with $F_n(x)$, $F(x)$ and $U_n(x)$ being the corresponding univariate quantities. We define

$$p_h(x, y) = p_{h(n),2}(x, y) \triangleq \mathrm{E}[\hat{p}(x, y)]$$

and note that

$$p_h(x, y) = \int k_h(x-u)k_h(y-v)p(u, v)\,\mathrm{d}u\,\mathrm{d}v$$

and similarly for $p_h(x)$. Using the abbreviations $\hat{A}(x, y) = \hat{p}(x, y) - \hat{p}(x)\hat{p}(y)$ and $A_h(x, y) = p_h(x, y) - p_h(x)p_h(y)$ we write $\hat{I}_4$ as

$$\begin{aligned}\hat{I}_4 = &\int A_h(x, y)w(x, y)\,\mathrm{d}F(x, y) + \int \{\hat{A}(x, y) - A_h(x, y)\}w(x, y)\,\mathrm{d}F(x, y)\\ &+ n^{-1/2}\int A_h(x, y)w(x, y)\,\mathrm{d}U_n(x, y)\\ &+ n^{-1/2}\int \{\hat{A}(x, y) - A_h(x, y)\}w(x, y)\,\mathrm{d}U_n(x, y).\end{aligned}$$

We do a 3-variate Taylor expansion (which in our case is an identity) of $\hat{A}(x, y) = \hat{p}(x, y) - \hat{p}(x)\hat{p}(y)$ around $\{p_h(x, y), p_h(x), p_h(y)\}$. Using the fact that $n^{1/2}\{\hat{p}(x, y) - p_h(x, y)\} = \int k_h(x-u)k_h(y-v)\,\mathrm{d}U_n(u, v)$ and $n^{1/2}\{\hat{p}(x) - p_h(x)\} = \int k_h(x-u)\,\mathrm{d}U_n(u)$ we obtain the following identity, which is basic for most of the subsequent analysis of this section,

$$\begin{aligned}\hat{I}_4 = &\int \{p_h(x, y) - p_h(x)p_h(y)\}w(x, y)\mathrm{d}F(x, y)\\ &+ n^{-1/2}\Big\{\int k_h(x-u)k_h(y-v)w(x, y)\mathrm{d}U_n(u, v)\mathrm{d}F(x, y)\\ &\quad - \int k_h(x-u)p_h(y)w(x, y)\,\mathrm{d}U_n(u)\,\mathrm{d}F(x, y)\\ &\quad - \int k_h(y-v)p_h(x)w(x, y)\mathrm{d}U_n(v)\,\mathrm{d}F(x, y)\\ &\quad + \int \{p_h(x, y) - p_h(x)p_h(y)\}w(x, y)\,\mathrm{d}U_n(x, y)\Big\}\\ &+ n^{-1}\Big\{-\int k_h(x-u)k_h(y-v)w(x, y)\mathrm{d}U_n(u)\,\mathrm{d}U_n(v)\,\mathrm{d}F(x, y)\\ &\quad + \int k_h(x-u)k_h(y-v)w(x, y)\,\mathrm{d}U_n(u, v)\,\mathrm{d}U_n(x, y)\\ &\quad - \int k_h(x-u)p_h(y)w(x, y)\mathrm{d}U_n(u)\,\mathrm{d}U_n(x, y)\end{aligned}$$

$$- \int k_h(y-v)p_h(x)w(x,y)\,\mathrm{d}U_n(v)\,\mathrm{d}U_n(x,y)\Big\}$$

$$-n^{-3/2}\int k_h(x-u)k_h(y-v)w(x,y)\,\mathrm{d}U_n(u)\,\mathrm{d}U_n(v)\,\mathrm{d}U_n(x,y). \tag{15.17}$$

Note that in the null situation of independence, since we use a product kernel, $p_h(x, y) \equiv p_h(x)p_h(y)$, so that the first term of the expansion (15.17) drops out.

15.5.1 Asymptotic bias

Our goal is to derive a test of independence, and it is necessary to evaluate $E(\hat{I}_4)$ when $\{X_t\}$ is i.i.d. The independence of the X_t's does not imply independence between the pairs (X_{t+1}, X_t), (X_t, X_{t-1}) and (X_{t-1}, X_{t-2}), and we need an extension of lemma 2.1 of Joe (1989).

Lemma 15.2

Let $\{X_t\}$ be a series of i.i.d. random variables, and let $F(x, y, z) = F_3(x, y, z) = F(x)F(y)F(z)$ be the joint cumulative distribution function for (X_{t+1}, X_t, X_{t-1}). Then for arbitrary functions $a_n(x)$, $a_n(x, y)$, $a_n(x,y; u)$, $a_n(x, y; u, v)$ and $b(x, y; u, v)$ for which the integrals below exist, we have

$$\mathrm{E}\Big\{\int a_n(x)\,\mathrm{d}U_n(x)\Big\} = \mathrm{E}\Big\{\int a_n(x,y)\,\mathrm{d}U_n(x,y)\Big\} = 0 \tag{15.18}$$

$$\mathrm{E}\Big\{\int a_n(x,y)\,\mathrm{d}U_n(x)\,\mathrm{d}U_n(y)\Big\} = \int a_n(x,x)\,\mathrm{d}F(x) - \int a_n(x,y)\,\mathrm{d}F(x,y) \tag{15.19}$$

$$\mathrm{E}\Big\{\int a_n(x,y;u,v)\,\mathrm{d}U_n(x,y)\,\mathrm{d}U_n(u,v)\Big\}$$

$$= \int a_n(x,y;x,y)\,\mathrm{d}F(x,y) + \int \{a_n(x,y;y,u) + a_n(y,u;x,y)\}\,\mathrm{d}F(x,y,u)$$

$$-3\int a_n(x,y;u,v)\,\mathrm{d}F(x,y)\,\mathrm{d}F(u,v) + \mathrm{O}(n^{-1}) \tag{15.20}$$

$$\mathrm{E}\Big\{\int a_n(x,y;u)\,\mathrm{d}U_n(x,y)\,\mathrm{d}U_n(u)\Big\}$$

$$= \int \{a_n(x,y;x) + a_n(x,y;y)\}\,\mathrm{d}F(x,y)$$

$$-2\int a_n(x,y;u)\,\mathrm{d}F(x,y,u) + \mathrm{O}(n^{-1}) \tag{15.21}$$

$$E\left\{\int b(x, y; u, v)\,dU_n(x, y)\,dU_n(u)\,dU_n(v)\right\} = O(n^{-1/2}) \tag{15.22}$$

Proof

The identities (15.18) and (15.19) follow directly from definitions and from Joe (1989, lemma 2.1). We indicate the proof of (15.20). Those of (15.21) and (15.22) are similar. Using the definition of U_n we have that (15.20) can be written

$$n^{-1}E\left\{\sum_t\sum_s a_n(X_t, X_{t-1}, X_s, X_{s-1})\right\} - E\left\{\sum_t \int a_n(X_t, X_{t-1}, u, v)\,dF(u, v)\right\}$$
$$- E\left\{\sum_s \int a_n(x, y, X_s, X_{s-1})\,dF(x, y)\right\} + n\int a_n(x, y; u, v)\,dF(x, y)\,dF(u, v). \tag{15.23}$$

The two middle terms and the first double sum term of (15.23) for $|s-t|>1$ give rise to $-2n+n(n-3)/n$ terms of the type given in the fourth term. Combining them yields the last term on the right hand side of (15.20). Letting $s=t$ and $|s-t|=1$ in the double sum we obtain the first term and $(n-1)/n$ times the second term on the right hand side of (15.20).

The lemma is used to obtain the bias of $\hat{I}_4$. We do this both for the leave-one-out estimates given in (15.13) and for the estimates

$$\hat{p}(X_t, X_{t-1}) = \frac{1}{n}\sum_s k_h(X_s - X_t)k_h(X_{s-1} - X_{t-1})$$
$$\hat{p}(X_t) = \frac{1}{n}\sum_s k_h(X_s - X_t) \tag{15.24}$$

Moreover, we use a product weight function with $w_2(x, y) = 1(x\in S)1(y\in S)$, where $1(z\in S)$ is an indicator function with support on the set S.

Proposition 15.3

Let $\{X_t\}$ be a series of i.i.d. random variables and let $p(x, y)$ and $p(x)$ be twice continuously differentiable. Assume that the previously stated conditions on k hold and that $h = h(n)\to 0$ and $nh^2\to\infty$ as $n\to\infty$. Then, when using the estimates (15.24), we have

$$E(\hat{I}_4) = n^{-1}h^{-2}k^2(0)\left\{\int p(x)w(x)\,dx\right\}^2$$
$$- 2n^{-1}h^{-1}k(0)\int p^2(x)w(x)\,dx\int p(x)w(x)\,dx$$

$$+ n^{-1}\left[2\left\{\int p^2(x)w(x)\,dx\right\}^2 - \int p^3(x)w(x)\,dx\right]$$

$$+ O(n^{-1}h) + O(n^{-2}h^{-2}). \tag{15.25}$$

If the leave-one-out estimates (15.13) are utilized, then the two first terms on the right hand side of (15.25) drop out.

Proof

Taking expectations in (15.17) we see from (15.18) that the $n^{-1/2}$ terms of (15.17) give zero contribution. Let the n^{-1} terms of (15.17) be denoted by $n^{-1}(-S_1 + S_2 - S_3 - S_4)$. Then using independence and (15.19)

$$E(S_1) = \int k_h(x-u)k_h(y-u)p(x)w(x)p(y)w(y)p(u)\,dx\,dy\,du$$

$$- \int k_h(x-u)k_h(y-v)p(x)w(x)p(y)w(y)p(u)p(v)\,dx\,dy\,du\,dv.$$

Introducing $z = h^{-1}(x-u)$ and similar substitutions for the arguments of the other kernel functions, Taylor expanding and using

$$p_h(x) = E[\hat{p}(x)] = p(x) + \tfrac{1}{2}h^2 p''(x) + o(h^2), \tag{15.26}$$

it is not difficult to show that

$$E(S_1) = \int p^3(x)w(x)\,dx - \left\{\int p^2(x)w(x)\,dx\right\}^2 + O(h^2).$$

In the same manner, using (15.20) and (15.21), we conclude that

$$E(S_2) = h^{-2}k^2(0)\left\{\int p(x)w(x)\,dx\right\}^2 + 2\int p^3(x)w(x)\,dx - 3\left\{\int p^2(x)w(x)\,dx\right\}^2$$

$$+ O(h^2) + O(n^{-1})$$

and

$$E(S_3) = E(S_4) = h^{-1}k(0)\int p^2(x)w(x)\,dx \int p(x)w(x)\,dx + \int p^3(x)w(x)\,dx$$

$$- 2\left\{\int p^2(x)w(x)\,dx\right\}^2 + O(h).$$

Moreover, using (15.22) the highest order term from the $n^{-3/2}$ term of (15.17) is given by $n^{-2}h^{-2}k^2(0)\int w(x)\,dx$. Collecting terms we obtain (15.25). If we use the leave-one-out method, this effectively means setting $k(0) = 0$, and the second part of the proposition follows.

In the Gaussian case with $X_t \sim N(0,1)$ and $w(x) \equiv 1$,

$$2\left\{\int p^2(x)\,dx\right\}^2 - \int p^3(x)\,dx = \frac{1}{2\pi}(1 - 3^{-1/2}) \approx 0.067.$$

In practice we have found that the bias of $\hat{I}_4$ can be neglected when we use the leave-one-out estimates (15.13) in constructing the test statistic for $n \geqslant 100$; i.e. we replace $E(\hat{I}_4)$ by zero. This is not the case if the estimators (15.24) are used.

15.5.2 Asymptotic variance

We essentially use the same method as in section 15.5.1, and we assume $w(x, y) = 1(x \in S)1(y \in S)$ where $S \subset R$ is a Lebesgue measurable set.

Proposition 15.4

Under the assumptions of proposition 15.3 we have both for the estimates (15.13) and (15.24)

$$\operatorname{var}(\hat{I}_4) = n^{-1}\left(\int p^3(x)w(x)\,dx - \left\{\int p^2(x)w(x)\,dx\right\}^2\right)^2 + O(n^{-1}h^2) + O(n^{-2}h^{-2}). \tag{15.27}$$

Proof

We start by finding the variance of the $n^{-1/2}$ term of (15.17). We note that in the null situation of independence this term can be rewritten as

$$B_n = n^{-1/2}\left\{\int p_{h,w}(x, y)\,dU_n(x, y) - 2\int p_h(x)p(x)w(x)\,dx \int p_{h,w}(x)\,dU_n(x)\right\}. \tag{15.28}$$

where

$$p_{h,w}(x, y) = \int k_h(x-u)k_h(y-v)p(u, v)w(u, v)\,du\,dv$$

and similarly for $p_{h,w}(x)$. Due to (15.18), we have $\operatorname{var}(B_n) = E(B_n^2)$, and we can use (15.19)–(15.21) to compute expected values of the integrals. We obtain, using $p_h(x, y) = p_h(x)p_h(y)$ and $p_{h,w}(x, y) = p_{h,w}(x)p_{h,w}(y)$,

$$E\left[\left\{\int p_{h,w}(x, y)\,dU_n(x, y)\right\}^2\right] = \left\{\int p_{h,w}^2(x)p(x)\,dx\right\}^2 + 2\frac{n-1}{n}\int p_{h,w}^2(x)p(x)\,dx\left\{\int p_{h,w}(x)p(x)\,dx\right\}^2 - 3\left\{\int p_{h,w}(x)p(x)\,dx\right\}^4,$$

$$\mathrm{E}\left[\int p_{h,w}(x,y)\,\mathrm{d}U_n(x,y)\int p_h(x)p(x)w(x)\,\mathrm{d}x\int p_{h,w}(x)\,\mathrm{d}U_n(x)\right]$$
$$=\int p_h(x)p(x)w(x)\,\mathrm{d}x\left[2\int p_{h,w}^2(x)p(x)\,\mathrm{d}x\int p_{h,w}(x)p(x)\,\mathrm{d}x\right.$$
$$\left.-2\left\{\int p_{h,w}(x)p(x)\,\mathrm{d}x\right\}^3\right],$$

$$\mathrm{E}\left[\left\{\int p_h(x)p(x)\,\mathrm{d}x\int p_{h,w}(x)\,\mathrm{d}U_n(x)\right\}^2\right]$$
$$=\left\{\int p_h(x)p(x)w(x)\,\mathrm{d}x\right\}^2\left[\int p_{h,w}^2(x)p(x)\,\mathrm{d}x-\left\{\int p_{h,w}(x)p(x)\,\mathrm{d}x\right\}^2\right]. \tag{15.29}$$

From (15.26) it follows that

$$\int p_{h,w}(x)p(x)\,\mathrm{d}x=\int p^2(x)w(x)\,\mathrm{d}x+\mathrm{O}(h^2)$$

with similar expressions being true for the other integrals involving $p_h(x)$ and $p_{h,w}(x)$. By straightforward calculation it is then shown that

$$\operatorname{var}(B_n)=n^{-1}\left(\int p^3(x)w(x)\,\mathrm{d}x-\left\{\int p^3(x)w(x)\,\mathrm{d}x\right\}^2\right)^2+\mathrm{O}(n^{-1}h^2).$$

The n^{-1} term and the $n^{-3/2}$ term of (15.17) can be treated likewise, but an extension of lemma 15.2 (and of lemma 2.1 (*iv*) of Joe, 1989) is needed. The details are somewhat messy and are omitted. The leading term of the variance of the n^{-1} term is given by

$$2n^{-2}h^{-2}\left\{\int k^2(x)\,\mathrm{d}x\right\}\left\{\int p^2(x)w(x)\,\mathrm{d}x\right\}^2, \tag{15.30}$$

whereas for the $n^{-3/2}$ term it is of order $\mathrm{O}(n^{-3}h^{-2})$. The covariances of the various terms of (15.17) are of order $\mathrm{o}(n^{-2}h^{-2})$, and thus (15.27) is obtained.

A discussion of the practical problems of implementing (15.27) in a test statistic is given in section 15.6. In the Gaussian case with $w(x)\equiv 1$

$$\left[\int p^3(x)\,\mathrm{d}x-\left\{\int p^2(x)\,\mathrm{d}x\right\}^2\right]^2=\frac{1}{4\pi^2}\left(\tfrac{1}{2}-3^{-1/2}\right)^2\approx 0.00015.$$

For the functional $\hat{I}_3$ the $n^{-1/2}$ term corresponding to (15.17) is given by, with $w(x)\equiv 1$,

$$n^{-1/2}\left[2\int\{p_h(x,y)-p_h(x)p_h(y)\}\,\mathrm{d}U_n(x,y)\right.$$
$$\left.-4\int p_h(x)k_h(y-v)p(x,y)\,\mathrm{d}x\,\mathrm{d}y\,\mathrm{d}U_n(v)+4\int p_h(x)p(x)\,\mathrm{d}x\int p_h(x)\,\mathrm{d}U_n(x)\right]$$

which is zero if $p_h(x, y) = p_h(x)p_h(y)$ and $p(x, y) = p(x)p(y)$. The dominating term of the variance for $\hat{I}_3$ is then of order $O(n^{-2}h^2)$, whereas it is of order $O(n^{-1})$ for $\hat{I}_1$ and $\hat{I}_{1,\gamma}$ defined in (15.5) and (15.6) for $\gamma > 0$. We also refer to Wahlen (1991) for a similar result for $\hat{I}_3$ in a different situation.

15.5.3 Asymptotic normality

We only analyse the null situation although asymptotic normality can be proved in the dependent case as well using a truncated Taylor expansion and a mixing theorem. We take as our starting point the expression (15.17) and denote the $n^{-1/2}$, n^{-1} and $n^{-3/2}$ terms by I, II and III, respectively, so that

$$\hat{I}_4 = n^{-1/2}I + n^{-1}II + n^{-3/2}III.$$

Here $n^{-1/2}I = B_n$ with B_n as in (15.28). From the preceding section we have $\mathrm{E}[\{n^{-1/2}I\}^2] = O(n^{-1})$, $n^{-2}\mathrm{E}[\{II - \mathrm{E}(II)\}^2] = O(n^{-2}h^{-2})$, $\mathrm{E}(n^{-3/2}III) = O(n^{-2}h^{-2})$ and $\mathrm{var}(n^{-3/2}III) = O(n^{-3}h^{-2})$. Hence $n^{-1/2}I = O_p(n^{-1/2})$, and under the assumptions of proposition 15.3, $n^{-1}\{II - \mathrm{E}(II)\} + n^{-3/2}III = o_p(n^{-1/2})$. It follows from standard results (see e.g. Brockwell and Davis 1987, p. 198) that $n^{1/2}\{\hat{I}_4 - n^{-1}\mathrm{E}(II)\}$ and $I = n^{1/2}B_n$ have the same asymptotic distribution.

Theorem 15.5

Let the assumptions of proposition 15.3 be fulfilled and let the weight function $w(x, y) = 1(x \in S)1(y \in S)$ have compact support S and $\int p^3(x)w(x)\,dx - \{\int p^2(x)\, w(x)\,dx\}^2 \neq 0$. Then

$$n^{1/2}\{\hat{I}_4 - n^{-1}\mathrm{E}(II)\} \xrightarrow{d} N\left(0, \left\{\int p^3(x)w(x)\,dx - \left\{\int p^2(x)w(x)\,dx\right\}^2\right\}^2\right),$$

where $\mathrm{E}(II)$ is given in proposition 15.3.

Proof

As explained, it is sufficient to prove asymptotic normality of $n^{1/2}B_n$ with B_n given by (15.28). Using the definition of U_n

$$\begin{aligned} n^{1/2}B_n &= n^{-1/2}\sum_t \left\{p_{h,w}(X_t, X_{t-1}) - \int p_{h,w}(x, y)p(x, y)\,dx\,dy\right\} \\ &\quad - 2n^{-1/2}\int p_h(x)p(x)w(x)\,dx \sum_t \left\{p_{h,w}(X_t) - \int p_{h,w}(x)p(x)\,dx\right\} \\ &= n^{-1/2}\sum_t \{C_{h,w}(X_t, X_{t-1}) - 2D_{h,w}(X_t)\} = n^{-1/2}\sum_t G_{h,w}(X_t, X_{t-1}), \end{aligned}$$

where

$$C_{h,w}(u,v) = p_{h,w}(u,v) - \int p_{h,w}(x,y)p(x,y)\,dx\,dy,$$

$$D_{h,w}(u) = \int p_h(x)p(x)w(x)\,dx\left\{p_{h,w}(u) - \int p_{h,w}(x)p(x)\,dx\right\}$$

and

$$G_{h,w}(u,v) = C_{h,w}(u,v) - 2G_{h,w}(u).$$

In obvious notation we can write this as

$$n^{-1/2}\sum_t G_{h,w}(X_t, X_{t-1}) = n^{-1/2}\sum_t G_w(X_t, X_{t-1})$$

$$+ n^{-1/2}\sum_t \{G_{h,w}(X_t, X_{t-1}) - G_w(X_t, X_{t-1})\}$$

where

$$n^{-1/2}\sum_t G_w(X_t, X_{t-1}) = \int p(x,y)w(x,y)\,dU_n(x,y)$$

$$-2\int p(x)w(x)\,dx \int p(x)w(x)\,dU_n(x).$$

Using the reasoning of section 15.5.2 and the assumptions in the theorem it follows that

$$\mathrm{E}\left[\left\{n^{-1/2}\sum_t G_w(X_t, X_{t-1})\right\}^2\right] \rightarrow \left[\int p^3(x)w(x)\,dx - \left\{\int p^2(x)w(x)\,dx\right\}^2\right]^2 > 0,$$

as $n \rightarrow \infty$. Since $\{X_t\}$ is i.i.d. and G_w is a measurable function on R^2, the process $\{Y_t\} = \{G_w(X_t, X_{t-1})\}$ is strictly stationary and 1-dependent. From the central limit theorem for m-dependent processes (Brockwell and Davis 1987, theorem 6.4.2)

$$n^{-1/2}\sum_t G_w(X_t, X_{t-1}) \xrightarrow{d} N\left(0, \left[\int p^3(x)w(x)\,dx - \left\{\int p(x)w(x)\,dx\right\}^2\right]^2\right).$$

The theorem will be proved if it can be shown that

$$n^{-1/2}\sum_t \{G_{h,w}(X_t, X_{t-1}) - G_w(X_t, X_{t-1})\} = o_p(1)$$

or

$$n^{-1}\mathrm{E}\left[\left(\sum_t \{G_{h,w}(X_t, X_{t-1}) - G_w(X_t, X_{t-1})\}\right)^2\right] = o(1).$$

Using the notation already established and lemma 15.2,

$$n^{-1}\mathrm{E}\left[\left(\sum_t \{G_{h,w}(X_t, X_{t-1}) - G_w(X_t, X_{t-1})\}\right)^2\right]$$

$$=\mathrm{E}\left[\left(\int\{C_{h,w}(x,y)-C_w(x,y)\}\,\mathrm{d}U_n(x,y)-2\int\{D_{h,w}(x)-D_w(x)\}\,\mathrm{d}U_n(x)\right)^2\right]$$

$$=\int\{C_{h,w}(x,y)-C_w(x,y)\}^2 p(x,y)\,\mathrm{d}x\,\mathrm{d}y$$

$$+2\frac{n-1}{n}\int\{C_{h,w}(x,y)-C_w(x,y)\}\{C_{h,w}(y,v)-C_w(y,v)\}p(x,y,v)\,\mathrm{d}x\,\mathrm{d}y\,\mathrm{d}v$$

$$-3\left(\int\{C_{h,w}(x,y)-C_w(x,y)\}p(x,y)\,\mathrm{d}x\,\mathrm{d}y\right)^2$$

$$+4\int\{D_{h,w}(x)-D_w(x)\}^2 p(x)\,\mathrm{d}x-4\left(\int\{D_{h,w}(x)-D_w(x)\}p(x)\,\mathrm{d}x\right)^2$$

$$-4\int\{C_{h,w}(x,y)-C_w(x,y)\}\{D_{h,w}(x)+D_{h,w}(y)-D_w(x)-D_w(y)\}p(x,y)\,\mathrm{d}x\,\mathrm{d}y$$

$$+8\int\{C_{h,w}(x,y)-C_w(x,y)\}p(x,y)\,\mathrm{d}x\,\mathrm{d}y\int\{D_{h,w}(x)-D_w(x)\}p(x)\,\mathrm{d}x.$$

Since $p_h(x,y)=\int k_h(x-u)k_h(y-v)p(u,v)\,\mathrm{d}u\,\mathrm{d}v$ it follows from the continuity of p that $p_h(x,y)\to p(x,y)$ for all (x,y) as $h=h(n)\to 0$. Since $w(x)$ has compact support S we have uniform continuity of p and hence uniform convergence $p_h(x,y)\to p(x,y)$ on $S\times S$ as $n\to\infty$. From the definition of $C_{h,w}$, C_w, $D_{h,w}$ and D_w it follows that $C_{h,w}(x,y)\to C_w(x,y)$ and $D_{h,w}(x)\to D_w(x)$ uniformly on $S\times S$ and S. (It is really necessary to redefine $C_w(x,y)$ and $D_w(x)$ at the boundary parts of $S\times S$ and S, but assuming that these have Lebesgue measure zero this contribution is negligible.) Since the integrals in (15.31) are over S, $S\times S$ and $S\times S\times S$, and p and p_h are bounded on S due to continuity, it follows that for a given $\varepsilon>0$ there exists an N such that for $n>N$, $|C_{h,w}(x,y)-C_h(x,y)|<\varepsilon$ and $|D_{h,w}(x)-D_h(x)|<\varepsilon$, and there is an M such that $p(x)\leqslant M$ for all $x\in S$ and $(x,y)\in S\times S$. Thus each integral is majorized by $\varepsilon^p M^q|S|^q$ where $p=1$ or 2, $q=1,2$ or 3 and $|S|$ is the Lebesgue measure of S. It follows that the expression (15.31) is of order o(1) as $n\to\infty$, and this completes the proof.

For the functional $\hat{I}_3$ the $n^{-1/2}$ term of a Taylor expansion analogous to (15.17) does not dominate in the null situation. Asymptotic distributions must then be obtained from the n^{-1} term. This can be done using an extension of Hall's (1984) U-statistic arguments. Under quite weak assumptions asymptotic normality can be established with an asymptotic variance of order $\mathrm{O}(n^{-2}h^{-2})$, whereas in the non-null situation it is of order $\mathrm{O}(n^{-1})$.

15.6 CAN THE ASYMPTOTIC THEORY BE USED IN PRACTICE?

Based on the results in the three preceding sections a natural statistic for testing of independence is

$$\hat{I}_4^* = \{\hat{I}_4 - \hat{\mathrm{E}}(\hat{I}_4)\}/\hat{S}D(\hat{I}_4), \tag{15.32}$$

where asymptotic expressions for $\mathrm{E}(\hat{I}_4)$ and $SD(\hat{I}_4)$ are given in (15.25) and (15.27), and where natural estimators in the leave-one-out case are given by

$$\hat{\mathrm{E}}(\hat{I}_4) = n^{-1}\left[2\left\{\frac{1}{n}\sum_t \hat{p}(X_t)w(X_t)\right\}^2 - \frac{1}{n}\sum_t \hat{p}^2(X_t)w(X_t)\right]$$

and

$$\hat{S}D(\hat{I}_4) = n^{-1/2}\left|\frac{1}{n}\sum_t \hat{p}^2(X_t)w(X_t) - \left\{\frac{1}{n}\sum_t \hat{p}(X_t)w(X_t)\right\}^2\right|. \tag{15.33}$$

For n large $\hat{I}_4^*$ is expected to be approximately standard normal under the null hypothesis. If we reject the null hypothesis of independence if $\hat{I}_4^* \geqslant u_{1-\alpha}$, where $u_{1-\alpha}$ is the upper $(1-\alpha)$ fractile in the standard normal distribution, then this should result in a test of approximate level α.

To examine this closer we look first at the asymptotic approximations (15.25) and (15.27) for the mean (leave-one-out case) and the standard deviation. These are plotted on Figure 15.3(a) as functions of n for a Gaussian i.i.d. process $\{e_t\}$ with $\mathrm{E}(e_t) = 0$ and $SD(e_t) = 1$ and with weight function

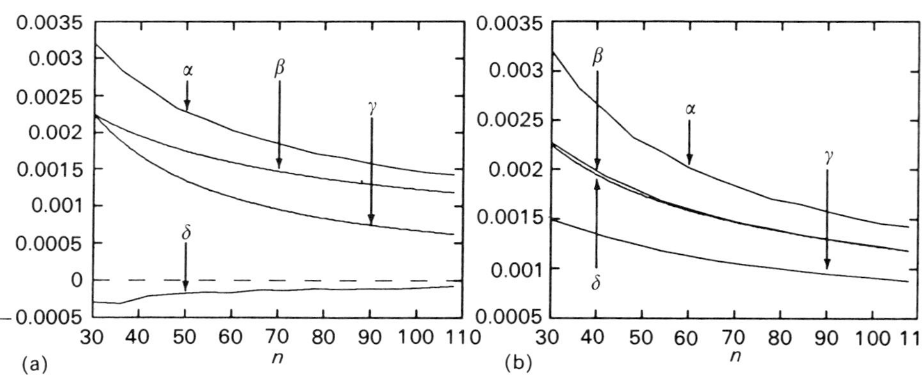

Figure 15.3. Performance of asymptotic formulae when n varies and X_t is i.i.d. $N(0, 1)$: (a) α = simulated $SD(\hat{I}_4)$, β = one-term asymptotic formula (15.27) for $SD(\hat{I}_4)$, δ = simulated $E(\hat{I}_4)$, γ = one-term asymptotic formula (15.25) for $E(\hat{I}_4)$; (b) α = simulated $SD(\hat{I}_4)$, β = one-term asymptotic formula (15.27) for $SD(\hat{I}_4)$, δ = expected value of $\hat{S}D(\hat{I}_4)_{\mathrm{corr}}$ given by (15.34), γ = expected value of $\hat{S}D(\hat{I}_4)$ given by (15.33).

$w(x) \equiv 1$. On the same figure are plotted the simulated mean and standard deviation of $\hat{I}_4$ based on averaging over 8000 realizations of $\{e_t\}$, each of length n. We have again used a bandwidth $h = n^{-1/6}$, but very similar results were obtained over a wide range of h-values.

Based on the mean value plot in Figure 15.3(a) we decided to replace $\hat{\mathrm{E}}(\hat{I}_4)$ by 0 in (15.32). The standard deviation is much more troublesome, however. For $n = 100$ it is seen that use of (15.27) leads to a clear underestimation. To some degree this situation persists for $n = 500$ where the simulated and asymptotic first order approximation for the standard deviation are given by 0.00059 and 0.00055, respectively. One may think that improvements can be obtained by including next order terms in the asymptotic expansion, which for $h = n^{-1/6}$ are the terms due to bias in $\hat{p}(x)$ (cf. 15.29) of order $\mathrm{O}(n^{-1/2}h)$ and the leading term of the standard deviation of the n^{-1}-term of (15.17), which is of order $n^{-1}h^{-1}$. For $n = 100$, $n^{-1/2}h = n^{-2/3} \approx 0.046$ and $n^{-1}h^{-1} = n^{-5/6} \approx 0.022$ so that there are relatively small differences in order between this term and the first term of order $n^{-1/2} = 0.1$. In addition the n-independent part of the higher order terms of (15.27) are larger than that of the first term. In fact for $n = 100$ the $n^{-1/2}h$ term is larger in absolute value than the $n^{-1/2}$ term and negative, thus resulting in a negative (!) variance if only these terms are included, whereas the $n^{-1}h^{-1}$ term is roughly of the same value. Thus these terms are useless as correction terms for $n = 100$. Moreover, since $h = n^{-1/6}$ decreases very slowly, n must be very substantial to obtain significantly better results.

In practice the integrals entering in (15.27) are estimated by (15.33). Implementation of this expression causes the quality of the approximation to deteriorate even more (as seen from Figure 15.3(b)). The reason is that in the standard normal case the values $\int p^3(x)\,dx$ and $\{\int p^2\,dx\}^2$ are very close to each other, so that the estimate of the difference between them is dominated by the bias of the estimates of the individual integrals. We can adjust for bias by using (15.26) so that $\hat{p}$ is replaced by $\hat{p} - \frac{1}{2}h^2\hat{p}''$ and $\hat{p}^2$ by $\hat{p}^2 - h^2\hat{p}\hat{p}''$, where $\hat{p}''(x)$ is the kernel estimate of $p''(x)$. This gives a bias corrected version of $\widehat{SD}(\hat{I}_4)$:

$$\widehat{SD}(\hat{I}_4)_{\mathrm{corr}} = n^{-1/2}\left|\frac{1}{n}\sum_t [\hat{p}^2(X_t) - h^2\hat{p}(X_t)\hat{p}''(X_t)]w(X_t) - \left\{\frac{1}{n}\sum_t [\hat{p}(X_t) - \tfrac{1}{2}h^2\hat{p}''(X_t)]w(X_t)\right\}^2\right|. \tag{15.34}$$

We then came very close to the first order term of (15.27) as shown in Figure 15.3(b).

The distributional approximation of $\hat{I}_4$ to normality is studied in Figure 15.4(a), where a plot of the simulated distribution of the standardized test statistic $\hat{I}_4^*$ is shown both when $SD(\hat{I}_4)$ is approximated by simulations and when $SD(\hat{I}_4)$ is estimated by (15.33) and by (15.34). Again 8000 realizations

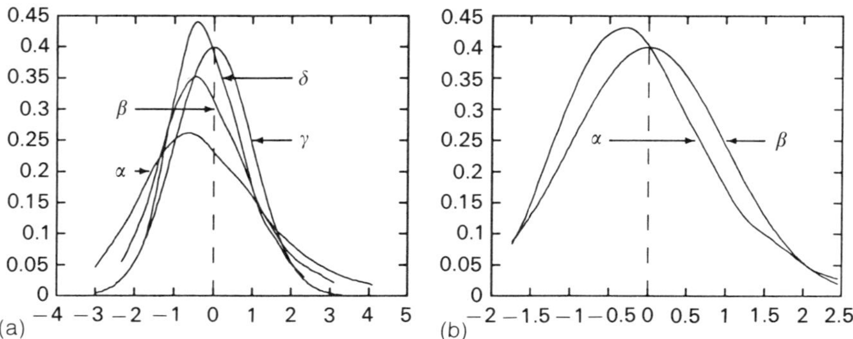

Figure 15.4. The null distribution of $\hat{I}_4$ for $n+1=100$ and $X_t \sim N(0,1)$: (a) density of $\hat{I}_4/SD$ with SD estimated by $\alpha = \widehat{SD}(\hat{I}_4)$ given by (15.33), $\beta = \widehat{SD}(\hat{I}_4)_{\text{corr}}$ given by (15.34), δ = simulated standard deviation of $\hat{I}_4$. The standard normal density is labeled γ. (b) α = density of $\hat{I}_4/SD$ with bootstrap SD, β = the standard normal density.

have been used, and a standard normal distribution is plotted for reference purposes. The approximation to normality is reasonably good, but it is clear that a normalization based on both (15.33) and (15.34) will lead to wrong level of the test if fractiles from the standard normal distribution are used. In experiments with significance levels of 0.01, 0.05 and 0.1 the simulated level for a test statistic based on $\hat{I}_4^*$ and (15.34) were 0.048, 0.093 and 0.135. Normalizing with the simulated standard deviation on the other hand gave 0.021, 0.059 and 0.096. This evidence and the fact that in the null situation $\{X_t\}$ is i.i.d. strongly suggest the use of the bootstrap methods. Since the normal approximation seems to work fairly well we chose to stick to that particular result of the asymptotic theory, and only bootstrapped the standard deviation in (15.32). This also makes it far easier to evaluate the bootstrap by simulation, since relatively few bootstrap replicas are needed for each realization. The simulated distribution for $\hat{I}_4^*$ for the Gaussian process $\{e_t\}$ with $\hat{\mathrm{E}}(\hat{I}_4)=0$ and $\widehat{SD}(\hat{I}_4)$ computed from 50 bootstrap replicas is shown in Figure 15.4(b). The approximation to the standard normal curve is seen to be better than that obtained using the asymptotic expansion for the standard deviation. Similar results can be obtained for the other functionals discussed in this paper, and, based on the evidence so far, introducing the bootstrap seems to result in a test that is superior to one based purely on asymptotic theory. Moreover, this appears to hold for quite a wide range of sample sizes.

At last we make a comparison of the $\hat{I}_4$ test and the correlation test. These, among others, were compared in Figure 15.1, but now we also take the problem of estimating critical values into consideration. The $\hat{I}_4$ test comes in two versions, one with standard deviation given by (15.34), resulting in a test criterion $\hat{I}_4^*$, and one using the bootstrap method discussed above, giving a test criterion denoted by $\hat{I}_4^\dagger$. The bootstrap method was also used in

Table 15.1 Power of tests (significance level 0.05) for the AR(1) process with $n+1=100$ and $e_t \sim N(0, 1)$

Test	$\alpha=0$	$\alpha=0.2$	$\alpha=0.5$
$\hat{I}_4^*$	0.092	0.216	0.908
$\hat{I}_4^\dagger$	0.062	0.159	0.875
$\hat{I}_c$	0.054	0.464	0.996

calculating the critical values for the correlation test based on $\hat{I}_c$. For the Gaussian process (15.14) with $n+1=100$ the power of these three tests is tabulated in Table 15.1 for three values of α, including $\alpha=0$ which gives the null-situation. It is seen that $\hat{I}_c$ is closest to the intended level 0.05. In accordance with Figure 15.1, $\hat{I}_c$ is the most powerful statistic in the alternatives $\alpha=0.2$ and $\alpha=0.5$.

REFERENCES

Brockwell, P.J. and Davis, R.A. (1987) *Time Series: Theory and Methods* Springer-Verlag, New York.

Chan, N.H. and Tran, L.T. (1992) Nonparametric tests for serial dependence. *J. Time Series Anal.*, **13**, 19–28.

Granger, C.W.J. and Lin, J.L. (1991) *Nonlinear Correlation Coefficients and Identification of Nonlinear Time Series Models*. Dept. of Economics, University of California, San Diego.

Hall, P. (1984) Central limit theorem for integrated square error of multivariate nonparametric density estimators. *J. Mult. Anal.*, **14**, 1–16.

Joe, H. (1989) Estimation of entropy and other functionals of a multivariate density. *Ann. Stat. Math.*, **41**, 683–697.

Robinson, P.M. (1991) Consistent nonparametric entropy-based testing. *Review of Economic Studies*, **58**, 437–453.

Rosenblatt, M. (1975) A quadratic measure of deviation of two-dimensional density estimates and a test of independence, *Ann. Statist.*, **3**, 1–14.

Silvermann, B.W. (1986) *Density Estimation for Statistics and Data Analysis*, Chapman and Hall, London.

Wahlen, B.E. (1991) *A Nonparametric Measure of Independence*, PhD thesis, Department of Mathematics, University of California, San Diego.

16

Measuring nonlinearity in time series

J. Pemberton

16.1 INTRODUCTION

A classical definition of a linear system is that the principle of superposition should hold and that the response to a single frequency sine wave input should be a sine wave of the same frequency possibly with a phase shift and scaled amplitude. Neither property holds for a nonlinear system (see e.g. Priestley, 1988, p. 27). This may be regarded as defining a nonlinear system. Nonlinear time series models such as bilinear and threshold autoregressive models certainly are nonlinear in this respect. However, when it comes to deciding if a single realization has been generated by a linear or nonlinear mechanism, the definition is not of much use directly. Instead we must concentrate on some manifestations of nonlinearity and test the series for whether these are present.

Some tests are designed to seek fairly general departures from nonlinearity, such as the bispectral tests of Subba Rao and Gabr (1980) and Hinich (1982), the test for non-additivity of Keenan (1985) and the so called BDS test for independence of Brock *et al.* (1986). Others are designed to test for a specific kind of nonlinearity as for example with the test of Petruccelli and Davies (1986). For a fairly up to date description and list of references on this problem we refer to section 5.3 of Tong (1990). All these tests are based on different manifestations of nonlinearity, and as such must be limited in their ability to detect nonlinearity as in the classical definition.

All the above tests essentially require that the model errors are independent. This requirement is of the utmost importance if we are to find some transformation of a time series which captures all the probabilistic structure, i.e. reduces it to strict white noise (see e.g. Priestley, 1988, p. 14). If a transformation merely reduces the series to uncorrelated noise, then there will still be structure left unaccounted for. Thus by a model for a time series we mean some transformation reducing the series to independent errors. A linear model is then a special case where the transformation is linear.

When we are interested in forecasting, however, a linear forecast could be optimal (in the sense of least-squares) even when the data-generating mechanism is nonlinear. This occurs when the linear prediction errors are a sequence of martingale differences. In this case we believe that the above tests may not always tell us. A test for this has been developed by An and Cheng (1991). Indeed, as we were preparing the final draft of this paper, Hinich and Patterson (1992) appeared in which they develop a bispectral test which they say will later be extended to include the trispectrum and even higher order cumulant spectra.

In this paper we compute quantities that indicate how close to having this feature certain nonlinear models can be. Related quantities were also used by Tong (1990, p. 177) in his second order index of nonlinearity.

16.2 MEASURING NONLINEARITY

Let $\{X_t\}$ be a strictly stationary, linearly non-deterministic time series with mean μ. As is well known, it will have a Wold representation in terms of an uncorrelated sequence $\{\varepsilon_t\}$ of zero mean random variables

$$X_t - \mu = \sum_{u=0}^{\infty} \theta_u \varepsilon_{t-u} \tag{16.1}$$

with $\theta_0 = 1$ and $\sum \theta_u^2 < \infty$. If we denote the best linear predictor of X_{t+m} in terms of $X_t, X_{t-1}, \dots,$ by $X(t+m|t)$ then we can take ε_t to be $X_t - X(t|t-1)$ and the representation is unique (see for example Hannan, 1970 or Priestley, 1981). A useful term for ε_t is linear innovations used by Hannan and Deistler (1988). Using $\mathscr{F}_t$ and $\mathscr{G}_t$ to denote the σ-fields $\sigma(X_t, X_{t-1}, \dots)$ and $\sigma(\varepsilon_t, \varepsilon_{t-1}, \dots)$ respectively, the best predictor of X_{t+m} is of course

$$\mathrm{E}[X_{t+m}|\mathscr{F}_t] = \mathrm{E}[X_{t+m}|\mathscr{G}_t]$$

which we denote by $X_t(\mathrm{m})$. The difference between $X_t(1)$ and $X(t+1|t)$ is easily seen to be $\mathrm{E}[\varepsilon_{t+1}|\mathscr{G}_t]$, so that the condition for these two predictors to be identical (for all m) is that $\{\varepsilon_t\}$ be a martingale difference sequence i.e.

$$\mathrm{E}[\varepsilon_t|\mathscr{G}_{t-1}] = 0 \tag{16.2}$$

One of the earliest uses (and derivations) of this in the statistical time series literature seems to have been by Hannan and Heyde (1972), repeated in Hall and Heyde (1980). It was first given as a definition of linearity by Hannan (1976) and Hannan (1986) refers to it as a minimal requirement for linearity.

In this paper we are concerned with how close nonlinear models can be to having (16.2) hold. The reason we are led to ask this question is twofold. First, we have observed poor comparative forecast performance of nonlinear models with linear approximations (see Davies *et al.* 1988). Secondly, both the An and

Cheng (*op. cit.*) test and our own (as yet incomplete and unpublished) preliminary study of a test for (16.2) suffer from a seeming lack of power for some nonlinear models. It is thus of interest to know if low power of a test of (16.2) only occurs when (16.2) holds approximately. We are also interested in finding out if some of the nonlinear models that have been fitted to data may also possess predictors that are 'almost linear'. Of course for this purpose the measure will have to be scaled somehow (unless it is shown that (16.2) holds exactly). To this end we calculate $\mathrm{E}[\varepsilon_t|\mathscr{G}_{t-1}]$ for some simple models and then use the mean of this conditional on X_{t-1} and compare it to the conditional standard deviation. This introduces a scaling to allow us at least a rough sense of what is meant by '(16.2) holds approximately' and 'almost linear'.

In the sequel we will refer to the degree of linearity/nonlinearity of a model as being measured by the difference between the linear and nonlinear one-step predictors.

16.3 LINEAR INNOVATIONS FOR SIMPLE THRESHOLD AUTOREGRESSIVE MODELS

We would hope that eventually an investigation of the type we are beginning here could be carried out for a first order nonlinear model of the form

$$X_t = \alpha(X_{t-1}) + \sigma(X_{t-1})a_t \tag{16.3}$$

or even to higher order models (replacing X_{t-1} by a vector of past values). Here $\{a_t\}$ is a sequence of independent and identically distributed random variables.

For now, we try for a more modest aim by studying the so called piecewise constant autoregressive models (PCM) of Pemberton (1990). These are a special case of threshold autoregressive models (see e.g. Tong, 1990) and are obtained from (16.3) by setting

$$\alpha(x) = \alpha_i,\ \sigma(x) = \sigma_i \quad \text{if} \quad x \in R_i,\ i = 1,2,\ldots,k,$$

where R_i are a partition of $\mathbb{R}$. Although it has to be admitted that these are trivial as examples of nonlinear models, they do offer some advantages in trying to gain insight into some aspects of nonlinearity. One such advantage is that the full probabilistic structure is obtainable in closed form when the parameters and the distribution of a_t are specified. More importantly for this study, where we use linear least-squares forecasts, is that these are also obtainable in closed form because under weak assumptions the autocorrelation structure of the model is the same as that of an ARMA(p,p) where $p \leqslant k-1$ (for these results, see Pemberton, 1990). In all the examples we consider, $p = k-1$.

Hence we can represent X_t as a causal and invertible ARMA process of the form

$$\phi(\mathrm{B})(X_t - \mu) = \theta(\mathrm{B})\varepsilon_t, \tag{16.4}$$

where ϕ and θ are both polynomials of degree p and B is the backshift operator. Because of invertibility ε_t are the linear innovations for the X_t process. The parameters of the model (16.4) are obtained as follows. The zeros of the AR operator are simply the inverses of the non-zero and non-unit eigenvalues of the transition matrix $\mathbf{P}$ of the finite Markov chain underlying $\{X_t\}$, that is the process $\{M_t\}$ whose value is j when $X_t \in R_j$. The MA coefficients are then simply obtained by matching the first p autocorrelations of X_t obtained from its known distributional structure with those of the ARMA model. From the ARMA representation we can obtain

$$\varepsilon_t = \sum_{u=0}^{\infty} \pi_u(X_{t-u} - \mu). \tag{16.5}$$

The linear least-squares one-step forecast of X_t is then an infinite series in the semi-infinite past of $\{X_t\}$, while the best forecast is easily seen to be α_i with $i = M_{t-1}$. Thus by subtracting the former from the latter (or equivalently taking conditional expectations of (16.5)), we obtain $\mathrm{E}[\varepsilon_t | \mathscr{G}_{t-1}]$.

Since $\mathrm{E}[\varepsilon_t | \mathscr{G}_{t-1}]$ is an infinite series in the semi-infinite realizations of $\{X_t\}$, its distribution over these realizations will be difficult to obtain. We therefore concentrate on the conditional mean and variance given X_{t-1},

$$\mathrm{E}[\mathrm{E}[\varepsilon_t | \mathscr{G}_{t-1}] | X_{t-1}] = \mathrm{E}[\varepsilon_t | X_{t-1}], \tag{16.6}$$

$$\mathrm{Var}(\mathrm{E}[\varepsilon_t | \mathscr{G}_{t-1}] | X_{t-1}). \tag{16.7}$$

In other words we consider the average behaviour at the origin, X_{t-1}, of a nonlinear forecast of X_t and the variability about that average. In general we would expect that their behaviour will depend on the forecast origin, so it is important that we do condition on this value. In particular we are interested if the average is close to zero with small variance, as this would indicate that a linear forecast could give approximately the same value as the nonlinear forecast. We would also expect a test to have low power. Conversely, for a model that has a large nonzero value of (16.6) we should hope to obtain high power.

For any model, if we can obtain the infinite AR representation (16.5), then (16.6) and (16.7) can be obtained as infinite series in terms of conditional means and conditional covariances of X_{t-r} given X_{t-1} where r runs from 0 to infinity. When the infinite AR is in fact derived from a finite ARMA(p, q) these infinite series are then a linear combination of power series in the reciprocals of the zeros of the MA operator. For the PCM examples we consider below, because of the form of the conditional means and covariances, these series have a closed form sum and are thus exactly computable.

We now look at some examples to illustrate the results that can be obtained. In each, $a_t \sim \mathrm{N}(0, 1)$, σ_i are all unity and $R_i = (r_{i-1}, r_i]$, a sequence of consecutive intervals with $r_0 = -\infty$, $r_k = \infty$, the r_i being called threshold values. The parameter values of the models are given below.

Model	*α values*	*Thresholds*
(16.3)	$\alpha, -\alpha (\alpha > 0)$	0
(16.4)	$1.75, 1.25, -1.25, -1.75$	$-1, 0, 1$
(16.5)	$2, -2, 0.5$	$-1, 1$

Our first example, model (16.3), is probably the simplest that could be constructed. It has the explicit form

$$X_t = \begin{cases} \alpha + a_t & \text{if} \quad X_{t-1} \leqslant 0 \\ -\alpha + a_t & \text{if} \quad X_{t-1} > 0. \end{cases}$$

We had initially thought that as α increased, the model would become more nonlinear. That this is true only as far as some value of α can be seen intuitively, but the value of α beyond which linearity begins to increase again is perhaps not so easy to see. The model has the ARMA(1, 1) representation

$$X_t - \beta X_{t-1} = \varepsilon_t - \theta \varepsilon_{t-1}, \tag{16.8}$$

where θ is the root of $\theta^2 + A(\alpha)\theta + 1 = 0$ which lies inside the unit circle, $\beta = 1 - 2\Phi(\alpha)$, $A(\alpha) = (\beta^2 - 2\beta\rho + 1)/(\rho - \beta)$, $\rho = -\alpha C(\alpha)/(1 + \alpha^2)$ is the lag one autocorrelation of X_t, $C(\alpha) = 2\phi(\alpha) - \alpha\beta$ and ϕ and Φ are the pdf and cdf of the standard normal distribution.

For this model (16.6) and (16.7) are given by

$$\mathrm{E}[\varepsilon_t | X_{t-1} = x] = (\theta - \beta)\{x + B(x)D(\alpha)\} - \mathrm{sign}(x)\alpha \tag{16.9}$$

$$\mathrm{Var}(\mathrm{E}[\varepsilon_t | \mathscr{G}_{t-1}] | X_{t-1} = x) = D(\alpha)\{\alpha\theta - B^2(x)D(\alpha)\} \tag{16.10}$$

where $B(x) = h(x)/f(x)$, $2h(x) = \phi(x + \alpha) - \phi(x - \alpha)$, $2f(x) = \phi(x + \alpha) + \phi(x - \alpha)$, ($f$ being the stationary pdf of X_t) and $D(\alpha) = (\theta - \beta)\theta C(\alpha)/(1 - \beta\theta)$. We must also define $\mathrm{sign}(x) = -1$ when $x = 0$. Surface plots of (16.9) and (16.10) against (x, α) in the positive quadrant are shown in Figures 16.1 and 16.2. The range of values for α is 0.1 through 8.0 and for x is 0.5 through 5.0. Both plots show a peak (somewhere between $\alpha = 2$ and $\alpha = 3$) and that the functions tend to zero as α increases (except at $x = 0$ where we can show that the mean tends to α and the variance to α^2). This is in line with our intuition which tells us that the process will essentially alternate between α and $-\alpha$ for large α, so becoming linear in the sense of predictability. (Note, however, that in the limit as $\alpha \to \infty$, both β and $\theta \to -1$ so that the ARMA model does not have a stationary solution.) Figures 16.3 and 16.4 show the mean ± 2 standard deviations (i.e. (16.6) and (16.6) ± 2 sqrt (16.7)) for $\alpha = 2$ and 5

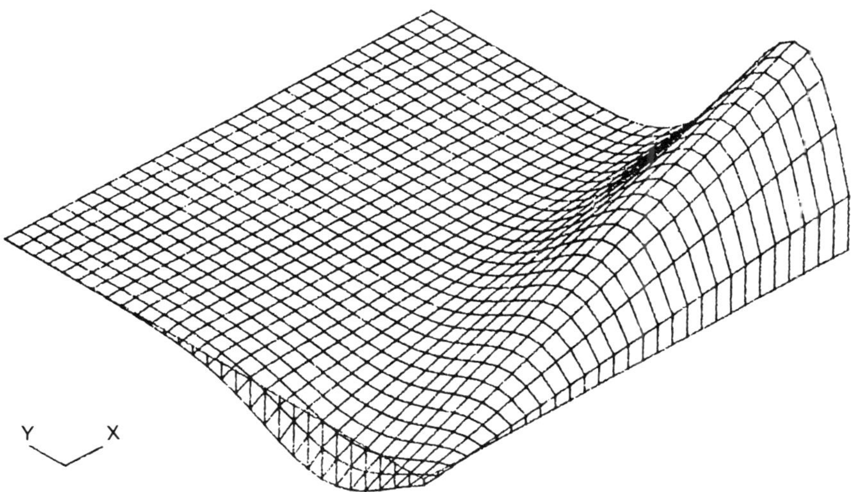

Figure 16.1. Conditional mean vs. α (Y) and $X(t-1)$ (X) for model (16.3).

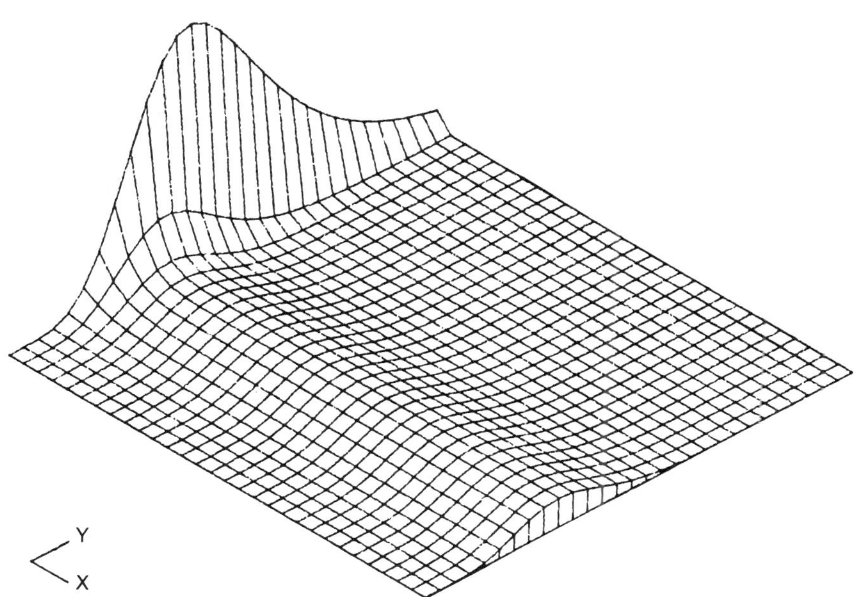

Figure 16.2. Conditional variance vs. α (Y) and $X(t-1)$ (X) for model (16.3).

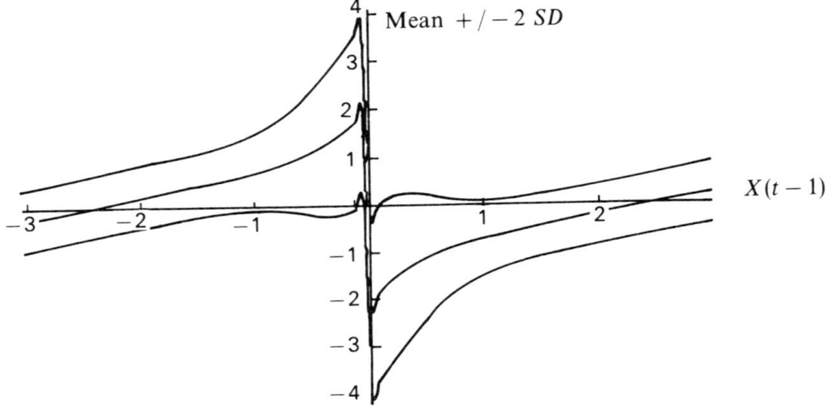

Figure 16.3. Conditional mean +/−2 standard deviations for model (16.3), $\alpha = 2$.

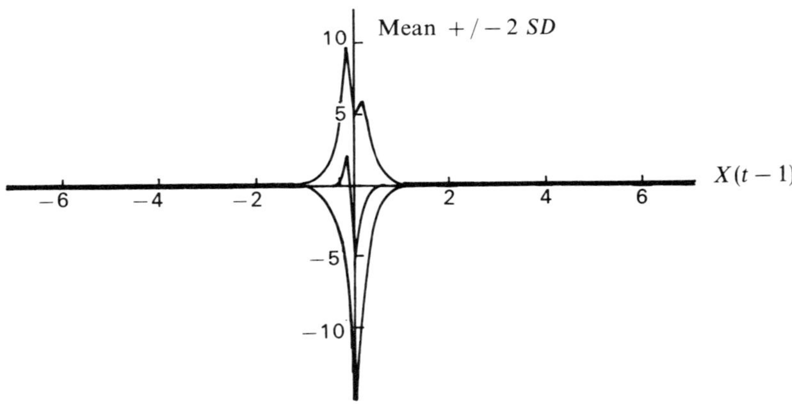

Figure 16.4. Conditional mean +/−2 standard deviations for model (16.3), $\alpha = 5$.

respectively. These are really scatter plots of the three functions, but the points have been joined by straight lines. Figures 16.5 and 16.6 give plots of the mean versus the stationary density of X_{t-1}. We can see that at $\alpha = 2$ there is still considerable nonlinearity while by the time $\alpha = 5$ it has all but gone. The nonlinearity represented by the larger values of the mean and variance can be seen to reside in regions of low probability for X_{t-1} at $\alpha = 5$, whereas for $\alpha = 2$ there is more nonlinearity associated with the larger probability. This seems to suggest how often regions where (16.6) is large will be visited. Hence, if we were to test for the condition (16.2) with a single

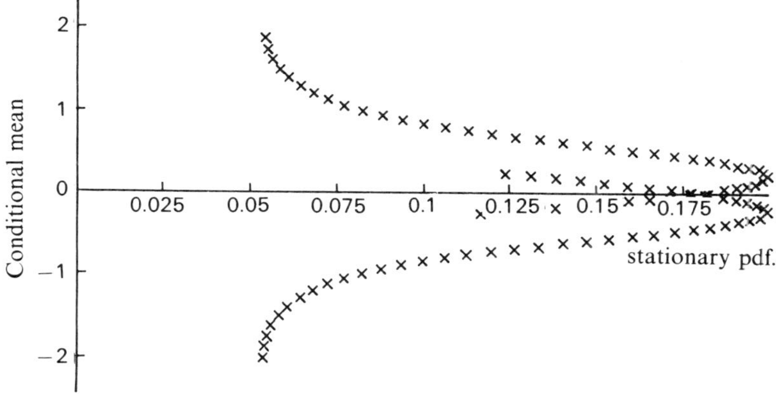

Figure 16.5. Conditional mean vs. stationary pdf for model (16.3), $\alpha = 2$.

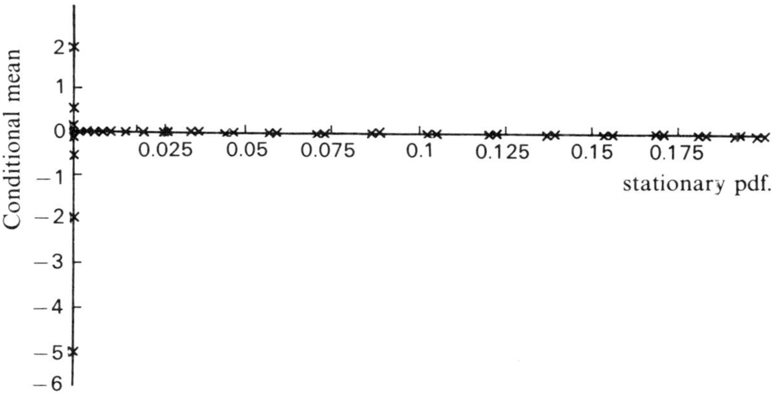

Figure 16.6. Conditional mean vs. stationary pdf for model (16.3), $\alpha = 5$.

realization, we would expect that the power of such a test would be largest when α is in the region of 2 or 3.

Our second example was a first attempt to approximate model 5 of An and Cheng (*op. cit.*) for which the power was low, but we really have little idea of how well it does approximate this model (in fact, we suspect that model (16.3) with $\alpha \approx 2$ may provide a better approximation!). Model (16.4) shows considerable nonlinearity in that (16.6) is not very close to zero as can be seen from Figures 16.7 and 16.8. Although there is symmetry about $X_{t-1} = 0$, we have plotted the functions for negative and positive values of X_{t-1} between -5 and 5. Figure 16.7 is again a scatter plot for the mean

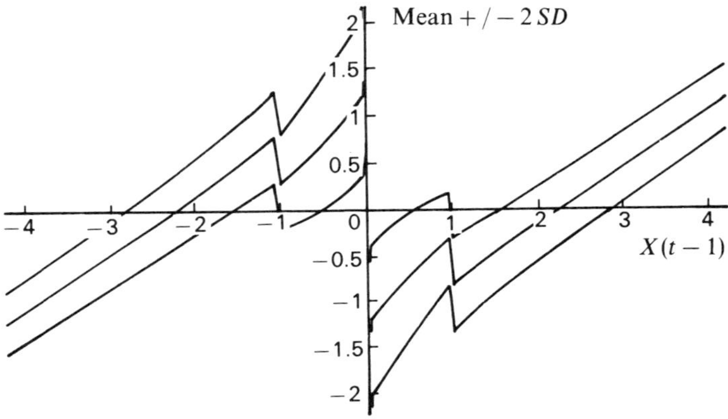

Figure 16.7. Conditional mean +/− 2 standard deviations for model (16.4).

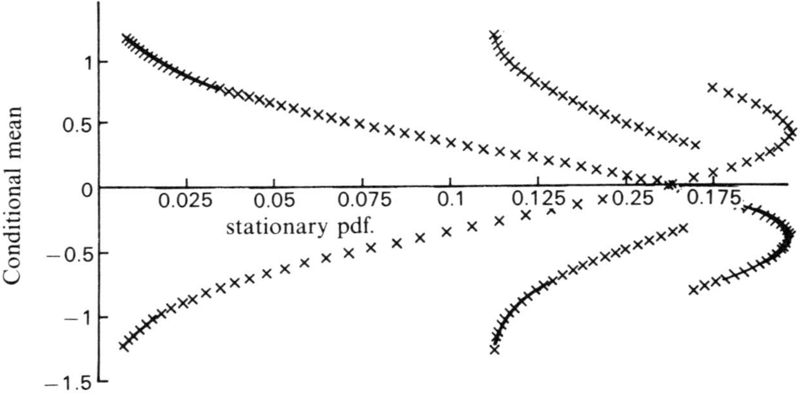

Figure 16.8. Conditional mean vs. stationary pdf for model (16.6).

and 2 standard deviations on either side of it (with points joined by straight lines again). Figure 16.8 shows (16.6) plotted against the stationary pdf for all values of X_{t-1} considered. The model has an odd regression function and hence with the symmetric noise distribution, symmetry appears in these figures. It will be interesting to see what power a test of (16.2) has for this model.

Figures 16.9 and 16.10 show the same quantities as above for the final example, model (16.5). These suggest that the model is quite nonlinear and hence we would expect that a test for linearity of prediction should have

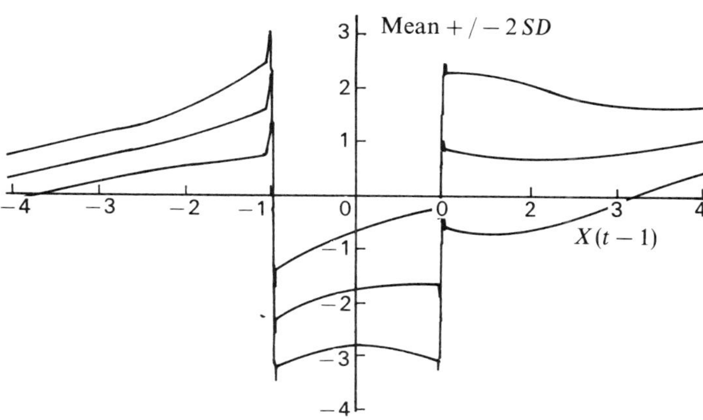

Figure 16.9. Conditional mean +/−2 standard deviations for model (16.5).

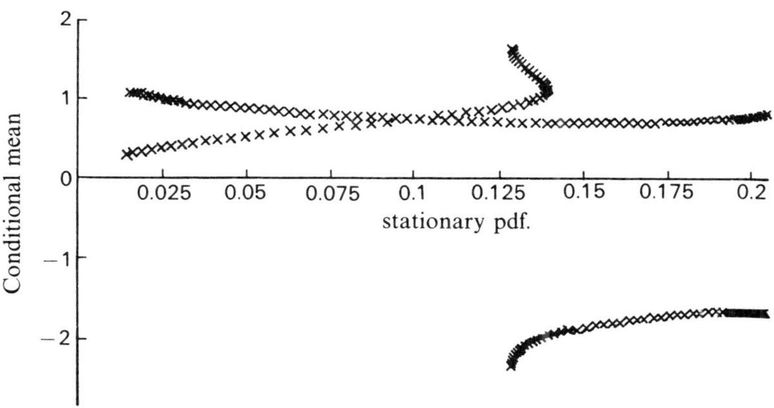

Figure 16.10. Conditional mean vs. stationary pdf for model (16.5).

reasonable power. We await with interest the results of such a test for this and indeed any models for which we can compute $\mathrm{E}[\varepsilon_t|\mathscr{G}_{t-1}]$.

16.4 DISCUSSION

We have employed a measure of nonlinearity that indicates how close the least-squares forecast is to being linear. This can actually be calculated for a simple class of nonlinear time series models and examples of models that appear to be fairly linear and others that seem quite nonlinear have been given. We hope that it can provide useful information with regard to the power of tests for this kind of nonlinearity although we have yet to complete the work on this problem. It will also be of interest to observe how well the measure can predict forecast performance of linear models for data generated by nonlinear models. We shall report the results of these further studies elsewhere.

REFERENCES

An H.-Z. and Cheng, B. (1991) A Kolmogorov–Smirnov type statistic with application to test for nonlinearity in time series. *International Statistical Review*, **59**, 287–307.

Brock, W.A., Dechert, W.D. and Scheinkman, J. (1986) A test for independence based on the correlation dimension. Unpublished manuscripts, University of Wisconsin and University of Chicago.

Davies, N., Pemberton, J. and Petruccelli, J.D. (988) An automatic procedure for identification, estimation and forecasting self-exciting threshold autoregressive models. *The Statistician*, **37**, 337–342.

Hall, P. and Heyde, C.C. (1980) *Martingale Limit Theory and its Application*, Academic Press.

Hannan, E.J. (1970) *Multiple Time Series*, John Wiley & Sons.

Hannan, E.J. (1976) The asymptotic distribution of serial covariances. *Annals of Statistics*, **4**, 396–399

Hannan, E.J. (1986) Rememberance of things past, in *The craft of probabilistic modelling*, (ed. J. Gani), Springer-Verlag, pp. 31–42.

Hannan, E.J. and Deistler, M. (1988) *The Statistical Theory of Linear Systems*, John Wiley & Sons.

Hannan, E.J. and Heyde, C.C. (1972) On limit theorems for quadratic functions of discrete time series. *Annals of Mathematical Statistics*, **43**, 2058–2066.

Hinich, M. (1982) Testing for Gaussianity and linearity of a stationary time series. *Journal of Time Series Analysis*, **3**, 169–176.

Hinich, M. and Patterson, D.M. (1992) A new diagnostic test of model inadequacy which uses the martingale difference criterion. *Journal of Time Series Analysis*, **13**, 233–252.

Keenan, D.M. (1985) A Tukey non-additivity-type test for time series nonlinearity. *Biometrika*, **72**, 39–44.

Pemberton, J. (1990) *Piecewise Constant Models for Univariate Time Series*. Technical report MCS-90-04, Department of Mathematics, University of Salford, Salford, UK.

Petruccelli, J.D. and Davies, N. (1986). A portmanteau test for self-exciting threshold autoregressive type nonlinearity in time series. *Biometrika*, **73**, 687–694.

Priestley, M.B. (1981) Spectral analysis and time series. Vols I and II. Academic Press.

Priestley, M.B. (1988) *Non-linear and Non-stationary Time Series Analysis*. Academic Press.

Subba Rao, T. and Gabr, M.M. (1980) A test for linearity of stationary time series. *Journal of Time Series Analysis*, **1**, 145–158.

Tong, H. (1990) *Non-linear Time Series*, Clarendon Press, Oxford.

17

A Chernoff–Savage result for serial signed rank statistics

J. Allal and M. Hallin

17.1 INTRODUCTION

A Chernoff–Savage technique is used to derive the joint asymptotic normality and strong consistency of a class of serial signed rank statistics under strong mixing assumptions. This class of statistics includes the signed-rank autocorrelation coefficients, the role of which has been shown (Hallin and Puri, 1991a and b) fundamental in the nonparametric analysis of time-series models with unspecified symmetric innovation density. The present result extends an earlier one by Tran (1990) on (unsigned) serial rank statistics, as well of course as Chernoff and Savage's (1958) classical central limit theorem for nonserial, unsigned two-sample statistics under independent observations. This type of result is essential in problems dealing with the power of signed rank tests under fixed (nonlocal) alternatives, as well as in the study of the asymptotic behaviour of R-estimators for time-series models.

Let $(X_t; t\in\mathbb{Z})$ be a strictly stationary, real-valued process defined on (Ω, A, P). Considering a finite realization $\mathbf{X}^{(n)} = (X_1,\ldots, X_n)$, denoted by $R_t^{(n)}$ the rank of X_t among $X_1,\ldots, X_n$, by $R_{+,t}^{(n)}$ the rank of $|X_t|$ among $|X_1|,\ldots, |X_n|$, by $s_t = 2\mathrm{I}[X_t \geqslant 0] - 1$ the sign of X_t, and let $\mathbf{R}^{(n)} = (R_1^{(n)},\ldots, R_n^{(n)})$, $\mathbf{R}_+^{(n)} = (R_{+,1}^{(n)},\ldots, R_{+,n}^{(n)})$, $\mathbf{s}^{(n)} = (s_1,\ldots, s_n)$.

Linear serial rank statistics, of the form

$$S^{(n)} = (n-p)^{-1} \sum_{t=p+1}^{n} a^{(n)}(R_t^{(n)}, R_{t-1}^{(n)},\ldots, R_{t-p}^{(n)}), \tag{17.1}$$

where $a^{(n)}(i_1,\ldots, i_{p+1})$ is a collection of scores defined over the set of all $(p+1)$ tuples of distinct integers in $\{1,\ldots, n\}$ were introduced in Hallin *et al.* (1985) and shown to be asymptotically normal, after adequate normalization and under adequate regularity conditions, for absolutely continuous, exchangeable sequences $\mathbf{X}^{(n)}$. Similar results have been obtained in Hallin and Puri (1991a) for signed rank analogues of (17.1), the linear serial signed rank statistics

$$S_{+}^{(n)} = (n-p)^{-1} \sum_{t=p+1}^{n} a_{+}^{(n)} (s_t R_{+,t}^{(n)}, s_{t-1} R_{+,t-1}^{(n)}, \ldots, s_{t-p} R_{+,t-p}^{(n)}). \quad (17.2)$$

Here the scores $a_{+}^{(n)}(i_1, \ldots, i_{p+1})$ are defined over all $(p+1)$ tuples $i_1, \ldots, i_{p+1}$ in $\{\pm 1, \ldots, \pm n\}$ such that the absolute values $|i_1|, \ldots, |i_{p+1}|$ are distinct. Most serial statistics of practical interest in the study of linear processes can be decomposed into linear combinations of simpler statistics, of the form

$$S^{(n)} = (n-p)^{-1} \sum_{t=p+1}^{n} a^{(n)}(R_t^{(n)}) b^{(n)}(R_{t-p}^{(n)}) \quad (17.3)$$

or

$$S_{+}^{(n)} = (n-p)^{-1} \sum_{t=p+1}^{n} a_{+}^{(n)}(s_t R_{+,t}^{(n)}) b_{+}^{(n)}(s_{t-p} R_{+,t-p}^{(n)}); \quad (17.4)$$

the runs statistic of order p for example is a serial rank statistic of the form $S^{(n)} = S_1^{(n)} + S_2^{(n)}$, with

$$S_1^{(n)} = (n-p)^{-1} \sum_{t=p+1}^{n} I[2R_t^{(n)} < n+1] I[2R_{t-p}^{(n)} > n+1]$$

and

$$S_2^{(n)} = (n-p)^{-1} \sum_{t=p+1}^{n} I[2R_t^{(n)} > n+1] I[2R_{t-p}^{(n)} < n+1]$$

when the runs are taken with respect to the median. It is a signed rank statistic of the form

$$S_{+}^{(n)} = (n-p)^{-1} \sum_{t=p+1}^{n} s_t s_{t-p}$$

when the runs are taken with respect to the origin (see Goodman 1958 or Dufour 1981)—in this latter case, it actually reduces to the autocorrelation coefficient of order p computed from the series of signs. Statistics of the form (17.3) or (17.4) will be called simple (unsigned or signed) serial statistics. The Wald–Wolfowitz (1943) autocorrelation coefficients, as well as the f-rank autocorrelation coefficients defined in Hallin et al. (1987) and Hallin and Puri (1988, 1991a, b) belong to this class of simple serial rank statistics.

Whereas only the null distribution (under the hypothesis of white noise or exchangeability) of the above statistics is needed in order to compute critical values in hypothesis testing problems, the study of the nonlocal power of the resulting tests, or the construction of R-estimates for time series parameters also requires distributional results under alternatives of serial dependence. The asymptotic distribution and strong consistency under strong mixing of simple serial rank statistics in the unsigned case (17.3) has been derived by

Tran (1990). His approach, of the Chernoff–Savage type, is considerably simpler than the generalized Pyke–Shorack approach considered in Harel and Puri (1990a and b), and applies under less restrictive conditions. Our purpose here is to derive an analogue of Tran's result for the signed case (17.4).

17.2 CONSISTENCY AND ASYMPTOTIC DISTRIBUTIONS

17.2.1. Strong consistency

The simple signed rank serial statistics considered throughout are of the form (17.4), with

$$a_+^{(n)}(i)b_+^{(n)}(j) = J_1^{(n)}(i/(n+1))\,J_2^{(n)}(j/(n+1)),$$

and

$$J_1^{(n)}(i/(n+1))\,J_2^{(n)}(j/(n+1)) = E[J_1(\text{sgn}(i)U_{(|i|)}^{(n)})]E[J_2(\text{sgn}(j)U_{(|j|)}^{(n)})|,$$
$$i \neq j \in \{\pm 1, \ldots, \pm n\},$$

or

$$J_1^{(n)}(i/(n+1))\,J_2^{(n)}(j/(n+1)) = E[J_1(\text{sgn}(i)U_{(|i|)}^{(n)})]E[J_2(\text{sgn}(j)U_{(|j|)}^{(n)})|,$$
$$i \neq j \in \{\pm 1, \ldots, \pm n\},$$

where $U_{(1)}^{(n)} \leqslant U_{(2)}^{(n)} \leqslant \cdots \leqslant U_{(n)}^{(n)}$ denote an ordered sample of IID rectangular $[0, 1]$ random variables. Moreover, we make the following two assumptions.

(A1) The score functions J_1 and J_2 are monotone over $(-1, 1)$, twice differentiable (denoted by $J_k^{(l)}$ the derivative of order l of $J_k = J_k^{(0)}$, $k = 1, 2$), and satisfy

$$|J_k^{(l)}(v)| \leqslant K[|v|/(1-|v|)]^{\alpha_k+1}, \quad l = 0, 1, 2, \quad k = 1, 2, \quad v \in (-1, 1),$$

with

$$\alpha_1 = (1-2\delta)/2q, \quad \alpha_2 = (1-2\delta)/2(1-q)$$

for some $K > 0, 0 < \delta < 1/2, 0 < q < 1$.

(A2) The process $(X_t; t \in \mathbb{Z})$ is strictly stationary and strongly mixing, with mixing rate $\alpha(n) = \mathrm{O}(\mathrm{e}^{-sn})$ for some $s > 0$ (see Rosenblatt (1956); recall that strong mixing is weaker than all other usual mixing conditions, such as φ-mixing or absolute regularity); the distribution of X_1 is absolutely continuous with respect to the Lebesgue measure, symmetric with respect to the origin, with distribution function F and probability density f.

Letting $F_+ = 2F - 1$, note that $F_+(|x|) = P(|X_1| \leqslant |x|)$, and $F_+(x) = \text{sgn}(x) \times F_+(|x|)$. Let also H_{i+1} denote the joint distribution function of (X_t, X_{t-i}), viz. $H_{i+1}(x, y) = P(X_t \leqslant x, X_{t-i} \leqslant y)$, and

$$m_p = \int_{-\infty}^{\infty} \int_{-\infty}^{\infty} J_1(F_+(x)) J_2(F_+(y)) dH_{p+1}(x, y). \tag{17.5}$$

Proposition 17.1

Under assumptions (A1) and (A2),

$$S_+^{(n)} = (n-p)^{-1} \sum_{t=p+1}^{n} J_1^{(n)}\left(\frac{s_t R_{+,t}^{(n)}}{n+1}\right) J_2^{(n)}\left(\frac{s_{t-p} R_{+,t-p}^{(n)}}{n+1}\right) \tag{17.6}$$

converges to m_p almost surely; in probability, its convergence is uniform with respect to H_{p+1}.

The proof of proposition 17.1 relies on four lemmas. Lemma 17.2 generalizes Tran's (1990) lemma 2.3.

Lemma 17.2

Let $r(v) = [|v|(1-|v|)]^{-1}$, $v \in (-1, 1)$, with the convention $r(0) = 0$, and

$V_+^{(n)}(v) = \text{sgn}(v) n^{-1} \sum_{t=1}^{n} I[F_+(|X_t|) \leqslant v]$, $\quad v \in (-1, 1)$. Under (A2), for all $\varepsilon > 0$,

$$\sup_{v \in (-1,1)} n^{1/2} r(v)^{(1/2)-\varepsilon} |V_+^{(n)}(|v|) - |v|| = O((\log n)^2) \quad \text{a.s..}$$

Proof

$$\begin{aligned}
&\sup_{v \in (-1,1)} n^{1/2} r(v)^{(1/2)-\varepsilon} |V_+^{(n)}(|v|) - |v|| \\
&\quad \leqslant \sup_{v \in (-1,0)} n^{1/2} [-v(1+v)]^{\varepsilon-(1/2)} |V_+^{(n)}(-v) + v| \\
&\qquad + \sup_{v \in (0,1)} n^{1/2} [v(1-v)]^{\varepsilon-(1/2)} |V_+^{(n)}(v) - v| \\
&\quad \leqslant 2 \sup_{v \in (0,1)} n^{1/2} [v(1-v)]^{\varepsilon-(1/2)} |V_+^{(n)}(v) - v|,
\end{aligned}$$

a quantity which, from Tran's lemma 2.3, is a.s. $O((\log n)^2)$.

Let

$$F_{|X|}^{(n)}(x) = \left[n^{-1} \sum_{t=1}^{n} I[|X_t| \leqslant x] \right], \quad F_+^{(n)}(x) = \text{sgn}(x) F_{|X|}^{(n)}(|x|),$$

$$F_{+,\theta}^{(n)}(x) = \theta \frac{n}{n+1} F_+^{(n)}(x) + (1-\theta) F_+(x), \quad \theta \in [0, 1],$$

and

$$H_{i+1}^{(n)}(x, y) = (n-i)^{-1} \sum_{t=i+1}^{n} I[X_t \leqslant x, X_{t-i} \leqslant y].$$

The following lemmas generalize Tran's lemmas 2.4 and 2.5.

Lemma 17.3

For all $\theta \in [0, 1]$, all $\lambda \in (0, 1)$, and all x such that $n^{-1+\lambda} \leqslant |F_+(x)| \leqslant 1 - n^{-1+\lambda}$,

$$\begin{aligned} 1 - |F_{+,\theta}^{(n)}(x)| &\geqslant [1 - |F_+(x)|][1 - \mathrm{o}(1)] \\ &= [1 - |F_+^{(n)}(|x|)|][1 - \mathrm{o}(1)] \quad \text{a.s., as } n \to \infty. \end{aligned}$$

Proof

Assume $x > 0$. Then

$$\begin{aligned} 1 - F_{+,\theta}^{(n)}(x) &= 1 - F_+(x) - \theta\left[\frac{n}{n+1} F_+^{(n)}(x) - F_+(x)\right] \\ &= [1 - F_+(x)] \\ &\quad \times \left\{1 - \theta\left[\left(\frac{n}{n+1} F_+^{(n)}(x) - F_+(x)\right) \Big/ (1 - F_+(x))F_+(x)\right] F_+(x)\right\} \\ &\leqslant [1 - F_+(x)] \\ &\quad \times \left\{1 - \left[\left|\frac{n}{n+1} F_+^{(n)}(x) - F_+(x)\right| \Big/ (1 - F_+(x))F_+(x)\right]\right\}. \end{aligned}$$

Now, for $n^{-1+\lambda} \leqslant |F_+(x)| \leqslant 1 - n^{-1+\lambda}$,

$$\begin{aligned} &\left|\frac{n}{n+1} F_+^{(n)}(x) - F_+(x)\right| \Big/ (1 - F_+(x))F_+(x) \\ &\leqslant n^{\lambda 2 - 1}\left|\frac{n}{n+1} F_+^{(n)}(x) - F_+(x)\right| [(1 - F_+(x))F_+(x)]^{(\lambda-1)/2} \\ &\leqslant \sup_{x>0} n^{1/2}\left|\frac{n}{n+1} F_+^{(n)}(x) - F_+(x)\right| [(1 - F_+(x))F_+(x)]^{(\lambda-1)/2} n^{\lambda 2 - 3/2} \\ &\leqslant \sup_{v \in (0,1)} n^{1/2} |V_+^{(n)}(|v|) - |v|| [v(1-v)]^{(\lambda-1)/2} n^{\lambda 2 - 3/2} = \mathrm{o}(1), \end{aligned}$$

as $n \to \infty$, for $\lambda \in (0, 1)$. For $x < 0$,

$$F^{(n)}_{+,\theta}(x) = -\theta \frac{n}{n+1} F^{(n)}_{+}(-x) + (1-\theta)F_{+}(x),$$

and the proof is similar.

Lemma 17.4

For all $\lambda \in (0, 1)$,

$$|F^{(n)}_{+}(F^{-1}_{+}(1 - n^{-1+\lambda})) - (1 - n^{-1+\lambda})| = \mathrm{O}(n^{-1+\lambda/2}(\log n)^2) \quad \text{a.s., as } n \to \infty,$$

and

$$|F^{(n)}_{+}(F^{-1}_{+}(n^{-1+\lambda} - 1)) - (n^{-1+\lambda} - 1)| = \mathrm{O}(n^{-1+\lambda/2}(\log n)^2) \quad \text{a.s., as } n \to \infty.$$

Proof

For n sufficiently large, $F^{-1}_{+}(1 - n^{-1+\lambda}) > 0$, so that

$$F^{(n)}_{+}(F^{-1}_{+}(1 - n^{-1+\lambda})) = F^{(n)}_{|X|}(F^{-1}_{+}(1 - n^{-1+\lambda}))$$

and

$$F^{(n)}_{+}(F^{-1}_{+}(n^{-1+\lambda} - 1)) = F^{(n)}_{|X|}(-F^{-1}_{+}(n^{-1+\lambda} - 1)) = -F^{(n)}_{|X|}(F^{-1}_{+}(1 - n^{-1+\lambda}));$$

the result then follows from applying Trans's lemma 2.5.

Consider the simple signed rank statistic

$$\begin{aligned} S^{(n)}_{p,+} &= (n-p)^{-1} \sum_{t=p+1}^{n} J^{(n)}_{1}\left(\frac{s_t R^{(n)}_{+,t}}{n+1}\right) J^{(n)}_{2}\left(\frac{s_{t-p} R^{(n)}_{+,t-p}}{n+1}\right) \\ &= (n-p)^{-1} \sum_{t=p+1}^{n} J^{(n)}_{1}\left(\frac{n \operatorname{sgn}(X_t) F^{(n)}_{|X|}(|X_t|)}{n+1}\right) J^{(n)}_{2}\left(\frac{n \operatorname{sgn}(X_{t-p}) F^{(n)}_{|X|}(|X_{t-p}|)}{n+1}\right) \\ &= \int_{-\infty}^{\infty}\int_{-\infty}^{\infty} J^{(n)}_{1}\left(\frac{n}{n+1} F^{(n)}_{+}(x)\right) J^{(n)}_{2}\left(\frac{n}{n+1} F^{(n)}_{+}(y)\right) \mathrm{d}H^{(n)}_{p+1}(x, y). \end{aligned} \tag{17.7}$$

Let

$$\begin{aligned} A^{(n)}_{p,+} &= \int_{-\infty}^{\infty}\int_{-\infty}^{\infty} J_1(F_{+}(x)) J_2(F_{+}(y)) \mathrm{d}H^{(n)}_{p+1}(x, y) \\ &= (n-p)^{-1} \sum_{t=p+1}^{n} J_1(F_{+}(X_t)) J_2(F_{+}(X_{t-p})), \end{aligned}$$

$$B^{(n)}_{p,+} = \int_{-\infty}^{\infty}\int_{-\infty}^{\infty} [F^{(n)}_{+}(x) - F_{+}(x)] J^{(1)}_{1}(F_{+}(x)) J_2(F_{+}(y)) \mathrm{d}H_{p+1}(x, y),$$

$$C_{p,+}^{(n)} = \int_{-\infty}^{\infty}\int_{-\infty}^{\infty} [F_+^{(n)}(y) - F_+(y)] J_1(F_+(x)) J_2^{(1)}(F_+(y)) \mathrm{d}H_{p+1}(x, y),$$

$$L_{p,+}^{(n)} = A_{p,+}^{(n)} + B_{p,+}^{(n)} + C_{p,+}^{(n)},$$

and define

$$R_{p,+}^{(n)} = S_{p,+}^{(n)} - L_{p,+}^{(n)}.$$

Lemma 17.5

$n^{1/2} R_{p,+}^{(n)}$ converges to zero, a.s. and, in probability, uniformly in H_{p+1}, as $n \to \infty$.

Proof

Let $a_1^{(n)}$, $b_1^{(n)}$, $c_1^{(n)}$ and $d_1^{(n)}$ be such that

$$F_+(a_1^{(n)}) = n^{-1+\delta_1}, \qquad F_+(b_1^{(n)}) = 1 - n^{-1+\delta_1},$$
$$F_+(c_1^{(n)}) = n^{-1+\delta_2}, \qquad F_+(d_1^{(n)}) = 1 - n^{-1+\delta_2},$$

where $\delta_1 = \delta/q$ and $\delta_2 = \delta/(1-q)$ (q and δ as in assumption A2). Also, for all $\xi \in (0, 1/4)$, let

$$\gamma_1 \in (0, \min(\xi(1+\alpha_1)^{-1}, (1-2\alpha_2)/8\alpha_1)),$$
$$\gamma_2 \in (0, \min(\xi(1+\alpha_2)^{-1}, (1-2\alpha_1)/8\alpha_2)),$$

and $a_2^{(n)}$, $b_2^{(n)}$, $c_2^{(n)}$ and $d_2^{(n)}$ be such that

$$F_+(a_2^{(n)}) = n^{-\gamma_1}, \qquad F_+(b_2^{(n)}) = 1 - n^{-\gamma_1},$$
$$F_+(c_2^{(n)}) = n^{-\gamma_2}, \qquad F_+(d_2^{(n)}) = 1 - n^{-\gamma_2}.$$

Denote by $\mathbb{R}_-$ and $\mathbb{R}_+$ the half-lines $(-\infty, 0)$ and $(0, \infty)$ respectively, and put

$$I_{1,1}^{(n)} = [a_1^{(n)}, b_1^{(n)}] \times [c_1^{(n)}, d_1^{(n)}] \qquad I_{1,5}^{(n)} = (\mathbb{R}_+ \times \mathbb{R}_+) \backslash I_{1,1}^{(n)}$$
$$I_{1,2}^{(n)} = [-a_1^{(n)}, -b_1^{(n)}] \times [-c_1^{(n)}, -d_1^{(n)}] \qquad I_{1,6}^{(n)} = (\mathbb{R}_- \times \mathbb{R}_-) \backslash I_{1,2}^{(n)}$$
$$I_{1,3}^{(n)} = [a_1^{(n)}, b_1^{(n)}] \times [-c_1^{(n)}, -d_1^{(n)}] \qquad I_{1,7}^{(n)} = (\mathbb{R}_+ \times \mathbb{R}_-) \backslash I_{1,3}^{(n)}$$
$$I_{1,4}^{(n)} = [-a_1^{(n)}, -b_1^{(n)}] \times [c_1^{(n)}, d_1^{(n)}] \qquad I_{1,8}^{(n)} = (\mathbb{R}_- \times \mathbb{R}_+) \backslash I_{1,4}^{(n)}.$$

Similarly define (from $a_2^{(m)}, b_2^{(m)}$, etc) $I_{2,1}^{(n)}, I_{2,2}^{(n)}, I_{2,3}^{(n)}, I_{2,4}^{(n)}, I_{2,5}^{(n)}, I_{2,6}^{(n)}, I_{2,7}^{(n)}$ and $I_{2,8}^{(n)}$: $R_{p,+}^{(n)}$ then decomposes into

$$R_{p,+}^{(n)} = \sum_{k=1}^{59} R_{p,k}^{(n)},$$

with

$$R_{p,1}^{(n)} = \int_{-\infty}^{\infty}\int_{-\infty}^{\infty} [J_1^{(n)}(F_+^{(n)}(x)) - J_1(nF_+^{(n)}(x)/(n+1))]$$
$$\times [J_2^{(n)}(F_+^{(n)}(y)) - J_2(nF_+^{(n)}(y)/(n+1))]\, \mathrm{d}H_{p+1}^{(n)}(x, y),$$

$$R_{p,2}^{(n)} = \int_{-\infty}^{\infty}\int_{-\infty}^{\infty} [J_1^{(n)}(F_+^{(n)}(x)) - J_1(nF_+^{(n)}(x)/(n+1))]$$
$$\times [J_2(nF_+^{(n)}(y)/(n+1))]\, \mathrm{d}H_{p+1}^{(n)}(x, y),$$

$$R_{p,3}^{(n)} = \int_{-\infty}^{\infty}\int_{-\infty}^{\infty} [J_1(nF_+^{(n)}(x)/(n+1))]$$
$$\times [J_2^{(n)}(F_+^{(n)}(y)) - J_2(nF_+^{(n)}(y)/(n+1))]\, \mathrm{d}H_{p+1}^{(n)}(x, y),$$

$$R_{p,3+k}^{(n)} = \iint_{I_{1,k}^{(n)}} [J_1(nF_+^{(n)}(x)/(n+1)) - J_1(F_+(x))]$$
$$\times [J_2(nF_+^{(n)}(y)/(n+1)) - J_2(F_+(y))]\, \mathrm{d}H_{p+1}^{(n)}(x, y), \quad k = 1, \ldots, 8,$$

$$R_{p,11+k}^{(n)} = \iint_{I_{1,k}^{(n)}} \{J_1(nF_+^{(n)}(x)/(n+1)) - J_1(F_+(x)) - J_1^{(1)}(F_+(x))$$
$$\times [(nF_+^{(n)}(x)/(n+1)) - F_+(x)]\} J_2(F_+(y))\, \mathrm{d}H_{p+1}^{(n)}(x, y), \quad k = 1, \ldots, 8,$$

$$R_{p,19+k}^{(n)} = \iint_{I_{1,k}^{(n)}} J_1(F_+(x))\{J_2(nF_+^{(n)}(y)/(n+1)) - J_2(F_+(y)) - J_2^{(1)}(F_+(y))$$
$$\times [(nF_+^{(n)}(y)/(n+1)) - F_+(y)]\}\, \mathrm{d}H_{p+1}^{(n)}(x, y), \quad k = 1, \ldots, 8,$$

$$R_{p,23+k}^{(n)} = -\iint_{I_{2,k}^{(n)}} J_1^{(1)}(F_+(x)) J_2(F_+(y))$$
$$\times [F_+^{(n)}(x) - F_+(x)]\, \mathrm{d}H_{p+1}(x, y), \quad k = 5, \ldots, 8,$$

$$R_{p,27+k}^{(n)} = -\iint_{I_{2,k}^{(n)}} J_1(F_+(x)) J_2^{(1)}(F_+(y))$$
$$\times [F_+^{(n)}(y) - F_+(y)]\, \mathrm{d}H_{p+1}(x, y), \quad k = 5, \ldots, 8,$$

$$R_{p,36+k}^{(n)} = -\iint_{I_{1,k}^{(n)} \cap I_{2,k+4}^{(n)}} J_1^{(1)}(F_+(x)) J_2(F_+(y))$$
$$\times [F_+^{(n)}(x) - F_+(x)]\, \mathrm{d}H_{p+1}^{(n)}(x, y), \quad k = 1, \ldots, 4,$$

$$R_{p,39+k}^{(n)} = -\iint_{I_{1,k}^{(n)} \cap I_{2,k+4}^{(n)}} J_1(F_+(x)) J_2^{(1)}(F_+(y))$$
$$\times [F_+^{(n)}(y) - F_+(y)]\, \mathrm{d}H_{p+1}^{(n)}(x, y), \quad k = 1, \ldots, 4,$$

$$R^{(n)}_{p,43+k} = \iint_{I^{(n)}_{2,k}} J^{(1)}_1(F_+(x))J_2(F_+(y))\,[F^{(n)}_+(x) - F_+(x)]\,\mathrm{d}[H^{(n)}_{p+1}(x,y) - H_{p+1}(x,y)], \quad k = 1,\ldots,4,$$

$$R^{(n)}_{p,47+k} = \iint_{I^{(n)}_{2,k}} J_1(F_+(x))J^{(1)}_2(F_+(y))\,[F^{(n)}_+(y) - F_+(y)]\,\mathrm{d}[H^{(n)}_{p+1}(x,y) - H_{p+1}(x,y)], \quad k = 1,\ldots,4,$$

$$R^{(n)}_{p,51+k} = -(n-1)^{-1}\iint_{I^{(n)}_{1,k}} J^{(1)}_1(F_+(x))J_2(F_+(y))F^{(n)}_+(x)\,\mathrm{d}H^{(n)}_{p+1}(x,y), \quad k = 1,\ldots,4,$$

$$R^{(n)}_{p,55+k} = -(n-1)^{-1}\iint_{I^{(n)}_{1,k}} J_1(F_+(x))J^{(1)}_2(F_+(y))F^{(n)}_+(y)\,\mathrm{d}H^{(n)}_{p+1}(x,y), \quad k = 1,\ldots,4.$$

Using Tran's lemma 2.2, and lemmas 17.2, 17.4 and 17.5 above, each of these 59 terms can be shown to be $o_P(n^{-(1/2)-v})$, $v > 0$, uniformly in the underlying joint distribution H_{p+1}; $n^{1/2}R^{(n)}_{p,k}$, $k = 1,\ldots,59$ thus converges to zero a.s., as $m \to \infty$.

The case of $R^{(n)}_{p,4}$ is treated here as an example. Let $\lambda = \eta + 1/2, 0 < \eta$. Then

$$\begin{aligned} n^\lambda|R^{(n)}_{p,4}| &\leqslant n^\lambda \iint_{I^{(n)}_{1,1}} |J_1(nF^{(n)}_+(x)/(n+1)) - J_1(F_+(x))| \\ &\quad \times |J_2(nF^{(n)}_+(y)/(n+1)) - J_2(F_+(y))|\,\mathrm{d}H^{(n)}_{p+1}(x,y) \\ &\leqslant n^\lambda \iint_{I^{(n)}_{1,1}} |(nF^{(n)}_+(x)/(n+1)) - F_+(x)|J^{(1)}_1(F^{(n)}_{+,\theta_1}(x)) \\ &\quad \times |(nF^{(n)}_+(y)/(n+1)) - F_+(y)|J^{(1)}_2(F^{(n)}_{+,\theta_2}(y))\,\mathrm{d}H^{(n)}_{p+1}(x,y). \end{aligned}$$

It then follows from assumption (A1) and lemma 17.3 that

$$\begin{aligned} n^\lambda|R^{(n)}_{p,4}| &\leqslant K^2 n^\lambda \iint_{I^{(n)}_{1,1}} |(nF^{(n)}_+(x)/(n+1)) - F_+(x)|\,[r(F^{(n)}_{+,\theta_1}(x))]^{\alpha_1+1} \\ &\quad \times |(nF^{(n)}_+(y)/(n+1)) - F_+(y)|\,[r(F^{(n)}_{+,\theta_2}(y))]^{\alpha_2+1}\,\mathrm{d}H^{(n)}_{p+1}(x,y) \\ &\leqslant C_1C_2K^2n^\lambda \iint_{I^{(n)}_{1,1}} |(nF^{(n)}_+(x)/(n+1)) - F_+(x)|\,[r(F_+(x))]^{\alpha_1+1} \\ &\quad \times |(nF^{(n)}_+(y)/(n+1)) - F_+(y)|\,[r(F_+(y))]^{\alpha_2+1}\,\mathrm{d}H^{(n)}_{p+1}(x,y) \\ &\leqslant C_1C_2n^\lambda \sup_{x>0}\{|(nF^{(n)}_+(x)/(n+1)) - F_+(x)|[r(F_+(x))]^{(2-\delta_1),4}\} \end{aligned}$$

$$\times \sup_{y>0} \{ |(nF_+^{(n)}(y)/(n+1)) - F_+(y)| [r(F_+(y))]^{(2-\delta_2)/4} \}$$

$$\times \iint_{I_{1,1}^{(n)}} [r(F_+(x))]^{\alpha_1 + (2+\delta_1)/4} [r(F_+(y))]^{\alpha_2 + (2+\delta_2)/4} \, \mathrm{d}H_{p+1}^{(n)}(x, y),$$

with positive constants C_1, C_2. Lemma 17.2 finally implies that, with probability one, the above quantity is less than

$$C_3 n^{\eta - 1/2} (\log n)^4 \iint_{I_{1,1}^{(n)}} [r(F_+(x))]^{\alpha_1 + (2+\delta_1)/4} [r(F_+(y))]^{\alpha_2 + (2+\delta_2)/4} \, \mathrm{d}H_{p+1}^{(n)}(x, y), \tag{17.8}$$

where $C_3 > 0$ and the integral, since (X_t, X_{t-p}) is ergodic, converges a.s. to the corresponding (finite) expectation. The announced strong consistency result thus follows from the fact that $n^{\eta - 1/2}(\log n)^4 = o(1)$. As for the uniform convergence in probability, we have, from (17.8),

$$\begin{aligned} E[n^\lambda |R_{p,4}^{(n)}| &\leqslant C_3 n^{\eta - 1/2} (\log n)^4 \iint_{I_{1,1}^{(n)}} [r(F_+(x))]^{\alpha_1 + (2+\delta_1)/4} \\ &\quad \times [r(F_+(y))]^{\alpha_2 + (2+\delta_2)/4} \, \mathrm{d}H_{p+1}^{(n)}(x, y) \\ &\leqslant C_3 n^{\eta - 1/2} (\log n)^4 [r(n^{\delta_2 - 1})]^{(2+\delta_2)/4} \iint_{I_{1,1}^{(n)}} [r(F_+(x))]^{\alpha_1 + (2+\delta_1)/4} \\ &\quad \times [r(F_+(y))]^{\alpha_2} \, \mathrm{d}H_{p+1}(x, y). \end{aligned}$$

As in Ruymgaart *et al.* (1972), we may find p_1 and q_1 such that $p_1^{-1} + q_1^{-1} = 1$, $0 < (\alpha_1 + (2+\delta_1)/4)p_1 < 1$ and $0 < \alpha_2 q_1 < 1$. Applying Hölder's inequality,

$$\begin{aligned} &\iint_{I_{1,1}^{(n)}} [r(F_+(x))]^{\alpha_1 + (2+\delta_1)/4} [r(F_+(y))]^{\alpha_2} \, \mathrm{d}H_{p+1}(x, y) \\ &\quad \leqslant \left[\int_0^1 [r(u)]^{(\alpha_1 + (2+\delta_1)/4)p_1} \, \mathrm{d}u \right]^{1/p_1} \left[\int_0^1 [r(u)]^{\alpha_2 q_1} \, \mathrm{d}u \right]^{1/q_1} = C_4, \end{aligned}$$

so that

$$\begin{aligned} E[n^\lambda |R_{p,4}^{(n)}|] &\leqslant C_5 n^{\eta - 1/2} (\log n)^4 [r(n^{\delta_2 - 1})]^{(2+\delta_2)/4} \\ &= C_5 n^{\eta - 1/2} (\log n)^4 [n^{\delta_2 - 1}(1 - n^{\delta_2 - 1})]^{-(2+\delta_2)/4} \\ &\leqslant C_6 n^{(\eta - (\delta_2 + \delta)_2^2/4)}, \end{aligned}$$

a quantity that does not depend on H_{p+1} and tends to zero provided that $0 < \eta < (\delta_2 + \delta_2^2)/4$.

Proof of proposition 17.1

Define

$$A^{(n)}_{p,t,+} = J_1(F_+(X_t))J_2(F_+(X_{t-p})),$$

$$B^{(n)}_{p,t,+} = \int_{-\infty}^{\infty}\int_{-\infty}^{\infty} [\operatorname{sgn}(x)\phi_{|X_t|}(|x|) - F_+(x)]J_1^{(1)}(F_+(x))J_2(F_+(y))\,\mathrm{d}H_{p+1}(x,y)$$

and

$$C^{(n)}_{p,t,+} = \int_{-\infty}^{\infty}\int_{-\infty}^{\infty} [\operatorname{sgn}(y)\phi_{|X_t|}(|y|) - F_+(y)]J_1(F_+(x))J_2^{(1)}(F_+(y))\,\mathrm{d}H_{p+1}(x,y),$$

where $\phi_{|X_t|}(z) = I[|X_t| \leqslant z]$. Then,

$$A^{(n)}_{p,+} = (n-p)^{-1}\sum_{t=p+1}^{n} A^{(n)}_{p,t,+}$$

$$B^{(n)}_{p,+} = n^{-1}\sum_{t=1}^{n} B^{(n)}_{p,t,+}$$

$$C^{(n)}_{p,+} = n^{-1}\sum_{t=1}^{n} C^{(n)}_{p,t,+}.$$

First, we show that $B^{(n)}_{p,+}$ and $C^{(n)}_{p,+}$ tend to zero, a.s. and, in probability, uniformly in H_{p+1}, as $n\to\infty$. The process $\{B^{(n)}_{p,t,+};\ t\geqslant p+1\}$ is ergodic and strongly mixing, with mean zero and mixing coefficient $\alpha(n) = \mathrm{O}(\mathrm{e}^{-sn})$, $s>0$, so that $B^{(n)}_{p,+}$ converges a.s. to zero. Moreover, there exist p_1 and q_1 such that $p_1^{-1}+q_1^{-1}=1$, $0<(\alpha_1+(2+\delta)/4)p_1<1$, $0<\alpha_2 q_1<1$, and

$$\begin{aligned}
|B^{(n)}_{p,+}| &\leqslant K^2 \sup_{x\in\mathbb{R}}\{|F^{(n)}_+(x) - F_+(x)|\}[r(F_+(x))]^{(2-\delta)/4} \\
&\quad\times \int_{-\infty}^{\infty}\int_{-\infty}^{\infty} [r(F_+(x))]^{\alpha_1+(2+\delta)/4}[r(F_+(y))]^{\alpha_2}\,\mathrm{d}H_{p+1}(x,y) \\
&\leqslant K^2 \sup_{x\in\mathbb{R}}\{|F^{(n)}_+(x) - F_+(x)|\}[r(F_+(x))]^{(2-\delta)/4} \\
&\quad\times\left[\int_0^1 [r(u)]^{(\alpha_1+(2+\delta)/4)p_1}\,\mathrm{d}u\right]^{1/p_1}\left[\int_0^1 [r(u)]^{\alpha_1 q_1}\,\mathrm{d}u\right]^{1/q_1}.
\end{aligned}$$

The desired convergence to zero of $B^{(n)}_{p,+}$ then follows from lemma 17.2; that of $C^{(n)}_{p,+}$ follows along the same lines. Next, we show that $A^{(n)}_{p,+}$ similarly converges to m_p (see (17.5)). Let

$$A^{*(n)}_{p,+} = A^{(n)}_{p,+} - m_p = (n-p)^{-1}\sum_{t=p+1}^{n}(A^{(n)}_{p,t,+} - m_p) = (n-p)^{-1}\sum_{t=p+1}^{n} A^{*(n)}_{p,t,+}.$$

Here again, $\{A^{*(n)}_{p,t,+};\ t \geqslant p+1\}$ is ergodic and strongly mixing, with mean zero and mixing coefficient $\alpha(n) = \mathrm{O}(\mathrm{e}^{-sn})$, $s > 0$, so that $A^{*(n)}_{p,+}$ converges a.s. to zero. The assumptions on the score functions J_1 and J_2 imply that $|A^{*(n)}_{p,t,+}|$ has a finite moment of order $2+\delta$, with δ given in assumption (A1). By Minkowski's inequality,

$$\{E[|A^{(n)}_{p,t,+}|^{2+\delta}]\}^{1/(2+\delta)} \leqslant \{E[|J_1(F_+(X_t))J_2(F_+(X_{t-p}))|^{2+\delta}]\}^{1/(2+\delta)} + |m_p|.$$

Now,

$$\begin{aligned} |m_p| &\leqslant E[|J_1(F_+(X_{p+1}))J_2(F_+(X_1))|] \\ &\leqslant K^2 E\{[r(F_+(X_{p+1}))]^{\alpha_1}[r(F_+(X_1))]^{\alpha_2}\} \\ &\leqslant K^2(E\{[r(F_+(X_{p+1}))]^{(1-2\delta)/2}\})^{1/p}(E\{[r(F_+(X_1))]^{(1-2\delta)/2}\})^{1/q} = K_1. \end{aligned}$$

Similarly,

$$\{E[|J_1(F_+(X_t))J_2(F_+(X_{t-p}))|^{2+\delta}]\}^{1/(2+\delta)} \leqslant K_2.$$

The desired convergence follows, which completes the proof of proposition 17.1.

17.2.2 Asymptotic normality

Letting

$$\mathbf{S}^{(n)}_+ = (S^{(n)}_{1,+}, \ldots, S^{(n)}_{p,+})', \qquad \mathbf{m}_+ = (m_{1,+}, \ldots, m_{p,+})',$$

$$\mathbf{L}^{(n)}_+ = (L^{(n)}_{1,+}, \ldots, L^{(n)}_{p,+})', \quad \text{and} \quad \mathbf{R}^{(n)}_+ = (R^{(n)}_{1,+}, \ldots, R^{(n)}_{p,+})', p \geqslant 1,$$

we have

$$n^{1/2}(\mathbf{S}^{(n)}_+ - \mathbf{m}_+) = n^{1/2}(\mathbf{L}^{(n)}_+ - \mathbf{m}_+) + n^{1/2}\mathbf{R}^{(n)}_+.$$

Denote by $\mathbf{\Gamma}^{(n)}_+ = (\gamma^{(n)}_{i,j,+})$ the covariance matrix of $n^{1/2}(\mathbf{S}^{(n)}_+ - \mathbf{m}_+)$.

Proposition 17.6

Under assumptions (A1) and (A2), there exists a real matrix $\mathbf{\Gamma}_+ = (\gamma_{i,j,+})$ such that $\gamma^{(n)}_{i,j,+} \to \gamma_{i,j,+}$ as $n \to \infty$. If moreover $\mathbf{\Gamma}_+$ is positive definite, $n^{1/2}(\mathbf{S}^{(n)}_+ - \mathbf{m}_+)$ is asymptotically multinormal, with mean zero and covariance matrix $\mathbf{\Gamma}_+$.

Proof

The proof is exactly the same as in Tran (1990) theorem 3.2.

REFERENCES

Chernoff, H. and Savage, I.R. (1958) Asymptotic normality and efficiency of certain nonparametric test statistics. *Annals of Mathematical Statistics*, **29**, 972–994.

Dufour, J.-M. (1981) Rank tests for serial dependence. *Journal of Time Series Analysis*, **2**, 117–128.

Goodman, L.A. (1958) Simplified runs tests and likelihood ratio tests for Markov chains. *Biometrika*, **45**, 181–197.
Hallin, M., Ingenbleek, J.-Fr. and Puri, M.L. (1985) Linear serial rank tests for randomness against ARMA alternatives. *Annals of Statistics*, **13**, 1156–1181.
Hallin, M., Ingenbleek, J.-Fr. and Puri, M.L. (1987) Linear and quadratic serial rank tests for randomness against serial dependence. *Journal of Time Series Analysis*, **8**, 409–424.
Hallin, M. and Puri, M.L. (1988) Optimal rank-based procedures for time series analysis: testing an ARMA model against other ARMA models. *Annals of Statistics*, **16**, 402–432.
Hallin, M. and Puri, M.L. (1991a) Time series analysis via rank-order theory: signed rank tests for ARMA models. *Journal of Multivariate Analysis*, **39**, 1–29.
Hallin, M. and Puri, M.L. (1991b) *Aligned Rank Tests for Linear Models with Autocorrelated Error Terms*, Technical Report, Department of Mathematics, Indiana University, Bloomington.
Harel, M. and Puri, M.L. (1990a) Weak convergence of serial rank statistics under dependence with applications in time series and Markov processes. *Annals of Probability*, **18**, 1361–1387.
Harel, M. and Puri, M.L. (1990b) Weak convergence of serial rank statistics with unbounded scores and regression constants under mixing conditions. *Journal of Statistical Planning and Inference*, **25**, 163–186.
Rosenblatt, M. (1956) A central limit theorem and a strong mixing condition. *Proceedings of the National Academy of Sciences USA*, **42**, 43–47.
Ruymgaart, Fr.H., Shorack, G.R. and van Zwet, W.R. (1972) Asymptotic normality of nonparametric tests for independence. *Annals of Mathematical Statistics*, **43**, 1122–1135.
Tran, L.T. (1990) Rank statistics for serial dependence. *Journal of Statistical Planning and Inference*, **24**, 215–232.
Wald, A. and Wolfowitz, J. (1943) An exact test for randomness in the nonparametric case based on serial correlation. *Annals of Mathematical Statistics*, **14**, 378–388.

Part Five

Nonlinear and Non-Gaussian Time Series Models

18

Non-Gaussian characteristics of exponential autoregressive processes

T. Ozaki

18.1 INTRODUCTION

It is more than ten years since the excellent time series book by Priestley was published (Priestley, 1981). That book, which ended by describing some interesting nonlinear time series models developed in late 1970s, was further supplemented by his later work (Priestley, 1988). As mentioned there, most nonlinear models of the late 1970s such as exponential AR models, threshold AR models and bilinear models, were introduced to answer such pragmatic concerns as (a) Can the model give a better prediction than linear models?; (b) Can we estimate the model from the data?; and (c) Can we produce a similar series to the original data by simulating the estimated model? A model which cannot be estimated from data, or one which cannot reproduce a similar series by simulation is certainly not interesting to statisticians although it may still interest some mathematicians. Nonlinear models which cannot give a better prediction performance than linear models are also unattractive and meaningless for applied time series analysts. Considering these three criteria, quite a lot of progress has been made in nonlinear time series modelling over the last ten years, and several books on nonlinear time series models have been published. However, as far as the exponential AR model is concerned, it must be admitted, to the author's regret as one of the originators, that the model, in spite of its potential, has not been sufficiently exploited either by applied time series analysts or by theoretical time series analysts. On the occasion of celebrating Professor Priestley's 60th birthday, I would like to reconsider and discuss the ExpAR model as a general nonlinear time series model by adding, to the above three criteria, a fourth question, i.e. Can the model characterize the non-Gaussian characteristics of the series?, which, I hope, may contribute to future developments of nonlinear time series analysis both in theory and practice.

18.2 EXPAR PROCESSES AND MARGINAL DISTRIBUTIONS

In this section we will see how the non-Gaussian marginal distribution of an ExpAR process is characterized by the first order ExpAR model,

$$x_{t+1} = \{\phi_1 + \phi_2 \exp(-\gamma x_t^2)\} x_t + n_{t+1}. \tag{18.1}$$

In (18.1), γ is some scaling constant and n_{t+1} is Gaussian white noise with variance σ^2.

18.2.1 Histograms

One of the simplest and most natural ways of checking the non-Gaussianity of time series is to check its histogram. Many nonlinear time series models have been introduced, but it is not known how the parameters of these models affect the shape of the distribution of the series. The ExpAR model is an exceptional example whose parameters clearly explain what kind of marginal distribution the generated time series is going to have.

Figures 18.1 (a)–(c) show the histograms of three sets of time series, each with 80 000 observations, generated from the following three ExpAR models with $\sigma^2 = 1$,

$$x_{t+1} = \{1 - 0.2 \exp(-x_t^2)\} x_t + n_{t+1} \tag{18.2}$$

$$x_{t+1} = \{0.8 + 0.2 \exp(-x_t^2)\} x_t + n_{t+1} \tag{18.3}$$

$$x_{t+1} = \{0.8 + 0.4 \exp(-x_t^2)\} x_t + n_{t+1} \tag{18.4}$$

The three histograms show three different non-Gaussian characteristics of the density functions. The first histogram clearly shows a distribution with long heavy tails like a stable distribution. This type of histogram is generated from any ExpAR model (18.1) with $\phi_1 = 1$ and $-1 < \phi_2 < 0$. The second histogram shows short light tails and a fairly concentrated centre, which is a characteristic of exponential family distributions. This type of histogram is generated from any ExpAR model with $\phi_1 + \phi_2 = 1, 0 < \phi_1 < 1$ and $0 < \phi_2 < 1$. The third histogram shows a bimodal distribution with light and short tails, which is also a characteristic of some distributions in the exponential family. This type of histogram is generated from any ExpAR model with $\phi_1 + \phi_2 > 1$, $0 < \phi_1 < 1$ and $0 < \phi_2 < 1$.

These examples of ExpAR models clearly show that we can generate time series with various types of marginal distributions by controlling the two parameters, ϕ_1 and ϕ_2, of the ExpAR model (18.1)

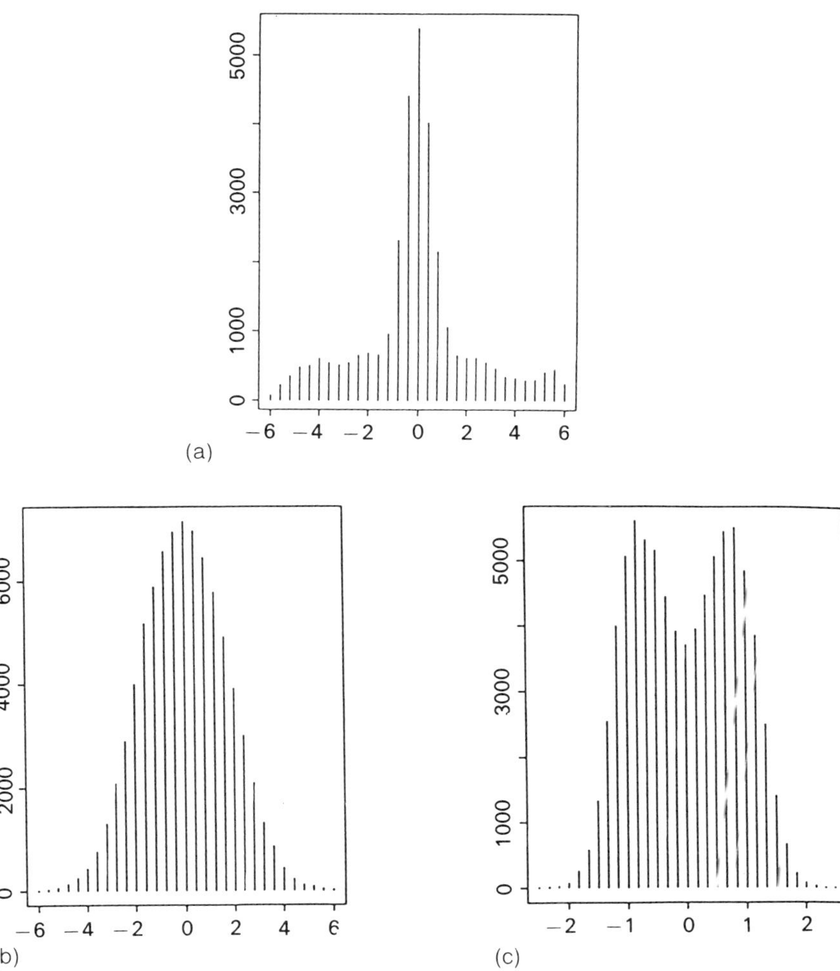

Figure 18.1. Histograms of the series $x_1, x_2, \ldots, x_{80-000}$: (a) of the model (18.2); (b) of the model (18.3); (c) of the model (18.4).

18.2.2 Exponential family and stochastic dynamical systems

The second and third histograms in the previous section imply that the marginal distribution of an ExpAR process may belong to the exponential family. Actually, 'Exp'AR models and the 'exponential' family of distributions are not only related phonetically. To show this we need to establish, first, the relationship between the exponential family and a class of diffusion processes. It was Wong (1963) who first pointed out a close relationship between a certain class of distributions and some class of diffusion processes. He pointed out that for any density function $W(x)$ belonging to the Pearson system,

$$\frac{\mathrm{d}W(x)}{\mathrm{d}x} = \frac{f_0 + f_1 x}{g_0 + g_1 x + g_2 x^2} W(x), \tag{18.5}$$

it is possible to give a diffusion process whose marginal density is $W(x)$. The result was extended (Ozaki, 1985a) to any density function $W(x)$ defined by proper analytic functions $f(x)$ and $g(x)$ in a distribution system given by,

$$\frac{\mathrm{d}W(x)}{\mathrm{d}x} = \frac{f(x)}{g(x)} W(x). \tag{18.6}$$

The corresponding Markov diffusion process is given (Ozaki, 1985a, 1992) by the following Fokker–Planck equation,

$$\frac{\partial p}{\partial t} = -\frac{\partial}{\partial x}\left[\left\{f(x) + \frac{\partial g(x)}{\partial x}\right\} p\right] + \frac{1}{2}\frac{\partial^2}{\partial x^2}[2g(x)p]. \tag{18.7}$$

When $g(x)$ is a constant and $f(x)$ is a polynomial, density functions $W(x)$ defined by (18.6) constitute the exponential family of distributions. On the other hand, the diffusion process in (18.7) has, when $g(x)$ is a constant g_0, the stochastic dynamical system representation,

$$\dot{x}(t) = f(x(t)) + n(t), \tag{18.8}$$

where $n(t)$ is a Gaussian white noise of variance $2g_0$. Thus, from a diffusion process $x(t)$, defined by (18.7), we have its marginal density function, $W(x)$, which satisfies $\mathrm{d}W(x)/\mathrm{d}x = \{f(x)/g_0\}W(x)$. Since a distribution of the exponential family is completely characterized by a finite number of its moments, the marginal distribution of the process $x(t)$ of (18.8) is also characterized by a finite number of its moments. We can see this in the following example. Let

$$\dot{x}(t) = ax(t) + bx(t)^3 + cx(t)^5 + n(t), \tag{18.9}$$

where $n(t)$ is a Gaussian white noise of variance σ^2.

The marginal density function $W(x)$ satisfies the distribution system

$$\frac{\mathrm{d}W(x)}{\mathrm{d}x} = \frac{(ax + bx^3 + cx^5)}{\frac{\sigma^2}{2}} W(x),$$

and the density function is given by

$$W(x) = W_0 \exp\left(\frac{\frac{a}{2}x^2 + \frac{b}{4}x^4 + \frac{c}{6}x^6}{2\sigma^2}\right), \tag{18.10}$$

where W_0 is a normalizing constant and σ^2 is the variance of the white noise $n(t)$.

Since $W(x)$ satisfies the relation

$$\frac{d}{dx}\{x^{2k-1}W(x)\} = (2k-1)x^{2k-2}W(x) + \left[\frac{2\{ax^{2k} + bx^{2k+2} + cx^{2k+4}\}}{\sigma^2}\right]W(x), \tag{18.11}$$

we have, for $k = 1, 2, \ldots$, the relations,

$$(2k-1)\mu_{2k-2} + \left(\frac{2}{\sigma^2}\right)(a\mu_{2k} + b\mu_{2k+2} + c\mu_{2k+4}) = 0. \tag{18.12}$$

The relations (18.12) can also be derived from the stochastic dynamical system (18.9). From the dynamical system relation we have

$$\frac{d}{dt}\left(\frac{x^{2k}}{2k}\right) = x^{2k-1}\{ax(t) + bx(t)^3 + cx(t)^5\} + \frac{1}{2}\{(2k-1)x^{2k-2}\}n(t)^2 + x^{2k-1}n(t) \tag{18.13}$$

for $k = 1, 2, \ldots$. This is because for any analytic function $F(\cdot)$, it holds that

$$F(x + \Delta x)) = F(x(t)) + F'(x(t))\Delta x(t) + \tfrac{1}{2}F''(x(t))[\Delta x(t)]^2 + \ldots,$$

and by Ito's stochastic calculus this implies that

$$\begin{aligned} F(x(t + \Delta t)) = F(x(t)) + F'(x(t))[\{ax(t) + bx(t)^3 + cx(t)^5\}\Delta t \\ + \Delta w(t)] + \tfrac{1}{2}F''(x(t))\Delta w(t)^2 + \ldots, \end{aligned} \tag{18.14}$$

where $w(t)$ is a Brownian motion process such that

$$\frac{\Delta w(t)}{(\Delta t)^{1/2}} \to n(t) \quad \text{for} \quad \Delta t \to 0.$$

(18.13) is obtained from (18.14) by using $F(x) = x^{2k}/2k$ and letting $\Delta t \to 0$. Taking the expectation of both sides of the equation (18.13), we have, for $k = 1, 2, \ldots$,

$$\begin{aligned} 0 &= \frac{d}{dt}E\left[\frac{x^{2k}}{2k}\right] \\ &= E[ax(t)^{2k} + bx(t)^{2k+2} + cx(t)^{2k+4}] + E\left[\left(\frac{2k-1}{2}\right)x^{2k-2}n(t)^2\right] \\ &\quad + E[x^{2k-1}n(t)] \\ &= a\mu_{2k} + b\mu_{2k+2} + c\mu_{2k+4} + \frac{(2k-1)\sigma^2}{2}\mu_{2k-2}. \end{aligned} \tag{18.15}$$

Since the process $x(t)$ is symmetric about $x(t) = 0$, μ_0 and all the odd order moments are zero. The equations show that μ_2, μ_4 and μ_6 are sufficient statistics since all the higher order moments, $\mu_8, \mu_{10}, \mu_{12}, \ldots$ may be calculated recursively from relation (18.12) and μ_2, μ_4 and μ_6. Process $x(t)$ of (9) is completely specified by the parameters a, b, c and σ^2, the marginal distribution, and so μ_2, μ_4 and μ_6, could also be explicitly written using those parameters. In spite of the importance of this type of distribution in nonlinear and non-Gaussian time series analysis, not much work has been devoted to the study of the exponential family, except for a general treatment (Barndorff–Nielsen, 1978), and even the moment generating function of distribution function (18.10) does not seem to have been discussed in the statistics literature.

18.2.3 Stochastic dynamical systems and ExpAR models

The general nonlinear stochastic dynamical system

$$\dot{x}(t) = f(x(t)\,|\,\underline{a}) + n(t) \tag{18.16}$$

and its discrete time approximate model have been studied by applied probabilists, physicists, geneticists and engineers. The main purpose of those studies has been to simulate model (18.16). It has been widely known in this field that when $f(x)$ is nonlinear, the simulation of the discretized model often ends up in a computational explosion when Δt is not small enough. This often happens even though the original continuous time process is known to be stationary. For example, $x(t)$ of

$$\dot{x}(t) = -x^3 + n(t) \tag{18.17}$$

is known to be stationary and has a stationary marginal distribution,

$$W(x) = W_0 \exp\left(-\frac{\dfrac{x^4}{4}}{2\sigma^2}\right)$$

where W_0 is a normalizing constant and σ^2 is the variance of the white noise $n(t)$. We can easily see this computational problem by simulating (18.17) with large σ^2 and Δt. To avoid the computational explosion in simulations people often take a small time interval Δt in the discretization.

An inductive use of discretized models of dynamical system (18.16) came from time series analysis in statistics, where researchers are interested in estimating the parameters of the model in order to characterize the non-Gaussian and nonlinear nature of the time series. Here there is a problem because we cannot choose both the time interval Δt and the white noise variance σ^2. Both are fixed by the sampling of the time series and it is often the case that the noise variance is quite large and the sampling interval is not very small. Naturally, special attention was paid to the computational stability of the discretization scheme used to derive the model with fairly large Δt and with very large σ^2. It was reported in Ozaki and Oda (1978) that when the Euler type discretization scheme is employed, simulations of the estimated model computationally explode very easily, even though the process defined by the model of the Euler type discretization scheme converges, in probability, to the original continuous time process $x(t)$ on the finite time interval $[0, T]$ when $\Delta t \to 0$. In fact, not only the Euler type scheme but also more sophisticated schemes, such as the Heun or Runge–Kutta types, are inappropriate for inductive use, even though all three types are consistent. This is because they all yield polynomial type discrete time models when $f(x)$ in (18.16) is of polynomial type as for example in (18.17) and such discrete time models driven by Gaussian white noise are known to be transient and computationally explosive in a finite time interval with probability one.

Ozaki (1985a, b, 1986, 1990a, b, 1992) has introduced a scheme (called the local linearization scheme) which yields a model which has both computational stability and consistency properties. From a nonlinear stochastic dynamical system (18.16), the local linearization yields the following nonlinear time series model:

$$
\begin{aligned}
x_{t+\Delta t} &= A(x_t|\underline{a})x_t + B(x_t|\underline{a})n_{t+\Delta t} \\
A(x_t|\underline{a}) &= \exp\{K(x_t|\underline{a})\Delta t\} \\
B(x_t|\underline{a}) &= \left(\frac{\exp\{2K(x_t|\underline{a})\Delta t\} - 1}{2K(x_t|\underline{a})}\right)^{1/2} \\
K(x_t|\underline{a}) &= \frac{1}{\Delta t}\log\{1 + J_t^{-1}(e^{J_t\Delta t} - 1)f(x_t|\underline{a})/x_t\} \\
J_t &= \left(\frac{\partial f(x|\underline{a})}{\partial x}\right)_{x=x_t}.
\end{aligned}
\tag{18.18}
$$

Since it holds that

$$A(x_t|\underline{a})x_t = x_t + \Delta t f(x_t|\underline{a}) + o(\Delta t)$$

and

$$B(x_t|\underline{a})x_t = (\Delta t + o(\Delta t)^{1/2},$$

the locally linearized model (18.18) is asymptotically equivalent to the Euler type model and shares the same asymptotic properties with regard to Δt.

When the original model (18.16) is specified, we can check if the locally linearized model (18.18) is a stationary Markov chain by using Tweedie's (1975) theorem. Ozaki (1985a) gave sufficient conditions for the function $f(x)$ in (18.16) to produce a stationary locally linearized model (18.17). The conditions are not very strict and are satisfied by most interesting nonlinear functions seen in defining non-Gaussian diffusion processes.

An interesting point in the local linearization is that the original continuous time function $f(x)$ can be recovered from the locally linearized discrete time model by a numerical method (see Ozaki (1985b, 1992)). Of course, since approximation is involved in the discretization, the recovered function is an approximation to the original $f(x)$. As is usual for any function approximation, the accuracy of the approximation depends on how it is approximated—i.e. in the local linearization case, on the time interval Δt of the discretization. Naturally the difference between the two functions becomes zero when $\Delta t \to 0$.

Another interesting point in the local linearization is that it yields, from the stochastic dynamical system model (18.17), the following $A(x_t)$,

$$A(x_t) = \tfrac{2}{3} + \tfrac{1}{3}\exp(-3x_t^2 \Delta t)$$

Since $B(x_t)$ is almost constant and is given by $B(x_t) \approx (\Delta t)^{1/2}$, the discretized model is almost equivalent to an ExpAR model. From the stochastic dynamical system,

$$\dot{x} = x - x^3 + n(t). \tag{18.19}$$

It yields

$$A(x_t) = \frac{-2x_t^2}{1-3x_t^2} + \frac{1-x_t^2}{1-3x_t^2}\exp\{(1-3x_t^2)\Delta t\}.$$

The function $A(x_t)$ is a positive valued smooth function and goes to a constant, 2/3, when x_t goes to infinity, and $1 < A(x_t)$ when $-1 < x_t < 1$. The shape of the function $A(x_t)$ is very similar to the $\phi(\cdot)$ function of the ExpAR model (18.3). Model (18.19) is known to have a bimodal distribution similar to ExpAR model (18.3). Looking at the above examples we can see that the local linearization scheme yields, from the class of general nonlinear stochastic dynamical systems,

$$\dot{x}(t) = a_1 x(t) + a_2 x(t)^2 + \cdots a_{2k+1} x(t)^{2k+1} + n(t), \quad (k = 1, 2, \ldots,) \tag{18.20}$$

a smooth and positive valued function $A(x_t)$ which approaches the constant values $2k/(2k+1)$ when $x_t \to \infty$. This type of function is exactly what Ozaki (1985a) considered in his extended version of the ExpAR model, i.e.

$$x_{t+1} = \phi(x_t)x_t + n_{t+1}, \tag{18.21}$$

where

$$\phi(x_t) = \phi_1 + \{\phi_2 + \phi_3 x_t + \cdots + \phi_r x_t^{r-2}\}\exp(-\gamma x_t^2).$$

By taking r of the $\phi(\cdot)$ function sufficiently large we can approximate any smooth function $A(x_t)$ which converges to a constant for $x_t \to \infty$. It means that the discrete time process obtained from the diffusion process (18.20) can be very closely approximated using the extended ExpAR model (18.21). It also means that if the $\phi(\cdot)$ function of an extended ExpAR model is given, it may be closely approximated by $A(x_t)$ which may be found from a function of the form

$$f(x) = a_1 x(t) + a_2 x(t)^2 + \cdots + a_{2k+1} x(t)^{2k+1},$$

and we can characterize the non-Gaussian behaviour of the series using the nice relationship between the diffusion process defined by (18.20) and the exponential family. Further work in this field, both empirical and theoretical, will certainly be worth doing in the future.

18.3 ARCH MODELS AND EXPAR MODELS

In nonlinear time series analysis, it is sometime argued that nonlinear models driven by homogeneous Gaussian white noise, like ExpAR models, are not appropriate for analysing time series whose marginal distribution has heavy tails. Such time series are found in financial time series data. For the analysis of such time series data, researchers often use models with Gaussian white noise whose variance is dependent on the value of the series. A well-known example is the ARCH model (Engel, 1982). One simple example of an ARCH model is given by

$$x_{t+1} = \phi x_t + (\alpha + \beta x_t^2)^{1/2} n_{t+1}. \tag{18.22}$$

This type of model looks very different from an ExpAR model. However, we have already seen in section 18.2 that ExpAR models can also generate time series having a marginal distribution with heavy tails. Further investigation of these two different types of nonlinear time series model reveals the interesting role of instantaneous transformations in non-Gaussian time series analysis.

We know that the model (18.22) can be obtained by discretizing a continuous time diffusion process model of the type

$$\dot{x} = -ax + (\alpha + \beta x^2)^{1/2} n(t). \tag{18.23}$$

Here we assume the Stratonovich type stochastic calculus. The process $x(t)$ defined by the model has a stable marginal distribution (see Guegan, 1992). The model (18.23) can be transformed into the form

$$\dot{y} = \frac{-a}{\beta^{1/2}} \tanh(\alpha^{1/2} y) + n(t), \tag{18.24}$$

by using the transformation

$$y = \frac{1}{\beta^{1/2}} \sinh^{-1}\left[\left(\frac{\beta}{\alpha}\right)^{1/2} x\right].$$

It is shown in Ozaki (1992) that the model (18.24) yields a discrete time model

$$x_{t+1} = A(x_t)x_t + B(x_t)n_{t+1},$$

where $B(x_t)$ is almost constant and equals $(\Delta t)^{1/2}$, while $A(x_t)$ is $0 < A(x_t) < 1$ and is a smooth function of x_t approaching 1 for $|x_t| \to \infty$. This means that the ExpAR model combined with an instantaneous variable transformation, i.e.

$$y_{t+1} = \{1 - \pi \exp(-\gamma y_t^2)\} y_t + n_{t+1}$$
$$x_{t+1} = \left(\frac{\alpha}{\beta}\right)^{1/2} \sin h(\beta^{1/2} y_{t+1}),$$

is a reasonable model for a heteroscedastic time series if the parameters α, β, γ and π are properly chosen. Thus ExpAR models combined with instantaneous variable transformations can characterize quite large varieties of non-Gaussian processes.

18.4 EXPAR(p) MODELS AND NONLINEAR STATE-SPACE REPRESENTATIONS

We have so far seen the non-Gaussian properties of ExpAR models of lag order 1. Such models are not appropriate for fitting non-Gaussian oscillatory time series whose spectrum has peaks. One example of non-Gaussian oscillatory time series is data generated from

$$\ddot{x}(t) + a\dot{x}(t) + b(x(t))x(t) = n(t). \tag{18.25}$$

Model (18.25) is known to be a model of hard spring type nonlinear oscillations if $b(x) = b_1 + b_2 x^2$ with $b_2 > 0$ and of soft spring type nonlinear oscillations if $b_2 < 0$. $x(t)$ is known to have the following non-Gaussian marginal distribution (Caughey, 1963; Ozaki, 1990b),

$$p(x) = p_0 \exp\left\{\frac{-2a\int^x b(\xi)\xi \, d\xi}{\sigma^2}\right\},$$

where σ^2 is the variance of the Gaussian white noise $n(t)$ and p_0 is the normalizing constant. In linear Gaussian cases, i.e. when $b_2 = 0$, AR(p) models, or ARMA(p, q) models, with $p > 1$ are usually considered for the analysis of such time series. To analyse non-Gaussian and, at the same time, oscillating time series, the second order ExpAR(2) model

$$x_{t+1} = \{\phi_{1,1} + \phi_{1,2} \exp(-\gamma x_t^2)\} x_t + \{\phi_{2,1} + \phi_{2,2} \exp(-\gamma x_t^2)\} x_{t-1} + n_{t+1},$$

or the pth lag order ExpAR(p) model

$$x_{t+1} = \{\phi_{1,1} + \phi_{1,2} \exp(-\gamma x_t^2)\} x_t + \cdots$$
$$+ \{\phi_{p,1} + \phi_{p,2} \exp(-\gamma x_t^2)\} x_{t-p+1} + n_{t+1},$$

have been introduced (Ozaki & Oda (1978), Haggan & Ozaki (1981)). Although ExpAR(p) models are useful for explaining many interesting nonlinear phenomena, they are not, in a sense, general enough; they share the same defects as the linear AR models, i.e., they are not appropriate for time series whose spectrum has sharp troughs. It is widely recognized in applied time series analysis that ARMA(p, q) models are much more efficient to model such series parsimoniously. Also, it is known that a discretization of the linear vibration model

$$\ddot{x}(t) + a\dot{x}(t) + bx(t) = n(t)$$

yields the ARMA(2, 1) model,

$$x_{t+1} = \phi_1 x_t + \phi_2 x_{t-1} + \theta n_t + n_{t+1}$$

where ϕ_1, ϕ_2 and θ are given as functions of a and b (Pandit and Wu, 1975). Naturally, a generalization of the ExpAR(2) model has been considered as a discrete time version for the nonlinear vibration model (18.25) so that the generalized model includes the ARMA(2, 1) model as a special case. Ozaki (1980) proposed the ExpARMA(2, 1) model

$$x_{t+1} = \{\phi_1 + \pi_1(\exp - \gamma x_t^2)\}x_t + \{\phi_2 + \pi_2 \exp(-\gamma x_t^2)\}x_{t-1} \\ + \{\theta_1 + \theta_2 \exp(-\gamma x_t^2)\}n_t + n_{t+1}$$

as a discrete version of the nonlinear vibration model (18.25), and the ExpARMA(p, $p-1$) model as a discrete version of the nonlinear stochastic differential equation model

$$x^{(p)}(t) + a_1(x(t))x^{(p-1)}(t) + \cdots + a_{p-1}(x(t))\dot{x}(t) + a_p(x(t))x(t) = n(t). \quad (18.26)$$

However, the problem of these generalizations is that the maximum likelihood method does not work properly in practice. For the estimation of the parameters of these models we need to use a nonlinear optimization method to maximize the likelihood, as in the linear ARMA cases, but in most cases the Hessian becomes almost singular. This may be because the model has too many parameters, twice as many as in the linear case.

In order to eliminate these numerical problems we need to generalize ExpAR models in a compact and parametrically parsimonious way. A useful idea arises from checking the nonlinear characteristics of the model introduced from the continuous time model (18.25) by the local linearization (Ozaki, 1986 and 1989). The nonlinear random vibration model (18.25) can be written down as the two-dimensional stochastic dynamical system

$$\dot{\underline{z}} = \underline{f}(\underline{z}|\underline{\alpha}) + \underline{n}(t), \quad (18.27)$$

where $\underline{n}(t) = \{n(t), 0\}'$, $z = (\dot{x}, x)'$ and $f(z|\underline{\alpha}) = \{-a(x|\underline{\alpha})\dot{x}, b(x|\underline{\alpha})x\}'$. The local linearization of (18.27) is obtained in the same way as in the scalar case as follows:

$$\underline{z}_{t+\Delta t} = \mathbf{A}(\underline{z}_t|\underline{\alpha})\underline{z}_t + \mathbf{B}(\underline{z}_t|\underline{\alpha})\underline{n}_{t+\Delta t}$$

$$\mathbf{A}(\underline{z}_t|\underline{\alpha}) = \text{Exp}\{\mathbf{K}(\underline{z}_t|\underline{\alpha})\Delta t\}$$

$$\mathbf{K}(\underline{z}_t|\underline{\alpha}) = \frac{1}{\Delta t}\text{Log}\{1 + J_t^{-1}(e^{J_t\Delta t} - 1)\mathbf{F}(\underline{z}_t|\underline{\alpha})\}$$

$$J_t = \left(\frac{\partial f(\underline{z}|\underline{\alpha})}{\partial \underline{z}}\right)_{\underline{z}=\underline{z}_1}. \quad (18.28)$$

$\mathbf{F}(\underline{z}|\underline{\alpha})$ is a matrix which satisfies $\mathbf{F}(\underline{z}_t|\underline{\alpha})z_t = f(\underline{z}_t|\underline{\alpha})$, Exp(·) is the matrix exponential function and Log(·) is the matrix logarithmic function. The elements of the matrix $\mathbf{B}(\underline{z}_t|\underline{\alpha})$ are explicitly given as functions of the eigenvalues of the matrix $\mathbf{K}(\underline{z}_t|\underline{\alpha})$ (see Appendix). In many applications we usually have observation records only of $x(t)$ and the record of $\mathrm{d}x/\mathrm{d}t$ is not available. Then we have a nonlinear state space representation model

$$\begin{aligned} \underline{z}_{t+\Delta t} &= \mathbf{A}(\underline{z}_t|\underline{\alpha})\underline{z}_t + \mathbf{B}(\underline{z}_t|\underline{\alpha})\underline{n}_{t+\Delta t} \\ x_t &= \mathbf{H}\underline{z}_t, \end{aligned}$$

where $\mathbf{H} = (0, 1)$. Each element of the matrix $\mathbf{A}(\underline{z}_t|\underline{\alpha})$ and $\mathbf{B}(\underline{z}_t|\underline{\alpha})$ for the model (18.28) is given as a function of the parameters of the original continuous model (18.25). By checking the elements of the matrix $\mathbf{A}(\underline{z}_t|\underline{\alpha})$ we know (see Figures 4.2 and 4.3 of Ozaki, 1986) that the (1,1)th and (1,2)th elements, $a_{1,1}$ and $a_{1,2}$, of the matrix $\mathbf{A}(\underline{z}_t|\underline{\alpha})$ are smooth functions of x_t, while the (2,1)th and (2,2)th elements are almost constant, i.e. $a_{2,1} = \Delta t$ and $a_{2,2} = 1$. This suggests that the state space model

$$\begin{aligned} \underline{z}_{t+1} &= \mathbf{F}\underline{z}_t + \mathbf{G}n_{t+1} \\ x_t &= \mathbf{H}\underline{z}_t \end{aligned} \tag{18.29}$$

is a reasonably parsimonious discrete time model for nonlinear random vibrations, where

$$\mathbf{F} = \begin{pmatrix} f_1(x_t) & f_2(x_t) \\ c & 1 \end{pmatrix}, \qquad \mathbf{G} = \begin{pmatrix} \sigma_1 \\ \sigma_2 \end{pmatrix}, \qquad \mathbf{H} = (0, 1),$$

$$f_1(x_t) = \phi_1 + \pi_1 \exp(-\gamma x_t^2) \qquad \text{and} \qquad f_2(x_t) = \phi_2 + \pi_2 \exp(-\gamma x_t^2).$$

n_{t+1} is a Gaussian white noise of variance σ^2. The two-dimensional state vector $\underline{z}_t$ is given by $\underline{z}_t = (\xi_t, x_t)'$, where the conceptual unobserved variable ξ_t is introduced to represent the velocity of the vibrations. Thus x_{t+1} of the model (18.29) is given by $x_{t+1} = x_t + c\xi_t$. The parameter c corresponds to the time interval of the time series data, which is fixed when the data is sampled, but we do not know which scaling unit we should use in the model (18.29) in real data analysis. When $\pi_1 = 0$ and $\pi_2 = 0$, obviously the model (18.29) is equivalent to the ARMA(2, 1) model

$$x_{t+1} = \lambda_1 x_t + \lambda_2 x_{t-1} + \mathbf{H}(\mathbf{F} - \lambda_1 I)\mathbf{G}n_t + \mathbf{HG}n_{t+1},$$

where λ_1 and λ_2 are the characteristic coefficients of the matrix $\mathbf{F}$, satisfying

$$\mathbf{F}^2 - \lambda_1 \mathbf{F} - \lambda_2 I = 0.$$

More generally, the eigenvalues of the matrix $\mathbf{F}$ are functions of x_t. Then the model yields a van der Pol type nonlinear oscillator if ϕ_1, ϕ_2, π_1 and π_2 are such that the absolute value of the roots of

$$\Lambda^2 - (1 + \phi_1 + \pi_1)\Lambda + \{(\phi_1 + \pi_1) - c(\phi_2 + \pi_2)\} = 0$$

are less than one, and the absolute value of the roots of

$$\Lambda^2 - (1 + \phi_1)\Lambda + (\phi_1 - c\phi_2) = 0$$

are greater than one. One such example is the model (18.29) with the **F**-matrix

$$\mathbf{F} = \begin{pmatrix} 0.94 + 0.26\exp(-x_t^2) & -0.2 \\ 0.1 & 1 \end{pmatrix}. \tag{18.30}$$

Figure 18.2 shows the typical limit cyclic behaviour obtained by simulating model (18.30) with an initial value $\underline{z}_t = (\xi_t, x_t)' = (0.001, 0.001)'$ and with zero input noise.

The idea is immediately extendible to the pth order nonlinear stochastic differential equation model (18.26), for which we have the p-dimensional state space representation model

$$\begin{aligned} \underline{z}_{t+1} &= \mathbf{F}\underline{z}_t + \mathbf{G}\underline{n}_{t+1} \\ x_t &= \mathbf{H}\underline{z}_t, \end{aligned} \tag{18.31}$$

where

$$\mathbf{F} = \begin{pmatrix} f_1(x_t) & . & . & . & .. & f_p(x_t) \\ c & 1 & 0 & 0 & .. & 0 \\ 0 & c & 1 & 0 & .. & 0 \\ . & . & . & . & .. & . \\ 0 & . & . & . & c & 1 \end{pmatrix},$$

$$\mathbf{G} = (\sigma_1, \sigma_2, \ldots, \sigma_p),$$
$$\mathbf{H} = (0, \ldots, 0, 1),$$
$$f_i(x_t) = \phi_i + \pi_i \exp(-\gamma x_t^2) \quad \text{for} \quad i = 1, 2, \ldots, p.$$

The state vector $\underline{z}_t$ is given by $\underline{z}_t = (\xi_p, \xi_{p-1}, \ldots, \xi_2, x_1)'_t$, where ξ_i $(i = 2, 3, \ldots, p)$ stands for the $(i-1)$th order differential of $x(t)$, i.e. $\mathrm{d}^{i-1}x/\mathrm{d}t^{i-1}$ at time point t.

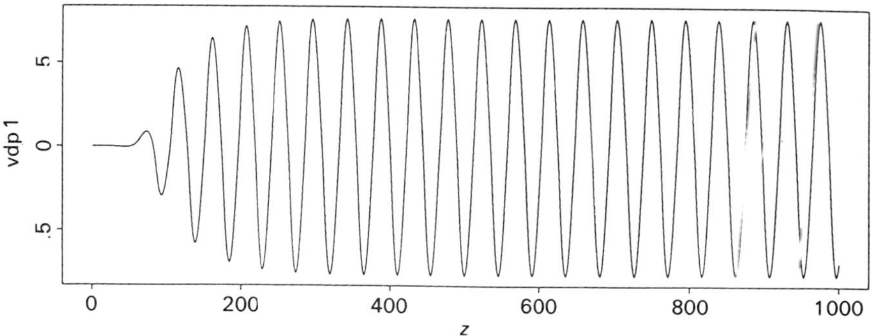

Figure 18.2. Limit cycles simulated by the model (18.29); (**F** is given by (18.30)).

Estimating such a model as (18.31) from data, $x_t (t = 1, 2, \ldots, N)$, may look difficult since the state variables $\xi_p, \xi_{p-1}, \ldots$ and ξ_2 are not observed. However, if we use nonlinear Kalman filtering we can transform the data into prediction error sequences, i.e. Gaussian white noise innovations. This allows us to write down the likelihood of the non-Gaussian distributed series x_t by using the Gaussian likelihood of the innovations and by maximizing the likelihood we can obtain the maximum likelihood estimates of the model parameters. Asymptotic properties, such as consistency and normality, of the estimates have been proved (Ljung, 1976; Caines, 1978). Effectiveness of the estimation method in practice can be checked using simulation data. Numerical studies of the estimation method will be given in a future paper.

18.5 CONCLUSION

It has been shown that a simple first lag order ExpAR model can generate time series with various types of marginal distributions by varying the two model coefficients. A close relationship between ExpAR models and the exponential family of distributions is pointed out. It is shown that for a given distribution of the exponential family we can have a special type of diffusion process whose diffusion coefficient is constant and whose marginal distribution is equivalent to the given distribution. Any such stationary homogeneous diffusion process with constant diffusion coefficient can be closely approximated by a discrete time nonlinear autoregressive model whose autoregressive coefficient is a smooth function of the process and approaches a constant for $|x_t| \to \infty$. This means that for a given distribution of the exponential family we can introduce an ExpAR model whose marginal distribution is very close to the given distribution.

For the analysis of time series with a stable marginal distribution, heteroscedastic nonlinear time series models such as ARCH models are used. It has shown that an ExpAR model combined with an instantaneous variable transformation is useful for the modelling of such heteroscedastic time series.

A nonlinear state space model was introduced as an extension of the ExpAR model. The model-set defined by the nonlinear state space representation includes ARMA$(p, p-1)$ models as a subset. Thus it can be regarded as a natural nonlinear extension of linear Markovian models for time series.

APPENDIX

Matrix $\mathbf{B}(\underline{z}_t | \underline{\alpha})$ in (18.28).

The variance – covariance matrix of the discrete time white noise

$$\underline{n}_{t+\Delta t} = \int_t^{t+\Delta t} \exp\{\mathbf{K}_t(t + \Delta t - u)\} n(u) \mathrm{d}u$$

of the model (18.27) is given by

$$E\left[\int_t^{t+\Delta t}\exp\{\mathbf{K}_t(t+\Delta t-u)\}n(u)\mathrm{d}u\int_t^{t+\Delta t}n(u)\exp\{\mathbf{K}_t'(t+\Delta t-u)\}\mathrm{d}u\right]$$

$$=\int_t^{t+\Delta t}\exp\{\mathbf{K}_t(t+\Delta t-u)\}\mathbf{\Sigma}\exp\{\mathbf{K}_t'(t+\Delta t-u)\}\mathrm{d}u$$

$$=\int_t^{t+\Delta t}\mathbf{V}_t\mathbf{M}\mathbf{V}_t^{-1}\mathbf{\Sigma}(\mathbf{V}_t^{-1})'\mathbf{M}\mathbf{V}_t'\mathrm{d}u$$

$$=\mathbf{V}_t\mathbf{S}\mathbf{V}_t',$$

where

$$\mathbf{M}=\begin{bmatrix}\exp\{\mu_1(t+\Delta t-u)\} & 0\\ 0 & \exp\{\mu_2(t+\Delta t-u)\}\end{bmatrix}.$$

$\mathbf{V}_t$ is a matrix which satisfies

$$\mathbf{V}_t^{-1}\mathbf{K}_t\mathbf{V}_t=\begin{bmatrix}\mu_1 & 0\\ 0 & \mu_2\end{bmatrix}\quad\text{and is given by}\quad \mathbf{V}_t=\begin{bmatrix}\mu_1 & \mu_2\\ 1 & 1\end{bmatrix}.$$

$\mathbf{\Sigma}$ is the variance–covariance matrix of the white noise $n(t)$ given by

$$\mathbf{\Sigma}=\begin{bmatrix}\sigma^2 & 0\\ 0 & 0\end{bmatrix}.$$

The matrix $\mathbf{S}$ is

$$\mathbf{S}=\mathbf{V}_t^{-1}\mathbf{\Sigma}(\mathbf{V}_t^{-1})'$$

$$=\begin{bmatrix}e_{11}s & e_{12}s\\ e_{21}s & e_{22}s\end{bmatrix},$$

where $s=\sigma^2/(\mu_1-\mu_2)^2$, $e_{11}=\{\exp(2\mu_1\Delta t)-1\}/(2\mu_1)$, $e_{12}=[\exp\{(\mu_1+\mu_2)\Delta t\}-1]/(\mu_1+\mu_2)$, $e_{21}=e_{12}$ and $e_{22}=\{\exp(2\mu_2\Delta t)-1\}/(2\mu_2)$.

The matrix $\mathbf{S}$ is diagonalized by the unitary matrix $\mathbf{U}_t$,

$$\mathbf{S}=\mathbf{U}_t\begin{bmatrix}\lambda_1 & 0\\ 0 & \lambda_2\end{bmatrix}\mathbf{U}_t',$$

where λ_1 and λ_2 are eigenvalues of the matrix $\mathbf{S}$ and are given as roots of

$$\Lambda^2-(e_{11}s+e_{22}s)\Lambda+e_{11}e_{22}s^2-e_{12}^2s^2=0.$$

The unitary matrix $\mathbf{U}_t$ is given by

$$\mathbf{U}_t=(e_{12}^2s^2+(\lambda_1-e_{11}s)^2)^{-1/2}\begin{pmatrix}e_{12}s & \lambda_2-e_{22}s\\ \lambda_1-e_{11}s & e_{12}s\end{pmatrix}.$$

Using $\mathbf{V}_t$, $\mathbf{U}_t$, λ_1 and λ_2 the matrix $\mathbf{B}(\underline{z}_t|\underline{\alpha})$ of (18.28) is given by

$$\mathbf{B}(\underline{z}_t|\underline{\alpha}) = \mathbf{V}_t\mathbf{U}_t\begin{pmatrix} (\lambda_1)^{1/2} & 0 \\ 0 & (\lambda_2)^{1/2} \end{pmatrix}.$$

REFERENCES

Barndorff-Nielsen, O. (1978) *Information and Exponential Families in Statistical Theory*, Wiley, New York.

Caines, P.E. (1978) Stationary linear and nonlinear system identification and prediction set completeness, *IEEE Transactions on Automatic Control*, **AC-23**, 583–594.

Caughey, T.K. (1963) Derivation and application of the Fokker–Planck equation to discrete nonlinear dynamic systems subjected to white random excitation. *J. Acoust. Soc. Amer.*, **35**, 1683–1692.

Engel, R.F. (1982) Autoregressive Conditional Heteroscedasticity with Estimates of the Variance of United Kingdom Inflation, *Econometrica*, **50**, 987–1007.

Guegan, D. (1992) *A Continuous Time ARCH Model*, Prepublications Mathematiques, **92**, (1), Universite Paris.

Haggan, V. and Ozaki, T. (1981) Modelling random vibrations using an amplitude-dependent autoregressive time series models, *Biometrika*, **68**, 189–196.

Ljung, L. (1976) On the consistency of prediction error identification methods, in (eds) R.K. Mehra and D.G. Lainiotis, *System Identification — Advanced Studies*, Academic Press, New York.

Ozaki, T. (1980) Nonlinear time series models for nonlinear random vibrations. *J. Appl. Probab.*, **17**, 84–93.

Ozaki, T. (1985a) Nonlinear time series models and dynamical systems, in (eds) E.J. Hannan, P. Krishnaiah, M.M. Rao, *Handbook of Statistics*, vol 5, North-Holland, pp. 25–83.

Ozaki, T. (1985b) Statistical identification of storage models with application to stochastic hydrology, *Water Resources Bulletin*, **21**, 663–675.

Ozaki, T. (1986) Local Gaussian modelling of stochastic dynamical systems in the analysis of nonlinear random vibrations, in (eds) J. Goni and M.B. Priestley, *Essays on Time Series and Allied Processes*, Applied Probability Trust, pp. 241–255.

Ozaki, T. (1989) Statistical identification of nonlinear random vibration systems. *J. of Appl. Mechanics*, **56**, 186–191.

Ozaki, T. (1990a) *A Local Linearization Approach to Nonlinear Filtering*, Research Memo, No. 378, Institute of Statistical Mathematics, Tokyo.

Ozaki, T. (1990b) Identification of nonlinearities and non-Gaussianities in time series, to appear in the proceedings of 1990 Summer International Interdisciplinary Workshop of Time Series Analysis at the IMA, July 2–27, 1990, Springer-Verlag.

Ozaki, T. (1992) A bridge between nonlinear time series models and nonlinear stochastic dynamical systems: a local linearization approach. *Statistica Sinica*, **2**, 113–135.

Ozaki, T. and Oda, H. (1978) Nonlinear time series model identification by Akaike's information criterion, in (ed) B. Dubuisson, *Information and Systems*, Pergamon Press, pp. 83–91.

Pandit, S. and Wu, S.M. (1975) Unique estimates of the parameters of a continuous stationary stochastic processes. *Biometrika*, **62**, 497–501.

Priestley, M.B. (1981) *Spectral Analysis and Time Series*, 2 vols, Academic Press, London.

Priestley, M.B. (1988) *Non-linear and Non-stationary Time Series Analysis*, Academic Press, London.

Tweedie, R.L. (1975) Sufficient conditions for ergodicity and stationarity of Markov chains on a general state space. *Stoch. Proc. Their Appl.*, **3**, 385–403.

Wong, E., (1963) The construction of a class of stationary Markoff process, *Proc. Amer. Math. Soc. Symp. Appl. Math*, **16**, 264–276.

19

Bispectrum based checking of linear predictability for time series

G. Terdik and J. Máth

19.1 INTRODUCTION

Finding the best predictor is a long standing problem. Prediction of a stationary time series usually means the following: take the Hilbert space generated by the elements of the series as vectors in L_2, then find the orthogonal projection of an element in the future to the subspace generated by the past of the series. The precise mathematical formulation and the solution of this problem is given by Kolmogorov (1941) in his pioneering work on stationary processes. This predictor is called linear because it is a linear combination of the values from the past. Wiener (1949) gave an independent solution for finding the best linear predictor.

It is a well known fact that the best least squares predictor is the conditional expectation with respect to the past of the process. Wiener himself, collaborating with Kallianpur and Masani, see Masani and Wiener (1959) and Kallianpur (1981), gave a method for finding the best predictor and has shown that under some circumstances the Hilbert space spanned by all the polynomials of the past is the same as the Hilbert space generated by the random variables with second moments, measurable with respect to the σ algebra generated by the past. Interesting particular cases were considered by Nelson and Van Ness (1973) from computational aspects of a quadratic predictor, by Granger and Andersen (1978) for transformed series, and by Hida and Kallianpur (1975) for squared Gaussian Markov processes in continuous time.

In this paper we consider cases not necessarily Gaussian when the linear predictor is as good as the linear and quadratic ones together. The assumption of the best linear predictability is concerned with the bispectrum of the innovation series originated from the best linear predictor. The bilinear realizable Hermite degree-2 process with separable kernel is an example of the situation when, although the process is non-Gaussian, the linear predictor is the best among all possible nonlinear ones.

19.2 QUADRATIC PREDICTOR

Suppose we are given a zero mean time series Y_t stationary to the third order with finite fourth moments. The spectrum and bispectrum are denoted by

$$S_Y(z) = \sum_{k=-\infty}^{\infty} c_Y(k) z^{-k},$$

$$B_Y(z_1, z_2) = \sum_{k,j=-\infty}^{\infty} c_{YY}(k,j) z_1^{-k} z_2^{-j},$$

respectively, where $z = e^{i2\pi\lambda}$, $z_1 = e^{i2\pi\lambda_1}$, $z_2 = e^{i2\pi\lambda_2}$, $c_Y(k) = EY_0 Y_k$ and $c_{YY}(k,j) = EY_0 Y_k Y_j$.

The quadratic predictor of one lag is of the form

$$\hat{Y}_Q(t+1) = \sum_{k=0}^{\infty} a_k Y_{t-k} + \sum_{j,k=0}^{\infty} a_{jk} Y_{t-j} Y_{t-k}, \tag{19.1}$$

and the coefficients $a_k, a_{j,k}$ are chosen such that the mean square error

$$\mathrm{E}|Y_{t+1} - \hat{Y}_Q(t+1)|^2,$$

be minimum.

The question we address is whether the contribution of the quadratic term in (19.1) is significant or not, i.e. whether the linear predictor $\hat{Y}_L(t+1)$ is the same (in mean square sense) as the quadratic one $\hat{Y}_Q(t+1)$.

The construction of the linear predictor is well known (see Priestley, 1981), and based on the spectrum S_Y of the process. One needs only the Szegö assumption, i.e.

$$\int_0^1 \log S_Y(z) \mathrm{d}\lambda > -\infty \tag{19.2}$$

to be fulfilled. Let

$$e_t = Y_t - \hat{Y}_L(t),$$

be the innovation process. Note that under the assumption (19.2), Y_t has moving average representation

$$Y_t = \sum_{k=0}^{\infty} \mathrm{d}_k e_{t-k}.$$

It is well known that if the process Y_t is Gaussian then the conditional expectation of Y_{t+1} with respect to $Y_t, Y_{t-1}, Y_{t-2}, \ldots$ is linear, i.e., $\hat{Y}_L(t)$ is the best predictor. However, if the process Y_t is non-Gaussian then it can happen that the variance of the error process according to the quadratic predictor, $Y_{t+1} - \hat{Y}_Q(t+1)$, is smaller than the variance of the linear innovation process, e_t. Recently it was shown by Terdik and Subba Rao (1989) that the variance of the best linear predictor of bilinear processes driven by Gaussian white noise u_t is greater than the variance of the noise process u_t.

We use Masani and Wiener's (1959) definition of the spectrum of a distribution which is often used in the measure theory.

By the spectrum of a distribution function F on Euclidean q-space we shall mean the set of points $\boldsymbol{x} = (x_1, x_2, \ldots, x_q)$ such that $\mu(I) > 0$ for every open interval I containing $\boldsymbol{x}$, where μ is the Lebesque-Stieltjes measure generated by F.

Now we are in a position to prove the following.

Theorem 19.1

Let the time series Y_1 be stationary in third order with spectral density S_Y. Moreover suppose that the fourth order moments of Y_t exist, and let S_Y fulfill the Szegö condition (19.2) and let all finite dimensional distributions of Y_t have positive spectrum. Then the necessary and sufficient condition for the equivalence of the linear $\hat{Y}_L(t)$ and the quadratic $\hat{Y}_Q(t)$ predictor is that the bispectrum $B_e(z_1, z_2)$ of the innovation process e_t has the form

$$B_e(z_1, z_2) = H(z_1) + H(z_2) + H(z_1^{-1} z_2^{-1}). \tag{19.3}$$

where

$$H(z) = \sum_{k=0}^{\infty} c_k z^k,$$

and

$$z = \mathrm{e}^{i2\pi\lambda}, \quad z_1 = \mathrm{e}^{i2\pi\lambda_1}, \quad z_2 = \mathrm{e}^{i2\pi\lambda_2}.$$

Proof

To prove that the members of the vector system

$$S = \{Y_{t-j}, Y_{t-k} Y_{t-l}, j = 0, 1, 2 \ldots, k, l = 0, 1, 2, \ldots\},$$

are linearly independent in the Hilbert space H_S generated by S we refer to the paper of Masani and Wiener (1959). Consider any polynomial P from S, i.e.

$$P(Y_{t-k_1}, \ldots, Y_{t-k_Q}) = \sum g_r Y_{t-k_r} + \sum g_{r,q} Y_{t-k_r} Y_{t-k_q},$$

then the norm square

$$\| P \|^2 = \int_{R^Q} |P(y_1, \ldots y_Q)|^2 \, \mathrm{d}F_{k_1,\ldots,k_Q}(y_1, \ldots, y_Q),$$

cannot be zero unless $g_r = g_{r,q} = 0$ because of the positive spectrum of the distribution $F_{k_1,\ldots,k_Q}$. So the subspaces H_{S_1} and H_{S_2} in H_S are disjoint where H_{S_1} and H_{S_2} are the Hilbert spaces generated by $\{Y_{t-k}, k = 0, 1, 2, \ldots\}$ and $\{Y_{t-j}\, Y_{t-k}, j, k = 0, 1, 2, \ldots\}$ respectively.

Now the necessary and sufficient assumption for $\hat{Y}_L(t) = \hat{Y}_Q(t)$ is that the error of the orthogonal projection of Y_t to the Hilbert space H_{S_1} be orthogonal to the H_{S_2} as well, i.e.,

$$\mathrm{E}(Y_t - \hat{Y}_L(t))\, Y_{t-j} Y_{t-k} = \mathrm{E}e_t\, Y_{t-j}\, Y_{t-k} = 0, k, j = 1, 2, \ldots. \tag{19.4}$$

Under the assumptions of the theorem there exists a linear relationship between the process Y_t and its innovation e_t therefore (19.4) is equivalent to

$$c_{ee}(-j, -k) = \mathrm{E}e_t e_{t-j} e_{t-k} = 0, \quad j, k = 1, 2, \ldots. \tag{19.5}$$

In view of the symmetry of the third order moments

$$\begin{aligned} c_{ee}(-j, -k) &= c_{ee}(-k, -j) = c_{ee}(j, j-k) = c_{ee}(j-k, j) \\ &= c_{ee}(k, k-j) = c_{ee}(k-j, k) \end{aligned}$$

the only possibility for $c_{ee}(j, k)$ not to be zero is when either $k = j$, or $k = 0$, $j > 0$, or $j = 0$, $k > 0$. Therefore,

$$B_e(z_1, z_2) = H(z_1) + H(z_2) + H(z_1^{-1} z_2^{-1}),$$

where

$$H(z) = \sum_{k=0}^{\infty} c_k z^k; \quad c_k = c_{ee}(k, k).$$

We note here that the assumption (19.3) is automatically fulfilled when the bispectrum of the process Y_t is zero for all frequencies because the bispectrum of the linearly filtered process e_t is given as a product of the bispectrum of the process Y_t and the filter. This implies that the bispectrum of the innovation process e_t is also zero. Therefore it may happen that although the linearity test fails, the best predictor is linear.

To decide whether the bispectrum of the innovation process e_t is of the form (19.3) the following theorem is useful, the multiplicative version of which is given by Sakaguchi (1991) and the proof of it is also similar.

Theorem 19.2

Let $B(\lambda_1, \lambda_2)$ be a bispectrum which is partially differentiable once with respect to λ_1. Then the following statements are equivalent.

(*a*) For any (a, b, c) the bispectrum $B(\lambda_1, \lambda_2)$ satisfies the relation

$$B(a,b)+B(c,0)+B(-a+c,-b-c)=B(b,c)+B(0,-a-b)+B(-a+c,-c) \quad (19.6)$$

(*b*) There exists a function $H(\lambda)$ such that

$$B(\lambda_1, \lambda_2) = H(\lambda_1) + H(\lambda_2) + H(-\lambda_1 - \lambda_2) \quad (19.7)$$

where $H(\lambda)$ is a complex valued function on R and satisfies

$$H(-\lambda) = H^*(\lambda),\ H(\lambda + 2\pi) = H(\lambda) \quad (19.8)$$

for any $\lambda \in R$, where H^* denotes the complex conjugate of H.

Proof

Let us use the following notation.

$$f_1(x, y) = \frac{\mathrm{d}}{\mathrm{d}\lambda} B(\lambda, y)\,|_{\lambda = x},$$

$$L(c, b) = f_1(c, -b - c),$$

$$G(c) = f_1(c, -c),$$

$$C(x) = \int_0^x G(y)\,\mathrm{d}y.$$

On the basis of the properties of the bispectrum it is easy to prove

$$f_1(-x, -y) = -f_1^*(x, y) \quad (19.9)$$

$$f_1(-x, 0) = -f_1(x, 0) \quad (19.10)$$

$$f_1(x, -x - y) = f_1(x, y) - f_1(y, x). \quad (19.11)$$

For L we can write

$$\int_0^x L(c, x + y - c)\mathrm{d}c = \int_0^x f_1(c, -x - y)\mathrm{d}c$$

$$= B(x, -x + y) - B(0, -x - y) = B(x, y) - B(0, -x - y). \quad (19.12)$$

If (*a*) holds, differentiating both sides of (19.6) with respect to a and setting a equal to zero and using (19.10), (19.11) we get

$$f_1(b, -b - c) = f_1(c, -c) - f_1(b, -b).$$

Using the above notation we have

$$L(c, b) = G(c) - G(b),$$

and hence

$$\int_0^x L(c, x+y-c)\mathrm{d}c = \int_0^x (G(c) - G(x+y-c))\mathrm{d}c$$

$$= C(x) + C(y) - C(x+y). \tag{19.13}$$

From (19.12) and (19.13) we obtain

$$B(x, y) = B(0, -x-y) + C(x) + C(y) - C(x+y). \tag{19.14}$$

Setting $x = -y = z$ and then $z = -x - y$,

$$B(0, z) = B(z, -z) = B(0,0) + C(z) + C(-z),$$

$$B(0, -x-y) = B(0,0) + C(-x-y) + C(x+y). \tag{19.15}$$

Substituting (19.15) in (19.14) we have the form

$$B(x, y) = B(0,0) + C(x) + C(y) + C(-x-y). \tag{19.16}$$

This is almost the final form we are looking for but $C(\lambda)$ is not periodic. It follows from the definition that $G(-\lambda) = -G^*(\lambda)$ and $C(-\lambda) = C^*(\lambda)$ and because of the periodicity of both $B(\lambda, 0)$ and $f_1(\lambda, -\lambda)$, we have

$$\int_\alpha^{\alpha+2\pi} G(\lambda)\mathrm{d}\lambda = \int_{-\pi}^{\pi} G(\lambda)\mathrm{d}\lambda$$

$$= 2i \int_0^\pi Im G(\lambda)\mathrm{d}\lambda = i\theta, \tag{19.17}$$

where θ is a real constant. As $C(\lambda + 2\pi) = C(\lambda) + i\theta$, we put

$$H(\lambda) = \frac{B(0,0)}{3} + C(\lambda) - \frac{i\theta\lambda}{2\pi}, \tag{19.18}$$

and $H(\lambda)$ satisfies (19.7) and (19.8).

If (*b*) holds it is easy to show that (*a*) also holds.

19.3 AN EXAMPLE— HOMOGENEOUS BILINEAR REALIZABLE TIME SERIES WITH HERMITE DEGREE-2

One of the nonlinear/non-Gaussian time series is the bilinear realizable time series with Hermite degree-2. The general form of a homogeneous process (see Brillinger (1965)), given by its Wiener–Ito representation is

$$Y_t = \int_0^1 \int_0^1 \mathrm{e}^{i2\pi(\omega_1 + \omega_2)t} g_2(z_1, z_2) W(\mathrm{d}\omega_1, \mathrm{d}\omega_2),$$

where W denotes the stochastic spectral measure with respect to the Gaussian white noise series ε_t. The process Y_t is bilinear realizable if and only if its transfer function g_2 is a rational function of two variables with particular form

$$g_2(z_1, z_2) = \frac{\gamma(z_1, z_1 z_2)}{\alpha_{22}(z_1 z_2)\alpha_{21}(z_1)},$$

(see Terdik, 1991), where the polynomials $\alpha_{21}(z)$, $\alpha_{22}(z)$ and $\gamma(z, v)$ are given by

$$\alpha_{21}(z) = \sum_{k=0}^{P_1} a_k^{(1)} z^{-k}; \quad a_0^{(1)} = 1,$$

$$\alpha_{22}(z) = \sum_{k=0}^{P_2} a_k^{(2)} z^{-k}; \quad a_0^{(2)} = 1,$$

$$\gamma(z, v) = \sum_{m=1, n=0}^{R,S} c_{m,m+n} z^{-n} v^{-m}.$$

In this case the process Y_t can be given by the following state space equations (see Terdik, 1991),

$$\sum_{k=0}^{P_1} a_k^{(1)} X_{t-k}^{(1)} = \varepsilon_t,$$

$$\sum_{k=0}^{P_2} a_k^{(2)} X_{t-k}^{(2)} = \sum_{m=1, n=0}^{R,S} c_{m,m+n} X_{t-m-n}^{(1)} \varepsilon_{t-m} + const.,$$

$$Y_t = X_{t+1}^{(2)}. \tag{19.19}$$

The process Y_t is called separable if the polynomial γ is the product of two polynomials of a single variable, i.e.

$$\gamma(z_1, z_1 z_2) = \gamma_0(z_1 z_2)\gamma_1(z_1).$$

As the spectrum and the bispectrum for bilinear realizable processes with Hermite degree-2 are explicitly given (Terdik and Meaux, 1991), theorem 19.3 follows.

Theorem 19.3

If the homogeneous bilinear realizable Hermite degree-2 process (19.19) is separable and the roots of γ_0 are inside the unit circle, then the best linear predictor is the best quadratic one as well.

Proof

In this case the spectrum has the form (Terdik and Meaux, 1991)

$$\sigma^4 \left| \frac{\gamma_0(z_1)}{\alpha_{22}(z_1)} \right|^2 \left[1 + \int_0^1 \left| \frac{\gamma_1(z)}{\alpha_{21}(z)} \right|^2 \mathrm{d}\lambda \right] = \sigma_e^2 \left| \frac{\gamma_0(z_1)}{\alpha_{22}(z_1)} \right|^2,$$

where σ_e^2 is the variance of the residual series of the best linear predictor. Assuming the roots of γ_0 are inside the unit circle the residual series takes the form

$$e_t = \frac{\alpha_{22}(L)}{\gamma_0(L)} Y_t,$$

where L is the backward shift operator, i.e. $LY_t = Y_{t-1}$ and the bispectrum of the residual series is also simple, that is

$$\Psi(z_1, z_2) = 6\sigma^6(h(z_1) + h(z_2) + h(z_1^{-1} z_2^{-1})),$$

where

$$h(z_1) = 1/3 + \int_0^1 \frac{|\gamma_1(z)|^2 \gamma_1(z^{-1} z_1^{-1})}{|\alpha_{21}(z)|^2 \alpha_{21}(z^{-1} z_1^{-1})} \, d\lambda.$$

As $\Psi(z_1, z_2)$ satisfies the necessary and sufficient condition of theorem 19.1, the proof is completed.

Moreover, we show that in the case of separability the best linear predictor is the best polynomial one as well. Put

$$e_t = \frac{\alpha_{22}(L)}{\gamma_0(L)} Y_t = \int_0^1 z_1^t z_2^t \frac{\gamma_1(z_1)}{\alpha_{21}(z_1)} W(d\lambda_1, d\lambda_2). \tag{19.20}$$

It can be seen from (19.18) that the state space equations of e_t are

$$\begin{aligned} \alpha_{21}(L) X_t^{(1)} &= \gamma_1(L)\varepsilon_t, \\ e_t &= X_t^{(1)} \varepsilon_t - \sigma^2, \end{aligned} \tag{19.21}$$

where ε_t is the Gaussian white noise series according to the stochastic spectral measure W. We have from (19.19)

$$\begin{aligned} &\mathrm{E} e_t (Y_{t-i_1}^{k_1} Y_{t-i_2}^{k_2} \cdots Y_{t-i_n}^{k_n} - \mathrm{E} Y_{t-i_1}^{k_1} Y_{t-i_2}^{k_2} \cdots Y_{t-i_n}^{k_n}) \\ &= \mathrm{E}(\varepsilon_t^2 - \sigma^2 + \varepsilon_t \phi_1(\varepsilon_1, \varepsilon_2 \ldots))\phi_2(\varepsilon_1, \varepsilon_2, \ldots) = 0, \end{aligned}$$

for any functions ϕ_1 and ϕ_2 and for all $n, k_1, k_2, \ldots, k_n, i_1, i_2, \ldots, i_n = 1, 2, \ldots$ because ε_t is an independent series.

The question whether the assumption of separability is necessary is still open. Until now we have not been able to find a counter example. So our conjecture is that the linear predictor for bilinear realizable Hermite degree-2 processes is the best if and only if it is separable.

19.4 TESTING HYPOTHESES

There are several ways to check whether the linear predictor contains all the information that is contained in a quadratic predictor as well for large samples. The first step is to filter the process by fitting a linear model and consider the residual series e_t. The residual series is uncorrelated therefore its spectrum is a constant and equal to σ_e^2, and the sample variance $\hat{\sigma}_e^2$ of e_t gives a consistent estimate of it.

Now, the third order periodogram $\hat{B}$ or its smoothed version taken at different frequencies (ω_k, ω_l) are distributed asymptotically as complex Gaussian with mean which is the true bispectrum and variance $\sigma_e^{3/2}$ under some circumstances, (Brillinger and Rosenblatt, 1967). Consider the statistics

$$T_m(\omega_1, \omega_2, \omega_3) = \hat{B}(\omega_1, \omega_2) + \hat{B}(\omega_3, 0) + \hat{B}(\omega_3 - \omega_1, -\omega_2 - \omega_3) \\ - \hat{B}(\omega_2, \omega_3) - \hat{B}(0, -\omega_1 - \omega_2) - \hat{B}(\omega_3 - \omega_1, -\omega_3).$$

Under the null hypothesis that the best predictor is linear the statistic $T_m(\omega_1, \omega_2, \omega_3)$ can be shown to be, approximately, complex normal with mean 0 and variance $4\sigma_e^2$ and $2\sigma_e^2$ respectively for the real and the complex part. In other words, the statistic $|T_m(\omega_1, \omega_2, \omega_3)|^2$ is distributed as central χ^2 and it is noncentral under the alternative hypothesis.

REFERENCES

Brillinger, D.R. (1985) An introduction to polyspectra, *Ann. Math. Statist.*, **36**, 1351–1374.

Brillinger, D.R. and Rosenblatt, M. (1967), Asymptotic theory of estimates of kth order spectra, in, (ed) B. Harris Spectral Analysis of Time Series, J. Wiley, New York, NY, pp. 153–188.

Granger, C.W.J. and Andersen, A.P. (1978) Nonlinear time series modeling, in (ed.) D.F. Findely, *Applied Time Series Analysis*, Academic Press, New York, pp. 25–38.

Hida, T. and Kallianpur, G. (1975), The square of a Gaussian Markov process and nonlinear prediction, *Journal of Multivariate Analysis*, **5**, pp. 451–461.

Kallianpur, G. (1981), Some ramifications of Wiener's ideas on nonlinear prediction in (ed.) P. Masoni, N.Wiener: *Collected Works*, vol. III, MIT Press, pp. 402–425.

Kolmogorov, A.N. (1941), Interpolation and extrapolation of stationary sequences. *Izvestiya Akad. Nauk, Math. Ser.*, **5**, 3–14.

Masani, P. and Wiener, N. (1959), Nonlinear prediction, The Harald Cramer Volume ed. by Grenander, Stockholm, pp. 190–212.

Nelson, J.Z. and Van Ness, J. (1973), Formulation of a nonlinear prediction. *Technometries*, **15**, (1) 1–17.

Priestley, M.B. (1981), *Spectral Analysis and Time Series*, vol. 1, Academic Press, New York.

Sakaguchi, F. (1991) A relation for 'Linearity' of the bispectrum. *Journal of Time Series Analysis*, **12**, (3).

Terdik, G. and Subba Rao, T. (1989) On Wiener–Ito representation and the best linear predictors for bilinear time series. *J. Appl. Prob.*, **26**, 274–286.

Terdik, G. and Meaux, L. (1991) The exact bispectra for bilinear realizable processes with Hermite degree-2. *Adv. Appl. Prob.*, **23**, 798–808.

Terdik, Gy. (1991). Bilinear state space realization for polynomial stochastic systems. *Computers Math. Applic.*, **22** (7) 69–83.

Wiener, N. (1949). *Extrapolation, Interpolation and Smoothing of Stationary Time Series with Engineering Applications*, MIT Press, Cambridge, Mass.

Wiener, N. (1958) *Nonlinear Problems in Random Theory*, John Wiley and Sons.

20

Maximum likelihood fitting of bilinear models to time series with missing observations

M.M. Gabr

20.1 INTRODUCTION

The method of maximum likelihood has been previously applied to the problem of estimating the parameters of linear time series models such as the stationary autoregressive moving average (ARMA) and the autoregressive integrated moving average (ARIMA) models. Calculation of the maximum likelihood estimates (MLE) may be carried out iteratively by means of a scoring equation that involves the gradient of the negative log likelihood function (LF) and the Fisher information matrix. Evaluation of the information matrix requires implementation of a Kalman filter and its derivative with respect to each parameter (see, for example, Akaike (1978), Pearlman (1980), Jones (1980), Harvey and Pierse (1984), Kohn and Ansley (1986) and also Harvey (1989)).

Recently, missing data problems have successfully been approached using the state space methodology. For example, Jones (1980) proposed a method for obtaining MLE of the parameters of stationary ARMA processes when some observations are missing. Harvey and Pierse (1984) and Kohn and Ansley (1986) have extended the method for the non-stationary ARIMA processes. The extension to some cases of nonlinear processes, such as the bilinear processes, raises some non-trivial problems and has not been dealt with before. Bilinear models were initially discussed by Granger and Anderson (1978) and studied and greatly developed by Subba Rao (1981), Gabr and Subba Rao (1981), Subba Rao and Gabr (1984), Pham Dinh (1985), Liu and Brockwell (1988), Liu (1989), Gabr (1988, 1991), Kim and Billard (1990) and Subba Rao and Silva (1992).

The general form of a bilinear time series $\{x_t, t = 0, \pm 1, \pm 2, \ldots\}$, denoted by BL($p, q, m, k$), is defined by

$$X_t + \sum_{i=1}^{p} a_i X_{t-i} = e_t + \sum_{j=1}^{q} c_j e_{t-j} + \sum_{i=1}^{m} \sum_{j=1}^{k} b_{ij} X_{t-i} e_{t-j}, \tag{20.1}$$

where $\{e_t\}$ is an i.i.d. sequence of random variables with zero mean and common variance σ^2. The parameter estimation of the bilinear time series models has not received considerable attention in the literature. Subba Rao (1981) and Gabr and Subba Rao (1981) have used the repeated least squares and Newton Raphson iterations in estimating the parameters of the full and subset BL models. Guegan and Pham Dinh (1989) show that the least squares estimators are strongly consistent. Kim and Billard (1990) obtained moment estimators for the parameters of BL(1, 0,1, 1) model and studied their asymptotic properties. Subba Rao and Silva (1992) have used the so called Yule–Walker type difference equations for higher order moments and cumulants for fitting the BL(p, 0, p, 1) model. Recently, Gabr (1991) proposed methods for the recursive estimation of BL models.

In this paper the problem of parameter estimation of the BL(p, 0, p ,1) model is considered. The BL model is first represented by a suitable state-space form. The Kalman filter is then applied to compute the LF efficiently. A nonlinear optimization program is then used to obtain the maximum likelihood estimates of the parameters. In the case when some observations are missing, we can carry out the same procedure obtaining approximate maximum likelihood estimates of the parameters. Once this has been done, the missing observations can be estimated by smoothing.

20.2 STATE SPACE FORMULATION AND KALMAN FILTER

There are different state space representations of the BL model (1.1) (see, e.g. Pham Dinh (1985), Guegan (1987) and Gabr (1991)). For simplicity, we restrict ourselves to the stationary and invertible BL(p, 0, p, 1) model, namely

$$X_t + \sum_{i=1}^{p} a_i X_{t-i} = e_t + \sum_{j=1}^{p} b_j X_{t-j} e_{t-1}. \tag{20.2}$$

The conditions for stationarity and invertability for the above model are given in Subba Rao and Gabr (1984) and Liu (1989). Gabr (1991) has used the following state space representation

$$\mathbf{X}(t) = \mathbf{F}(t)\mathbf{X}(t-1) + \mathbf{C}e_t \tag{20.3}$$

$$y_t = \mathbf{H}\mathbf{X}(t), \tag{20.4}$$

where

$$\mathbf{X}(t)=\begin{bmatrix} X_t \\ X_{t-1} \\ \vdots \\ X_{t-p+1} \\ e_t \end{bmatrix}, \mathbf{C}=\begin{bmatrix} 1 \\ 0 \\ \vdots \\ 0 \\ 1 \end{bmatrix}, \quad \mathbf{F}(t)=\begin{bmatrix} -a_1 & -a_2 & \cdots & -a_p & \sum_{j=1}^{p} b_j \ X_{t-j} \\ & & & 0 & 0 \\ & \mathbf{I} & & \vdots & \\ 0 & \cdots & & 0 & 0 \end{bmatrix}$$

and $\mathbf{H}=[1 \quad 0\ldots0]$. $\mathbf{X}(t)$ is $(p+1)\times 1$ column vector representing the state of the process of time t. $\mathbf{F}(t)$ is $(p+1)\times(p+1)$ state transition matrix defining how the process progresses from one time point to the next. $\mathbf{C}$ is $p \times 1$ column vector and $\mathbf{H}$ is $1 \times p$ row vector both constant in time. $\{e_t\}$ is a sequence of normally distributed independent random variables with mean zero and variance σ^2, i.e. $e_t \sim \text{NID}(0,\sigma^2)$. In the measurement equation (20.4), y_t is the true observed value (without observational error).

$$y_t = \mathbf{H}\mathbf{X}(t). \tag{20.5}$$

This state space formulation of the BL model allows us to compute the LF of observations when some of these observations are missing and to estimate the missing values.

Although the transition matrix $\mathbf{F}(t)$ is stochastic (depends on observations up to and including y_{t-1}), it may be regarded as fixed once we are at time $t-1$. The distribution of y_t, conditional on $\{y_{t-1}, y_{t-2}, \ldots\}$ is normal for all $t=1,2,\ldots,N$, and therefore this model is considered a conditionally Gaussian model (see Jazwinski (1970), Liptser and Shiryayev (1978), Anderson and Moore (1979), Priestley (1988) and Harvey (1989)). Therefore, the derivation of the Kalman filter and the computation of the LF can be carried out exactly as in linear state space case but with a different interpretation.

Define $\alpha(t/t-1)$ and $\mathbf{P}(t/t-1)$ as the mean vector and covariance matrix of $\mathbf{X}(t)$, conditional on the information at time $t-1$. Here, $\alpha(t/t-1)$ is viewed as an estimator for $\mathbf{X}(t)$ and $\mathbf{P}(t/t-1)$ is regarded as its conditional error covariance, or mean square error (MSE) matrix. Given $\hat{\mathbf{X}}(t-1)$, the optimal estimator of the state vector at time $t-1$, together with its MSE matrix $\mathbf{P}(t-1)$, defined by

$$\mathbf{P}(t-1) = E[\{\hat{\mathbf{X}}(t-1) - \mathbf{X}(t-1)\}\{\hat{\mathbf{X}}(t-1) - \mathbf{X}(t-1)\}^T],$$

the optimal estimator of $\mathbf{X}(t)$ is given by

$$\alpha(t/t-1) = \mathbf{F}(t)\hat{\mathbf{X}}(t-1), \tag{20.6}$$

where the covariance matrix of the estimation error is given by

$$\mathbf{P}(t/t-1) = \mathbf{F}(t)\mathbf{P}(t-1)\mathbf{F}^T(t) + \sigma^2\mathbf{C}\mathbf{C}^T. \tag{20.7}$$

The updating equations, given a new observation y_t, are

$$\hat{\mathbf{X}}(t) = \alpha(t/t-1) + \mathbf{P}(t/t-1)\mathbf{H}^T[y_t - \mathbf{H}\alpha(t/t-1)]/S_t, \tag{20.8}$$

$$\mathbf{P}(t) = \mathbf{P}(t/t-1) - \mathbf{P}(t/t-1)\mathbf{H}^T\mathbf{H}\mathbf{P}(t/t-1)/S_t, \tag{20.9}$$

where

$$S_t = p_{11}(t/t-1) = \mathbf{H}\mathbf{P}(t/t-1)\mathbf{H}^T. \tag{20.10}$$

Note that $p_{11}(t/t-1)$ is the upper left-hand element of the $\mathbf{P}(t+1/t)$. The prediction error is given by

$$v_t = y_t - \mathbf{H}\alpha(t/t-1), \tag{20.11}$$

(for details see, e.g. Ljung and Söderström (1983) and Harvey (1989)).

Given N observations $\{y_1, y_2, \ldots, y_N\}$, one seeks the MLE of the parameters $\theta^T = [-a_1 \ldots -a_p b_1 \ldots b_p]$ in (20.2). Let $\mathbb{Y}(t-1) = \{y_{t-1}, y_{t-2}, \ldots, y_1\}$, then the conditional probability density function of y_t, conditional on $\mathbb{Y}(t-1)$ is normal for all $t = 1, 2, \ldots, N$. Therefore, the LF

$$\mathbf{L}(y_1, y_2, \ldots, y_N; \theta, \sigma^2) = \prod_{t=1}^{N} f(y_t/\mathbb{Y}(t-1))$$

can be constructed by the prediction error decomposition, yielding

$$\log \mathbf{L} = -\frac{N}{2}\log 2\pi - \frac{N}{2}\log\sigma^2 - \frac{1}{2}\sum_{t=1}^{N}\log S_t - \frac{1}{2\sigma^2}\sum_{t=1}^{N} v_t^2/S_t. \tag{20.12}$$

The parameter σ^2 cannot be removed completely from (20.12) as in the linear ARMA models case (see Jones (1980) and Harvey (1989)). The reason is that initial values of the elements of the variance covariance matrix $\mathbf{P}(t)$ contain different orders of σ^2 which makes it impossible to remove σ^2 completely as a common factor from all terms of $\mathbf{P}(t)$ simultaneously. The exact expression for these values will be given in the next section. Still it is more convenient to simplify (20.12) by scaling $\mathbf{P}(t/t-1)$ and S_t and dividing both sides of equations (20.7), (20.9) and (20.10) by σ^2. A numerical nonlinear optimization search procedure can be used to find the maximum of log $\mathbf{L}$ with respect to the unknown parameters θ. This gives the MLE of θ, and when this is completed the MLE of σ^2 can be obtained directly.

20.3 THE INITIAL ESTIMATES

In order to start the recursions, an initial estimator $\alpha(1/0)$ of the state $\mathbf{X}(0)$ is needed, together with the associated matrix $\mathbf{P}(1/0)$. In principle, these starting values for the Kalman recursions, are given by the mean vector and covariance matrix of the unconditional distribution of the state vector. Since the series is stationary, these initial conditions can be chosen as the unconditional expectations

$$\alpha(1/0) = \mathbf{E}[\mathbf{X}(t)],$$

$$\mathbf{P}(1/0) = \mathbf{E}[\{\mathbf{X}(t) - \alpha(1/0)\}\{\mathbf{X}(t) - \alpha(1/0)\}^T]/\sigma^2.$$

From the results given by Subba Rao and Gabr (1984) concerning this model, and after some algebraic manipulations, it can be shown that

$$\alpha(1/0) = c\sigma^2[1\,1\ldots1\,0]^T; \quad c = b_1/(1 + \sum_{i=1}^{p} a_i)$$

$$\mathbf{P}(1/0) = \begin{bmatrix} & & & 0 \\ & \mathbf{Q} & & \vdots \\ & & & 0 \\ 0 & \cdots & 0 & 1 \end{bmatrix},$$

where $\mathbf{Q}$ is a $p \times p$ matrix given by

$$\text{vec}(\mathbf{Q}) = [\mathbf{I} - \mathbf{A} \otimes \mathbf{A} - \mathbf{B} \otimes \mathbf{B}]^{-1} \text{vec}(D).$$

In the above equations $\otimes$ is the Kronecker product, the vec$(\cdot)$ operator is the single column vector obtained by stacking the columns of the matrix one on top of another in order from left to right and

$$\mathbf{A} = \begin{bmatrix} -a_1 & -a_2 & \cdots & -a_p \\ & & & 0 \\ & \mathbf{I} & & \vdots \\ & & & 0 \end{bmatrix}, \quad \mathbf{B} = \begin{bmatrix} b_1 & b_2 & \cdots & b_p \\ 0 & 0 & \cdots & 0 \\ \vdots & \vdots & \vdots & \vdots \\ 0 & 0 & \cdots & 0 \end{bmatrix}$$

and

$$\mathbf{D} = \begin{bmatrix} d_{11} & d_{12} & c^2\sigma^2 & \cdots & c^2\sigma^2 \\ d_{21} & 0 & 0 & \cdots & 0 \\ c^2\sigma^2 & 0 & 0 & \cdots & 0 \\ \vdots & \vdots & \vdots & \vdots & \vdots \\ c^2\sigma^2 & 0 & 0 & \cdots & 0 \end{bmatrix},$$

where

$$d_{11} = 1 + 2c\sigma^2(b_1 - a_1\sum b_i) - c^2\sigma^4\{1 - (\sum a_i)^2 - (\sum b_i)^2\sigma^2\},$$

$$d_{12} = d_{21} = c\sigma^2(\sum b_i).$$

It is seen clearly that we cannot remove σ^2 completely from the LF.

20.4 MISSING OBSERVATIONS

It is very often the case in practice that the values of the time series are recorded at unequally spaced times, through failure to observe one or more values of the series. As in the linear model case (see Jones (1980) and Harvey and Pierse (1984)), the prediction errors associated with the non-missing observations can be obtained simply by skipping the Kalman filter updating equations at the points where the observations are missing. Thus, when an observation $y(t)$ is missing, the Kalman recursion skips equations (20.6), (20.7), (20.10) and (20.11) and equations (20.8) and (20.9) are simply replaced by

$$\hat{\mathbf{X}}(t) = \alpha(t/t-1)$$

$$\mathbf{P}(t) = \mathbf{P}(t/t-1).$$

We have to replace the missing values $X(t)$ in the expression for $\mathbf{F}(t+1)$ by its estimate $\hat{X}(t)$. Thus, the model now is no longer exactly conditionally Gaussian but approximately so. Hence, the corresponding term in log $\mathbf{L}$ given by (20.12), is omitted from the likelihood. Thus, the approximate LF is of the form (20.12) with the summations covering only those values of t for which the variable is actually observed.

Once the parameters of the BL model have been estimated, the approximate mean square error estimates of missing observations can be calculated by smoothing. The most straightforward of the smoothing algorithms, known as the fixed-point smoother, can be applied by augmenting the state space model and applying the Kalman filter. Full details can be found in Anderson and Moore (1979) and Harvey (1989).

20.5 NUMERICAL ILLUSTRATION

In order to examine the performance of the above algorithm and its convergence, some simulations were carried out. The following BL (1, 0, 1, 1) model has been used as an example,

$$X_t + a_1 X_{t-1} = e_t + b_{t-1} X_{t-1} e_{t-1}. \tag{20.13}$$

where the $\{e_t\}$ are zero mean pseudo normal variates with $E[e_t^2] = \sigma^2 = 1$. The parameter values used are $a_1 = -0.4$ and $b_1 = 0.4$.

Two schemes for missing observations, as in Dunsmuir and Robinson (1981), were considered:

scheme I —5% and 10% of the full data are omitted, periodically, at $t = 14, 34, 54, \ldots,$ and $t = 14, 24, 34, \ldots;$

scheme II—5% and 10% of the full data are omitted, randomly, according to Bernoulli sampling.

Three sample sizes $N = 100, 200$ and 500 were used. For each sample size, 50 series were generated, and between successive series 600 observations of the white noise $\{e_t\}$ were discarded to guarantee the independence of replications. From each series the first 50 observations were discarded to avoid initialization effects.

The above method for maximum likelihood estimation of the parameters a_1, b_1 and σ^2 was applied to the full data set and to the two subsets of it (with 5% and 10% missing observations).

To maximize the LF we have used the same optimization algorithm used

Table 20.1 Maximum likelihood estimates of the parameters a_1 and b_1 of the model (20.13)

	$N = 100$			$N = 200$			$N = 500$		
	a_1	b_1	σ^2	a_1	b_1	σ^2	a_1	b_1	σ^2
True	−0.4	0.4	1.0						
Full data	−0.409 (.025)	0.391 (.022)	1.023 (.021)	−0.406 (.022)	0.394 (.017)	1.015 (.014)	−0.402 (.012)	0.396 (.007)	0.966 (.009)
5% Randomly missing data	−0.414 (.034)	0.388 (.027)	1.025 (.023)	−0.407 (.031)	0.390 (.022)	1.016) (.017)	−0.405 (.019)	0.393 (.012)	0.997 (.011)
10% Randomly missing data	−0.418 (.043)	0.382 (.032)	1.029 (.028)	−0.411 (.037)	0.385 (.026)	1.018 (.020)	−0.408 (.023)	0.388 (.018)	1.009 (.014)
5% Regularly missing data	−0.412 (.035)	0.386 (.026)	1.026 (.024)	−0.408 (.032)	0.391 (.023)	1.017 (.017)	−0.404 (.021)	0.391 (.013)	1.003 (.012)
10% Regularly missing data	−0.421 (.044)	0.383 (0.33)	1.031 (.029)	−0.413 (.039)	0.381) (.025)	1.021) (.019)	−0.410 (.025)	0.389 (.019)	1.011 (.014)

by Harvey and Pierse (1984) in dealing with the ARIMA models, namely, the Gill–Murray–Pitfield algorithm given in the UK NAG library routine E04JBF. This routine is a comprehensive quasi-Newton algorithm that calculates the derivatives numerically and allows simple bounds to be placed on the parameters.

The ML estimation results for the full and the two subsets of data giving estimates of the parameters together with their standard errors, as well as an estimate of the residual variance are shown in Table 20.1. The results reported in this table represent the average and the sample standard errors (given in parentheses) of the parameter estimates over the 50 realizations. The results of simulations show that the parameter estimates, in all cases, are quite close to the true values. As expected, both the bias and the standard errors are increasing as the number of missing observations increases and the sample size N decreases. The missing observations, in some series, are estimated by the fixed-point smoothing algorithm. The estimates are very close to the actual values. The optimal mean square error predictions of future observations with their conditional MSEs can also be obtained by repeated application of Kalman recursion prediction equations. Once the LF is calculated, the model order p can be selected easily by using the AIC (Akaike information criterion).

Generalization of the algorithm and the computer program to the general BL (p, q, m, k) model (20.1) and application to real data, such as the well known sunspot and Canadian lynx data are still to be investigated and will be the subject of a subsequent publication.

REFERENCES

Akaike, H. (1978) *Covariance Matrix Computation of the State Variable of a Stationary Gaussian Process*, Research Memorandum No. 139, The Institute of Statistical Mathematics, Tokyo.

Anderson, B.D.O. and Moore, J.B. (1979) *Optimal Filtering*, Prentice-Hall, Englewood Cliffs.

Dunsmuir, W. and Robinson, P.M. (1981) Estimation of Time Series Models in the Presence of Missing Data. *Journal of the American Statistical Association*, **76**, 560–567.

Gabr, M.M. (1986) A recursive (On-line) identification of bilinear systems. *Int. J. Control*, **44**(4), pp. 911–917.

Gabr, M.M. (1988) On the third-order moment structure and bispectral analysis of some bilinear time series. *J. of Time Ser. Anal.*, **9**(1), 11–20.

Gabr, M.M. (1991) *Recursive Estimation of Bilinear Time Series Models*, Technical Report No. 207, Dept. of Math., UMIST.

Gabr, M.M. and Subba Rao, T. (1981) The estimation and prediction of subset bilinear time series models with applications. *J. of Time Ser. Anal.*, **2**(3), 153–171.

Granger, C.W.J. and Andersen, A.P. (1978) *An Introduction to Bilinear Time Series Analysis*, Vandenhoeck & Ruprecht, Gottingen.

Guegan, D. (1987) Different representations for bilinear models. *J. of Time Ser.*, **8**(4), 389–408.

Guegan, D. and Pham. D.T. (1989) A note on the estimation of the parameters of the diagonal bilinear models by the method of least squares. *Scand. J. Statist.*, **16**, 129–136.

Harvey, A.C. (1989) *Forecasting, Structural Time Series Models and the Kalman Filter*, Cambridge University Press, Cambridge.

Harvey, A.C. and Pierse, R.G. (1984) Estimating missing observations in economic time series. *J. of the American Statistical Association*, **79**, 125–131.

Jazwinski, A.H. (1970) *Stochastic Processes and Filtering Theory*, New York, Academic Press.

Jones, R.H. (1980) Maximum likelihood fitting of ARMA models to time series with missing observations. *Technometrics*, **22**, 389–395.

Kim, W.K. and Billard, L. (1990) Asymptotic properties for the first-order bilinear time series model. *Commun. Statist.-Theory Meth.*, **19**(4), 1171–83.

Kohn, R. Ansley, C.F. (1986) Estimation, prediction and interpolation for ARIMA models with missing data. *J. of the American Statistical Association*, **81**, 751–61.

Liptser, R.S. and Shiryayev, A.N. (1978). *Statistics of Random Processes II: Applications*, Trans. A.B. Aries, Springer-Verlag, New York.

Liu, J. (1989) A simple condition for the existence of some stationary bilinear time series. *J. of Time Ser. Anal.*, **10**(1), 33–39.

Liu, J. and Brockwell, P.J. (1988) On the general bilinear time series model. *Stoch. Proc. Appl.*, **20**, 617–627.

Ljung, L., and Söderström, T. (1983) *Theory and Practice of Recursive Identification*, MIT Press, Cambridge.

Pearlman, J.G. (1980) An algorithm for the exact likelihood of a high-order Auto-regressive-Moving Average process. *Biometrika*, **67**, 232–3.

Pham Dinh, T. (1985) Bilinear markovian representation and bilinear models. *Stoch. Proc. Appl.*, **20**, 295–306.

Priestley, M.B. (1988) *Nonlinear and Non-Stationary Time Series Analysis*. Academic Press, London.

Subba Rao, T. (1981) On the theory of bilinear time series models, *J. Royal Statist. Soc.*, B **28**, pp 244–255.

Subba Rao, T. and Gabr, M.M. (1984) *An Introduction to Bispectral Analysis and Bilinear Time Series Models*, Lecture Notes in Statistics, **24**, Springer-Verlag, Berlin.

Subba Rao, T. and Silva, M.E. (1992) Identification of Bilinear time series models BL $(p, 0, p, 1)$. *Statistica Sinica*, **2**(2), 464–478.

Part Six

Time and Frequency Analysis of Time Series—Applications

21

Time series models for multivariate series of count data

K. Ord, C. Fernandes and A.C. Harvey

21.1 INTRODUCTION

An earlier paper, Harvey and Fernandes (1989), denoted subsequently as HF, proposed various time series models for count data, that is, observations consisting of non-negative integers. These models led to forecasts based on the exponentially weighted moving average (EWMA) with the parameter determining the rate of discounting being computed by maximum likelihood (ML). This paper considers a method for extending such models to cope with multivariate time series of count observations. In a Bayesian context, an univariate treatment of count data has been developed by West, Harrison and Migon (1985).

The models proposed by HF can be regarded as falling within the class of structural time series models (Harvey, 1989). These are models which are set up directly in terms of components of interest. The simplest structural model, the local level plus noise, takes the form

$$y_t = \mu_t + \varepsilon_t, \quad t = 1, \ldots, T, \tag{21.1}$$

$$\mu_t = \mu_{t-1} + \eta_t, \tag{21.2}$$

where μ_t is a permanent or level component, which can move up or down because of the disturbance term η_t, and ε_t is a transitory disturbance term. If both η_t and ε_t are normally distributed, with zero means and variances σ_η^2 and σ_ε^2 respectively, then the forecasts are an EWMA. However, the model is inappropriate for count data. Following Smith (1979) and Smith and Miller (1986), HF specify ε_t in such a way that the distribution of y_t conditional on μ_t is Poisson or negative binomial. The stochastic process governing the evolution of μ_t is then defined implicitly so as to have certain desirable properties and to allow the distribution of y_t given past observations to be obtained. This is the basis, not only for making predictions, but also for forming the likelihood function. For a general comparison of alternative Bayesian forecasting models see Smith (1992).

A multivariate version of (21.1), (21.2) can be set up for Gaussian observations. In this model, y_t, μ_t, η_t and ε_t are all $N \times 1$ vectors, and η_t and ε_t have covariance matrices $\mathbf{\Sigma}_\eta$ and $\mathbf{\Sigma}_\varepsilon$ respectively. In the special case when these two matrices are proportional, the series are said to be homogeneous (Fernandez and Harvey, 1990). The forecasts for individual series can then be computed from separate EWMA's, each with the same smoothing constant.

One way of trying to develop a multivariate count data model would be to assume a multivariate Poisson distribution for the observations (Taillie *et al.*, 1979). However, such an approach turns out not to be particularly attractive, one reason being that the bivariate Poisson distribution can only be defined for variables which are positively correlated. Instead we set up a model in which the total number of events recorded in each period follows a Poisson distribution and the split into the individual series is determined by a binomial, or multinomial, distribution. Both of these mechanisms may be made dynamic in the way suggested in HF. Combining the predictive distributions for each mechanism leads to a joint predictive distribution for the series, from which predictions may be made and a likelihood function constructed.

Section 21.2 reviews the relevant univariate models from HF. Section 21.3 then shows how these models may be brought together in the way outlined in the previous paragraph. The properties of the implied joint distributions and joint predictive distributions of the observations are then derived in section 21.4. Section 21.5 describes how explanatory variables may be incorporated into the model and section 21.6 gives an application. One of the examples in HF concerned the modelling of the series of goals scored by England against Scotland in football matches at Hampden Park, Glasgow, and section 21.6 estimates a multivariate model which considers the goals scored by both teams.

21.2 COUNT DATA MODELS

21.2.1 Univariate Poisson-gamma model

Suppose that the observation at time t is drawn from a Poisson distribution,

$$p(y_t | \mu_t) = \frac{\mu_t^{y_t} e^{-\mu_t}}{y_t!}. \tag{21.3}$$

This corresponds to the measurement equation of (21.1).

Let $p(\mu_{t-1} | Y_{t-1})$ denote the pdf of μ_{t-1} conditional on the information at time $t-1$. Suppose that this distribution is gamma, that is it is given by

$$p(\mu | a, b) = e^{-b\mu} \mu^{a-1} b^a / \Gamma(\mathrm{a}), \quad \mathrm{a, b} > 0 \tag{21.4}$$

with $\mu = \mu_{t-1}$, $a = a_{t-1}$ and $b = b_{t-1}$ where a_{t-1} and b_{t-1} are computed from the first $t-1$ observations, Y_{t-1}. Following HF we assume that $p(\mu_t | Y_{t-1})$ is gamma distributed with parameters $a_{t|t-1}$ and $b_{t|t-1}$ such that

$$a_{t|t-1} = \omega a_{t-1} \tag{21.5}$$

$$b_{t|t-1} = \omega b_{t-1} \tag{21.6}$$

and $0 < \omega \leqslant 1$. Then

$$E(\mu_t | Y_{t-1}) = a_{t|t-1}/b_{t|t-1} = a_{t-1}/b_{t-1} = E(\mu_{t-1} | Y_{t-1}),$$

while

$$\text{Var}(\mu_t | Y_{t-1}) = a_{t|t-1}/b^2_{t|t-1} = \omega^{-1}\,\text{Var}(\mu_{t-1} | Y_{t-1}).$$

The stochastic mechanism governing the transition of μ_{t-1} to μ_t is therefore defined implicitly rather than explicitly. However it is possible to show that it is formally equivalent to a multiplicative transition equation of the form

$$\mu_t = \omega^{-1}\mu_{t-1}\eta_t,$$

where η_t has a beta distribution, of the form (21.18), with parameters ωa_{t-1} and $(1-\omega)a_{t-1}$ (Smith and Miller, 1986). As recently demonstrated by Shephard (1993), if $\omega < 1$, $\mu_t \to 0$ almost surely, as $t \to \infty$. Shephard (1993) gives the necessary modifications needed on the transition equation to eliminate this problem.

Once the observation y_t becomes available, the posterior distribution $p(\mu_t | Y_t)$ is given by a gamma distribution with parameters

$$a_t = a_{t|t-1} + y_t \tag{21.7}$$

$$b_t = b_{t|t-1} + 1. \tag{21.8}$$

The initial prior gamma distribution, that is the distribution of μ_t at time $t = 0$, tends to become diffuse, or non-informative, as $a, b \to 0$. However, none of this prevents the recursions (21.5), (21.6), (21.7) and (21.8) being initialized at $t = 0$ with $a_0 = b_0 = 0$. A proper distribution for μ_t is then obtained at time $t = \tau$ where τ is the index of the first non-zero observation. It follows that, conditional on Y_τ, the joint density of the observations $y_{\tau+1}, \ldots, y_T$ is

$$p(y_{\tau+1}, \ldots, y_T; \omega) = \prod_{t=\tau+1}^{T} p(y_t | Y_{t-1}). \tag{21.9}$$

The predictive pdf at time t is given by the negative binomial distribution

$$p(y_t | Y_{t-1}) = \int_0^\infty p(y_t | \mu_t) p(\mu_t | Y_{t-1}) \mathrm{d}\mu_t \tag{21.10}$$

$$= \binom{a + y_t - 1}{y_t} b^a (1 + b)^{-(a + y_t)}, \tag{21.11}$$

where $a = a_{t|t-1}$ and $b = b_{t|t-1}$ and

$$\binom{a + y_t - 1}{y_t} = \frac{\Gamma(a + y_t)}{\Gamma(y_t + 1)\Gamma(a)},$$

although since y_t is an integer, $\Gamma(y_t + 1) = y_t!$. The log likelihood function for the unknown hyperparameter ω is

$$\log L(\omega) = \sum_{t=\tau+1}^{T} [\log\Gamma(a_{t|t-1} + y_t) - \log y_t! - \log\Gamma(a_{t|t-1})$$

$$+ a_{t|t-1}\log b_{t|t-1} - (a_{t|t-1} + y_t)\log(1 + b_{t|t-1})]. \tag{21.12}$$

It follows from the properties of the negative binomial that the mean and variance of the predictive distribution of y_{T+1} given Y_T are respectively

$$\tilde{y}_{T+1|T} = \mathrm{E}(y_{T+1}|Y_T) = \frac{a_{T+1|T}}{b_{T+1|T}} = \frac{a_T}{b_T}, \tag{21.13}$$

and

$$\mathrm{Var}(y_{T+1}|Y_T) = \frac{a_{T+1|T}(1 + b_{T+1|T})}{b^2_{T+1|T}}$$

$$= \omega^{-1}\mathrm{Var}(\mu_T|Y_T) + E(\mu_T|Y_T). \tag{21.14}$$

Repeated substitution from (21.5), (21.6), (21.7) and (21.8) shows that the one-step-ahead prediction is given by

$$\tilde{y}_{T+1|T} = a_T/b_T = \frac{\sum_{j=0}^{T-1}\omega^j y_{T-j}}{\sum_{j=0}^{T-1}\omega^j}. \tag{21.15}$$

In large samples the denominator of (21.15) is approximately equal to $1/(1-\omega)$ when $\omega < 1$ and the forecasts can be obtained recursively by the EWMA scheme

$$\tilde{y}_{t+1|t} = (1-\lambda)\tilde{y}_{t|t-1} + \lambda y_t, \quad t = 1, \ldots, T \tag{21.16}$$

where $y_{1|0} = 0$ and $\lambda = 1 - \omega$ is the smoothing constant. When $\omega = 1$, the right hand side of (21.15), is equal to the sample mean. Regarding this as an estimate of μ, the choice of zeroes as initial values for a and b in the filter is seen to be justified insofar as it yields the classical solution. It is also worth noting that, unlike the Gaussian case, no approximations are involved in the use of a diffuse prior in this model.

A model based on a negative binomial, rather than a Poisson, for the observations may also be constructed. The relevant conjugate prior distribution in this case is the beta distribution; see HF (section 21.5).

21.2.2 Binomial-beta and multinomial-Dirichlet distributions

If the observations at time t are generated from a binomial distribution then

$$p(y_t|\pi_t) = \binom{y_t}{n_t}\pi_t^{y_t}(1-\pi_t)^{n_t - y_t}, \quad y_t = 0,\ldots,n_t, \tag{21.17}$$

where π is the probability that y_t is unity when n_t is one. The value of n_t is assumed to be fixed and known.

The conjugate prior for the binomial distribution is the beta distribution

$$p(\pi|c,d) = [\mathrm{B}(c,d)]^{-1}\pi^{c-1}(1-\pi)^{d-1}, \tag{21.18}$$

where the beta function is

$$\mathrm{B}(c,d) = \frac{\Gamma(c)\Gamma(d)}{\Gamma(c+d)}.$$

Let $p(\pi_{t-1}|Y_{t-1})$ have a beta distribution with parameters c_{t-1} and d_{t-1}. The updating step from $\pi_{t-1}|Y_{t-1}$ to $\pi_t|Y_{t-1}$ does not preserve conjugacy. Therefore, following HF, we assume that $p(\pi_t|Y_{t-1})$ is also beta with parameters given by equations exactly the same as those in (21.5), (21.6). This again ensures that the mean of $\pi_t|Y_{t-1}$ is the same as that of $\pi_{t-1}|Y_{t-1}$ but the variance increases. Specifically, $c_{t|t-1} = \omega c_{t-1}$ and $d_{t|t-1} = \omega d_{t-1}$, so that

$$\begin{aligned}\mathrm{E}(\pi_t|Y_{t-1}) &= \frac{c_{t|t-1}}{c_{t|t-1} + d_{t|t-1}} \\ &= \frac{c_{t-1}}{c_{t-1} + d_{t-1}}\end{aligned}$$

and

$$\mathrm{Var}(\pi_t|Y_{t-1}) = \frac{c_{t-1}d_{t-1}}{(c_{t-1} + d_{t-1})^2(\omega c_{t-1} + \omega d_{t-1} + 1)}.$$

This approach is similar in spirit to that of Harrison and Stevens (1976), who used a moments-preserving approximation in their multistate model.

Once the tth observation becomes available, the distribution of $\pi_t|Y_t$ is beta with parameters

$$c_t = c_{t|t-1} + y_t \tag{21.19}$$

$$d_t = d_{t|t-1} + n_t - y_t. \tag{21.20}$$

The predictive distribution, $p(y_t|Y_{t-1})$ is beta-binomial

$$p(y_t|Y_{t-1}) = \frac{1}{n_t+1}\,\frac{\mathrm{B}(c+y_t, d+n_t-y_t)}{\mathrm{B}(y_t+1, n_t-y_t+1)\mathrm{B}(c,d)}, \tag{21.21}$$

where $c = c_{t|t-1}$ and $d = d_{t|t-1}$. The likelihood function is again (21.9) with τ

defined as the first time period for which

$$0 < \sum_{t=1}^{\tau} y_t < \sum_{t=1}^{\tau} n_t. \tag{21.22}$$

This condition ensures that a_τ and b_τ are strictly positive, although again there is nothing to prevent us starting the recursions (21.5), (21.6), (21.19) and (21.20) at $t = 1$ with $c_0 = d_0 = 0$; see the comments in Lehmann (1983, p. 243).

From the properties of the beta-binomial distribution, the mean and variance of y_{T+1} conditional on the information at time T are

$$\tilde{y}_{T+1|T} = E(y_{T+1} | Y_T) = \frac{n_{T+1} c_T}{c_T + d_T}, \tag{21.23}$$

$$\operatorname{Var}(y_{t+1} | Y_t) = \frac{n_{T+1} c_T d_T (c_T + d_T + \omega^{-1} n_{T+1})}{(c_T + d_T)^2 (c_T + d_T + \omega^{-1})} \tag{21.24}$$

By substituting repeatedly from the recursive equations (21.19), (21.20) it can be seen that, for n_t constant $\tilde{y}_{T+1|T}$ is effectively an EWMA.

When there are more than two categories, the observations are said to be polytomous and the multinomial distribution is appropriate. Let there be N possible categories, and suppose that the probability that, at time t, an object belongs to the ith category is π_{it}. If there are n_t trials and the number of objects in the ith category is y_{it}, then

$$p(y_{1t}, \ldots, y_{Nt}) = \binom{n_t}{y_{1t}, \ldots, y_{Nt}} \prod_{i=1}^{N} \pi_{it}^{y_{it}} \tag{21.25}$$

with

$$\sum_{i=1}^{N} y_{it} = n_t \quad \text{and} \quad \sum_{i=1}^{N} \pi_{it} = 1.$$

The conjugate prior for the multinomial distribution is the multivariate beta or Dirichlet distribution

$$p(\pi_1, \ldots, \pi_N | c_1, \ldots, c_N) = \frac{\Gamma(\sum c_i)}{\prod \Gamma(c_i)} \prod_{i=1}^{N} \pi_i^{c_i - 1} \tag{21.26}$$

where the summations are from $i = 1$ to N. (When $N = 2$ this collapses to the beta distribution with $c_1 = c$ and $c_2 = d$). Proceeding as in the previous section, it is not difficult to show that the recursive equations corresponding to (21.19), (21.20) become

$$c_{i,t|t-1} = \omega c_{i,t-1} \tag{21.27a}$$

$$c_{i,t} = c_{i,t|t-1} + y_{it}, \quad i = 1, \ldots, N. \tag{21.27b}$$

The likelihood for ω is as in (21.9) with τ the first value of t which yields $c_{i,t} > 0$ for all $i = 1, \ldots, N$. The predictive distribution in this case is known as the multinomial-Dirichlet. The forecasts can again be expressed in terms of EWMAs.

21.3 THE MULTIVARIATE COUNT DATA MODEL

Suppose we have N series of count data observations. Let the number in the ith series at time t be y_{it}, $i = 1, \ldots, N$, $t = 1, \ldots, T$, and let the aggregate over all series be y_t, that is

$$y_t = \sum_{i=1}^{N} y_{it}, \quad t = 1, \ldots, T. \tag{21.28}$$

We assume that y_t can be modelled by the Poisson-gamma model of section 21.2.1, with hyperparameter ω_1. For a given value of y_t, the split into individual series is then assumed to be such that it can be modelled by the multinomial-Dirichlet scheme of section 21.2.2 with hyperparameter ω_2. This model implies a particular joint distribution of $y_{1t}, \ldots, y_{Nt}$ conditional on a set of stochastic parameters, $\mu_{1t}, \ldots, \mu_{Nt}$, and it implies a particular joint distribution for $\mu_{1t}, \ldots, \mu_{Nt}$. The properties of these distributions are explored in the next section. This section concentrates on the statistical treatment of the proposed model which is remarkably simple.

The joint predictive density function for $\{y_{1t}, \ldots, y_{Nt}\}$ is the same as that for $\{y_{1t}, \ldots, y_{N-1,t}, y_t\}$ with y_t given by the sum in (21.28). Thus

$$p(y_{1t}, \ldots, y_{Nt} | Y_{t-1}; \omega_1, \omega_2) = p(y_{1t}, \ldots, y_{N-1,t} | y_t, Y_{t-1}; \omega_2) p(y_t | Y_{t-1}; \omega_1), \tag{21.29}$$

where Y_t denotes all the observations on all the series up to and including time t. Thus, (21.29) is the product of the negative binomial predictive distribution for y_t, (21.11) and a multinomial-Dirichlet predictive distribution for $y_{1t}, \ldots, y_{N-1,t}$. For $N = 2$, this latter distribution is (21.21). The updating equations used to obtain a_t, b_t, c_t and d_t are exactly as in (21.7), (21.8) and (21.26). The log-likelihood function is obtained by summing the logarithms of the joint predictive distributions from $\tau + 1$ to T where τ is defined as the first value of t for which all the series have had at least one non-zero observation; compare (21.22). Of course, unless a restriction such as $\omega_1 = \omega_2$, is placed on the model, ω_1 appears only in the predictive distribution for y_t and ω_2 appears only in the predictive distribution for $y_{1t}, \ldots, y_{N-1,t}$. Hence the overall likelihood function may be maximized by maximizing two separate likelihood functions, one with respect to ω_1 and the other with respect to ω_2.

The joint density function of one-step ahead predictions is given by evaluating (21.29) for $t = T + 1$. However, the expected values of the individual series at time $T + 1$ can be written down immediately since

$$E(y_{i,T+1} | y_{T+1}, Y_T) = \frac{c_{iT}}{c_T} y_{T+1},$$

where $c_T = \sum c_{iT}$, (compare (21.23) in the case $N = 2$) and so

$$E(y_{i,T+1} | Y_T) = \frac{c_{iT}}{c_T} \frac{a_T}{b_T}, \quad i = 1, \ldots, N. \tag{21.30}$$

From (21.15), the conditional expectation of y_{T+1} is an EWMA with weights determined by the hyperparameter ω_1. Let this be denoted as $\mathrm{EWMA}_1(y_t)$. Furthermore c_{iT} is proportional to an EWMA of the y_{it} terms with hyperparameter ω_2, denoted $\mathrm{EWMA}_2(y_{it})$, and c_T is proportional to a similar EWMA for the sum of the y_{it} terms. Thus

$$E(y_{i,T+1}|Y_T) = \mathrm{EWMA}_2(y_{iT})\cdot\frac{\mathrm{EWMA}_1(y_t)}{\mathrm{EWMA}_2(y_t)}, \quad i=1,\ldots,N. \qquad (21.31)$$

In the special case when $\omega_1 = \omega_2$, (21.31) reduces to an EWMA of the observations in the ith series. Hence there is a parallel with the homogeneous case of the Gaussian multivariate local level model described in section 21.1. However, as will be seen in the next section, setting $\omega_1 = \omega_2$ implies that the observations in the different series are independent of each other, something which is not necessarily the case for a homogeneous Gaussian local level model. An interesting corollary of the independence of the series is that the likelihood function for $\omega_1 = \omega_2$ is given by the product of the likelihood functions for the individual series.

A likelihood ratio test for the hypothesis that $\omega_1 = \omega_2$ can be carried out for $0 < \omega_1,\ \omega_2 < 1$. If the null hypothesis is accepted, the series should be forecast separately.

21.4 PROPERTIES OF JOINT DISTRIBUTIONS IMPLIED BY THE MODEL

In this section we explore the structure of the joint distributions introduced in section 21.3; for notational ease, the t subscript will be dropped unless needed explicitly. The joint model for $\mathbf{y} = (y_1, \ldots, y_N)'$ conditional on $\boldsymbol{\lambda} = (\lambda_1, \ldots, \lambda_N)'$ may be written as

$$p_1(\mathbf{y}|\boldsymbol{\lambda}) = \prod_{i=1}^{N} p_i(y_i|\lambda_i),$$

where $p_i(\cdot|\cdot)$ denotes the Poisson probabilities for y_i given λ_i, since these are conditionally independent. The joint prior distribution for $\boldsymbol{\lambda}$ is $p(\boldsymbol{\lambda})$ which we assume may be factorized into two parts as

$$p(\boldsymbol{\lambda}) = p_2(\mu|a,b)\,p_3(\boldsymbol{\pi}|\mathbf{c}),$$

where $\mathbf{c} = (c_1, \ldots, c_N)'$, $\boldsymbol{\pi} = (\pi_1, \ldots, \pi_N)'$, $\lambda_i = \mu\pi_i$, p_2 is the gamma prior for μ given in (21.4) and p_3 is the Dirichlet prior for π given in (21.26). Thus

$$p(\mathbf{y}) = \int_{(N)} p_1(\mathbf{y}|\mu,\boldsymbol{\pi})\,p_2(\mu|a,b)\,p_3(\boldsymbol{\pi}|\mathbf{c})\,\mathrm{d}\mu \prod_{i=1}^{(N-1)} \mathrm{d}\pi_i. \qquad (21.32)$$

The subscript (N) on the integral sign denotes that integration takes place over μ and the N-dimensional simplex $\{\sum\pi_i = 1,\ \pi_i \geqslant 0\}$. Reversing the argument, expression (21.32) may be rewritten as

$$p(\mathbf{y}) = \int_{(N)} p_1(\mathbf{y}|\boldsymbol{\lambda}) p_2^*(\boldsymbol{\lambda}|a, b, \mathbf{c}) \prod_{i=1}^{N} \mathrm{d}\lambda_i, \tag{21.33}$$

where

$$p_2^*(\boldsymbol{\lambda}) \equiv p_2^*(\boldsymbol{\lambda}|a, b, \mathbf{c}) = \frac{\Gamma(c)\mu^{a-c}}{\Gamma(a)b^{-a}} \prod_{i=1}^{N} \left(\frac{\lambda_i^{c_i - 1} \mathrm{e}^{-\lambda_i b}}{\Gamma(c_i)} \right) \tag{21.34}$$

where $\mu = \sum \lambda_i$, $c = \sum c_i$. When $a = c$, (4.3) splits into N distinct factors and the λ_i are independent gamma (c_i, b) random variables, as is well known (Johnson and Kotz, 1972, pp. 231–233). When $a \neq c$, the variates are dependent and we have a multivariate gamma distribution for which the sum is always gamma distributed, although the individual elements are gamma if and only if they are independent. These multivariate models are of interest in diverse areas such as marketing where μ represents total sales and the $\{\lambda_i\}$ market shares (*cf.* Goodhart, Ehrenberg and Chatfield, 1984) or the relative abundance of species (cf. Taillie *et al.*, 1979).

21.4.1 The multivariate gamma distribution

The moments of the $\{\lambda_i\}$ are readily shown to be

$$E(\lambda_i) = \frac{ac_i}{bc} \tag{21.35}$$

$$\mathrm{Var}(\lambda_i) = \frac{ac_i}{b^2 c^2 (c+1)} [(c_i + 1)c + a(c - c_i)] \tag{21.36}$$

$$\mathrm{Cov}(\lambda_i, \lambda_j) = \frac{ac_i c_j}{b^2 c^2 (c+1)} (c - a) \tag{21.37}$$

As expected, these reduce to the Dirichlet moments when $a \to \infty$ with (a/b) fixed and to those of the gamma when $c \to \infty$ with (c_i/c) fixed. It is apparent from (21.37) that zero correlations among all pairs imply $a = c$ and hence independence. Considering the case $c_i = c/N$, the range of possible correlations is found to be

$$-(N-1)^{-1} < \rho < 1, \tag{21.38}$$

the lower and upper bounds being approached, respectively, for the Dirichlet and gamma limiting cases. This range is the largest possible for equally correlated variables and reduces to $(-1, +1)$ for $N = 2$.

As regards other gamma models, several different multivariate gamma models have been suggested over the years, mostly based upon either addition such as $Y_i = X_1 + X_3$, $Y_2 = X_2 + X_3$, or minimization, i.e. $Y_1 = \min(X_1, X_3)$, $Y_2 = \min(X_2, X_3)$; see Stuart and Ord (1987, section 5.52), Patil *et al.* (1985). The addition law preserves gamma marginals provided the scale factors are equal, as in (21.34), but does not provide a tractable bivariate density. The minimization law is attractive for reliability applications, particularly when the distributions are exponential. However, both have restricted ranges for the correlation parameters and, in particular, both imply that $\rho > 0$.

21.4.2 The multivariate negative binomial distribution

The predictive distribution given in (21.29) may be rewritten in symmetric form as

$$p(\mathbf{y}) = p(y_1, \ldots, y_N) = \frac{\Gamma(a+s)}{\Gamma(a)} \frac{b^a}{(1+b)^{a+s}} \frac{\Gamma(c)}{\Gamma(c+s)} \prod_{i=1}^{N} \frac{\Gamma(c_i + y_i)}{y_i! \Gamma(c_i)}, \quad (21.39)$$

where $s = y_1 + \cdots + y_N$. Thus

$$p(\mathbf{0}) = b^a (1+b)^{-a} \quad (21.40)$$

and

$$p(y_1 + 1, \ldots, y_N) = \frac{(a+s)(c_1 + y_1)}{(1+b)(c+s)(y_1+1)} p(\mathbf{y}), \quad (21.41)$$

allowing ready evaluation of the probabilities. The moments are

$$\mathrm{E}(y_i) = \mathrm{E}(\lambda_i) = ac_i/bc \quad (21.42)$$

$$\mathrm{Var}(y_i) = \mathrm{Var}(\lambda_i) + E(\lambda_i) \quad (21.43)$$

$$\mathrm{Cov}(y_i, y_j) = \mathrm{Cov}(\lambda_i, \lambda_j). \quad (21.44)$$

Although the correlation between two ys is less in absolute value than that for the corresponding λs, the limits in (21.38) may still be approached. As before, the sign of all the covariances is the same and is given by the sign of $(c - a)$. It follows from (21.7) and (21.27) that

$$a_t = y_t + \omega_1 y_{t-1} + \omega_1^2 y_{t-1} + \ldots$$

and

$$c_t = y_t + \omega_2 y_{t-1} + \omega_2^2 y_{t-1} + \ldots.$$

Given the same initial values, it follows that $c - a > 0$ (< 0) when $\omega_2 >$ $(<)$ ω_1. The correlation is zero when $\omega_1 = \omega_2$ and, again, zero correlation implies independence, because with c set equal to a, expression (21.39) is seen to be equal to the product of N negative binomial predictive distributions.

Although the above set-up may be restrictive for larger N, it does provide a very flexible bivariate distribution, particularly compared to many previous suggestions (*cf.* Taillie *et al.*, 1979). The Poisson-lognormal model of Aitchison and Ho (1990) provides an equally flexible bivariate scheme, but requires numerical quadrature for evaluation of the probabilities. For our purpose the lack of conjugacy would make updating a major problem.

21.5 EXPLANATORY VARIABLES

In HF, explanatory variables were introduced into the Poisson model by means of the link function

$$\mu_t^+ = \mu_t \exp(\mathbf{x}_t'\boldsymbol{\delta}) \tag{21.45}$$

using the GLIM framework (McCullagh and Nelder, 1983). For explanatory variables which have an impact on the overall sum, this approach may be used without modification. Therefore, we concentrate upon those variables that affect the relative shares, i.e. the random variables $\{\pi_{it}\}$. Note however, that since the two analyses proceed independently, the two sets of explanatory variables may be overlapping. More commonly, we may use sums like $\mathbf{x}_t$ in (21.45) and proportions like $\mathbf{x}_{it}/\mathbf{x}_t$ for the relative shares model. Our discussion is now restricted to the case $N = 2$ when the natural (GLIM) link function is the logit

$$\text{logit}(\pi_t^+) = \log\left\{\frac{\pi_t^+}{(1-\pi_t^+)}\right\} = \text{logit}(\pi_t) + \mathbf{x}_t'\boldsymbol{\delta} \tag{21.46}$$

or

$$\pi^+ = \frac{\pi u}{(1 - \pi + \pi u)}, \tag{21.47}$$

where $u = \exp(\mathbf{x}'\boldsymbol{\delta})$ and the subscripts are to be understood from the context $\pi_2 = 1 - \pi$ since $N = 2$. We note that

$$1 > \pi^+ > \pi \quad \text{for } u > 1 \quad \text{and} \quad 0 < \pi^+ < \pi \quad \text{for } u < 1.$$

In order to proceed with the model development, we must evaluate the integral

$$J = J(y, s, c, d, u) = \int_0^1 (\pi^+)^y (1-\pi^+)^{s-y} \pi^{c-1} (1-\pi)^{d-1}\, \mathrm{d}\pi \tag{21.48}$$

since the predictive distribution for $y_{1t}|s_t$, where $s_t = y_{1t} + y_{2t}$, is

$$p(y_1|s) = \binom{s}{y_1} \frac{J(y_1, s, c, d, u)}{\beta(c, d)}. \tag{21.49}$$

Since (21.48) is a single integral, it could clearly be evaluated numerically; however, this option rapidly becomes infeasible as N increases, recalling that J must be evaluated for each time period and for each iteration of the likelihood maximization search routine. Instead, using (21.47) we may rewrite (21.48) as as

$$J = u^y \int_0^1 \pi^{c+y+1} (1-\pi)^{s+d-y-1} (1 - \pi + \pi u)^{-s}\, \mathrm{d}\pi. \tag{21.50}$$

We may always code the two series such that $u < 1$; if $u = 1$, J reduces to the beta function. Then expanding the term in brackets and integrating term-by-term, J is given by the convergent series expansion

$$J = u^y \sum_{i=0}^{\infty} \binom{s+i-1}{i} \beta(c+y+i, d+s-y)(1-u)^i. \tag{21.51}$$

Constants apart, the sum is the hypergeometric series $F = {}_2F_1(s, c+y; c+d+s; 1-u)$; the ratio of the $(n+1)$th term to the nth is

$$\begin{aligned} \frac{A_{n+1}}{A_n} &= \frac{(c+y+n)(s+n)(1-u)}{(c+d+s+n)(n+1)} \\ &= \alpha_n \quad \text{say,} \end{aligned} \tag{21.52}$$

so that the sum may be approximated by its partial sum to n terms, F_n say, plus a geometric series approximation for the remainder:

$$F = F_n + \frac{A_n}{1-\alpha_n}. \tag{21.53}$$

In practice, terms are summed until the remainder is sufficiently small.

21.5.1 A modal approximation

The series expansion approach is generally adequate for $N = 2$, but becomes tedious for large N or u near zero. A rapid, but more approximate procedure is to replace the terms in π^+ in (21.48) by the term

$$\pi^z (1-\pi)^{A-z} \tag{21.54}$$

where (A, z) are selected so that the mode of (5.10), $\pi_M = z/A$ agrees with the mode of the π^+ function, $\pi_M^+ = y_1/s$. Since there are two parameters to be specified, we set $A = s$ so that the approximation involves a reallocation of the 'observations' between the two populations, but retains the overall sum. Also, this approach allows direct extension to $N > 2$ series. It follows that

$$z = sy_1u/[uy_1 + s - y_1] \tag{21.55}$$

and $0 \leqslant z \leqslant y_1$ if $u \leqslant 1$, $y_1 \leqslant z \leqslant s$ if $u \geqslant 1$. J in (21.48) now reduces to a beta function and the approximation for (21.49) becomes (with y_1 in (21.55) in place of y)

$$p(y_1|s) = K \binom{s}{y_1} [\beta(c+z, d+s-z)/\beta(c,d)], \tag{21.56}$$

the constant K being selected to make the probabilities sum to 1.0. Since this approach will be most useful for small s, the computational effort required to obtain the predictive distribution is fairly modest. Also, the posterior beta distribution is given by the updating

$$c_t = \omega_2 c_{t-1} + z_t \tag{21.57}$$

$$d_t = \omega_2 d_{t-1} + s_t - z_t. \tag{21.58}$$

For $N > 2$, we may consider the link functions

$$\log(\pi_i^+ / \pi_N^+) = \log(\pi_i / \pi_N) + \mathbf{x}_i' \boldsymbol{\delta}_i \tag{21.59}$$

and use (21.55) to define z_i, $i = 1, \ldots, N$ in terms of y_i and $u_i = \exp(\mathbf{x}_i' \boldsymbol{\delta}_i)$.

21.6 GOALS SCORED BY ENGLAND AND SCOTLAND

Harvey and Fernandes (1989) fitted the Poisson-gamma model of section 21.2 to the number of goals socred by England in international football matches played against Scotland at Hampden Park in Glasgow. Apart from the war years these matches were played in Glasgow every other year, starting in 1872. (The year 1985 is also an exception; the match should have been played at Wembley, but was played in Scotland). Treating the observations as though they were evenly spaced, estimation of the Poisson-gamma model gave $\tilde{\omega} = 0.844$. The variance of the standardized residuals is 1.269 and a plot of them shows no indication of misspecification. A post-sample predictive test carried out over the last five observations gave no hint of model breakdown with $\xi(5) = 4.54$. The forecasted value for the mean of future observations is 0.82. The multivariate model of section 21.3 can be used to formulate a model in which the goals scored by England are modelled jointly with those scored by Scotland. (Although football theory is somewhat vague on the likely correlation between the goals scored by two teams in a match, a model which would only allow positive correlation seems too restrictive).

Given that the football matches have been played either in England (mostly at Wembley) or Scotland (at Hampden Park), the match venue is the natural explanatory variable for the proportion of goals scored by the teams. Since we are interested in predicting the goals scored by England we investigate how this dummy affects England's proportion. The dummy variable x_t is defined such that

$$x_t = +1 \text{ for matches played in England,}$$

$$-1 \text{ for matches played in Scotland.}$$

The above dummy has also been used for the total of goals. We have found that according to standard goodness of fit criteria the best specifications were given by:

M_1—a model in which we have assumed at the outset the constraint $\omega_1 = \omega_2$, i.e. independence between the two series of goals. The dummy is used both for the overall sum and England's relative share;

M_2—an unconstrained model where the dummy is used in both mechanisms.

Table 21.1 Bivariate model fitted to series of goals by England and Scotland

	Estimates				Goodness-of-fit			
	$\tilde{\omega}_1$	$\tilde{\omega}_2$	$\tilde{\delta}_1$	$\tilde{\delta}_2$	ML	AIC	BIC	Theil's U
M_1	0.885	$=\tilde{\omega}_1$	0.136	0.203	−157.95	323.35	331.29	0.677
M_2	0.844	0.930	0.139	0.203	−158.67	323.907	334.485	0.678

δ_1 and δ_2 are the dummy hyperparameters associated with μ and π respectively. Note that for both specifications the series expansion has been the best technique to introduce the dummy for the relative share π.

The selected specifications indicate that the venue is a relevant factor in explaining both the total number of goals and the share of England in this total, Table 21.1. Model M_2 seems to suggest some sort of dependence between the two series, although the improvement in the fit is barely affected if independence is assumed at the outset, by setting $\omega_1 = \omega_2$. In fact the likelihood ratio test statistic is 1.44, so that the null hypothesis this restriction is valid seems to be supported by the data. Hence we are led to believe that the goals scored by the two teams are independent and as a result they should be forecasted independently.

REFERENCES

Aitchison, J. and Ho, C.H. (1990) The multivariate Poisson-lognormal distribution. *Biometrika*, **76**, 643–653.

Fernandez, F.J. and Harvey, A.C. (1990). Seemingly unrelated time series equations and a test for homogeneity. *Journal of Business and Economic Statistics*, **8**, 71–81.

Goodhart, G.J., Ehrenberg, A.S.C. and Chatfield, C. (1984) The Dirichlet: A comprehensive model of buying behaviour. *Journal of the Royal Statistical Society, Series A*, **147**, 621–655.

Harrison, P.J. and Stevens, C.F. (1976) Bayesian Forecasting. *Journal of the Royal Statistical Society, Series B*, **38**, 205–247.

Harvey, A.C. (1989) *Forecasting, Structural Time Series Models and the Kalman Filter*, Cambridge University Press, Cambridge.

Harvey, A.C. and Fernandes, C. (1989). Time series models for count or qualitative observations. *Journal of Business and Economic Statistics*, **7**, 407–422.

Johnson, N.L. and Kotz, S. (1972) *Distributions in Statistics: Continuous Multivariate Distributions*, Houghton Mifflin, New York.

Lehmann, E.L. (1983) *Theorey of Point Estimation*. John Wiley and Sons Inc., New York.

McCullagh, P. and Nelder, J.A. (1983) *Generalised Linear Models*, Chapman and Hall, London.

Ord, J.K. (1972) Families of Frequency Distributions, Griffin, London.

Patil, G.P., Boswell, M.T. and Ratnaparkhi (1985) *Dictionary of Classified Bibliography of Statistical Distributions in Scientific Work*, Vol 3, International Co-operative Publishing House, Burtonsville, Maryland.

Shephard, N.G. (1990) *A Local Scale Model: an Unobserved Component Alternative to Integrated GARCH Processes.* STICERD Discussion Paper EM/220, London School of Economics.

Smit, J.Q. (1979) A Generalization of the Bayesian steady forecasting model. *Journal of the Royal Statistical Society, Series B*, **41**, 375–387.

Smith, J.Q. (1992) A comparison of the characteristics of some Bayesian forecasting models. *International Statistical Review*, **60**, 75–87.

Smith, R.L. and Miller, J.E. (1986) A non-Gaussian state space model and application to prediction of records. *Journal of the Royal Statistical Society, Series B*, **48**, 79–88.

Stuart, A. and Ord, J.K. (1987) *Kendall's Advanced Theory of Statistics*, vol. I. Griffin, London.

Taillie, C., Ord, J.K., Mosimann, J.E. and Patil, G.P. (1979) Discrete multivariate distributions, in (ed.) Ord, J.K., Patil, G.P. and Taillie, C. *Statistical Distributions in Ecological Work*, ICPH, Burtonsville, Maryland, pp. 159–178.

West, M., Harrison, P.J. and Migon, H.S. (1985) Dynamic generalized linear models and Bayesian forecasting. *Journal of the American Statistical Association*, **80**, 73–97.

22

Conditional maximum likelihood estimates for INAR(1) processes and their application to modelling epileptic seizure counts

J. Franke and T. Seligmann

22.1 INTRODUCTION

Daily seizure counts are a prime tool in investigating the epileptic disease and in evaluating the usefulness of drugs. To make a detailed analysis of such data, we have to consider them as time series of counts, i.e. as non-negative, integer-valued stochastic processes in discrete time. Models and procedures of conventional time series analysis have primarily been designed for continuously valued and, in particular, Gaussian processes. Therefore, they do not seem to be adequate tools for analyzing processes assuming only few different values.

The larger part of the literature on discrete-valued time series relies on a regression-like approach, whereas in Zeger and Qaqish (1988) and references therein the mean and variance of the outcome at time t are up to some unknown parameters given functions of past values of that outcome and of some covariates. This approach can even be extended to categorical time series as in Fahrmeir and Kaufmann (1987) and Kaufmann (1987). A related, but different approach is discussed by Zeger (1988) who considers a regression model for count data where the correlation is assumed to arise from an unobservable latent time series added to the conditional mean in a log-linear model. A more specific class of discrete-valued time series has been proposed by Jacobs and Lewis (1983) under the name of DARMA processes. Here, the outcome at time t is a certain random mixture of past outcomes and independent exogenous impulses.

In this paper, we investigate a class of parametric models, the integer-valued autoregressive schemes, which makes allowance for the discreteness of the data while retaining some of the properties of the conventional autoregressive models. These so-called INAR models have been discussed by Al-Osh and Alzaid (1987) and by Du and Li (1991). The same types of processes have been introduced under a different name by McKenzie (1985–1988) who derived various properties for the case of specific marginal distributions and extended the concept to ARMA-like models. Here, we prove some asymptotic results for the conditional maximum likelihood parameter estimates of INAR(1) processes. Then, we illustrate the usefulness of the model for a special case and apply the estimation procedure to some data coming from an extensive study concerned with the effectiveness of a certain antiepileptic drug. Here, a large number of patients recorded for approximately half the year the daily numbers of epileptic seizures, first with only standard medication being applied, then with additional use of the new drug in the test group and a placebo in the control group. Comparison of the total number of fits before and after treatment and between test and control group showed the general effectiveness of the drug against particular kinds of seizures. The time-ordering of the data was, however, not used in the analysis, and the question was, if it contains more information about how the reduction of the number of fits is achieved.

There is some previous literature on models for epileptic seizure counts. Milton *et al.* (1987) investigated the possibility that epileptic seizures are simply generated by a Poisson process and, therefore, the daily seizure counts are i.i.d. Poisson random variables. For about half the data sets considered, this simple mechanism could be rejected, and the authors suggested that the availability of longer records would lead to even more evidence against the Poisson model. On the other hand, there are enough hints in the data that independent Poisson variables might be useful as building blocks of more complex models for the seizure counts.

Hopkins *et al.* (1985) postulated that there are two states of seizure susceptibility corresponding to times of stress and times of calmness. The transition between the states should be governed by a two-state Markov chain, whereas the seizure counts are assumed to be i.i.d. Poisson variables with mean depending on the current state. However, Hopkins *et al.* do not discuss methods for inference and estimation. Albert (1991) fills this gap by developing estimates for both the transition probabilities of the Markov chain as well as for the two Poisson means. As the states cannot be observed directly, the use of the EM type estimation algorithm is quite involved, and theoretical results like asymptotic normality of the estimates are not available. Nevertheless, the procedure applied to simulated data and to epileptic seizure counts does quite well compared to fitting quasi-likelihood regression models (Wedderburn, 1974, McCullagh and Nelder 1983).

We follow Albert's approach in so far as we use Poisson variables as the basic building block of a model and incorporate dependence in time. To get a simpler estimation theory, we use a class of models discussed in detail by

Al-Osh and Alzaid (1987)—the so-called INAR(1) process (integer-valued autoregressions of order 1). To define them we introduce the notation

$$p \circ X = \sum_{j=1}^{X} Y_j,$$

where X is a random variable with values in $\mathbb{N}_0 = \{0, 1, 2, \dots\}$, $0 \leqslant p \leqslant 1$, and $Y_1, Y_2, \dots$ are i.i.d. Bernoulli-variables, independent of X, with

$$p = pr(Y_j = 1) = 1 - pr(Y_j = 0).$$

A straightforward calculation shows that $q \circ (p \circ X)$ is distributed as $(pq) \circ X$, and

$$\begin{aligned} \mathrm{E}(p \circ X) &= p\mathrm{E}X \\ \mathrm{var}(p \circ X) &= p^2 \,\mathrm{var}\, X + p(1-p)\mathrm{E}X \end{aligned} \tag{22.1}$$

Using this notation, Al-Osh and Alzaid (1987) introduced the INAR(1) model:

Definition 22.1

An $\mathbb{N}_0$-valued time series $\{X_t, -\infty < t < \infty\}$ is called an INAR(1) process, i.e. an integer-valued autoregressive process of order 1, if

$$X_t = p \circ X_{t-1} + \varepsilon_t, \quad -\infty < t < \infty, \tag{22.2}$$

for some $0 \leqslant p \leqslant 1$ and i.i.d. $\mathbb{N}_0$-valued random variables ε_t, $-\infty < t < \infty$. Such a process has an intuitive interpretation: X_t may represent the number of individuals of a population in the tth generation. Each individual generates an individual in the following generation with probability p independently of all other individuals, which amounts to a total number $p \circ X_t$ of offspring. Additionally, a random number ε_{t+1} of individuals enter the population from outside. Analogously, X_t may represent the number of certain events at time t (e.g. epileptic fits), where each one has a chance p to give rise to a like event at time $t+1$. A number ε_{t+1} of events is generated from independent sources. A more detailed discussion of the model for the epileptic seizure counts is given in section 22.3.

Du and Li (1991) generalized the INAR(1) model to the INAR(m) model, $m \geqslant 1$, and proved stationarity conditions which are completely analogous to the familiar AR(m) model. For $m = 1$, their results imply the existence of a unique weakly stationary $\mathbb{N}_0$-valued time series $\{X_t\}$ satisfying (22.2) for any i.i.d. sequence $\{\varepsilon_t\}$ with finite variance provided $p < 1$. Then, $\mathrm{cov}(X_s, \varepsilon_t) = 0$ for $s < t$ such that the ε_t can be called 'innovations'. They even provide an analogue to the MA(∞) representation of an AR(1) process

$$X_t \stackrel{d}{=} \sum_{j=0}^{\infty} p^j \circ \varepsilon_{t-j}$$

(compare (2) of Al-Osh and Alzaid, 1987) where '$\stackrel{d}{=}$' stands for equality in distribution. Also, the autocovariance sequence of a stationary INAR(1) process decreases exactly as for the AR(1) model (compare section 3 of Al-Osh and Alzaid, 1987):

$$r_k = \text{cov}(X_t, X_{t-k}) = p^k \,\text{var}\, X_t, \quad k \geqslant 1.$$

In the following section, we present a special INAR(1) model and the corresponding results on estimation and inference. This quite simple model allows a more detailed discussion of epileptic seizure counts by incorporating the two-state postulate of Hopkins *et al.* (1985). The application of the model to some of the real data and the interpretation of the model parameters is given in section 22.3. Finally, the Appendix provides the necessary theory for the general INAR(1) processes like asymptotic normality of conditional maximum likelihood estimates.

Finally, let us remark that the INAR(1) process is a special case of the Galton–Watson process with immigration studied e.g. by Seneta (1969), Venkataraman (1982) and Venkataraman and Nanthi (1982). In particular the last paper also studies maximum likelihood estimates of parameters. However, the authors assume that the immigration component, corresponding to ε_t in (22.2), is observed, which is not the case in the application which we have in mind.

22.2 THE SINAR(1) MODEL

Al-Osh and Alzaid (1987) define the INAR(1) model in a general context, but they discuss in detail only Poisson-distributed ε_t. For this case, they compare several estimates of the model parameters by simulations, and the conditional maximum likelihood (CML) estimate performs best. The Poisson assumption, however, implies that $\text{E}X_t = \text{var}\, X_t$ in the stationary state which makes it unsuitable for modelling seizure counts. Our data show a distinct tendency to overdispersion, i.e. $\text{E}X_t < \text{var}\, X_t$, which has also been observed by Milton *et al.* (1987) and by Albert (1991). Keeping the two-state postulate of Hopkins *et al.* (1985) in mind, we, therefore, generalize the INAR(1) model with Poisson innovations:

Definition 22.2

An INAR(1) process

$$X_t = p \circ X_{t-1} + \varepsilon_t, \quad -\infty < t < \infty,$$

is called a SINAR(1) process if

$$\varepsilon_t = Q_t \varepsilon_t^{(1)} + (1 - Q_t)\varepsilon_t^{(2)}, \quad -\infty < t < \infty,$$

where, for some $0 \leqslant q < 1$ and $0 < \lambda_1 < \lambda_2$, Q_t, $\varepsilon_t^{(1)}$, $\varepsilon_t^{(2)}$, $-\infty < t < \infty$, are independent, and the $\varepsilon_t^{(i)}$ are Poisson-distributed with parameters λ_i, $i = 1,2$, and the Q_t are Bernoulli-variables with parameter $q = pr(Q_t = 1)$.

SINAR stands for switching INAR as the innovations ε_t correspond to two states represented by Poisson variables with intensities λ_1, λ_2 where switching from one state to the other is controlled by the $0-1$ variables Q_t. Note, that the SINAR(1) model with $q = 0$ is just the INAR(1) model with Poisson innovations and intensity λ_2. Some simulated data from SINAR models are given in Figure 22.1(a)–(f).

Using (22.1) and the uncorrelatedness of X_{t-1} and ε_t, we get the mean and variance of a stationary SINAR(1) process:

$$\mu = \mathrm{E}X_t = \frac{\mathrm{E}\varepsilon_t}{1-p} = \frac{q\lambda_1 + (1-q)\lambda_2}{1-p}$$

$$\sigma^2 = \operatorname{var} X_t = \frac{p\mu}{1+p} + \frac{\operatorname{var}\varepsilon_t}{1-p^2}$$
$$= \mu + \frac{q(1-q)(\lambda_2 - \lambda_1)^2}{1-p^2}.$$

Let $\vartheta = (p, q, \lambda_1, \lambda_2)^T$ denote the parameter vector of the SINAR(1) model. We always assume

$$\vartheta \in \Theta = \{(p, q, \lambda_1, \lambda_2)^T;\ 0 < p < 1,\ 0 \leqslant q < 1,\ 0 < \lambda_1 < \lambda_2\}.$$

To show consistency and asymptotic normality of the estimates defined below, we use that a SINAR(1) process $\{X_t\}$ is an irreducible, aperiodic Markov chain (compare theorem 22.2) with transition probabilities

$$P_\vartheta(m, n) = pr_\vartheta(X_t = n / X_{t-1} = m)$$
$$= qA_p(m, n; \lambda_1) + (1-q)A_p(m, n; \lambda_2), \quad m, n \geqslant 0,$$

where, with $m \wedge n = \min\{m, n\}$,

$$A_p(m, n; \lambda) = \mathrm{e}^{-\lambda} \sum_{j=0}^{m \wedge n} \frac{\lambda^{n-j}}{(n-j)!} \binom{m}{j} p^j (1-p)^{m-j}, \quad m, n \geqslant 0$$

are the transition probabilities of an INAR(1) process with Poisson innovations.

The likelihood of the sample $x = (x_0, \ldots, x_N)$ is

$$L_N(x, \vartheta) = pr_\vartheta(X_0 = x_0) \cdot \prod_{t=1}^{N} P_\vartheta(x_{t-1}, x_t).$$

We neglect the dependency on the initial value and consider as an estimate for ϑ the conditional maximum likelihood estimate given $X_0 = x_0$ which we get by maximizing the conditional log-likelihood

$$\ell_N(X, \vartheta / X_0) = \sum_{t=1}^{N} \log P_\vartheta(X_{t-1}, X_t)$$

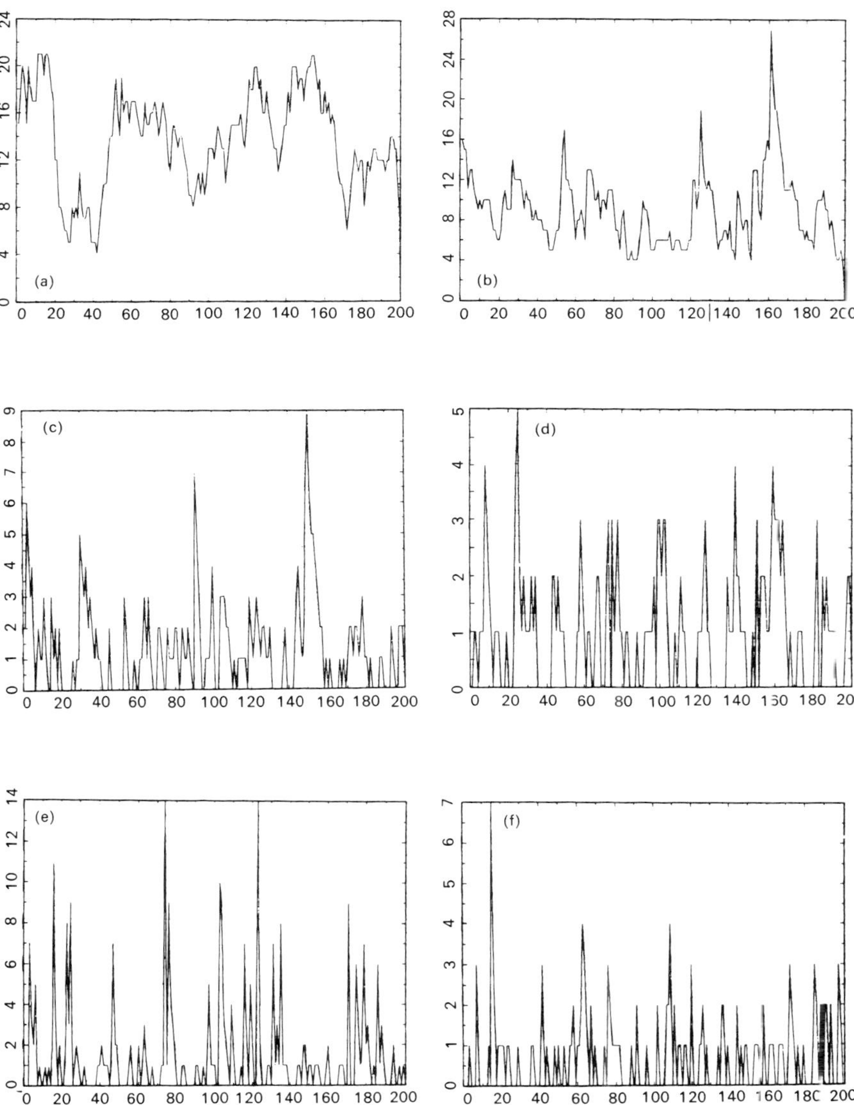

Figure 22.1. Typical sample paths of length 200 of a SINAR(1) process. (a) $p = 0.9$, $q = 0.5$, $\lambda_1 = 0.5$, $\lambda_2 = 2.0$; (b) $p = 0.9$, $q = 0.9$, $\lambda_1 = 0.5$, $\lambda_2 = 8.0$; (c) $p = 0.5$, $q = 0.9$, $\lambda_1 = 0.5$, $\lambda_2 = 2.0$; (d) $p = 0.5$, $q = 0.15$, $\lambda_1 = 0.5$, $\lambda_2 = 4.0$; (e) $p = 0.1$, $q = 0.9$, $\lambda_1 = 0.5$, $\lambda_2 = 8.0$; (f) $p = 0.1$, $q = 0.9$, $\lambda_1 = 0.5$, $\lambda_2 = 2.0$.

over $\mathbf{\Theta}$, where $X = (X_0, \ldots, X_N)$ is the vector of available data. We denote the CML estimate by $\hat{\vartheta}_N = (\hat{p}, \hat{q}, \hat{\lambda}_1 \hat{\lambda}_2)^T$, and we obtain it as a solution of the equations

$$\frac{\partial \ell}{\partial p} = \frac{1}{p(1-p)}\left[\sum_{t=1}^{N} (X_t - pX_{t-1}) - q\lambda_1 S_{\vartheta,1}^{-} - (1-q)\lambda_2 S_{\vartheta,2}^{-}\right] = 0 \quad (22.3)$$

$$\frac{\partial \ell}{\partial q} = S_{\vartheta,1} - S_{\vartheta,2} = 0 \quad (22.4)$$

$$\frac{\partial \ell}{\partial \lambda_1} = q[S_{\vartheta,1}^{-} - S_{\vartheta,1}] = 0 \quad (22.5)$$

$$\frac{\partial \ell}{\partial \lambda_2} = (1-q)[S_{\vartheta,2}^{-} - S_{\vartheta,2}] = 0, \quad (22.6)$$

with

$$S_{\vartheta,i} = \sum_{t=1}^{N} A_p(X_{t-1}, X_t; \lambda_i)/P_\vartheta(X_{t-1}, X_t), \quad i = 1, 2$$

$$S_{\vartheta,i}^{-} = \sum_{t=1}^{N} A_p(X_{t-1}, X_t - 1; \lambda_i)/P_\vartheta(X_{t-1}, X_t), \quad i = 1, 2,$$

where we use $A_p(m, -1; \lambda) = 0$ as a convention.

Note that $\hat{\vartheta}$ is not necessarily the unique solution of (22.3–22.6). Theorem 22.4, however, shows that we have at least asymptotic uniqueness and, also, consistency of the CML estimate if we exclude the case $q = 0$. Furthermore, we even have asymptotic normality and some other asymptotic results useful for testing hypotheses on the parameter.

Theorem 22.1

Let $\{X_t\}$ be a stationary SINAR(1) process with parameter

$$\vartheta \in \mathbf{\Theta}_0 = \{(p, q, \lambda_1, \lambda_2)^T; 0 < p, q < 1, 0 < \lambda_1 < \lambda_2\}.$$

(a) The CML estimate $\hat{\vartheta}_N$ based on the sample $X_0, \ldots, X_N$ is asymptotically normal:

$$\sqrt{N}(\hat{\vartheta}_N - \vartheta) \xrightarrow[\mathscr{L}]{} \mathscr{N}(0, \Sigma^{-1}(\vartheta)) \quad \text{for } N \to \infty,$$

where $\sigma(\vartheta)$ is the non-singular 4×4 Fisher information matrix, i.e.

$$\sum_{ij}(\vartheta) = \mathrm{E}\left(\frac{\partial}{\partial \vartheta_i} \log P_\vartheta(X_1, X_2) \frac{\partial}{\partial \vartheta_j} \log P_\vartheta(X_1, X_2)\right), \quad i, j = 1, \ldots, 4,$$

where $\vartheta_1 = p, \vartheta_2 = q, \vartheta_3 = \lambda_1, \vartheta_4 = \lambda_2.$

(b) $$2\{\ell_N(X,\hat{\vartheta}_N/X_0) - \ell_N(X,\vartheta/X_0)\} \underset{\mathscr{L}}{\longrightarrow} \chi_4^2 \quad \text{for } N \to \infty$$

and

$$2\{\ell_N(X,\hat{\vartheta}_N/X_0) - \ell_N(X,\vartheta/X_0)\} - N(\hat{\vartheta}_N - \vartheta)^T \Sigma(\vartheta)(\hat{\vartheta}_N - \vartheta) \underset{p}{\longrightarrow} 0$$

for $N \to \infty$.

The theorem follows from the more general theorem 22.5 which holds for any INAR(1) process satisfying certain rather weak conditions on the law of the ε_t. We postpone the verification that the SINAR(1) process satisfies those assumptions to the appendix. We conclude this section with the remark that an analogue of theorem 22.1 holds for the INAR(1) process with Poisson innovations, too, which has been discussed by Al-Osh and Alzaid (1987) provided that we assume $0 < p < 1$.

22.3 APPLICATION TO EPILEPTIC SEIZURE COUNTS

There were 126 patients, registered as outpatients for approximately half a year, and daily numbers of different types of seizures of these outpatients were recorded. During the first three months, only standard medication was applied. Then, for two weeks, half of the standard dose of the new drug or the placebo was administered. Afterwards, during a treatment period of 76 days, the patients in the test group got the new drug, whereas the patients in the control group used the placebo. During the whole study, daily numbers of seizures of different kinds were recorded. In our investigations we neglected the data from the first two weeks as the patients are getting used to the drug. Furthermore, we added the counts of all four types of partial seizures following the terminology of the Commission on Classification and Terminology of the International League Against Epilepsy (1981). The drug was presumed to be particularly effective against those types of seizures. Also, we wanted to get a time series which is not 0 on a substantial number of days. However, there may be problems with combining counts of different types of seizures. Ferryanto (1991) investigated the same data using Walsh–Fourier spectral analysis, he found a more prominent treatment effect for the time series corresponding to one particular type of partial seizures than for the combined series.

We applied the SINAR(1) model of section 22.2 to the epileptic fit counts of several patients from the test group and from the control group separately for the treatment phase and for the period before treatment. The CML estimates of the parameters $p, q, \lambda_1, \lambda_2$ were calculated by iterative maximization of the conditional log-likelihood using the form of the partial derivatives from (22.3)–(22.6) for judging the size of the gradient. The influence of the initial values turned out to be negligible except for the parameter p. Here, the difference of the value used for starting the iteration and the true parameter value should not be larger than about 0.15 to avoid problems with the maximization procedure. Fortunately, the Yule–Walker estimate of p (Al-Osh and Alzaid, 1987) provides a good initial value which can be easily calculated from the data. More details on the numerical considerations, in particular for estimates close to the boundary of the parameter set $\boldsymbol{\Theta}$, are given by Seligmann (1991).

A small simulation, presented in Table 22.1, shows that the CML estimates give reliable quantitative results for medium-sized samples, i.e. for N equal to 250 or larger. As the time series of seizure counts have a length of 86 (before treatment) and 76 (treatment phase), the estimates can only be interpreted qualitatively, e.g. with respect to changes caused by the treatment

Table 22.1 Mean and standard deviation of CML parameter estimates taken from 100 Monte Carlo runs of a SINAR(1) process with sample size $N = 250$

Parameter	True value	Mean of estimates	Standard deviation
p	0.20	0.195	0.060
q	0.50	0.677	0.168
λ_1	0.50	0.480	0.186
λ_2	2.00	2.109	0.522

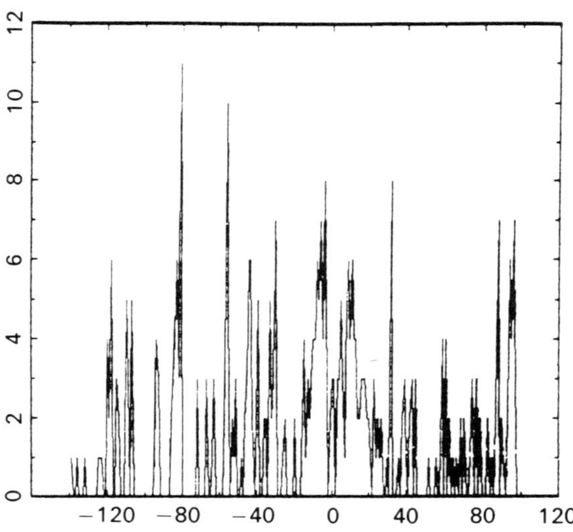

Figure 22.2. Daily epileptic seizure counts of patient no. 1 before treatment (up to day 0) and afterwards.

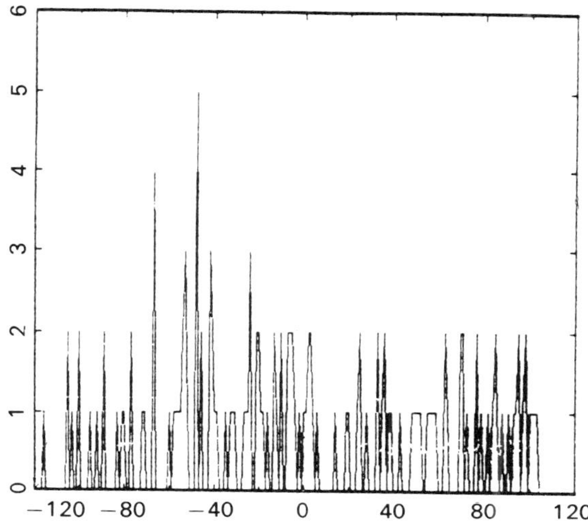

Figure 22.3. Daily epileptic seizure counts of patient no. 2 before treatment (up to day 0) and afterwards.

with drug or placebo. A smaller simulation with $N = 100$ showed that for such sample sizes the parameter estimates still show substantial variability.

Figures 22.2 and 22.3 present the time series for two patients from the test group where negative numbers correspond to the days before treatment. They serve as illustrations for the larger number of data sets which we have investigated. Table 22.2 contains the CML parameter estimates for both persons calculated separately for the period with and without application of the drug. In both cases, the drug led to a considerable reduction of the parameter λ_2 and a corresponding smaller increase of λ_1, and the parameter p as a measure of serial dependence decreased. These observations are typical for patients from the test group, whereas in the control group where a placebo replaced the drug we could not observe a similar consistent pattern.

Table 22.2 SINAR(1) parameter estimates for two patients from the test group before treatment and during the application of the drug

	Patient No. 1		Patient No. 2	
Parameter	Before treatment	During treatment	Before treatment	During treatment
p	0.068	0.000	0.059	0.016
q	0.557	0.493	0.705	0.489
λ_1	0.039	0.142	0.297	0.583
λ_2	3.315	2.498	1.354	0.590

We checked the model fit by applying a χ^2 test comparing the observed frequencies of days with given seizure counts with the marginal distribution of seizure counts (calculated by Monte Carlo simulation) for a SINAR(1) process with parameters equal to the CML estimates from the data. Those tests and an additional look at residuals made the class of SINAR(1) models appear as a reasonable description of the data (compare Seligmann, 1991, for the details).

With some caution, the parameters of the SINAR(1) process can even be interpreted. p measures the tendency to 'bursts' of seizure activity on consecutive days, referred to as clustering (Rodin, 1968). One explanation would be that the occurrence of an epileptic fit itself makes the patient more prone to get another fit in the near future (Hopkins *et al.*, 1985). On the other hand, the serial correlation of seizure counts may also be due to some unobservable underlying factor, for example, physiological or meteorological cycles usually influence the likelihood of epileptic fits.

From a pragmatic point of view, the parameters q, λ_1, λ_2 just describe a simple class of innovation laws which allow for the overdispersion observed in the data. On the other hand, the SINAR(1) model fits quite well into the philosophy of two states of seizure susceptibility (Hopkins *et al.*, 1985). λ_1, λ_2 give the intensity of seizures in the two states, and $q/(1-q)$ describes their relative proportion over time.

If one is willing to accept this kind of interpretation, then the data analysis shows that the drug under investigation reduces the intensity in the state of higher susceptibility or even causes the distinction between both states to vanish at all. Simultaneously, the serial correlation described by p is reduced by the drug. Both patients of Table 22.2 show the drug effects in a certain extreme sense: for patient 1 we have $\hat{p}=0$ in the treatment phase, and for patient 2 we have $\hat{\lambda}_1=\hat{\lambda}_2$, i.e. the drug causes the two states to merge. The reduction of the mean number of daily seizures caused by the drug can now be described in more detail as the reduction of clusters of consecutive days with many seizures (p decreases) and as the reduction of peak seizure counts (corresponding to the decrease of λ_2). To confirm this conjecture, a more extensive analysis of the data has, however, yet to be done.

We thank Dr. W. Sauermann, Goedecke A.G., Freiburg, for providing the data from a clinical study of Gabapentin (Neurontin®) which we used for testing the applicability of the SINAR(1)-model (See UK Gabapentin Study Group, 1990).

APPENDIX

Asymptotic behaviours of CML estimates for INAR(1) parameters

We consider an INAR(1) process $\{X_t, -\infty < t < \infty\}$ given by (22.2) where we assume $0 < p < 1$. Let

$$q^{\beta}(k) = pr_{\beta}(\varepsilon_t = k), \quad k \geqslant 0,$$

denote the weights of the law of the innovations ε_t which depend on some parameter $\beta \in B$ where B is an open subset of $\mathbb{R}^d$. Let

$$P(m,n) = \sum_{j=0}^{m \wedge n} q^{\beta}(n-j)\binom{m}{j} p^j (1-p)^{m-j}, \quad m, n \geqslant 0 \tag{22.7}$$

be the transition probabilities of the Markov chain $\{X_t\}$. To simplify notation, we do not stress their dependence on p and β explicitly.

Du and Li (1991) have shown the existence of a weakly stationary solution of (22.2) under the assumption of a second moment condition on the law of ε_t. We are interested in strict stationarity. For sake of simplicity, we assume $q^{\beta}(0) > 0$; otherwise, we would subtract $\min\{k; q^{\beta}(k) > 0\}$ from the X_t and study this shifted Markov chain. Also, we exclude the degenerate case $q^{\beta}(0) = 1$. Then, we have:

Theorem 22.2

Let $0 < p < 1$, $0 < q^{\beta}(0) < 1$. Then, any solution of (22.1) is an irreducible and aperiodic Markov chain on N_0. If, additionally, $\mathrm{E}\varepsilon_t < \infty$, then there exists a strictly solution of (22.2).

Proof

The irreducibility and aperiodicity of $\{X_t\}$ follows immediately from (22.7) as we have assumed $0 < q_{\beta}(0) < 1$. Now, let

$$Q(s) = \sum_{k=0}^{\infty} q^{\beta}(k) s^k, \quad 0 \leqslant s \leqslant 1,$$

be the generating function of the weights $q^{\beta}(k)$. Then, for any $j \geqslant 0$, $-\infty < t < \infty$, we have

$$pr(p^j \circ \varepsilon_s = 0) = \sum_{k=0}^{\infty} q^{\beta}(k)(1-p^j)^k = Q(1-p^j).$$

As $p^j \circ \varepsilon_{t-j}$, $j \geqslant 0$, are independent and non-negative, we conclude

$$pr\left(\sum_{j=0}^{n-1} p^j \circ \varepsilon_{t-j} = 0\right) = \prod_{j=0}^{n-1} Q(1-p^j).$$

Using $\mathrm{E}\varepsilon_t < \infty$ and (22.1),

$$\sum_{j=0}^{n-1} p^j \circ \varepsilon_{t-j} \longrightarrow \sum_{j=0}^{\infty} p^j \circ \varepsilon_{t-j}$$

in the mean, and, as pointed out by Al-Osh and Alzaid (1987), the right-hand side has the same distribution as X_t. By our assumptions on p and $q^\beta(0)$, $pr(X_t = 0) > 0$, and, therefore, we have

$$\lim_{t \to \infty} \prod_{j=0}^{t-1} Q(1 - p^j) > 0. \tag{22.8}$$

By theorem 1.2.1 of Rosenblatt (1971) and the remarks before the statement of this result, it remains to show that 0 (and then any other state) is positive recurrent if (22.8) holds. Let

$$P^{(t)}(m, n) = pr(X_t = n / X_0 = m)$$

denote that t-step transition probabilities. As usual, we have

$$P^{(t+1)}(m, n) = \sum_{k=0}^{\infty} P(m, k) P^{(t)}(k, n), \quad t \geqslant 1. \tag{22.9}$$

Below, we prove

$$P^{(t)}(0, 0) = \prod_{j=0}^{t-1} Q(1 - p^j), \quad t \geqslant 1. \tag{22.10}$$

As $0 \leqslant Q(s) \leqslant 1$ for all $0 \leqslant s \leqslant 1$, $P^{(t)}(0, 0)$ does not increase with t. Therefore, (22.8) immediately implies that

$$\sum_{t=0}^{\infty} P^{(t)}(0, 0) = \infty,$$

i.e. 0 is a recurrent state. Let μ_0 be the mean recurrence time for 0. Theorem 1.2.2 of Rosenblatt (1971) implies that $1/\mu_0$ is the limit of (22.8) and, therefore, $\mu_0 < \infty$, i.e. 0 is positive recurrent.

It remains to prove (22.10). For this purpose, let

$$B_m(s) = \sum_{j=0}^{m} \binom{m}{j} p^j (1 - p)^{m-j} s^j = \{1 - (1 - s)p\}^m$$

be the generating function of the bionomial weights, and let

$$F_m(s) = \sum_{n=0}^{\infty} P(m, n) s^n$$

be the generating function of the transition probabilities. As $P(m, n)$ is the convolution of binomial weights and the weights $q^\beta(k)$, we have $F_m(s) = B_m(s) Q(s)$.

These relations, (22.9) and

$$P(m, 0) = (1 - p)^m q^\beta(0) = (1 - p)^m Q(0)$$

imply by induction the following generalization of (22.10):

$$P^{(t)}(m, 0) = (1 - p^t)^m \prod_{j=0}^{t-1} Q(1 - p^j), \quad t \geqslant 1.$$

In the following, we are interested in estimating the parameter $\vartheta = (p, \beta)$ from the data $X = (X_0, \ldots, X_N)$. We consider the conditional log-likelihood

$$\ell_N(X, \vartheta/X_0) = \sum_{t=1}^{N} \log P(X_{t-1}, X_t),$$

and define the conditional maximum-likelihood (CML) estimates $\hat{\vartheta} = (\hat{p}, \hat{\beta})$ as a solution of

$$\frac{\partial}{\partial p} \ell_N(X, \vartheta/X_0) = 0$$

$$\frac{\partial}{\partial \beta_u} \ell_N(X, \vartheta/X_0) = 0, \quad u = 1, \ldots, d. \tag{22.11}$$

First, we remark that $P(m, n)$ and its partial derivative with respect to p satisfy the useful recurrence relations given in lemma 22.3 where, as a convention, we define $P(m, n) = 0$ for $m = -1$ or $n = -1$.

Lemma 22.3

(a) $$P(0, n) = q^{\beta}(n) \quad \text{for } n \geqslant 0,$$

$$P(m, n) = pP(m-1, n-1) + (1-p)P(m-1, n) \quad \text{for } m \geqslant 1,\ n \geqslant 0.$$

(b) $$\frac{\partial}{\partial p} P(m, n) = \frac{m}{1-p}\{P(m-1, n-1) - P(m, n)\} \quad \text{for } m, n \geqslant 0.$$

Proof

(a) For $m \geqslant 1$, let $Y_1, \ldots, Y_m$ be i.i.d. Bernoulli-variables with $p = pr(Y_j = 1)$. Then

$$\begin{aligned} P(m, n) &= pr\left(\sum_{j=1}^{m} Y_j + \varepsilon_t = n\right) \\ &= pr\left(Y_m = 1, \sum_{j=1}^{m-1} Y_j + \varepsilon_t = n - 1\right) + pr\left(Y_m = 0, \sum_{j=1}^{m} Y_j + \varepsilon_t = n\right) \\ &= pP(m-1, n-1) + (1-p)P(m-1, n). \end{aligned}$$

(b) $$\begin{aligned} \frac{\partial}{\partial p} P(m, n) &= -\frac{m}{1-p} P(m, n) + \sum_{j=1}^{m \wedge n} q^{\beta}(n-j)\binom{m}{j} j\{p^{j-1}(1-p)^{m-j} \\ &\quad + p^j(1-p)^{m-j-1}\} \\ &= -\frac{m}{1-p} P(m, n) + \sum_{j=1}^{m \wedge n} q^{\beta}(n-j)\binom{m-1}{j-1} mp^{j-1}(1-p)^{m-j-1} \\ &= \frac{m}{1-p}\{P(m-1, n-1) - P(m, n)\}. \end{aligned}$$

If additionally $p > 0$, an immediate consequence of lemma 22.3 is

$$-\frac{m}{1-p} \leqslant \frac{\partial}{\partial p} \log P(m,n) \leqslant \frac{m}{p}, \tag{22.12}$$

for all m, n with $P(m,n) > 0$.

As an abbreviation, we denote in the following partial derivatives with respect to β_u by a lower index u, e.g.

$$\frac{\partial}{\partial \beta_u} P(m,n) = P_u(m,n), \quad \frac{\partial^2}{\partial \beta_u \partial \beta_v} = q^\beta(k) = q^\beta_{uv}(k), \text{ etc.}$$

We want to apply results of Billingsley (1961) on estimates for the parameters of Markov processes. For this purpose we have to impose some regularity conditions on the weights $q^\beta(k)$:

$$\{k; q^\beta(k) > 0\} \text{ does not depend on } \beta; \tag{C1}$$

$$\mathrm{E}\varepsilon_t^3 = \sum_{k=0}^{\infty} k^3 q^\beta(k) < \infty; \tag{C2}$$

for any k, $q^\beta(k)$ is three times continuously differentiable on B; (C3)

for any $\beta' \in B$, there exists a neighbourhood U of β' such that

$$\sum_{k=0}^{\infty} \sup_{\beta \in U} q^\beta(k) \quad < \infty$$

$$\sum_{k=0}^{\infty} \sup_{\beta \in U} |q^\beta_u(k)| < \infty, \qquad u = 1, \ldots, d,$$

$$\sum_{k=0}^{\infty} \sup_{\beta \in U} |q^\beta_{uv}(k)| < \infty, \quad u, v = 1, \ldots, d; \tag{C4}$$

for $u, v, w = 1, \ldots, d$ and any $\beta' \in B$ there exists a neighbourhood U of β' and increasing sequences $\psi_u(n)$, $\psi_{uv}(n)$, $\psi_{uvw}(n)$, $n \geqslant 0$ (depending on β' and U) such that for all $\beta \in U$ and all $k \leqslant n$ with nonvanishing $q^\beta(k)$

$$|q^\beta_u(k)| \leqslant \psi_u(n) q^\beta(k)$$
$$|q^\beta_{uv}(k)| \leqslant \psi_{uv}(n) q^\beta(k)$$
$$|q^\beta_{uvw}(k)| \leqslant \psi_{uvw}(n) q^\beta(k)$$

and with respect to the stationary distribution of the INAR(1) process $\{X_t\}$

$$\mathrm{E}\psi_u^3(X_1) < \infty, \qquad \mathrm{E}X_1 \psi_{uv}(X_2) < \infty,$$
$$\mathrm{E}\psi_u(X_1)\psi_{vw}(X_1) < \infty, \quad \mathrm{E}\psi_{uvw}(X_1) < \infty; \tag{C5}$$

let $\Sigma(\vartheta) = (\sigma_{uv}(\vartheta))_{u,v=0,\ldots,d}$ denote the Fisher information matrix, i.e.

$$\sigma_{00}(\vartheta) = \mathrm{E}\left(\frac{\partial}{\partial p}\log P(X_1, X_2)\right)^2$$

$$\sigma_{0u}(\vartheta) = \mathrm{E}\left(\frac{\partial}{\partial p}\log P(X_1, X_2)\frac{\partial}{\partial \beta_u}\log P(X_1, X_2)\right) = \sigma_{u0}(\vartheta), \quad u = 1, \ldots, d,$$

$$\sigma_{uv}(\vartheta) = \mathrm{E}\left(\frac{\partial}{\partial \beta_u}\log P(X_1, X_2)\frac{\partial}{\partial \beta_v}\log P(X_1, X_2)\right), \quad u, v = 1, \ldots, d, \qquad \text{(C6)}$$

$\Sigma(\vartheta)$ is nonsingular.

Condition (C2) implies $\mathrm{E}X_t^3 < \infty$ for the stationary solution of (22.2). This can be shown completely analogous to the proof of theorem 2.1 of Du and Li (1991) where, among other things, the existence of the second moment of X_t is concluded from $\mathrm{E}\varepsilon_t^2 < \infty$. Note that conditions (C4) and (C5) are automatically satisfied for any innovation law with bounded support, i.e. with only finitely many nonvanishing weights $q^\beta(k)$.

Theorem 22.4

Let $\{X_t\}$ be an INAR(1) process satisfying the assumptions of theorem 22.3 and, additionally, (C1)–(C6). Then, there exists a consistent solution $\hat{\vartheta} = (\hat{p}, \hat{\beta})$ of (A5) which is a local maximum of $\ell_N(X, \vartheta/X_0)$ with probability going to 1. Moreover, any other consistent solution of (A5) coincides with $\hat{\vartheta}$ with probability going to 1.

Theorem 22.5

Under the assumptions of Theorem 22.4, the CML estimate $\hat{\vartheta} = (\hat{p}, \hat{\beta})$ is asymptotically normal, i.e.

$$\sqrt{N}(\hat{\vartheta} - \vartheta) \xrightarrow[\mathscr{L}]{} \mathcal{N}(0, \Sigma^{-1}(\vartheta)) \quad \text{for } N \to \infty.$$

Furthermore, for $N \to \infty$

$$2\{\ell_N(X, \hat{\vartheta}/X_0) - \ell_N(X, \vartheta/X_0)\} \xrightarrow[\mathscr{L}]{} \chi^2_{d+1}$$

and

$$2\{\ell_N(X, \hat{\vartheta}/X_0) - \ell_N(X, \vartheta/X_0)\} - N(\hat{\vartheta} - \vartheta)^T \Sigma(\vartheta)(-\vartheta) \xrightarrow[p]{} 0.$$

Proof of theorems 22.4 and 22.5

Both theorems are special cases of theorems 2.1 and 2.2 of Billingsley (1961). We only have to check that (C1)–(C6) imply the conditions of those general results.

(a) By (C1), (C3) and the explicit representation (22.7), $P(m,n)$ is three times continuously differentiable with respect to $p, \beta_1, \ldots, \beta_d$, and for any m, $\{n; P(m,n) > 0\}$ does not depend on p and β. Therefore, $\log P(m,n)$ is well-defined except on a set of $P(m,.)$-measure 0 which does not depend on the parameter values.

(b) For $n \geqslant m$, we have

$$P(m,n) = \sum_{j=0}^{m} q^{\beta}(n-j)\binom{m}{j} p^j (1-p)^{m-j} \leqslant \sum_{j=0}^{m} q^{\beta}(n-j).$$

The first relation of (C4), therefore, implies that for each $\vartheta' = (p', \beta')$ there exists a neighbourhood V such that for any fixed $m \geqslant 0$

$$\sum_{n=0}^{\infty} \sup_{\vartheta \in V} P(m,n) < \infty.$$

By lemma 22.3(b), the same summability condition holds for $\frac{\partial}{\partial p} P(m,n)$ and $\frac{\partial^2}{\partial p^2} P(m,n)$ and then, using the second and third relation of (C4), for all first and second derivatives of $P(m,n)$ with respect to $p, \beta_1, \ldots, \beta_d$, too.

(c) From (22.12) we know that in the stationary state

$$\mathrm{E}\left|\frac{\partial}{\partial p}\log P(X_1, X_2)\right|^2 \leqslant C \cdot \mathrm{E}X_1^2 < \infty$$

for a suitable constant C. Similarly, we have from (C5)

$$\left|\frac{\partial}{\partial \beta_u}\log P(m,n)\right| \leqslant \frac{1}{P(m,n)} \sum_{j=0}^{m \wedge n} |q_u^{\beta}(n-j)| \binom{m}{j} p^j (1-p)^{m-j}$$
$$\leqslant \psi_u(n)$$

and, therefore,

$$\mathrm{E}\left|\frac{\partial}{\partial \beta_u}\log P(X_1, X_2)\right|^2 \leqslant \mathrm{E}\psi_u^2(X_2) < \infty.$$

Therefore, the Fisher information matrix $\Sigma(\vartheta)$ is well-defined, and, by (C6), it is nonsingular.

(d) We have to show that local suprema of all third order derivatives of $\log P(X_1, X_2)$ have a finite mean. For this purpose, we use the abbreviations

$$d_i^p = \frac{\partial}{\partial p}\log P(m-i, n-i),$$

$$d_i^u = \frac{\partial}{\partial \beta_u}\log P(m-i, n-i), \quad i = 0, 1, \quad u = 1, \ldots, d.$$

From (22.12) and the first part of condition (C5) we know

$$|d_i^p| \leqslant \text{const} \cdot m, |d_i^u| \leqslant \psi_u(n) \tag{22.13}$$

where here and in the following 'const' stands for a generic constant which can be chosen independently of (p, β) in a suitable neighbourhood of any (p', β'), $0 < p < 1$, $\beta' \in B$. Using lemma 22.3 (b), a straightforward calculation shows

$$\frac{\partial^2}{\partial p^2} \log P(m, n) = \frac{1}{1-p} d_0^p + \left\{ d_0^p + \frac{m}{1-p} \right\} \{ d_1^p - d_0^p \}$$

and, therefore, using (22.13)

$$\left| \frac{\partial^2}{\partial p^2} \log P(m, n) \right| \leqslant \text{const} \cdot m^2.$$

Analogously, we get

$$\left| \frac{\partial^3}{\partial p^3} \log P(m, n) \right| \leqslant \text{const} \cdot m^3.$$

Therefore, for any $\vartheta' = (p', \beta')$ there exists a neighbourhood V such that in the stationary state

$$\mathrm{E} \sup_{\vartheta \in V} \left| \frac{\partial^3}{\partial p^3} \log P(X_1, X_2) \right| \leqslant \text{const}\, \mathrm{E} X_1^3 < \infty. \tag{22.14}$$

The same integrability condition has to hold for all third order derivatives of $\log P$. Again using Lemma 22.3 (b), we have for $u = 1, \ldots, d$

$$\frac{\partial^2}{\partial \beta_u \partial p} \log P(m, n) = \left\{ d_0^p + \frac{m}{1-p} \right\} \{ d_1^u - d_0^u \}$$

and, therefore, by (22.13) and monotonicity of ψ_u

$$\left| \frac{\partial^2}{\partial \beta_u \partial p} \log P(m, n) \right| \leqslant \text{const}\; m \psi_u(n).$$

Using the second inequality of (C5) and (22.13) we get

$$\left| \frac{\partial^2}{\partial \beta_u \partial \beta_v} \log P(m, n) \right| \leqslant \psi_{uv}(n) + \psi_u(n) \psi_v(n).$$

Analogously, we get bounds on the third order derivatives for $u, v = 1, \ldots, d$

$$\left| \frac{\partial^3}{\partial \beta_u \partial p^2} \log P(m, n) \right| \leqslant \text{const}\, m^2 \psi_u(n)$$

$$\left| \frac{\partial^3}{\partial \beta_u \partial \beta_v \partial p} \log P(m, n) \right| \leqslant \text{const}\, m \{ \psi_{uv}(n) + \psi_u(n) \psi_v(n) \}.$$

As, by condition (C5) and $\mathrm{E} X_1^3 < \infty$, $\mathrm{E} X_1^2 \psi_u(X_2) < \infty$, $\mathrm{E} X_1 \psi_u(X_2) \psi_v(X_2) < \infty$ and $\mathrm{E} X_1 \psi_{uv}(X_2) < \infty$, we have the same property as in (22.14) for all third order derivatives including at least one derivative with respect to p. As for all $u, v, w = 1, \ldots, d$

$$\left|\frac{\partial^3}{\partial\beta_u \partial\beta_v \partial\beta_w} \log P(m, n)\right|$$

$$\leqslant \psi_{uvw}(n) + \psi_u(n)\psi_{vw}(n) + \psi_v(n)\psi_{uw}(n) + \psi_w(n)\psi_{uv}(n) + 2\psi_u(n)\psi_v(n)\psi_w(n)$$

by condition (C5), we can also conclude that a property like (22.14) is shared by all third order derivatives of $\log P$.

(e) Points (a)–(d) together just imply that condition 1.1 of the above-mentioned theorems of Billingsley (1961) is satisfied. Theorem 1.3 of Billingsley (1961) and the proof of theorem 22.2 immediately imply the first part of condition 1.2 of Billingsley, and the second part follows from (c) and the footnote to this condition.

The conditions of theorem (22.2) and (C1)–(C6) are not restrictive. As an example and as a step towards the proof of theorem 22.1 we show that they are satisfied for Poisson innovations ε_t, i.e. for $\beta = \lambda$ and

$$q^{\beta}(k) = \frac{\lambda^k}{k!} e^{-\lambda} = e(k, \lambda).$$

Conditions (C1)–(C3) are obviously satisfied for Poisson weights. For any open interval $U \subseteq (0, \infty)$, $\sup_{\lambda \in U} e(k, \lambda)$ is summable, and together with

$$\frac{\mathrm{d}}{\mathrm{d}\lambda} e(k, \lambda) = e'(k, \lambda) = e(k-1, \lambda) - e(k, \lambda) \tag{22.15}$$

$$\frac{\mathrm{d}^2}{\mathrm{d}\lambda^2} e(k, \lambda) = e''(\lambda) = e(k-2, \lambda) - 2e(k-1, \lambda) + e(k, \lambda)$$

we have condition (C4). As also

$$e'(k, \lambda) = \left(\frac{k}{\lambda} - 1\right) e(k, \lambda), \tag{22.16}$$

the first inequality of (C5) is satisfied with $\psi_1(n) = \text{const} \cdot n$ where 'const' again stands for a suitable constant. Analogously we get by taking further derivatives with respect to λ that we can choose $\psi_{11}(n) = \text{const} \cdot n^2$ and $\psi_{111}(n) = \text{const} \cdot n^3$ in (C5). As the validity of (C2) implies $\mathrm{E}X_t^3 < \infty$, we also have the moment conditions of (C5).

Condition (C6) guarantees that the parameters of the INAR(1) process are not redundant. It is, e.g., satisfied if the matrix with entries

$$\sum_m(\vartheta) = \mathrm{E}\left(\frac{\partial}{\partial\beta_u} \log P(X_1, X_2) \cdot \frac{\partial}{\partial\beta_v} \log P(X_1, X_2) / X_1 = m\right), \quad u, v = 0, \ldots, d$$

(with $\beta_0 \equiv p$) is nonsingular for a set of m with positive measure under the stationary distribution. For Poisson innovations ε_t we calculate analogous to the proof of lemma 22.3(b):

$$\frac{\partial}{\partial\lambda}P(m,n)=\left(\frac{n}{\lambda}-1\right)P(m,n)-\frac{mp}{\lambda}P(m-1,n-1).$$

Using this relation and lemma 22.3(b), we get, with

$$D(m,n)=\frac{P(m-1,n-1)}{P(m,n)},$$

$$\frac{\lambda^2(1-p)^2}{m^2}\det\sum_m(p,\lambda)=\operatorname{var}\{(D(m,X_2)-1)(X_2-\lambda-mpD(m,X_2))\}.$$

This is positive except for the degenerate case

$$(D(m,n)-1)(n-\lambda-mpD(m,n))=\text{const for all } n\geqslant 0,$$

which can be true for at most a few values of m. Therefore (C6) is satisfied for the Poisson innovation law.

Proof of theorem 22.1

The weights of the innovations ε_t are with $\beta=(\lambda_1,\lambda_2,q)$

$$q^{\beta}(k)=qe(k,\lambda_1)+(1-q)e(k,\lambda_2). \tag{22.17}$$

Theorem 22.2 guarantees the existence of a strictly stationary SINAR(1) process with parameters λ_1,λ_2,q. (C1)–(C3) are, again, obviously satisfied. (22.15) and (22.17) imply (C4) as above for the Poisson weights. A straightforward calculation, using (22.16) and (22.17), shows that (C5) is also satisfied with

$$\psi_1(n)=\text{const}\cdot n,\ \psi_2(n)=\text{const}\cdot n,\ \psi_3(n)=\text{const},$$
$$\psi_{11}(n)=\text{const}\cdot n^2,\ \psi_{22}(n)=\text{const}\cdot n^2,\ \psi_{33}(n)=0,$$
$$\psi_{12}(n)=0,\ \psi_{13}(n)=\text{const}\cdot\psi_1(n),\ \psi_{23}(n)=\text{const}\cdot\psi_2(n),$$
$$\psi_{111}(n)=\text{const}\cdot n^3,\ \psi_{222}(n)=\text{const}\cdot n^3,\ \psi_{113}(n)=\text{const}\cdot\psi_{11}(n),$$
$$\psi_{223}(n)=\text{const}\cdot\psi_{22}(n)\text{ and all other }\psi_{uvw}(n)=0.$$

Finally, some elementary but tedious calculation shows that (C6) is satisfied, too (compare Seligmann, 1991).

REFERENCES

Albert, P.S. (1991) A two-state mixture model for analyzing time series count data. *Biometrics*, **47**, 1371–1381.

Al-Osh, M.A., and Alzaid, A.A. (1987) First-order integer-valued autoregressive (INAR(1)) processes. *J. Time Ser. Anal.*, **8**, 261–275.

Billingsley, P. (1961) *Statistical Inference for Markov processes.* University of Chicago Press, Chicago.

Commission on Classification and Terminology of the International League Against Epilepsy (1981) Proposal for revised clinical and electroencephalographic classification of epileptic seizures. *Epilepsia*, **22**, 489–501.

Du J.-G., and Li Y. (1991) The integer-valued autoregressive (INAR(p)) model. *J. Time Ser. Anal.*, **12**, 129–142.

Fahrmeir, L., and Kaufmann, H. (1987) Regression models for non-stationary categorical time series. *J. Time Ser. Anal.*, **8**, 147–160.

Ferryanto (1991) *The Effects of Drugs and Placebo on Epileptic fit-State Cycling—An Application of the Walsh-Fourier Analysis.* Masters Thesis, University of Kaiserslautern.

Hopkins, A., Davies, P., and Dobson, C. (1985) Mathematical models of patterns of seizures: Their use in the evaluation of drugs. *Archives of Neurology*, **42**, 463–467.

Jacobs, P.A., and Lewis, P.A.W. (1983) Stationary discrete autoregressive-moving average time series generated by mixtures. *J. Time Ser. Anal.*, **4**, 18–36.

Kaufmann, H. (1987) Regression models for non-stationary categorical time series: Asymptotic estimation theory. *Ann. Statist.* **15**, 79–98.

McCullagh, P., and Nelder, J.A. (1983) *Generalized Linear Models.* Chapman and Hall, London.

McKenzie, E. (1985) Discussion of Modelling and Residual Analysis of Nonlinear Autoregressive Time Series in Exponential Variables by A.J. Lawrence and P.A.W. Lewis. *JRSS Ser. B*, **47**, 187–188.

McKenzie, E. (1986) ARMA processes with negative binomial and geometric marginal distributions. *Adv. Appl. Prob.*, **18**, 679–705.

McKenzie, E. (1987) Innovation distributions for Gamma and negative binomial autoregressions. *Scand. J. Statist.*, **14**, 79–85.

McKenzie, E. (1988) Some ARMA models for dependent sequences of Poisson counts. *Adv. Appl. Prob.* **20**, 822–835.

Milton, J.G., Gotman, J., Remillard, G.M., and Andermann, F. (1987) Timing of seizure recurrence in adult epileptic patients: A statistical analysis. *Epilepsia*, **28**, 471–478.

Rodin, E.A. (1968) *The Prognosis of Patients with Epilepsy*, Charles C. Thomas, Springfield.

Rosenblatt, M. (1971) *Markov Processes, Structure and Asymptotic Behaviour*, Springer, Berlin-Heidelberg-New York.

Seligmann, Th. (1991) *Der SINAR-Prozeβ-ein stochastisches Modell für Epilepsiedaten*, Diploma Thesis, University of Kaiserslautern (in German).

Seneta, E. (1969) Functional equations and the Galton–Watson process. *Adv. Appl. Prob.*, **1**, 1–42.

UK Gabapentin Study Group (1990) Gabapentin in partial epilepsy. *Lancet*, **335**, 1114–1117.

Venkataraman, K.N. (1982) A time series approach to the study of the simple subcritical Galton-Watson process with immigration. *Adv. Appl. Prob.*, **14**, 1–20.

Venkataraman, K.N., and Nanthi, K. (1982) A limit theorem on subcritical Galton–Watson process with immigration. *Ann. Probab.*, **10**, 1069–1074.

Wedderburn, R.W.M. (1974) Quasi-likelihood functions, generalized linear models and the Gauss-Newton method. *Biometrika*, **61**, 439–447.

Zeger, S.L. (1988). A regression model for time series of counts. *Biometrika*, **75**, 621–629.

Zeger, S.L., and Qaqish, B. (1988) Markov regression models for time series: a quasi-likelihood approach *Biometrics*, **44**, 1019–1031.

23

An application of statistics to seismology: dispersion and modes

D.R. Brillinger

23.1 INTRODUCTION

Maurice Priestley has made substantial contributions to the analysis of time series. I mention, in particular, his work with the conceptualization of the idea of spectrum of a non-stationary time series (see Priestley 1965, his discussion of Loynes 1968 and Priestley and Tong (1973)) and with processes having discontinuous spectra (see Priestley 1962a,b; 1964). This present paper contains elements of both of these topics. Professor Priestley's work is seminal and broadly applicable. It has certainly influenced my own work.

23.2 OSCILLATIONS OF FINITE BODIES

Suppose that there is a finite body and suppose for the moment that it is linear, lying above the x-axis. Let $Y(x,t)$ denote the height at time t, of the body above coordinate x on the x-axis. In a simple case the motion could be given by

$$Y(x,t) = a\cos(kx - \omega t + \phi). \qquad (23.1)$$

The Fourier transform of the complex signal associated to $Y(x,t)$, with respect to x and t, is

$$a\frac{\mathrm{e}^{i\phi}}{(2\pi)^2}\iint \mathrm{e}^{i(kx-\omega t)}\mathrm{e}^{i(Kx+\Omega t)}\,\mathrm{d}x\,\mathrm{d}t = a\mathrm{e}^{i\phi}\delta(\Omega-\omega)\delta(K+k)$$

with $\delta(\cdot)$ the Dirac delta function. This transform is concentrated at a point in the $\Omega - K$ plane. Here ω, Ω refer to frequency and K, k to wavenumber.

In the case of a finite body, discreteness occurs, that is for given ω, k is restricted to a discrete set of possible values,

$$k = k_n(\omega) \qquad (23.2)$$

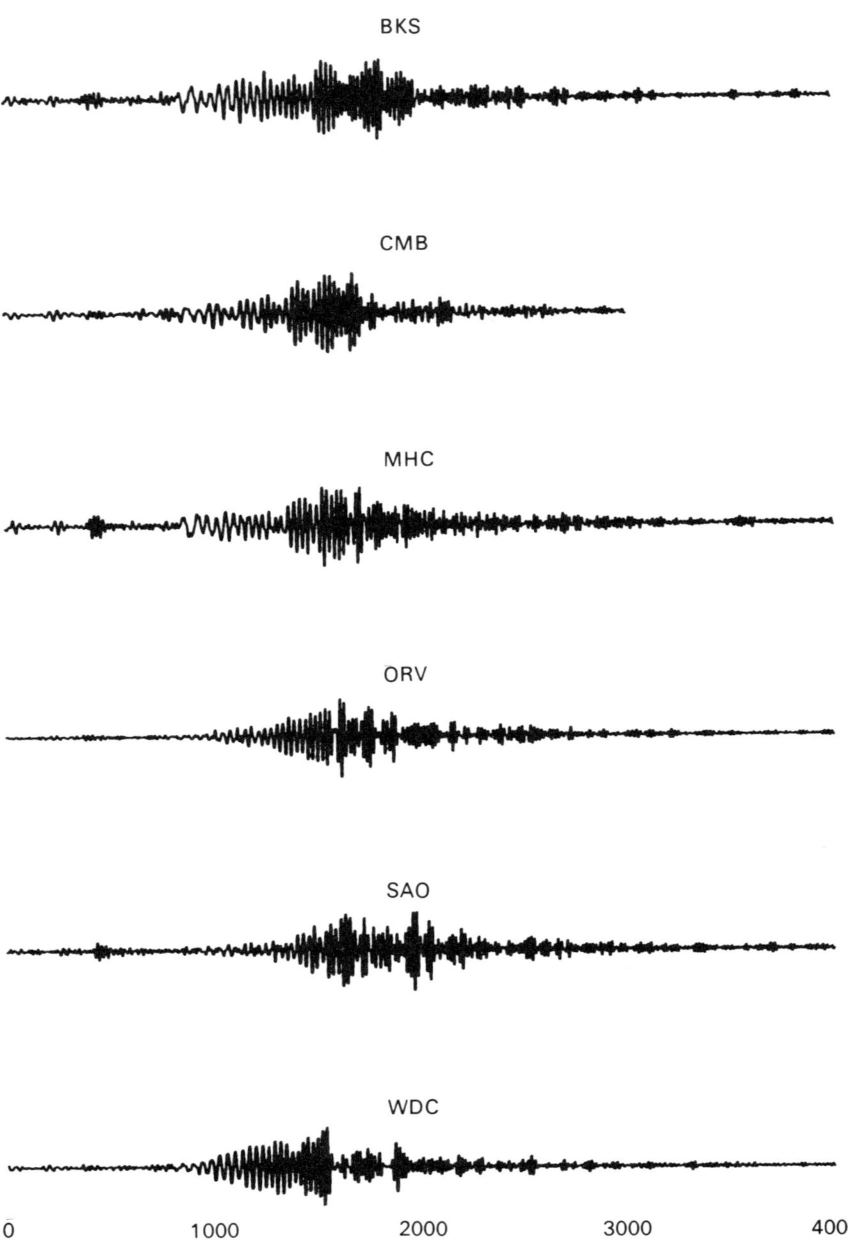

Figure 23.1. Vertical velocity records of the Armenian earthquake recorded at six stations of the Berkeley network. The concern is with Rayleigh (or surface) waves. These are seismic waves whose energy is trapped near the Earth's surface. Station locations indicated in text.

for discrete $n = 0, 1, \ldots$. The two dimensional transform of data $Y(x, t)$ of such a circumstance would lie on a sequence of curves. The allowable oscillations are referred to as modes with the case of $n = 0$ called the fundamental mode.

The (phase) velocity of the signal (23.1) is ω/k. In the case of (23.2), the ratio ω/k will generally not be constant in ω, and 'waves' of different frequencies travel with different velocities. This phenomenon is called dispersion.

Dispersion is observable in nature. One example is provided by the edge waves of oceanography. These are sea surface waves that can travel along, sidewise to the shore, in bays of certain geometry. Munk *et al.* (1964) analyze data collected in a linear array of 44 bottom pressure sensors spaced .73 km apart off the coast of southern California with Y now referring to pressure. In the above notation the data would be $Y(x_j, t)$ with $x_j = .73j$ and $j = 1, \ldots, 44$. The mass of the two-dimensional Fourier transform Munk *et al.* compute is seen to fall near curves. These curves turn out to be very close to the theoretical curves predicted by a model of the geometry of the bay concerned. A second example of the occurence of two-dimensional Fourier transforms lying close to curves is provided by the case of helioseismology. This is the subject concerned with the internal oscillations of the Sun. The cover of *Science* magazine for 6 September 1985 provides a striking example of the occurrence of such curves.

23.3 THE ARMENIAN EARTHQUAKE

The particular data that will be addressed in this paper was generated by the Armenian earthquake of 1988. This earthquake took place on 7 December at 07 hours, 41 minutes, 24.2 seconds Greenwich mean time in northern Armenia. It had an estimated magnitude of $M_S = 7.0$ and depth of 10 km. Approximately 25 000 people were killed and 19 000 injured. The signal travelled around 11 000 km along the surface of the Earth to the array of digital seismometers operated by the University of California, Berkeley in northern California.

Figure 23.1 presents records of the vertical velocity recorded at the stations BKS (Berkeley), ORV (Oroville), WDC (Whiskeytown), SAO (San Andreas), MHC (Mount Hamilton), CMB (Columbia). Their locations are scattered about northern California.

23.4 A STRUCTURAL MODEL

The frequency components of various seismic waves travel with differing velocities. Given a structural model, the dependence of frequency on velocity, and vice-versa, may be computed. Figure 23.2 presents an example of such a model. This particular model is referred to as a homogeneous layer over a homogeneous half-space. The parameters α refer to compressional velocities, β to shear velocities, ρ to densities, each for the top and lower

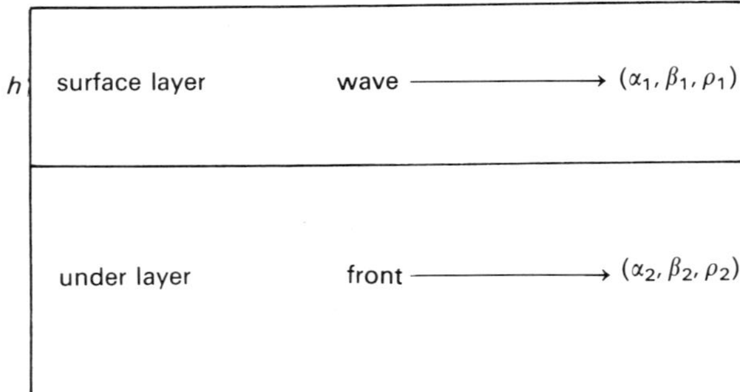

Figure 23.2. Earth model of a single layer over a half space.

layers respectively and the parameter h refers to thickness of the top layer. In the figure, the wave is envisaged as travelling from left to right.

Let x refer to distance from the origin of the earthquake to the observatory. The signal $Y(x, t)$ observed may be viewed as the motion (e.g. displacement, velocity or acceleration of the particle) at x at time t following an impulse at $x = 0$ and $t = 0$. Now it may be shown that the equations of motion have a solution only if a particular determinental criterion vanishes, see Bolt and Butcher (1960) and Bullen and Bolt (1985). The corresponding solution is

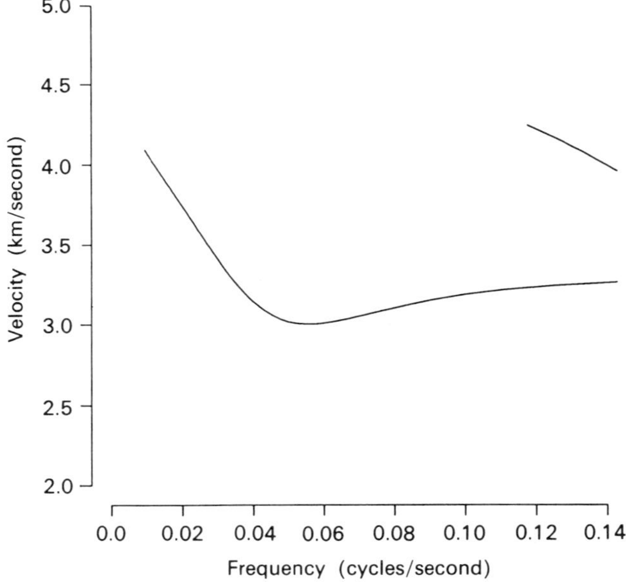

Figure 23.3. Velocities with which the frequency components particular surface waves (Rayleigh waves) travel for a specified Earth model.

called a dispersion relation. It gives the velocities with which the various frequencies travel. Higher modes correspond to further solutions. Figure 23.3 gives an example of such curves for one particular Earth model. In this case a fundamental and one higher mode occur.

23.5 DYNAMIC SPECTRUM ANALYSIS

By a dynamic spectrum will be meant a display of estimated intensity as a function of time and frequency. A naive method to derive such a display is

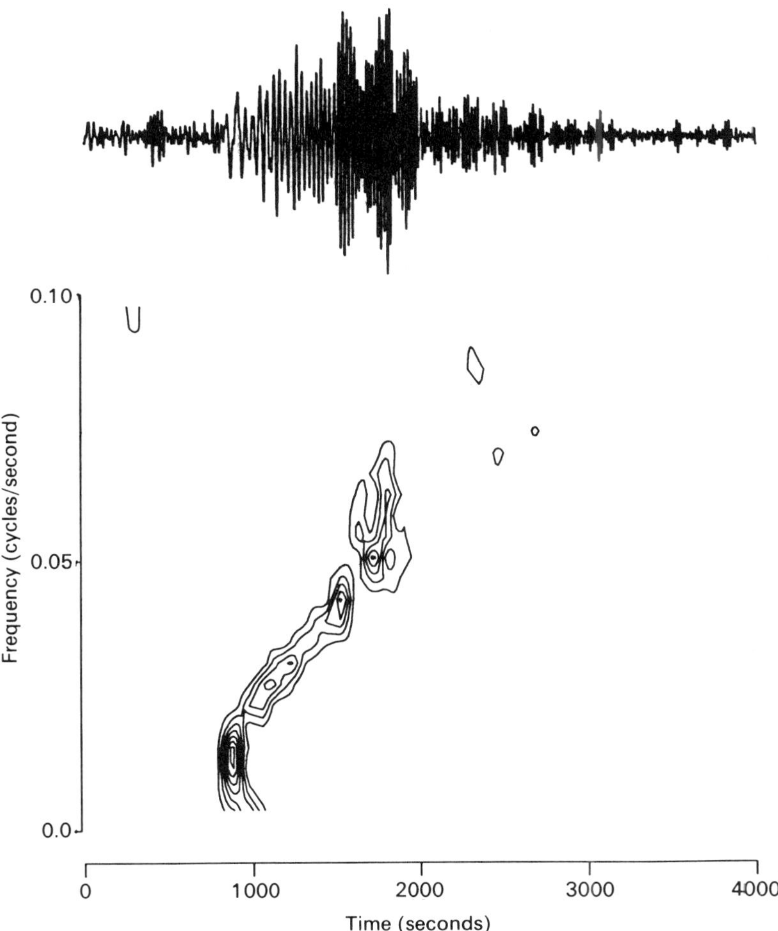

Figure 23.4. 1988 Armenian earthquake–top graph is vertical as recorded at Berkeley. Bottom graph is the estimated signal intensity as a function of frequency and time.

via a sliding Fourier analysis. Specifically, given a stretch of the time series $Y(t)$, one can compute

$$\left[\sum_v Y(t-v)\cos\omega v\right]^2 + \left[\sum_v Y(t-v)\sin\omega v\right]^2$$

with t referring to time, ω frequency and v summing over a restricted time interval. Given a contour plot of such an intensity as a function of (t, ω) one can look for ridges. In the seismological case these correspond to frequency components travelling with differing velocities.

Figure 23.4 presents part of the seismic trace of the Armenian earthquake, as recorded at the Berkeley station, and a parallel dynamic spectrum. The intensity display has the appearance of a ridge. Different frequency components do appear to be arriving at different times. The part of the seismic record analyzed, corresponds to what are called surface waves. These are waves whose energy is trapped near the Earth's surface. They provide information about the upper part of the mantle.

Dynamic spectrum analyses of earthquake records are presented in Dziewonski *et al.* (1969) and Levshin *et al.* (1972), for example.

What has usually been done, when seismologists have been concerned with fitting an Earth model, is that the parameter values of the model have been guessed until curves following the ridges of a plot like Figure 23.4 have been found. An intent of this present work is to automate that procedure.

23.6 PARAMETER ESTIMATION

When working empirically with dispersion curves and dynamic spectra, it is convenient to convert the time axis of the spectrum to velocity. This is a simple transformation $v = x/t$, the distance x to the earthquake source being known. Figure 23.5 presents the corresponding plots for the six stations of the Berkeley Digital Seismic Network. In each case there is a strong suggestion of ridges and of frequency components travelling with different velocities.

A basic problem of statistical concern is that of determining parameters of an Earth model leading to curves close to the ones suggested by Figure 23.5.

In the present case the following naive estimation procedure was employed. For station i and frequency ω, the velocity, $\hat{U}_i(\omega)$, where the intensity of the dynamic spectrum was largest, was determined. Next, $U(\omega|\theta)$ denotes the velocity at frequency ω of the fundamental mode for a particular Earth. Here θ denotes the unknown parameters. The estimate $\hat{\theta}$ is then determined by minimizing

$$\sum_{i=1}^{6} \sum_{\omega} [\hat{U}_i(\omega) - U(\omega|\theta)]^2 \tag{23.3}$$

as a function of θ. Figure 23.6 gives the $\hat{U}_i(\omega)$ and fitted curve $U(\omega|\hat{\theta})$ for

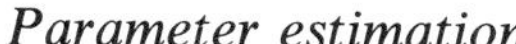

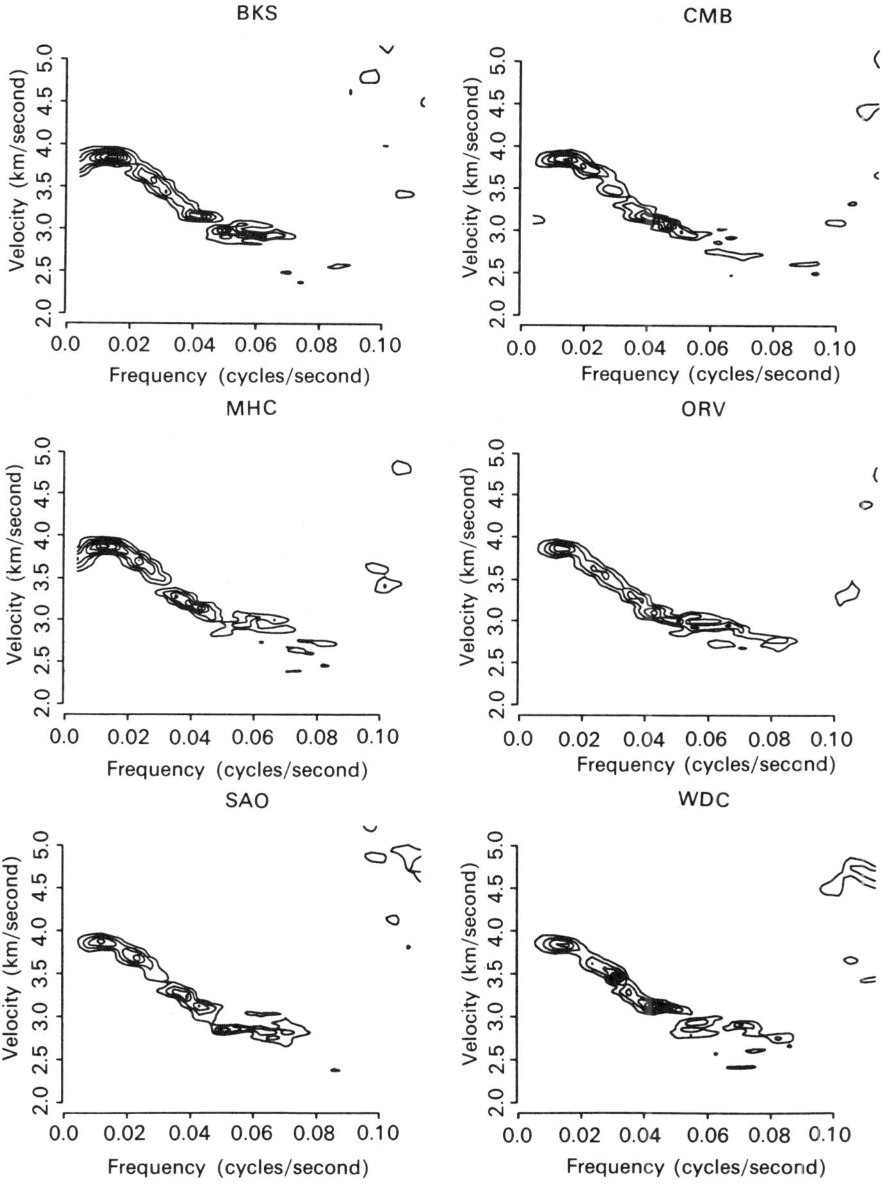

Figure 23.5. Estimated signal intensities as a functions of frequency and velocity.

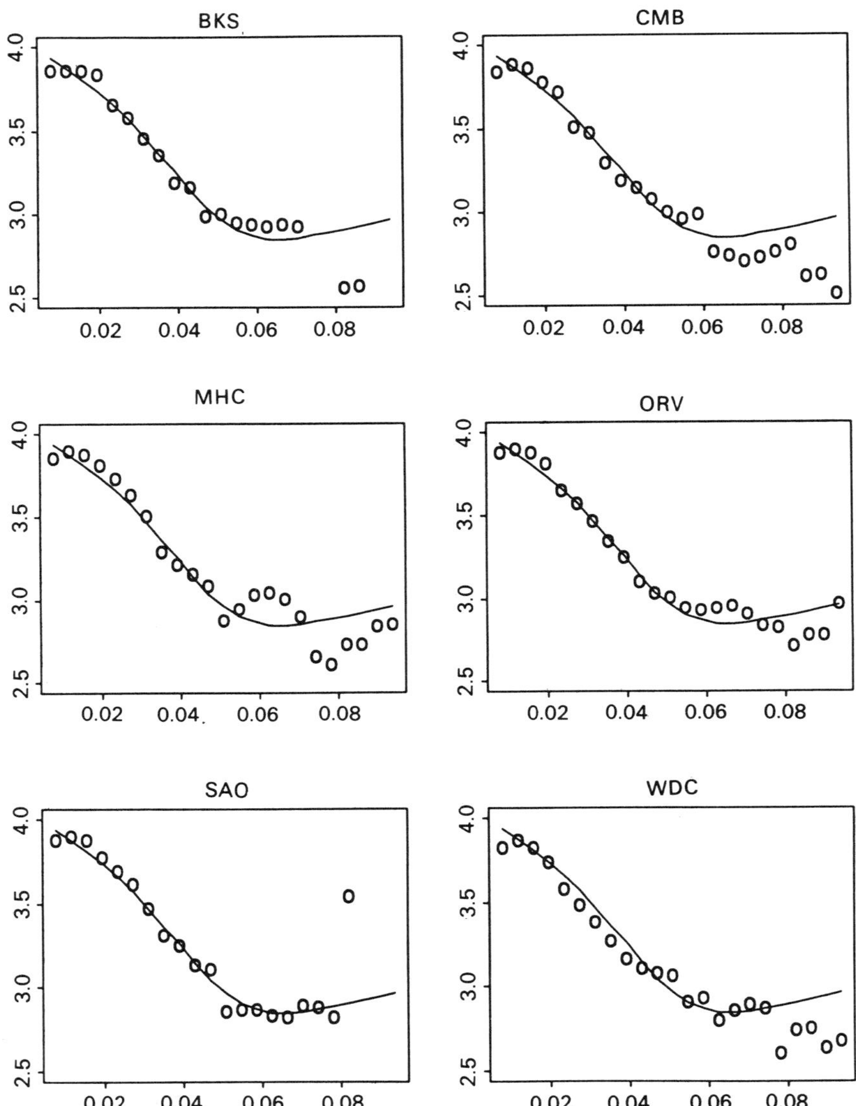

Figure 23.6. For given frequency, the '0' indicates the velocity at which the intensity was largest for the given station. The curve is the result of the fitting. The same curve is plotted for each station. The vertical axis is velocity in *k*/sec; the horizontal axis is frequency in cycles/sec.

the six stations. The fits appears reasonable, particularly at the lower frequencies where the signal to noise ratio is greatest.

In the model the ratio of densities, ρ_2/ρ_1 was taken to be 1.2, a figure derived from independent studies. The estimated parameter values are as follows:

$$\hat{h} = 22.61\ \text{km}$$

$$\hat{\alpha}_1 = 4.94\ \text{km/sec}$$

$$\hat{\beta}_1 = 3.88\ \text{km/sec}$$

$$\hat{\alpha}_2 = 6.62\ \text{km/sec}$$

$$\hat{\beta}_2 = 4.62\ \text{km/sec.}$$

Uncertainty is estimated via the jack-knife, events are dropped in turn from the criterion (23.3). Approximate 95% confidence intervals, taking the traces to be independent and the errors to be normal of constant variance, are

$$10.11 < h < 50.56$$

$$4.48 < \alpha_1 < 5.06$$

$$3.23 < \beta_1 < 4.65$$

$$1.17 < \alpha_2 < 37.37$$

$$1.01 < \beta_2 < 21.26.$$

23.7 DISCUSSION

The example of this paper shows the usefulness of the concept of spectrum for a nonstationary signal. It further illustrates the novel case where a two-dimensional Fourier transform is neither continuous, nor concentrated at points, but rather lies on curves.

The present approach makes no specific use of higher modes. In any case these may not have been excited in the present event. An alternate procedure, making use of higher modes and other data sets, is under development in joint research with B.A. Bolt.

REFERENCES

Bolt, B.A. and Butcher, J.C. (1960) Rayleigh wave dispersion for a single layer on an elastic half space. *Australian J. Physics*, **13**, 498–504.

Bullen, K.E. and Bolt, B.A. (1985) An Introduction to the Theory of Seismology. Cambridge University Press, Cambridge.

Dziewonski, A., Bloch, S. and Landisman, M. (1969) A technique for the analysis of transient seismic signals. *Bull. Seismol. Soc. America* **59**, 427–444.

Levshin, A.L., Pisarenko, V.F. and Pogrebiny, G.A. (1972) On a frequency-time analysis of oscillations. *Ann. Geophys.*, **28**, 211–218.

Loynes, R.M. (1968) On the concept of the spectrum for nonstationary processes. *J.R. Statist. Soc. B*, **30**, 1–30.

Munk, W., Snodgrass, F. and Gilbert, F. (1964) Long waves on the continental shelf: an experiment to separate trapped and leaky modes. *J. Fluid Mech.*, **20**, 529–554.

Priestley, M.B. (1962a) Analysis of stationary processes with mixed spectrum—*I.J.R. Statist. Soc. B*, **24**, 215–233.

Priestley, M.B. (1962b) Analysis of stationary processes with mixed spectrum—II. *J.R. Statist. Soc. B*, **24**, 511–529.

Priestley, M.B. (1964) The analysis of two-dimensional processes with discontinuous spectra. *Biometrika*, **51**, 195–217.

Priestley, M.B. (1965) Evolutionary spectra for non-stationary processes. *J.R. Statist. Soc. B*, **27**, 204–229.

Priestley, M.B. and Tong, H. (1973) On the analysis of bivariate non-stationary processes. *J.R. Statist. Soc. B*, **35**, 153–166.

24

On periodogram-based spectral estimation for replicated time series

P.J. Diggle and I. Al-Wasel

24.1 INTRODUCTION

Figure 24.1 shows time series consisting of measurements of the concentration of luteinizing hormone (LH) in blood samples taken at intervals of 5 minutes from each of 8 apparently healthy men. LH is secreted in a pulsatile manner involving complicated feed-back mechanisms in the endocrine system (Lincoln *et al.* 1985). Endocrinologists are interested in characterizing the frequency characteristics of this pulsatile process. The pattern of variation over time is complex, and spectral analysis is a natural technique to use in an attempt to characterize the contributions to the overall variation from different frequency ranges (Murdoch *et al.* 1985). Clearly, the sampling regime limits the range of frequencies which can be detected. Figure 24.2 shows a second set of data, taken from the same subjects, but in which each series consists of measurements from blood samples taken at intervals of 1 minute. In this second set of data, the objective is to discover whether there are any high-frequency patterns of variation superimposed on the low-frequency effects which are clearly visible in the first set of data. With this in mind, the low-frequency variation has been filtered out by subtracting from the original data a weighted 7-point moving average with weights proportional to 1, 3, 6, 7, 6, 3, 1.

Although spectral analysis is a very highly developed methodology, almost all of this development has been in the context of a single, long time series $\{x_t : t = 1, \ldots, n\}$. See, for example, Priestley (1981). This perhaps reflects the origins of the subject in signal processing and the physical sciences. However, the usefulness of time series methodology is becoming more widely accepted in the biomedical sciences, where replicated experiments are the rule rather than the exception.

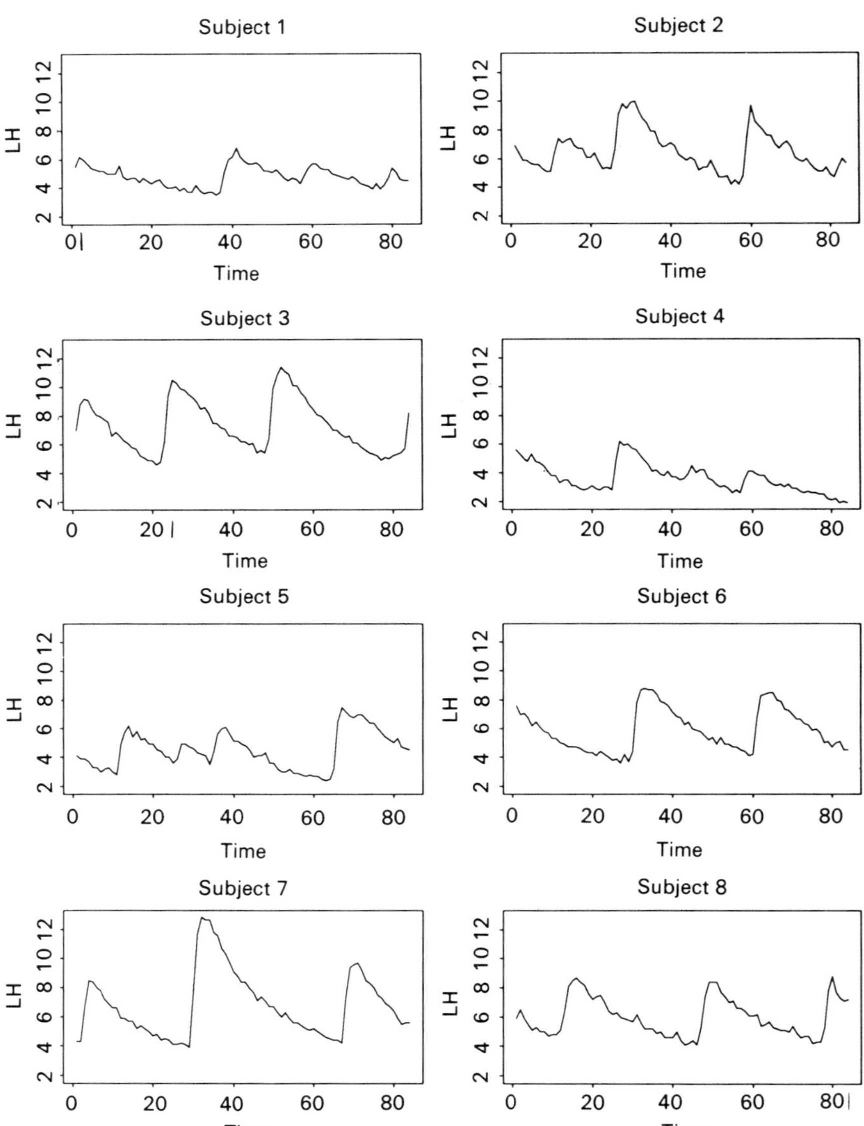

Figure 24.1. Replicated time series of LH in blood samples taken at 5-minute intervals.

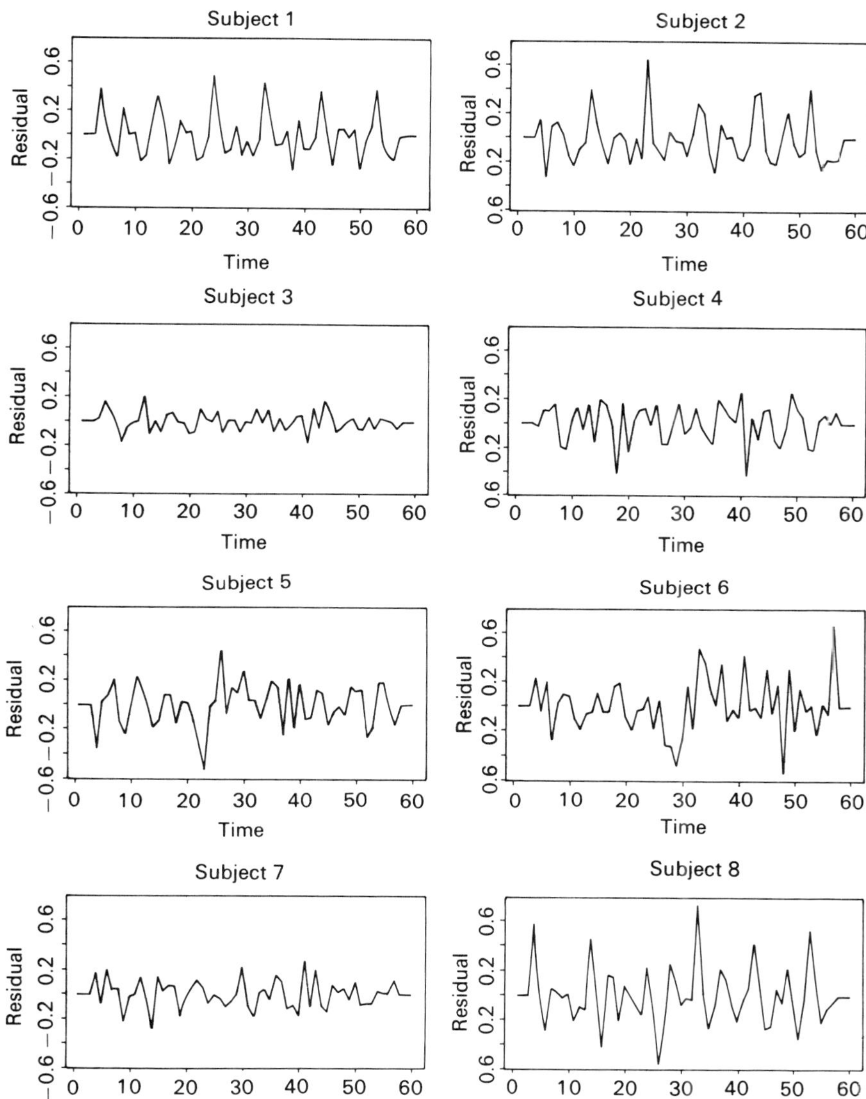

Figure 24.2. Replicated time series of LH blood samples taken at 1-minute intervals, after filtering to remove low-frequency variation.

This paper considers the application of spectral analysis to replicated time series such as those in Figures 24.1 and 24.2. In section 24.2 we review the standard distribution theory of periodogram-based estimators for the spectrum. In section 24.3 we return briefly to the LH data to motivate the need for a non-trivial treatment of replication. This need arises from the fact that the structure of biomedical time series typically includes a component of stochastic variation between subjects, implying a need to distinguish between **population** and **subject-specific** spectra. In section 24.4 we give a very simple model for the variation between subject-specific spectra, derive an interval estimation procedure for the population spectrum and apply the method to the LH data. Section 24.5 gives a brief outline of work in progress on a class of more realistic random effects models for the variation between subject-specific spectra.

24.2 BASIC DISTRIBUTION THEORY FOR THE PERIODOGRAM

Let $\{X_t : t \in \mathbb{Z}\}$ be a stationary stochastic process with absolutely continuous spectral density function $f(\omega)$, and $\{x_t : t = 1, \ldots, n\}$ a partial realization of $\{X_t\}$. To estimate $f(\omega)$ from the data $\{x_t\}$ we proceed as follows.

First, define the periodogram of $\{x_t\}$ to be the function

$$I(\omega) = n^{-1} \left| \sum_{t=1}^{n} x_t \exp(-i\omega t) \right|^2 .$$

Secondly, define the Fourier frequencies $\omega_j = 2\pi j/n$, for $j = 1, \ldots, [(n-1)/2]$. The asymptotic distribution theory for the periodogram ordinates $I(\omega_j)$ is remarkably simple, and is summarized by the results:

(a) $I(\omega_j) \sim f(\omega_j)\chi_2^2/2$;
(b) the set of periodogram ordinates $\{I(\omega_j) = j = 1, \ldots, [(n-1)/2]\}$ are asymptotically independent.

For a typically lucid discussion of the precise sense in which these results hold, see Priestley (1981, Chapter 6).

The first of these results establishes that $I(\omega_j)$ is an asymptotically unbiased, but inconsistent estimator for $f(\omega_j)$, since $\mathrm{E}[\chi_2^2] = 2$ and $\mathrm{Var}\ (\chi_2^2) = 4$, so that $\mathrm{E}[I(\omega_j)] \to f(\omega_j)$ and $\mathrm{Var}\{I(\omega_j)\} \to \{f(\omega_j)\}^2$ as $n \to \infty$. To obtain a consistent estimator, define the periodogram average of order $2k+1$,

$$I^k(\omega_j) = (2k+1)^{-1} \sum_{p=-k}^{k} I(\omega_{j+p}). \qquad (24.1)$$

Then, it follows readily from earlier results that $I^k(\omega_j)$ is a consistent estimator for $f(\omega_j)$ provided that $k = k(n)$ is such that $k(n) \to \infty$ and $n^{-1}k(n) \to 0$ as $n \to \infty$.

In practice, we use (24.1) with fixed k to give an estimator for $f(\omega_j)$ which is somewhat biased but with a substantially smaller variance than $I(\omega_j)$. The optimal choice of k depends on the underlying spectrum $f(\omega)$: a small value of k is optimal for a highly peaked $f(\omega)$, a larger value for a more slowly varying $f(\omega)$. See, for example, Priestley (1981, Chapter 7).

Now, suppose that our data consist of r independent partial realizations of $\{X_t\}$, and write the ith such replicate as $\{x_{it}:t=1,\ldots,n\}$. Let $I_i(\omega)$ be the periodogram of $\{x_{it}\}$. Then, a natural estimator for $f(\omega_j)$ is

$$\bar{I}(\omega_j)=r^{-1}\sum_{i=1}^{r} I_i(\omega_j). \tag{24.2}$$

Then, it follows immediately from the asymptotic distribution theory of the periodogram ordinates $I_i(\omega_j)$ that

$$\bar{I}(\omega_j)\sim f(\omega_j)\chi^2_{2r}/(2r),$$

$$\bar{I}(\omega_j) \text{ and } \bar{I}(\omega_k) \text{ are independent for } j\neq k.$$

Averaging across replicates in this way would seem to offer the same benefits as the smoothed periodogram estimator (24.1) with regard to reduction in variance, but without introducing any bias. Of course, if r is small, further smoothing of the $\bar{I}(\omega_j)$ across a band of frequencies may still be desirable so long as the trade-off between variance and bias remains favourable.

The striking simplicity of the asymptotic distribution theory for these periodogram-based estimators makes it relatively easy to develop inferential procedures. There is a formal connection with generalized linear models (McCullagh and Nelder, 1989) which apparently was first noted explicitly by Diggle (1985) following earlier (despite the publication date!) related work by Coates and Diggle (1986). This connection makes the necessary computations easily accessible through a range of standard statistical software packages (Diggle, 1985; Cameron and Turner, 1987). See also Diggle (1990, chapter 4).

24.3 HORMONE CONCENTRATIONS IN SERIAL BLOOD SAMPLES

Figure 24.3 shows the periodograms corresponding to the replicated time series in Figure 24.1. As we would expect in view of the theoretical results quoted in section 24.2, these periodograms show substantial random variation and it is hard to see any overall pattern. Figure 24.4 plots the empirical coefficients of variation of the sets of periodogram ordinates $I_i(\omega_j)$: $i=1,\ldots,8$ against ω_j. These fluctuate randomly about a value close to 1.0, which is consistent with the standard chi-squared distribution theory. We therefore

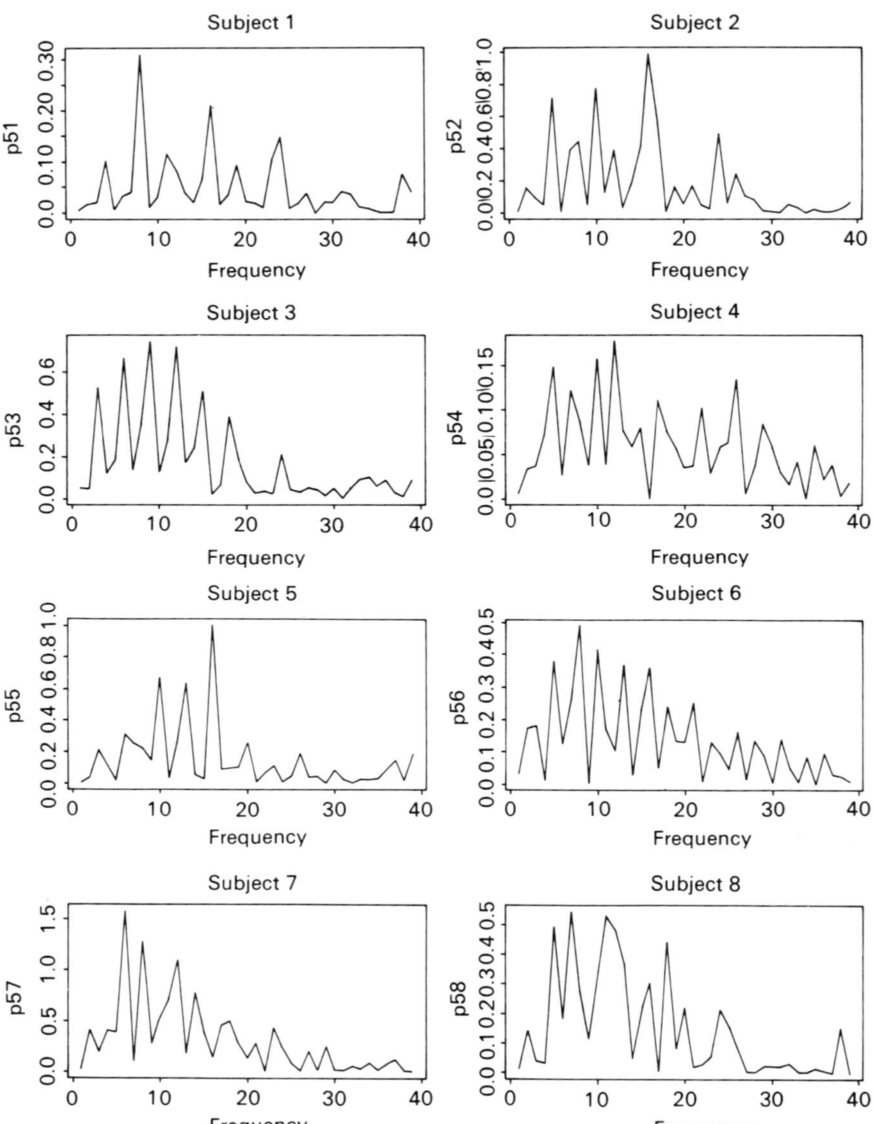

Figure 24.3. Periodograms of the time series of LH in blood samples taken at 5-minute intervals.

proceed to estimate $f(\omega_j)$ by a three-point moving average smoothing of the $\bar{I}(\omega_j)$ and attach point-wise confidence limits assuming that the sampling distribution of each smoothed estimate is proportional to χ^2_{48}. Figure 24.5 shows the resulting estimates and 90% confidence limits. These show a broad

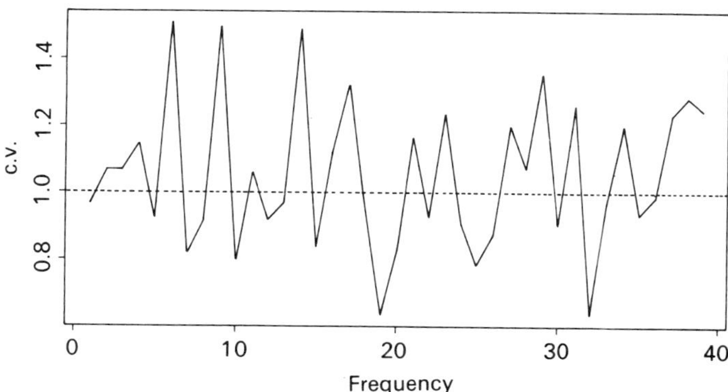

Figure 24.4. Empirical coefficients of variation of the periodogram ordinates (5-minute sampling interval).

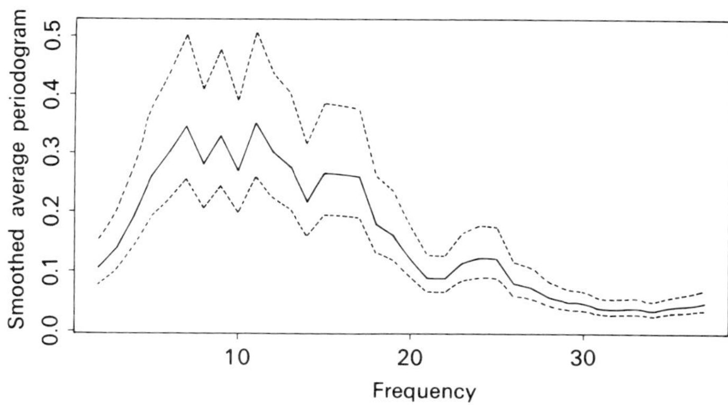

Figure 24.5. Smoothed estimate of the spectrum (solid line) and pointwise 90% confidence limits (dashed lines), for LH sampled at 5-minute intervals.

concentration of power over the low frequency range, confirming the visual impression gained from the raw data in Figure 24.1.

Figure 24.6 plots the empirical coefficients of variation for periodogram ordinates calculated from the replicated time series in Figure 24.2, for which the sampling interval was 1 minute rather than 5 minutes. In contrast to Figure 24.4, these coefficients of variation are almost all greater than 1.0. This behaviour is inconsistent with the standard distribution theory, and whilst smoothed versions of the $\bar{I}(\omega_j)$ may still be sensible point estimators for the $f(\omega_j)$, we need a different approach to interval estimation.

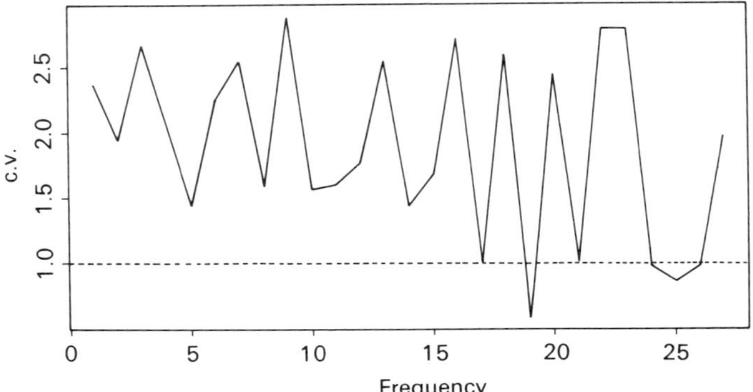

Figure 24.6. Empirical coefficients of variation of the periodogram ordinates (1-minute sampling interval).

24.4 ALLOWING FOR VARIATION BETWEEN SUBJECT-SPECIFIC SPECTRA

In the remainder of this paper, we write $Y_{ij} = I_i(\omega_j)$ for the periodogram ordinate of the ith replicate at the jth Fourier frequency, for $i = 1, \ldots, r$ and $j = 1, \ldots, m$ where $m = [(n-1)/2]$. Also, we assume that the standard asymptotic distribution theory for the periodogram ordinates holds exactly. Thus,

$$Y_{ij} = f_i(\omega_j)U_{ij}: i = 1, \ldots, r; \quad j = 1, \ldots, m,$$

where $f_i(\omega)$ is the subject-specific spectrum for the ith subject and the U_{ij} are mutually independent, unit mean exponential variates with common pdf e^{-u}, $u \geqslant 0$. Suppose now that the r units are selected at random from a given population. Then we regard the $f_i(w)$ as independent realizations of a random function $R(\omega)$, and set $f(\omega) = \mathrm{E}[R(\omega)]$ where the expectation is defined with respect to the population of subjects. We call $f(\omega)$ the population spectrum.

Note that this approach to the description of differences amongst the subject-specific spectra $f_i(\omega)$ is analogous to the use of random effects in the analysis of classically designed experiments.

24.4.1 Ad hoc methods

At each frequency ω_j, the $Y_{ij}: i = 1, \ldots, r$ are mutually independent, identically distributed random variables with common expectation $f(\omega_j)$ and unknown variance σ_j^2. Let y_{ij} denote the observed value of Y_{ij},

$$\bar{y}_j = r^{-1} \sum_{i=1}^{r} y_{ij},$$

the sample mean of the y_{ij} corresponding to frequency ω_j, and

$$s_j^2 = (r-1)^{-1} \sum_{i=1}^{r} (y_{ij} - \bar{y}_j)^2,$$

the sample variance corresponding to frequency ω_j. Then a natural ad hoc estimation procedure is to use

$$\hat{f}(\omega_j) = \bar{y}_j$$

with

$$\mathrm{Var}\{\bar{f}(\omega_j)\} = \sigma_j^2/r \simeq s_j^2/r.$$

For large r, we can base inferences for $f(\omega_j)$ at a specific frequency ω_j on the asymptotic result that $\{\bar{y}_j - f(\omega_j)\}(r/s_j^2)^{1/2} \sim N(0,1)$. However, this approximation will be poor for small r. For example, nominal 90% pointwise confidence limits calculated as $\bar{y} \pm 1.64(s_j^2/r)^{1/2}$ can easily extend to the physically meaningless region of negative $f(\omega_j)$.

One way to improve the normal approximation, and to avoid the problem of non-positive confidence limits, is to apply the above procedure to log-periodogram ordinates. However, there are two objections to this. First, the expectation of the log-periodogram ordinates at frequency ω_j is the function $g(\omega_j) = \log\{f(\omega_j)\} + c(\omega_j)$, where the correction term $c(\omega_j)$ depends on the unknown distribution of the Y_{ij}. Thus, for small r, estimates based on sample means of log-periodogram ordinates may be seriously biased to an unknown extent. Secondly, the method does not reduce to the standard, optimal estimator $\bar{y}_j$ when the stochastic variation between replicates turns out to be negligible.

We conclude that neither ad hoc method is satisfactory, and seek instead a method based on an explicit model for the stochastic variation between replicates.

24.4.2 A white noise model

We assume that

$$Y_{ij} = f(\omega_j) Z_i(\omega_j) U_{ij} : i = 1, \ldots, r; \quad j = 1, \ldots, m, \tag{24.3}$$

where $\{Z_i(\omega_j)\}$: $i = 1, \ldots, r$ are independent copies of a stochastic process $\{Z(\omega_j)\}$ with $\mathrm{E}[Z(\omega_j)] = 1$, for all ω_j. The simplest such process would be a white noise process, in which $\mathrm{Var}\{Z(\omega_j)\} = \sigma^2$ for all ω_j, and $Z(\omega_j)$ and $Z(\omega_k)$ are independent for $j \neq k$. Whilst this model is physically implausible, we shall see that it can provide a simple means of achieving our first objective—the construction of valid point-wise confidence limits for the $f(\omega_j)$: $j = 1, \ldots, m$.

We now have to decide on an appropriate distributional form for $Z_i(\omega_j)$ in (24.3). An obvious pragmatic strategy is to achieve a tractable form for the distribution of the product $W_{ij} = Z_i(\omega_j) U_{ij}$. Since we have no physical basis for choosing any particular distributional form, we pursue this pragmatic course, using the following result.

Proposition 24.1

If Z and U are independent random variables, with respective pdfs

$$f(z) = \{\Gamma(\kappa)\Gamma(1-\kappa)\}^{-1}\kappa(\kappa z)^{\kappa-1}(1-\kappa z)^{-\kappa} : 0 \leqslant z \leqslant \kappa^{-1},$$

for some $0 < \kappa < 1$, and

$$g(u) = \mathrm{e}^{-u} : u \geqslant 0,$$

then $W \equiv ZU$ has pdf

$$h(\omega) = \{\Gamma(\kappa)\}^{-1}\kappa^{\kappa}\omega^{\kappa-1}\exp(-\kappa\omega) : \omega \geqslant 0.$$

Proof

The given pdfs $f(\cdot)$ and $g(\cdot)$ are equivalent to the statements that κZ has a beta distribution with parameters κ and $1-\kappa$, and $\kappa^{-1}U$ a gamma distribution with scale parameter κ and shape parameter 1. Now, if X_1 and X_2 are independent gamma-distributed random variables with common scale parameter λ and shape parameters α_1 and α_2 respectively, it is well known that:

- $(X_1 + X_2)$ is gamma-distributed with scale parameter λ and shape parameter $\alpha_1 + \alpha_2$;
- $X_1/(X_1 + X_2)$ is beta-distributed with parameters α_1 and α_2;
- $(X_1 + X_2)$ and $X_1/(X_1 + X_2)$ are independent.

(Moran, 1968, Chapter 7). The proposition follows by equating $\alpha_1 = \kappa$, $\alpha_2 = 1-\kappa$, $X_1 + X_2 = \kappa^{-1}U$ and $X_1/(X_1 + X_2) = \kappa Z$, to give $W = X_1$ with the stated distribution.

24.4.3 Inference

Proposition 24.1 justifies a model for the periodogram ordinates Y_{ij} of the form

$$Y_{ij} = f(\omega_j)\,W_{ij} : i = 1, \ldots, r; \quad j = 1, \ldots, m, \tag{24.4}$$

where the random variables W_{ij} are mutually independent, gamma-distributed with scale parameter κ and shape parameter κ, for some value of $\kappa \leqslant 1$. Thus, using f_j to denote $f(\omega_j)$, the pdf of Y_{ij} is

$$h_{ij}(y) = \{\Gamma(\kappa)\}^{-1}(\kappa/f_j)^{\kappa}y^{\kappa-1}\exp(-\kappa y/f_j) : y \geqslant 0,$$

and the log-likelihood for data y_{ij} is

$$\begin{aligned} \ell = {} & mr\{\kappa\log\kappa - \log\Gamma(\kappa)\} \\ & - \kappa r\sum_{j=1}^{m}\log f_j + (\kappa-1)\sum_{i=1}^{r}\sum_{j=1}^{m}\log y_{ij} - \kappa\sum_{j=1}^{m} f_j^{-1}\sum_{i=1}^{r} y_{ij}. \end{aligned}$$

Differentiating with respect to the parameters f_j and κ we obtain

$$\frac{\partial \ell}{\partial f_i} = -\kappa r/f_j + \kappa r \bar{y}_j/f_j^2 \tag{24.5}$$

and

$$\frac{\partial \ell}{\partial \kappa} = mr\{\log \kappa + 1 - \gamma(\kappa)\} - r\sum_{j=1}^{m} \log f_j + \sum_{i=1}^{r}\sum_{j=1}^{m} \log y_{ij} - r\sum_{j=1}^{m} \bar{y}_j/f_j, \tag{24.6}$$

where $\gamma(\cdot)$ is the digamma function. From (24.3), setting $\partial\ell/\partial f_j = 0$ we obtain the maximum likelihood estimator

$$\hat{f}_j = \bar{y}_j,$$

irrespective of the value of κ. Substitution of $\hat{f}_j$ into (24.4) gives the maximum likelihood estimator $\hat{\kappa}$ as the solution of

$$mr\{\log \kappa - \gamma(\kappa)\} - r\sum_{j=1}^{m} \log \bar{y}_j + \sum_{i=1}^{r}\sum_{j=1}^{m} \log y_{ij} = 0,$$

or

$$\log \kappa - \gamma(\kappa) - c = 0,$$

where

$$c = m^{-1}\sum_{j=1}^{m} \log \bar{y}_k - (mr)^{-1}\sum_{i=1}^{r}\sum_{j=1}^{m} \log y_{ij}.$$

In our implementation of this model, we have used a Newton–Raphson iteration for $\hat{\kappa}$, with initial value

$$\hat{\kappa} = m^{-1}\sum_{j=1}^{m} \bar{y}_j^2/s_j^2,$$

motivated by the fact that the coefficient of variation of the gamma distribution with shape parameter κ is $\kappa^{-1/2}$.

Under the model (24.4), $\hat{f}_j = \bar{y}_j$ has a gamma sampling distribution with mean f_j and shape parameter $r\kappa$. To construct approximate pointwise confidence intervals for the f_j, we use this result with the maximum likelihood estimate $\hat{\kappa}$ in place of the unknown κ.

Al-Wasel (1993) describes a simulation experiment to investigate the small-sample properties of these approximate confidence intervals. Note that there are two distinct methodological issues to be addressed. Firstly, even when $\kappa = 1$ the standard theory is asymptotic, and we need to establish the extent to which it holds for relatively short series. Secondly, using the maximum likelihood estimate $\hat{\kappa}$ in place of the unknown κ may affect the coverage properties of the confidence intervals.

With regard to the first issue, Al Wasel applied the method to simulated realisations of first-order Gaussian autoregressive processes. For moderate values of the autoregressive parameter ρ, say $\rho \leqslant 0.5$, the standard asymptotic theory works well for series of length $n = 60$ and $r = 10$ replicates, but for ρ as large as 0.9 the actual coverage probabilities are too small—for example, nominal 90% intervals had actual coverage probabilities of about 85%, improving slowly as the values of n and r increased. With regard to the second issue, the results were generally encouraging; using $\hat{\kappa}$ in place of κ had little effect on the coverage properties of the confidence intervals for the range of situations considered ($n \geqslant 60$, $r \geqslant 10$).

24.4.4 Application to LH data

For the LH data shown in Figure 24.2, the maximum likelihood estimate of κ is $\hat{\kappa} = 0.534$. Using this estimate, we obtain pointwise 90% confidence limits for the $f(\omega_j)$. The estimated spectrum and associated confidence limits shown in Figure 24.7 are derived from a 3-point moving average smooth of the maximum likelihood estimates $\hat{f}_j = \bar{y}_j$. The point estimates of $f(\omega_j)$ show a peak at the 10th frequency, corresponding to a six-minute cycle. However, the confidence limits around this peak are wide, and the peak may not be significant. In fact, it is natural to ask whether the estimated spectrum is consistent with the data being a filtered version of white noise. The spectrum induced by the weighted moving average filter of the data takes the form

$$g(\omega) = \sigma^2 |A(\omega)|^2$$

where $\sigma^2 > 0$ is to be estimated,

$$A(\omega) = (1 - w_0) - 2 \sum_{j=1}^{3} w_j \cos(j\omega)$$

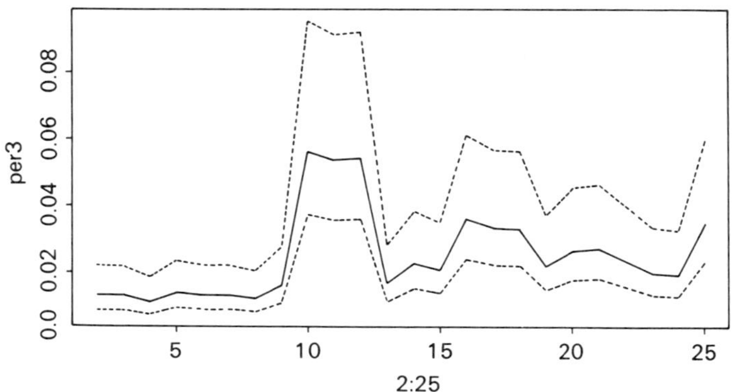

Figure 24.7. Smoothed estimate of the spectrum (solid line) and pointwise 90% confidence limits (dashed lines), for LH sampled at 1-minute intervals.

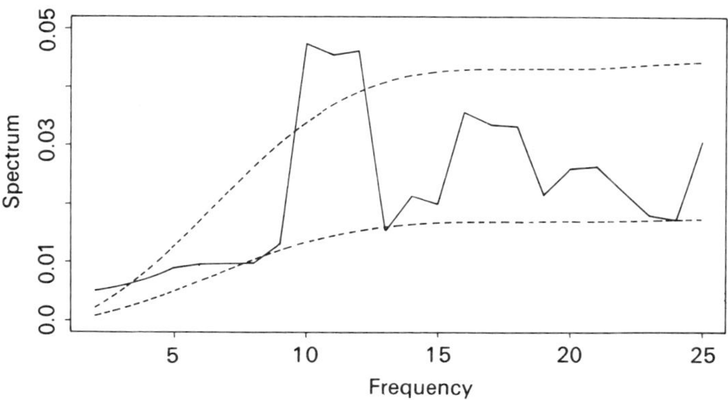

Figure 24.8. Smoothed estimate of the spectrum (solid line) for LH sampled at 1-minute intervals, and pointwise 90% tolerance limits (dashed lines) for filtered white noise.

and the weights w_j are those of the weighted moving average filter. Figure 24.8 shows the estimates $\hat{f}(\omega_k)$ together with pointwise 90% tolerance limits assuming that $f(\omega) = g(\omega)$, with σ^2 estimated so as to match the total areas under $\hat{f}(\omega)$ and $g(\omega)$. The peak in $\hat{f}(\omega)$ lies well outside the tolerance limits, and we conclude that for these data there **is** significant high-frequency variation superimposed on the basic, low-frequency pattern.

24.5 DISCUSSION

Although the white-noise model described in section 24.4 is a gross over-simplification, it achieves the limited aim of providing a simple method for calculating valid point-wise confidence intervals for a population spectrum in the presence of stochastic variation between subject-specific spectra. The model assumes that the coefficient of variation of the periodogram ordinates is the same at all frequencies, as can be checked by a plot of empirical coefficients of variation against frequency.

The basic model is amenable to various forms of generalization. One is to make κ a smooth function of ω. However, we feel that a more fruitful generalization is to build smoothness directly into the subject-specific spectra. In terms of our model (24.3), this implies incorporating positive autocorrelation structure into the process $\{Z(\omega)\}$. Amongst the various ways in which this can be done, one which appeals to us is a class of random-coefficient log-linear models,

$$Z(\omega) = \exp\left\{\sum_{j=0}^{p} A_j \phi_j(\omega)\right\}. \tag{24.6}$$

In (24.6), the $\phi_j(\omega)$ are a set of orthogonal basis functions, for example $\phi_{2k}(\omega) = \cos(k\omega)$ and $\phi_{2k+1}(\omega) = \sin(k\omega)$, whilst the A_j are independent Gaussian random variables with variances σ_j^2 and expectations constrained to that $\mathrm{E}[Z(\omega)] = 1$, for all ω. One of the attractions of this class of models is that it is possible to choose the basis functions and the nuisance parameters σ_j^2 to correspond to physically interpretable forms of subject-specific frequency modulation; for example, a sharpening or flattening of a spectral peak, a general shift of the power from lower to higher frequencies, etc. From a technical point of view, (24.6) is tractable to the extent that the associated likelihood for the population spectrum $f(\omega)$ and the nuisance parameters σ_j^2 is expressible as a $(p+1)$-dimensional integral, whilst the covariance structure of the array of periodogram ordinates Y_{ij} is expressible in closed form. We are currently investigating maximum likelihood and other methods of estimation for this class of models.

In conclusion, we believe that spectral analysis has much to offer in the interpretation of replicated time series in biomedicine and other, non-traditional fields of application. However, before it can be used as a formal inferential tool in this context we need to develop realistic models, and tractable methods, for describing and estimating stochastic variation between subject-specific spectra. We have made a start in this direction, but a great deal of work remains to be done.

REFERENCES

Al-Wasel, I. (1993). Unpublished Ph.D Thesis, Lancaster University.

Cameron, M.A. and Turner, T.R. (1987) Fitting models to spectra using regression packages. *Appl. Statist.*, **36**, 47–57.

Coates, D.S. and Diggle, P.J. (1986) Tests for comparing two estimated spectral densities. *J. Time Series Anal.*, **7**, 7–20.

Diggle, P.J. (1985) Comparing estimated spectral densities using GLIM, in (eds) R. Gilchrist, B. Francis and J. Whittaker, *Generalized Linear Models: proceedings, Lancaster 1985*, Springer-Verlag, Berlin, pp. 35–43.

Diggle, P.J. (1990) *Time Series: A Biostatistical Introduction*, Oxford University Press, Oxford.

Lincoln, D.W., Fraser, H.M., Lincoln, G.A., Martin, G.B. and McNeilly, A.S. (1985) Hypothalmic pulse generators. *Recent Prog. Horm. Res.*, **41**, 369–419.

McCullagh, P. and Nelder, J.A. (1989) *Generalized Linear Models*, Chapman and Hall, London.

Moran, P.A.P. (1968) *An Introduction to Probability Theory*, Oxford University Press, Oxford.

Murdoch, A.P., Diggle, P.J., Dunlop, W. and Kendall-Taylor, P. (1985) Determination of the frequency of pulsatile luteinizing hormone secretation by time series analysis. *Clin. Edocrin.*, **22**, 341–6.

Priestley, M.B. (1981) *Spectral Analysis and Time Series*, Academic Press, London.

25

The prediction of time–frequency spectra using covariance-equivalent models

J.K. Hammond, R.F. Harrison, Y.H. Tsao and J.S. Lee

25.1 INTRODUCTION

Non-stationary random processes having a 'frequency modulated' form often arise in the physical world. Examples include descriptions of the motion of a vehicle over rough terrain as it accelerates over the surface and the acoustical case of the Doppler effect of the sound perceived by an observer as a source travels past the observer. The second example is more complex than the first as it also includes range and directionality effects. This chapter is primarily directed at developing two-dimensional (time–frequency) spectra for this class of process and pays particular attention to the evolutionary spectral density due to Priestley (1965a, b, 1967).

The material presented here is a summary of aspects of studies carried out by the authors and their co-workers over a number of years. The first author's own interest in this field was initiated by Priestley (1965a, 1967) which led to applying the method to problems in acoustics and vibrations (Hammond, 1968, 1971, 1973). However, the explicit analytical treatment of problems having a frequency modulated structure came later and constitutes the main part of this chapter.

It should be noted that time–frequency descriptions are very attractive to the engineering world and empirical analysis of non-stationary processes using 'moving window spectral analysis' has been commonplace for a long time. Furthermore, there has been a recent upsurge of interest in a variety of descriptions other than evolutionary spectra and the conventional spectrogram. Accordingly this presentation will begin with a very brief review of some time–frequency distributions to help put the work in some context.

25.2 TIME–FREQUENCY REPRESENTATIONS

The aim of this section is to summarize the basic considerations underlying time–frequency analysis.

25.2.1 The uncertainty principle

Fourier analysis provides a decomposition of a time history in the frequency domain, i.e., the pair

$$x(t) = \int_{-\infty}^{\infty} e^{j\omega t} X(\omega) d\omega$$

$$X(\omega) = \frac{1}{2\pi} \int_{-\infty}^{\infty} x(t) e^{-j\omega t} dt \tag{25.1}$$

provide the interpretation of $x(t)$ as the sum of sines and cosines with 'amplitudes' $X(\omega)d\omega$ and the amplitude density $X(\omega)$ is given by (25.1). For the moment it is convenient to think of $x(t)$ as a deterministic signal for which $X(\omega)$ is defined.

Questions such as 'when does frequency component ω make its maximum contribution to the signal?' are difficult to answer, and raise the concept of introducing both time **and** frequency into the picture simultaneously. The difficulty is summarized in the uncertainty principle which begins by defining two 'widths' of the signal in the time domain T (say) and the frequency domain B (say) for which (Zadeh and Desoer, 1963; Hammond and Moss, 1991; Kodera *et al.*, 1978; Boashash *et al.*; 1989)

$$BT \geqslant \frac{1}{2}, \tag{25.2}$$

i.e., the bandwidth-time product of a signal is never less than 1/2. In Priestley's (1965) discussion of time-frequency spectra, the principle is fundamental to the notion that one cannot obtain **simultaneously** arbitrarily fine resolution of a signal in both time and frequency.

It is interesting to note that a Gaussian shaped pulse, say $x(t) = e^{-at^2}$ has a BT product of 1/2 and a function $e^{-\pi\gamma t^2} e^{(j2\pi\beta t^2)\frac{1}{2}}$ (Kodera *et al.*, 1978) has a BT product

$$\sim \frac{1}{2}\left(1 + \frac{\beta^2}{\gamma^2}\right)^{1/2}.$$

Note that in the latter case γ controls the amplitude modulation of a chirp signal whose instantaneous frequency, section 25.2.2, is $2\pi\beta t$. It is apparent, therefore, that the more cycles of the chirp are included (i.e., as β/γ increases) the larger is the BT product. This illustrates that the more like an amplitude modulated oscillation a signal appears, the larger is the BT product.

25.2.2 Instantaneous frequency and group delay

The time-frequency analysis of 'non-stationary' signals has, in the engineering literature, come to include the analysis of both random and deterministic processes. A commonly used example of a non-stationary deterministic signal is a 'swept sine wave' or chirp of fixed duration (or with decaying envelope) as in the second illustration above. Obviously this may be treated as a transient and conventional Fourier methods may be applied. However if, for example, it is an acoustic signal sweeping, say, from 100 Hz to 1000 Hz in 1 second, we easily perceive a 'changing frequency' and speak of the 'instantaneous frequency' and 'instantaneous amplitude' of the process. These terms are perhaps most easily understood if the signal is modelled not as the sum of sines and cosines in the Fourier sense but as

$$x(t) = A(t)\cos\phi(t), \tag{25.3}$$

where the emphasis is on amplitude modulation $A(t)$ and frequency modulation ($\phi(t)$ is the instantaneous phase and $\dot{\phi}(t)$ is the instantaneous frequency). An unambiguous way of defining $A(t)$ and $\phi(t)$ for a real valued signal $x(t)$ is to form the 'analytic signal' $\sigma_x(t) = x(t) + j\hat{x}(t)$, where $\hat{x}(t)$ is the Hilbert transform of $x(t)$. This complex valued signal (also sometimes called the pre-envelope signal) may be expressed in polar form, i.e.

$$\sigma_x(t) = A(t)e^{j\phi(t)}, \tag{25.4}$$

from which it is apparent that $x(t)$ is defined by (25.3) and that the Hilbert transform $\hat{x}(t)$ is

$$\hat{x}(t) = A(t)\sin\phi(t). \tag{25.5}$$

From this it is clear that a plot of $\dot{\phi}(t)$ versus time t essentially gives a description of the variation of the instantaneous frequency with time, i.e. a time–frequency pattern.

Time-frequency patterns also arise from considerations of conventional Fourier analyses. If $x(t)$ has Fourier transform $X(\omega)$, as in (25.1), then if $X(\omega)$ is written $|X(\omega)|e^{j\arg X(\omega)}$, the group delay $\tau_g(\omega)$ is

$$\tau_g(\omega) = -\frac{d}{d\omega}\arg X(\omega). \tag{25.6}$$

The quantity $\tau_g(\omega)$ may be interpreted as the time at which frequency component ω makes its maximum contribution to the signal. We may conceive of plotting $\tau_g(\omega)$ versus ω to obtain a time-frequency pattern.

We appear therefore to have two possible candidates for time–frequency laws based on the analytic signal (and the associated instantaneous frequency) and the Fourier transform (and the associated group delay). The question is, are they the same—or, more correctly, are they equivalent in the sense of being inverses of each other?

These aspects are discussed in Kodera *et al.* (1978), and Boashash *et al.* (1989), and it is argued and demonstrated that in general these two relationships are different. Conditions that must be met for them to be equivalent are that the signal should be asymptotic, i.e. the BT product should be 'large' and the signal should be 'monocomponent', i.e. the instantaneous frequency law should be invertible.

Having addressed some basic aspects for signals we now turn to dealing primarily with non-stationary random processes, though some of the descriptions are used for deterministic signals also.

25.2.3 Time–frequency representations

We shall now give a brief description of basic approaches to the definition of time–frequency spectra. The emphasis is on the fundamental definitions and the problem of estimation of time–frequency spectra is not discussed.

Evolutionary spectra

Priestley (1965a, b, 1967, 1988) has given a full description of a time–frequency representation for a class of non-stationary random processes termed 'oscillatory'. The representation of a stationary process as the sum of sines and cosines, i.e.

$$x(t) = \int_{-\infty}^{\infty} e^{j\omega t}\, dZ_x(\omega), \tag{25.7}$$

with $Z_x(\omega)$ an orthogonal process, is generalized. (N.B. We use the notation proposed by Priestley.) The essence of Priestley's approach is that the basic 'building blocks' in the representation, namely, the sines and cosines $e^{j\omega t}$ are replaced by amplitude modulated sines and cosines $A_t(\omega)e^{j\omega t}$ so that

$$x(t) = \int_{-\infty}^{\infty} A_t(\omega) e^{j\omega t}\, dZ_x(\omega) \tag{25.8}$$

where $Z_x(\omega)$ is still orthogonal, i.e. $E[dZ_x^*(\omega_1)dZ_x(\omega_2)] = 0, \omega_1 \neq \omega_2$; leading to the evolutionary spectral density

$$S_t^x(\omega) = |A_t(\omega)|^2 S_{xx}(\omega) \tag{25.9}$$

where

$$S_{xx}(\omega) = \frac{|dZ_x(x)|^2}{d\omega}. \tag{25.10}$$

The term 'oscillatory process' arises since $A_t(\omega)$ must be slowly varying (as a function of t) compared to $e^{j\omega t}$, i.e. each component $A_t(\omega)e^{j\omega t}$ considered as a function of t must have a 'large' BT product.

The autocorrelation form

The power spectral density of a stationary random process may be obtained by Fourier transforming the autocorrelation function,

$$S(\omega) = \int_{-\infty}^{\infty} R(\tau)\mathrm{e}^{-j\omega\tau}\mathrm{d}\tau. \tag{25.11}$$

It is natural to extend this to the time-varying case, where an appropriately defined time-dependent autocorrelation function $R(t, \tau)$ is used,

$$S(t, \omega) = \int_{-\infty}^{\infty} R(t, \tau)\mathrm{e}^{-j\omega\tau}\mathrm{d}\tau. \tag{25.12}$$

This defines another class of time-frequency spectral densities $S(t, \omega)$ which are different from Priestley's evolutionary spectral density $S_t(\omega)$.

There are many ways in which the correlation function may be defined and this approach has attracted a great deal of attention recently. Cohen (1989) is a good review of the subject, showing the richness of this class (referred to as the Cohen class) which includes the Wigner–Ville spectrum, the Choi–Williams distribution, the spectrogram and others. An exception is the evolutionary spectral density. As this paper is aimed mainly at demonstrating the use of the evolutionary spectral density, we shall not discuss how each attempts to satisfy various 'desirable properties' of time–frequency distributions. Readers are referred to Cohen (1989), Classen and Mecklenbrauker (1980), and Martin (1984) for this. However, we will formally relate evolutionary spectra to Cohen-class distributions later.

To conclude this section it is noted that the spectrogram (or moving window spectral estimation method) is well worthy of mention since it is so widely used. It is also of interest to point out a modification of the spectrogram due to Kodera *et al.* (1978) who utilized group delay and instantaneous frequencies to 'sharpen-up' the time–frequency patterns, and the usefulness of this is described by Hammond and Moss (1991), and Moss *et al.* (1989).

Finally, any description of time–frequency methods should include Gabor's (1946) classic work using two-dimensional representations based on Gaussian pulse modulated sines and cosines. Numerous generalizations and extensions have been given, all incurring the problem of complexity of calculation of the coefficients, (John, 1991).

25.3 COVARIANCE-EQUIVALENT MODELS

The aim of this section is to develop the analytical basis for calculating evolutionary spectra for the class of frequency modulated process described in section 25.1. The fundamental question that needs to be addressed is how representation (25.8), which says that a signal is the sum of amplitude modulated components, may describe frequency modulated processes.

A series of papers and PhD theses has been directed at this and we shall summarise the essential ideas here (Lee and Hammond, 1987; Hammond and Harrison 1981, 1984, 1985; Harrison and Hammond, 1986b; Hammond *et al.*, 1983; Tsao, 1983; Harrison, 1983; Lee, 1989). The first step is to recall an interpretation of the evolutionary spectral density as noted by Priestley (1965a), namely, that an oscillatory process may be regarded as the response of a time-varying filter to a stationary excitation, i.e. if $z(t)$ is a stationary process which is operated upon by a variable filter such that

$$x(t) = \int_{-\infty}^{\infty} h(t,u)z(t-u)\mathrm{d}u \tag{25.13}$$

then we may regard $x(t)$ as an oscillatory process with representation (25.8) where

$$A_t(\omega) = \int_{-\infty}^{\infty} h(t,u)\mathrm{e}^{-j\omega u}\mathrm{d}u. \tag{25.14}$$

Such interpretations of non-stationary processes in terms of time-variable 'shaping filters' are widely used, e.g. speech modelling.

Our concern is how we might develop analytical evolutionary spectral density models for frequency modulated processes and our approach is as follows. We shall start with a stationary process and then distort the independent variable so that the process is 'stretched out' or 'compressed' so as to create a non-stationary process. The term 'frequency modulated' for such a process is appropriate since the amplitude remains unchanged. An elementary version of this is to consider a pure sine wave which is expressed as a function of independent variable s (say), i.e. $\sin(2\pi ps)$ and now to let s be a function of t, $s(t)$ and so form $\sin[2\pi ps(t)]$ which, now regarded as a function of time, is frequency modulated in form depending on the form of $s(t)$, e.g. if $s(t) = \alpha t^2/2$, then the time history is a 'swept sine' increasing in frequency.

For the purposes of this presentation which deals with random processes, first define a process that is 'stationary' when viewed as a function of some independent variable say s(s is not, in general, time). We shall call this $y(s)$. Obtain a shaping filter representation for $y(s)$ in terms of filtered white noise $w(s)$. Now let s become a function of time, $s = s(t)$, such that $s(t)$ is not, in general, constant. This means that $y(s)$ may now be regarded as a function of time, $y[s(t)] = \tilde{y}(t)$ say, and the 'dilation' of the s axis nonlinearly results in a process that is frequency modulated in form. The question is whether the s-domain constant shaping filter representation transforms to a form analogous to (25.13). We shall show that this is indeed so, but in order to achieve it we introduce the concept of 'covariance-equivalence', using the following method. Full details may be found in the references cited.

As noted above, we begin by defining a process $y(s)$ that is not time-dependent but a function of another variable, say s (which may be a space variable, for example), $y(s)$. We shall assume that $y(s)$ is a stochastic process that is stationary (i.e. homogeneous) in the s-domain, having zero mean, variance σ_y^2, and autocovariance function (ACVF) $E[y(s_1)y(s_2)] = R_{yy}(s_2 - s_1)$. To create a frequency modulated process let us now regard s as a (deterministic) function of another variable t (time). Our aim now is to describe y not as a function of s, but as a function of t, i.e. we create $\tilde{y}(t) = y[s(t)]$ were $\tilde{y}$ is regarded as a function of t. The functional dependence of s on t will be described by $\dot{s}(t)$. If $\dot{s}$ is constant, then $\tilde{y}(t)$ is a stationary process, but if $\dot{s}$ is not constant $\tilde{y}(t)$ is obviously a non-stationary process but with the properties

$$E[\tilde{y}(t)] = E[y(s)] = \text{constant (assumed zero)},$$

$$E[y^2(t)] = E[y^2(s)] = \sigma_y^2 \text{ (constant)},$$

we assume $\dot{s} > 0$.

Even though the mean and variance are constant, the 'frequency structure' varies in accord with our notions of frequency modulation. This is, in turn, reflected in the ACVF for $\tilde{y}(t)$, i.e.

$$R_{\tilde{y}\tilde{y}}(t_1, t_2) = E[\tilde{y}(t_1)\tilde{y}(t_2)] = R_{yy}(s(t_2) - s(t_1)) \qquad (25.15)$$

which is a function of t_1 and t_2 and not simply $(t_2 - t_1)$ only (unless $\dot{s}$ is constant).

25.3.1 Shaping filter models

In order to be able to develop evolutionary spectral forms for such processes, we shall now assume that $y(s)$ can be described as the output of a shaping filter that is driven by white noise. This is a common model employed in time series analysis and, whilst imposing some restrictions, is of sufficient generality to be useful. It is convenient to use a state form to describe this filter. Let $y(s)$ be expressed as

$$y(s) = \mathbf{c}^T \mathbf{x}(s), \qquad (25.16)$$

where $\mathbf{c}^T$ is a (constant) vector having n components. The superscript T denotes the transpose. $\mathbf{x}(s)$ is an n-dimensional 'state vector' which is assumed to satisfy the shaping filter equation

$$\frac{\mathrm{d}}{\mathrm{d}s}\mathbf{x}(s) = \mathbf{A}\mathbf{x}(s) + \mathbf{b}w(s). \qquad (25.17)$$

$\mathbf{A}$ is an $(n \times n)$ constant matrix (and we shall assume the system to be stable), $\mathbf{b}$ is an $(n \times 1)$ constant vector, and $w(s)$ is a scalar white noise process with

$$E[w(s_1)w(s_2)] = \delta(s_1 - s_2) \qquad (25.18)$$

Since $\mathbf{c}$, $\mathbf{A}$ and $\mathbf{b}$ are constant, and if the system (25.17) is stable, then in the steady state $y(s)$ is a homogeneous (stationary in s) process. Quoting from standard results, we can write the ACVF of $y(s)$ as follows (for $s_2 > s_1$, say)

$$R_{yy}(s_2 - s_1) = \mathbf{c}^T R_{\mathbf{xx}}(s_2 - s_1)\mathbf{c}, \tag{25.19}$$

and

$$R_{\mathbf{xx}}(s_2 - s_1) = R_{\mathbf{xx}}(0)\Phi_A^T(s_2 - s_1). \tag{25.20}$$

$R_{\mathbf{xx}}$ is the covariance matrix for state vector $\mathbf{x}$; $\Phi_A(s_2 - s_1)$ is the state transition matrix for the system in (25.17).

To create the non-stationary process $\tilde{y}(t)$, we introduce $s(t)$ and it follows immediately from equations (25.19) and (25.20) that (for $t_2 > t_1$ and for $\dot{s} > 0$)

$$R_{\tilde{y}\tilde{y}}(t_1, t_2) = \mathbf{c}^T R_{\mathbf{xx}}(0)\phi_A^T(s(t_2) - s(t_1))\mathbf{c}. \tag{25.21}$$

We wish to obtain a time variable shaping filter form for $\tilde{y}(t)$ if we are to obtain the evolutionary spectral form. We now do this as follows. From (25.16) we see that

$$\tilde{y}(t) = \mathbf{c}^T \mathbf{x}[s(t)] = \mathbf{c}^T \tilde{\mathbf{x}}(t), \tag{25.22}$$

where $\tilde{\mathbf{x}}(t)$ denotes the vector $\mathbf{x}[s(t)]$ regarded as a function of time. To convert (25.17) to a form amenable to describe $\tilde{\mathbf{x}}(t)$ we note that

$$\frac{\mathrm{d}}{\mathrm{d}t}\tilde{\mathbf{x}}(t) = \frac{\mathrm{d}}{\mathrm{d}s}\mathbf{x}(s)|_t \dot{s}, \tag{25.23}$$

where $|_t$ denotes the evaluation of $\mathrm{d}/(\mathrm{d}s)\mathbf{x}(s)$ as a function of time. Using (25.17) in (25.23) gives

$$\frac{\mathrm{d}}{\mathrm{d}t}\tilde{\mathbf{x}}(t) = \dot{s}A\tilde{\mathbf{x}}(t) + \dot{s}\mathbf{b}w[s(t)]. \tag{25.24}$$

We remark that there may be situations that s may be regarded as a state element and then this model is related to the class of state dependent parameter models. This interpretation is not pursued here. Furthermore, the treatment is restricted to a single variable $\dot{s}$ modulating $\mathbf{A}$ and $\mathbf{b}$. The case of signals and delayed versions has been considered in Lee (1989), and Harrison and Hammond (1986a).

Equation (25.24) shows $\tilde{\mathbf{x}}(t)$ to be the solution of a time variable differential equation driven by process $w[s(t)]$. This independent variable dilated white process must be replaced by a function of time only in order to be able to proceed. The treatment of this problem may be approached formally as follows, noting a property of the delta function (Zadeh and Desoer, 1963), namely, if $g(t)$ is a function with simple zeros at $t = t_i$, then $\delta[g(t)]$ is equivalent to $\delta(t - t_i)/|\dot{g}(t_i)|$. Generalizing this slightly and applying it to the covariance of the white noise $w[s(t)]$ results in

$$E\{w[s(t_1)]w[s(t)]\} = \frac{\delta(t_1 - t)}{|\dot{s}(t)|}. \tag{25.25}$$

We will assume $\dot{s} > 0$ and so dispense with the modulus sign in (25.25); this is not a significant restriction.

It is equation (25.25) that we now use. An 'equivalent' covariance function would arise if we conceive of another white noise process, written as $w_1(t)[\dot{s}(t)]^{-1/2}$, where $w_1(t)$ is stationary with

$$E[w_1(t_1)w_1(t_2)] = \delta(t_1 - t_2) \tag{25.26}$$

so that

$$E\left[\frac{w_1(t_1)}{[\dot{s}(t_1)]^{1/2}}\frac{w_1(t)}{[\dot{s}(t)]^{1/2}}\right] = \frac{\delta(t_1 - t)}{\dot{s}(t)} \tag{25.27}$$

The process $w_1(t)[\dot{s}(t)]^{-1/2}$ is non-stationary in that it is a modulated white process, having an ACVF which is indistinguishable from the required form in (25.25). Accordingly we shall use $w_1(t)[\dot{s}(t)]^{-1/2}$ in place of $w[s(t)]$ in equation (25.24) and so produce a vector process which we shall call $\mathbf{x}_1(t)$, satisfying

$$\frac{\mathrm{d}}{\mathrm{d}t}\mathbf{x}_1(t) = \dot{s}\mathbf{A}\mathbf{x}_1(t) + \sqrt{\dot{s}}\mathbf{b}w_1(t). \tag{25.28}$$

Associated with (25.22) we write

$$y_1(t) = \mathbf{c}^T\mathbf{x}_1(t). \tag{25.29}$$

We use the notation y_1 rather than $\tilde{y}$ since it is apparent that y_1 and $\tilde{y}$ must differ in some respects. But, in view of the fact that equations (25.24) and (25.28) are both driven by excitations that are 'covariance-equivalent' (i.e. $w[s(t)]$ and $w_1(t)[\dot{s}(t)]^{-1/2}$), then it is reasonable to expect that $\tilde{y}(t)$ and $y_1(t)$ are also covariance-equivalent, i.e. $R_{\tilde{y}\tilde{y}}(t_1, t_2) = R_{y_1y}(t_1, t_2)$. That this is indeed so can easily be demonstrated.

We remark it can also be demonstrated that $\tilde{y}(t)$ and $\tilde{y}_1(t)$ are 'higher distribution equivalent', but we are only concerned with second order properties here. Furthermore, the equivalence of $w[s(t)]$ and $w_1(t)[\dot{s}(t)]^{-1/2}$ is a manifestation of 'self similarity, i.e. temporal scaling reveals a similar structure (Mandelbrot, 1983).

25.3.2 Evolutionary spectral forms for covariance-equivalent models

Evolutionary spectral forms for frequency modulated processes follow directly from the results of the previous section. The important point is that we shall use $y_1(t)$ in place of $\tilde{y}(t)$ and so will use equations (25.28) and (25.29). Let us formally express the stationary process $w_1(t)$ as

$$w_1(y) = \int_{-\infty}^{\infty} \mathrm{e}^{j\omega t}\,\mathrm{d}W(\omega), \tag{25.30}$$

with power spectral density for $w_1(t)$ written $S_{w_1w_1}(\omega) = 1$, then the solution of (25.28) may be written

$$\mathbf{x}_1(t) = \int_{-\infty}^{\infty} \int_{\infty}^{t} \Phi_{\dot{s}A}(t, t_1)[\dot{s}(t_1)]^{1/2}\mathbf{b}e^{j\omega t_1}\,dW(\omega)dt_1. \tag{25.31}$$

Using the substitutions $t - t_1 = \tau$ in (25.31), $y_1(t)$ may be expressed as

$$y_1(t) = \int_{-\infty}^{\infty} e^{j\omega t} A_t(\omega) dW(\omega), \tag{25.32}$$

where

$$A_t(\omega) = \mathbf{c}^T \int_0^{\infty} \Phi_{\dot{s}A}(t, t-\tau)[\dot{s}(t-\tau)]^{1/2} e^{-j\omega\tau}\,d\tau\mathbf{b}. \tag{25.33}$$

The evolutionary spectral density for $y_1(t)$ and hence (by covariance-equivalence) for $\tilde{y}(t)$, is

$$S_{yy,t}(\omega) = |A_t(\omega)|^2. \tag{25.34}$$

Note that $\Phi_{\dot{s}A}(t_2 - t_1) = \exp A[s(t_2) - s(t_1)]$ is the state transition matrix for the system in (25.28).

We note that analogous arguments may be put forward using the impulse response function rather than state space methods but the state space approach has proved convenient for the computations carried out in the applications. The next section illustrates how these ideas may be employed.

25.4 EXAMPLES OF COVARIANCE-EQUIVALENT PROCESSES

Examples of non-stationary random processes having a frequency modulated form have appeared in the literature cited and full descriptions are given in Tsao (1983), Harrison (1983) and Lee (1989). We also note that these concepts have been used in the context of control (Narayanan and Raju, 1992). In this section we will briefly describe three examples.

25.4.1 Vehicle motion over rough terrain

Let us consider the motion of the mass of a vehicle accelerating over rough ground (see Figure 25.1, Harrson (1983)).

If the ground is modelled as $h(s)$ having a covariance structure

$$R_{hh}(\xi) = E[h(s)h(s+\xi)] = \sigma^2 e^{-\alpha|\xi|} \tag{25.35}$$

then the spatial shaping filter for the ground is

$$\frac{dh}{ds} + \alpha h = \sigma(2\alpha)^{1/2} w(s) \tag{25.36}$$

where $w(s)$ is white.

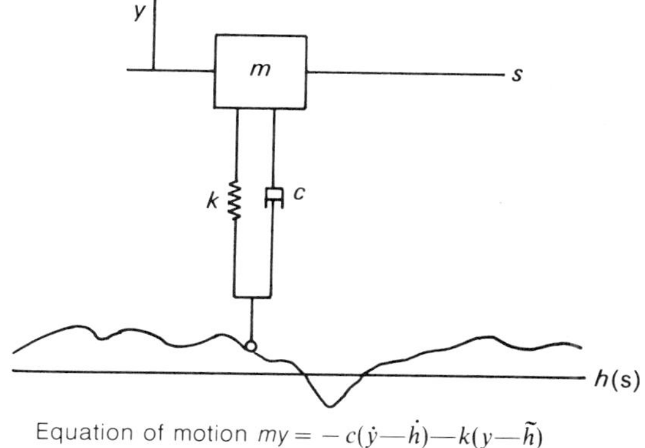

Figure 25.1. Simple vehicle model.

If s a function of time, then $h(s) \to h(s(t)) = \tilde{h}(t)$. Combining this with the dynamic equations $\left(\text{with } \omega_0^2 = \dfrac{k}{m};\ 2\zeta\omega_0 = \dfrac{c}{m}\right)$ yields

$$\frac{\mathrm{d}}{\mathrm{d}t}\begin{bmatrix} y \\ \dot{y} \\ \tilde{h} \end{bmatrix} \begin{bmatrix} 0 & 1 & 0 \\ -\omega_0^2 & -2\zeta\omega_0 & (\omega_0^2 - 2\alpha\zeta\omega_0\dot{s}) \\ 0 & 0 & -\alpha\dot{s} \end{bmatrix} \begin{bmatrix} y \\ \dot{y} \\ \tilde{h} \end{bmatrix} + \begin{bmatrix} 0 \\ 2\zeta\omega_0 \\ 1 \end{bmatrix} \sqrt{2\alpha\sigma\dot{s}}\, w[s(t)]. \tag{25.37}$$

Now using the result $w_1[s(t)] = w_1(t)(\dot{s})^{-1/2}$ and the other results given earlier in this section, yields the evolutionary spectral density shown in Figure 25.2.

Figure 25.2 is a clear indication of how the spectral density of the excitation broadens with time as the vehicle accelerates to excite the resonant behaviour

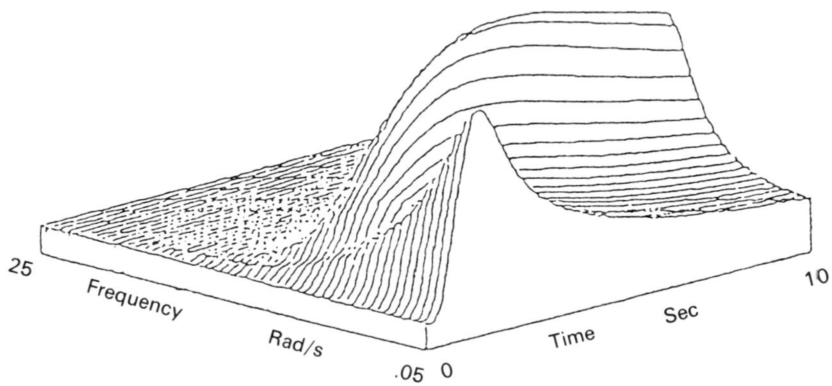

Figure 25.2. Evolutionary spectral density of mass displacement for an accelerating vehicle.

of the response. Note that the frequency variable ω runs from 0.05 and not zero. This is because at $t = 0$ (up to which time the vehicle is at rest) the 'frequencies' perceived by the vehicle are zero and the spectral density of the mass is concentrated at $\omega = 0$ represented as a delta function whose integral is σ^2.

Extensions to the above that have been described include general velocity variations $\dot{s}$, multi-wheel vehicles and inclusion of nonlinear dynamics (Harrison, 1983, and Harrison and Hammond, 1986a).

We emphasise that analytical and computational approaches are presented here to provide a sound basis for (empirical) analysis of data. The nature of the formulation is such that one may be able to take a physical situation and analyse the process/dynamics, treating both as stationary/constant. Only when motion is imposed does the process become non-stationary in the appropriate reference frame and then these procedures allow one to predict the time–frequency spectra that will arise. These predictions may in turn be used to 'validate' empirical analysis of recorded non-stationary data.

In the vehicle case, records of rough ground profiles are available, and assuming spatial homogeneity, constant parameter models may be fitted (e.g. constant AR models). The vehicle dynamics are modelled as constant coefficient differential equations. Only when the vehicle accelerates over the ground does the process becomes non-stationary. Harrison (1983), and Harrison and Hammond (1985) show how real data has been incorporated into this formulation.

25.4.2 Propagating acoustic sources

We shall briefly describe the formulation required for the determination of the evolutionary spectral density of the acoustic signal perceived by a fixed observer when a moving acoustic source emitting a random signal passes by (Tsao, 1983, Lee, 1989). The non-stationarity in this situation arises owing to range, Doppler and directivity effects. We restrict discussions to the case of a monopole moving at constant speed and Figure 25.3 depicts the geometry in three dimensions for the source travelling at constant velocity v_0. The signal received by the observer at time t is due to that generated by the source some time earlier (written as τ). We shall characterize the source distribution in the $y - \tau$ reference system as $4\pi q_0(\tau)\delta(y - v_0\tau)$ where $q_0(\tau)$ denotes the monopole volume strength which is assumed to be a stationary process. To obtain the pressure time history at R at time t requires the solution of the wave equation using the above source distribution. The (far) free-field solution is

$$p(t) = \frac{q(\tau)\dfrac{\mathrm{d}\tau}{\mathrm{d}t}}{R_e(1 - m_0 \cos\theta)}, \tag{25.38}$$

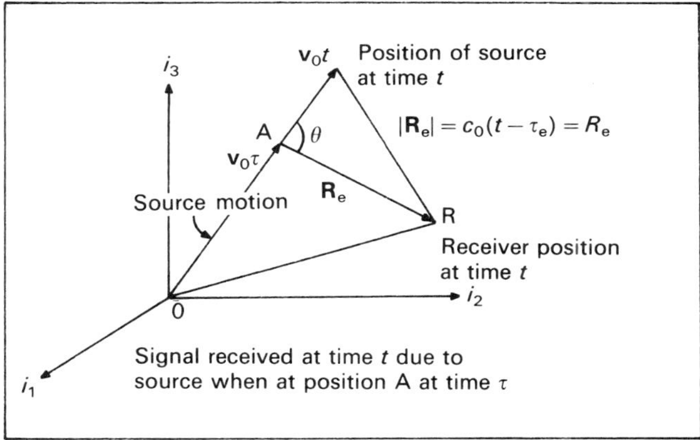

Figure 25.3. Geometry in three dimensions for a source travelling at a constant velocity v_0.

where

$$q(\tau) = \frac{\mathrm{d}q_0(\tau)}{\mathrm{d}\tau};$$

$\tau = t - R_e(t)/c_0$ is the so-called retarded time;

$$m_0 = \frac{v_0}{c_0}$$

is Mach number, assumed less than unity (c_0 is the speed of sound); R_e is the distance between the source and receiver; θ is the angle subtended by the distance vector with the source motion. Note that R_e and θ refer to the source position at the time when the signal received was generated.

It is possible to write down analytic forms for R_e and θ for a simple geometry and we see that equation (25.38) expresses the non-stationary signal $p(t)$ as

$$p(t) = m(t)q[\tau(t)], \tag{25.39}$$

where $m(t)$ accommodates the 'uniform modulation' (25.38) and $\tau(t)$ is the name given to the variable we previously called $s(t)$.

If we assume $q(\tau)$ (stationary in τ, i.e. the reference frame of the source) has a shaping filter representation, we can then conceive of a process $p_1(t)$ which is covariance-equivalent to $p(t)$ for which we can obtain the evolutionary spectral density.

For the case of a simple geometry, i.e. a source moving straight and level over an observer, a contour plot for the theoretical evolutionary spectral density is given in Figure 25.4. The source is assumed to be dominated by a single mode which is apparent from the figure as a 'high' frequency as the

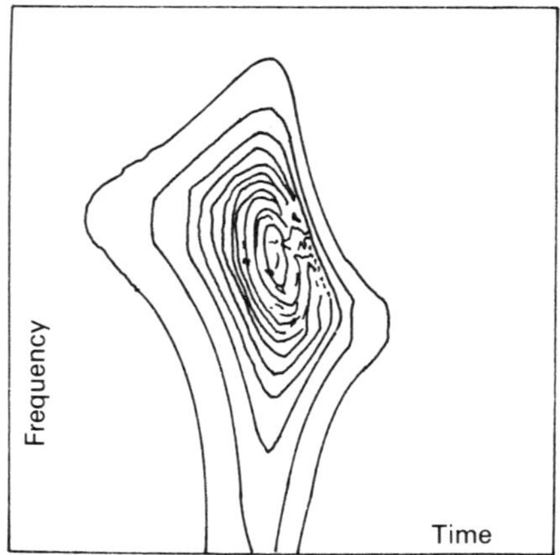

Figure 25.4. Wigner–Ville distribution, contour plot.

source approaches and is 'low' as it recedes. The 'flyover' point is apparent where the spectral density 'broadens' when rates of change are greatest.

More elaborate source structures and geometries, etc. were considered in Tsao (1983).

25.4.3 Directionality patterns of moving sources

The equation (25.39) was generalized in Lee (1989) to include directionality effects and may be written

$$p(t) = m(t)D[\psi(t)]f[\tau(t)] \tag{25.40}$$

The additional term $D[\psi(t)]$ accommodates the directionality of the source and $\psi(t)$ is the radiation angle relative to the observer. Lee considered the problem of estimating the source directionality pattern from sound measure-

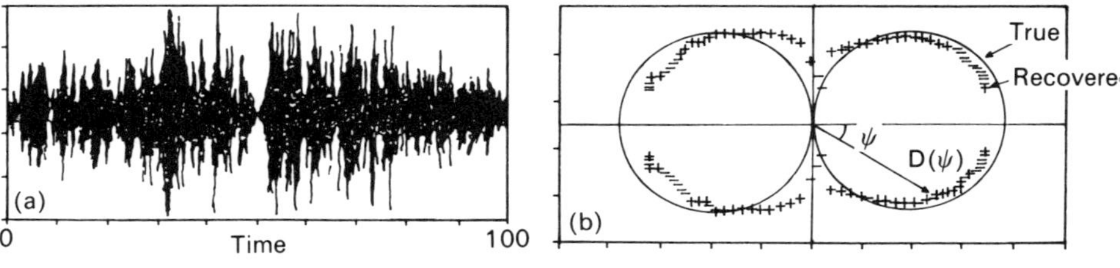

Figure 25.5. (a) Time history due to propagating dipole. (b) True and recovered directionality patterns.

ments which are both amplitude and frequency modulated. The application related to underwater sources and so reflected signals (such as off the water–air interface) were also included. Limitations on space preclude analytical details but Figures 25.5(a) and (b) show a simulation of a pressure measurement due to a dipole source and the true and recovered directionality patterns. Lee compared the evolutionary spectral approach with other time-frequency descriptions.

25.5 RELATIONSHIP OF THE EVOLUTIONARY SPECTRAL DENSITY WITH OTHER TIME–FREQUENCY DISTRIBUTIONS

As noted above, there has been a great deal of activity in time–frequency analysis centred around the so-called Cohen class of distributions and so it is of interest to relate these distributions to the evolutionary spectral density. This was first done by Hammond and Harrison (1986) where the Wigner–Ville distribution and evolutionary spectral density were related. Specifically, it was shown that, for real valued $x(t)$, if $W(t, \nu)$ is the Wigner–Ville spectral density and $S_t(\omega)$ the evolutionary spectral density, then

$$W(t, \nu) = \frac{1}{2\pi}\int_{-\infty}^{\infty} V(t, \nu, \omega) S_t(\omega) \mathrm{d}\omega, \tag{25.41}$$

where

$$V(t, \nu, \omega) = \frac{\displaystyle\int_{-\infty}^{\infty} A^*_{t-\tau/2}(\omega) A_{t+\tau/2}(\omega) \mathrm{e}^{-j\tau(\nu-\omega)} \mathrm{d}\tau}{|A_t(\omega)|^2}, \tag{25.42}$$

i.e. $W(t, \nu)$ is a weighted version of $S_t(\omega)$.

Figure 25.6 shows the two spectra for a uniformly modulated narrow band process and emphasises the significant differences (including negative values for $W(t, \nu)$). We note that these spectra were also compared for the overflying acoustic case described in the last section and it was shown that the Wigner–Ville spectral form was so similar as to be visually indistinguishable from the evolutionary spectral density using the flyover parameters that gave Figure 25.4, although detailed comparisons showed differences at times when rates of change were greatest.

A general relationship between the Cohen class of functions and evolutionary spectra can also be obtained (Hammond, 1992). In section 25.2 we stated a general definition of a frequency distribution as the Fourier transform of $R(t, \tau)$, i.e. equation (25.12). In Cohen (1989) this is written in a different form as

$$S(t, \omega) = \frac{1}{4\pi^2}\iint_{-\infty}^{\infty} \mathrm{e}^{-j\theta t - j\tau\omega + j\theta u} \phi(\theta, \tau) x^*\left(u - \frac{\tau}{2}\right) x\left(u + \frac{\tau}{2}\right) \mathrm{d}u\, \mathrm{d}\tau\, \mathrm{d}\theta. \tag{25.43}$$

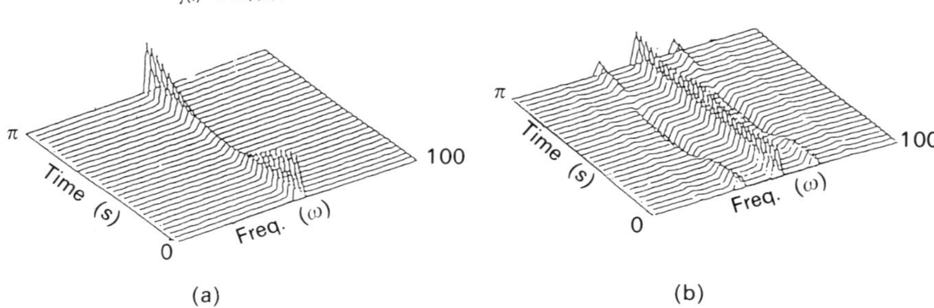

Figure 25.6. (a) Uniformly modulated case, evolutionary spectral density. (b) Uniformly modulated case, Wigner–Ville distribution.

$\phi(\theta, \tau)$ is referred to as the kernel and different choices of ϕ lead to different distributions, e.g. if $\phi = 1$, we obtain the Wigner distribution. Since our interest is in the stochastic case we take expectations of (25.43) to obtain a time–frequency distribution for a random signal and use $E[x^*(u - \tau/2)x(u + \tau/2)] = R_{xx}(u - \tau/2, u + \tau/2)$. Hammond (1992) shows that if $x(t)$ has representation (25.8), then

$$S(t, \omega) = \int_{-\infty}^{\infty} S_t(\omega') \frac{V_A(t, \omega - \omega')}{|A(t, \omega')|^2} \mathrm{d}\omega', \tag{25.44}$$

where

$$V_A(t, \omega - \omega') = \frac{1}{4\pi^2} \iiint_{-\infty}^{\infty} \mathrm{e}^{-j\theta t} \mathrm{e}^{-j\tau(\omega - \omega')} \mathrm{e}^{j\theta u} \phi(\theta, \tau) \times A^*_{u-\tau/2}(\omega') A_{u+\tau/2}(\omega') \mathrm{d}u \, \mathrm{d}\tau \, \mathrm{d}\theta \tag{25.45}$$

V_A is the Cohen class distribution for $A_t(\omega)$. This generalizes equations (25.41) and (25.42).

25.6 CYCLOSTATIONARY PROCESSES AND EVOLUTIONARY SPECTRA

This chapter has concentrated on processes having a frequency modulated structure, but we also note the current interest in so-called cyclostationary processes. There is a very extensive bibliography on the subject and only one reference is given here (Hardin and Miamee, 1990), in which a general class of processes called correlation autoregressive (CAR) is shown to include cyclostationary processes. Specifically such processes have covariance functions that satisfy a linear relationship of the type

$$R_{xx}(t_1, t_2) = \sum_{j=1}^{N} a_j R_{xx}(t_1 + \tau_j, t_2 + \tau_j). \tag{25.46}$$

In Hammond (1992) it is shown that if $x(t)$ is a CAR process described by (25.46), then the corresponding evolutionary spectral density is also autoregressive related, i.e.

$$S_t(\omega) = \sum_{j=1}^{N} a_j S_{t+\tau_j}(\omega). \tag{25.47}$$

Similar relationships are developed by Hammond (1992) for the Cohen class of distributions.

25.7 CONCLUDING REMARKS

Time–frequency distributions have become an indispensible tool for practical signal analysis in science and engineering. The conceptual complications and difficulties have not inhibited extensive empirical analyses by practitioners who are often puzzled by the caution with which the subject is treated. These different attitudes are very stimulating to signal analysts who must bridge both the fundamentals and carry out practical signal processing. The evolutionary spectral density, members of the Cohen class of spectra, Gabor distributions and its relatives (including wavelets (Rioul and Vetterli, 1991)) should become part of the standard armoury of practitioners as special purpose signal analysers increase in sophistication. The papers of Gabor (1946), Priestley (1965a) and, more recently, Cohen (1989), provide essential reading for those who really seek to understand what they are doing.

REFERENCES

Boashash, B., Jones, G. and O'Shea, P. (1989) Instantaneous frequency of signals: concepts, estimation techniques and applications. *Proc. Int. Conference on Advanced Algorithms and Architectures for Signal Processing*, 1152, SPIE 89.

Bozich, D. (1984) *The Analysis of a Class of Signals having Time-Variable Amplitudes and Frequencies.* PhD Thesis, University of Southampton.

Claasen, T.A.C.M. and Mecklenbrauker, W.F.G. (1980) The Wigner distribution – a tool for time–frequency analysis, Parts I, II, III. *Philips J. Res.*, **35**, 217–249; 276–301, 372–389.

Cohen, L. (1989) Time–frequency distributions – A review. *Proc. IEEE.* **77**(7), 941–981.

Gabor, D. (1946) Theory of communication. *J. IEEE, London*, **93**(III), 429–457.

Hammond, J.K. (1968) On the response of single and multi-degree of freedom systems to non-stationary random excitations. *Journal of Sound and Vibration*, **7**(3).

Hammond, J.K. (1971) *Frequency–time methods in vibrations.* PhD Thesis, University of Southampton.

Hammond, J.K. (1973) Evolutionary spectra in random vibrations. *Journal of the Royal Statistical Society*, **B35**, 167–188.

Hammond, J.K. (1992) Analytic time–frequency spectra for acoustic and vibration signal. *Proc. ICA*, Beijing.

Hammond, J.K. and Harrison, R.F. (1981) Non-stationary response of vehicles on rough ground—a state approach. *Trans. ASME Journal of Dynamic Systems, Measurement and Control*, **103**, 245–250.

Hammond, J.K. and Harrison, R.F. (1984) Modelling and deconvolution of non-stationary acoustic signals from moving sources using a covariance equivalent formulation. *Proc. ICASSP*, San Diego. 28B.4.1–28B.4.4.

Hammond, J.K. and Harrison, R.F. (1985) Wigner–Ville and evolutionary spectra for covariance equivalent non-stationary random processes. *Proc. ICASSP*, Tampa, Florida, 1025–1028.

Hammond, J.K. and Moss, J.C. (1991) Time–frequency spectra for nonstationary signals. *Proc. of the Workshop on Nonstationary Stochastic Processes and their Applications*, Hampton University, VA. World Publishing Co.

Hammond, J.K., Tsao, Y.H. and Harrison, R.F. (1983) Evolutionary spectral density models for random processes having a frequency modulated structure. *Proc. ICASSP*, Vol 1, Boston, 261–264.

Hardin, J.C. and Miamee, A.G. (1990) Correlation autoregressive processes with application to helicopter noise. *J. Sound Vib.*, **142**(2), 191–202.

Harrison, R.F. (1983) *The Non-stationary Response of Vehicles on Rough Ground.* PhD Thesis, University of Southampton.

Harrison, R.F. and Hammond, J.K. (1985) A systems approach to the characterisation of rough ground. *J. Sound Vib.*, **99**(3).

Harrison, R.F. and Hammond J.K. (1986a) Analysis of the non-stationary response of vehicles with multiple wheels. *Transactions of the American Society of Mechanical Engineers*, **108**, 69–73.

Harrison, R.F. and Hammond, J.K. (1986b) Evolutionary (frequency/time) spectral analysis of the response of vehicles moving on rough ground by using 'covariant equivalent' modelling. *Journal of Sound and Vibration*, **107**(1), 29–38.

John, R.Y. (1991) *Adaptive Filtering and the Identification of Tones in Broad Band.* PhD Thesis, University of Southampton.

Kodera, K., Gendrin, R. and de Villedary, C. (1978) Analysis of time-varying signals with small BT values. *Trans. on Acoustics, Speech and Signal Processing*, **ASSP-26**(1), 64–76.

Lee, J.S. (1989) *Time-varying Filter Modelling and Times–frequency Characterisation of Non-stationary sound fields due to a moving source.* PhD Thesis, University of Southampton.

Lee, J.S. and Hammond, J.K. (1987) Estimation of the directionality pattern of a moving acoustic source. *Proc. ICASSP*, Dallas, 1752–1756.

Mandelbrot, B.B. (1983) *The Fractal Geometry of Nature*, W.H. Freeman and Co., New York.

Martin, W. (1984) Spectral analysis of non-stationary processes. *Sixth Int. Conf. on Analysis and Optimisation of Systems* (special session on non-stationary processes), Nice, France.

Moss, J.C., Lee, J.S., Hammond, J.K. and Adamopoulos, P.G. (1989) Time–frequency spectra for non-stationary acoustic signals – the Wigner–Ville distribution – the evolutionary spectrum, the modified spectrogram and their inter-relationships. *Proc. of the 117th Meeting of the Acoustical Society of America*, Syracuse.

Narayanan, S. and Raju, G.V. (1992) Active control of non-stationary response of vehicles with nonlinear suspensions. *Vehicle System Dynamics* **21**(2).

Priestley, M.B. (1965a) Evolutionary spectra and nonstationary processes. *J. Roy. Stat. Soc.*, **B27**, 204–237.

Priestley, M.B. (1965b) Design relations for nonstationary processes. *J. Roy. Stat. Soc.*, **B28**, 228–240.

Priestley, M.B. (1967) Power spectral analysis of non-stationary processes. *Journal of Sound and Vibration*, **6**(1).

Priestley, M.B. (1988) *Nonlinear and Non-stationary Time-Series Analysis*. Academic Press, London.

Rioul, O. and Vetterli, M. (1991) Wavelets and signal processing. *IEEE Signal Processing Magazine*. **Oct**, 14–36.

Tsao, Y.H. (1983) *Aspects of Evolutionary Spectral Analysis with Applications to Problems in Acoustics*, PhD Thesis, University of Southampton.

Zadeh, L. and Desoer, C. (1963) *Linear System Theory*, McGraw Hill.

26

Time variable and state dependent modelling of non-stationary and nonlinear time series

P. Young

26.1 INTRODUCTION

Maurice Priestley has made many important contributions to the subject of time series analysis and his classic text on the subject (Priestley, 1981) is an indispensable asset for any student, research worker and practitioner who studies or utilizes time series methods for the advancement of science. Of his many contributions, the most innovative, in the view of the present author, are those which tackle the many problems on non-stationary and nonlinear time series analysis. In honour of Maurice Priestley, therefore, this chapter presents a unified approach to non-stationary and nonlinear time series analysis based on time variable and state dependent parameter estimation; an approach which relates directly to Priestley's own research in this area (e.g. Priestley, 1980, 1988a, b; Haggan *et al.*, 1984).

The methodological basis for the proposed approach is optimal recursive estimation. In particular, it utilizes the recursive filtering and fixed interval smoothing algorithms which derive directly from that best known of all recursive estimation algorithms, the Kalman filter. Indeed, the factor which most differentiates the Kalman filter from the prior recursive estimation algorithms of Gauss and Plackett (see e.g. Young, 1984) is its inherent ability to handle non-stationary systems described by stochastic state space (Gauss–Markov) models; i.e. systems where any, or all, of the parameters in the model may exhibit temporal variation over the observation interval. These include the parameters of the state space model itself, as well as the statistical parameters associated with the stochastic disturbances that are assumed to perturb the model, either as state disturbances or observational noise.

But, as Kalman admitted in his seminal paper (1960), the optimal state estimator (or filter, as it is known in the control and systems literature) has some limitations in practice. In particular, the state space model and its variable parameters need to be known **exactly** by the analyst, prior to the application of the filter, in order to exploit the many advantages of the formulation. In the light of this limitation in the Kalman filter, there has been much interest in the development of a more general procedure for handling models with unknown parameters that may vary over time. This was, indeed, one of the major motivations for the development of recursive techniques for time variable parameter (TVP) estimation, in which the object is to 'model the parameter variations' (Young, 1969a, b; 1984) by some form of stochastic state space model.

Such TVP models have been in almost continual use in the control and system's field since the early 1960s, when Kopp and Orford (1963) and Lee (1964) pioneered their use in the wake of the seminal Kalman (1960) and Kalman and Bucy (1961) papers. Interestingly, these two early but important contributions demonstrate rather different approaches to TVP estimation. Kopp and Orford recognized the nonlinearity of the state parameter estimation problem caused by the multiplication of the state variables by the unknown parameters, and introduced a method which is now universally known as the extended Kalman filter. Here the unknown parameters are also considered as state variables and are adjoined to the state, to form a composite state parameter vector. This composite state vector, which now characterizes a nonlinear system because of the product terms between the parameters and the state variables, is then estimated by a suboptimal linearization procedure applied at each recursion.

Lee, on the other hand, realized that, by allowing the system model to appear only in the 'observation equation' of the state space system, with the parameter variations alone being described by the Gauss–Markov state equations, it was possible to estimate the parameters using a time variable version of the earlier recursive least squares (RLS) estimation algorithm of Gauss and Plackett. In other words, Lee reversed the roles of the states and the parameters, with the states appearing only in an 'observation space' spanned by the measured variables in the model; and with the parameters defining a 'parametric' state space of dimension greater than, or equal to, the number of unknown parameters. This introduced some limitations on the approach, however, since the dynamic model for the system had to be of a type which would allow it to be considered from the standpoint of the observation equation alone. We shall have much more to say on this approach to TVP estimation later in the paper.

One of Lee's proposals was that the parameter variation should be characterized by a first order vector random walk (RW) model which, because of its unity roots, would allow for wide temporal variability in the parameters over any finite observation interval. The present author made liberal use of this same device in the 1960s within the context of self adaptive control design (Young, 1969a, 1970, 1971a, 1981), and proposed an extension to the idea if *a priori* information was available about the nature of the parametric time variability (Young, 1969b). Later, in the early 1970's, he also reminded a statistical audience of the extensive system's literature on recursive estimation and its application to TVP estimation (see Young, 1971b, 1975a; also the comments of W.D. Ray on the paper by Harrison and Stevens (1976)).

Another area where TVP modelling has been influential is adaptive digital signal processing. In 1979, the book by Willsky emphasized the close relationship between digital signal processing and control and estimation theory and, in subsequent years, there has been a continuing interplay of ideas on TVP estimation between the two research areas. One of the first and most famous adaptive algorithms for signal processing was the least mean square (LMS) algorithm of Widrow and Hoff (1960), a simple stochastic gradient algorithm which was introduced in the same year that Kalman's paper on optimal recursive estimation was published. The subsequent literature on adaptive signal processing is vast and too extensive to review here. It includes, for instance, the development of adaptive algorithms in areas such as signal equalization, noise cancellation, linear predictive coding (LPC) and spectral estimation. In general, these adaptive algorithms are based on TVP versions of common regression models and tend to concentrate on the development of fast versions of the algorithms, such as ladder and lattice methods (see e.g. Ljung and Soderstrom, 1983), which are desirable for 'on-line' implementation. Except for some of the recent research on wavelet transform methods (e.g. Rioul and Vetterli, 1991), however, there appears to be much less emphasis on 'off-line' methods, such as the fixed interval smoothing procedures which are of particular importance in the present chapter.

Since the early 1970s, TVP models have also been proposed and studied extensively in the statistical and econometrics literatures. For example, a major line of development has been linked to the well known 'structural' or 'component' time series model (e.g. Harrison and Stevens 1971, 1976; Kitagawa, 1981; Harvey, 1984). The term 'structural' has been used in other connections in both the statistical and economics literatures and so we will employ the former term. Here, the approach is an extension of the Lee procedure (although this is not overtly acknowledged by the authors), in which the parameter variations are described by a higher dimensional, vector random-walk type model termed the 'linear growth equation' by Harrison and Stevens. In some of these references, the potential importance of recursive smoothing is also highlighted and the methodology can be compared with that proposed in the systems literature by Norton (1975) and pursued in more detail by Jakeman and Young (1979, 1984).

The latter reference also shows how the recursive state-space algorithms are closely related and, in some cases yield equivalent results, to other smoothing procedures based on the optimization technique known as 'regularization', in which the smoothed estimate is obtained by minimizing (non-recursively) a least squares criterion function which includes constraints on the rates of change of the estimated variables (see Young, 1991). Recent research in the economic literature (e.g. Kalaba and Tesfatsion, 1988), which refers to this approach as 'flexible least squares' also uses this kind of optimization technique. However, we feel that the state space smoothing procedures used in the present paper provide a more elegant and flexible method of fixed interval smoothing estimation.

In the wider econometrics literature, there have been numerous contributions involving the concept of TVP estimation and Engle *et al.* (1988), for example, present a recent brief review of this topic and discuss an interesting application to electricity sales forecasting, in which the model is a time variable parameter regression plus an adaptive trend described by an RW model. Of considerable importance, particularly in the economics context, is the work of Sims and his co-workers (e.g. Doan *et al.*, 1984) on Bayesian vector autoregressive modelling and forecasting (BVAR). Here the vector autoregressive (VAR) model is extended so that its potentially time-variable parameters are each assumed to be described by random walk models. The model is then considered within a Bayesian framework, somewhat similar to that used by Harrison and Stevens, but with the Bayesian 'hyper parameters' estimated via maximum likelihood using special methods of numerical optimization.

Recent research by the present author and his collaborators (e.g. Young, 1988, 1989; Young and Ng, 1989; Ng and Young, 1990; Ng *et al.*, 1988; Young *et al.*, 1989, 1991c) has also been concerned with the component type of time series model and, like the earlier contributions in this context, employs the standard Kalman filter-type recursive filtering and smoothing algorithms. Except in the final forecasting and smoothing stages of the analysis, however, the justification for using these algorithms is not based on either a Bayesian interpretation (Harrison and Stevens, 1976) or 'optimality' in a prediction error or maximum likelihood (ML) sense (Harvey, 1984). Rather, the spectral properties of the algorithms are exploited in a manner which allows for straightforward and effective spectral decomposition of the time series into quasi-orthogonal components. A unifying element in this analysis is the modelling of non-stationary state variables and time variable parameters by a class of second order random walk models which are able to handle abrupt changes, or even discontinuities, in the states or parameters, so extending its range of applicability.

Finally, a number of previous papers by Maurice Priestley and the present author have attempted to consider the use of TVP estimation in a more general context; namely the identification and estimation of nonlinear stochastic, dynamic systems. Young (1978) and Young and Runkle (1989) approach this problem from an engineering standpoint, noting that normal

Taylor series linearization of nonlinear dynamic systems usually produces linearized, time variable coefficient models which can be estimated by TVP versions of the various recursive parameter estimation algorithms discussed above. In this manner, the nature of the nonlinearity can then be inferred and the model can either be useful in its own right, or as a prelude to nonlinear estimation based on the identified nonlinear structure and using techniques such as maximum likelihood. Priestley (1980, 1988a, b), Priestley and Heravi (1985), and Haggan *et al.* (1984) use a more formal approach which considers various linearized forms of the nonlinear models, including Volterra series expansions. However, their basic approach, as demonstrated in the paper by Haggan *et al.* (1984) and the book by Priestley (1988a), is very similar to that of Young: it also uses a Taylor series expansion of a particular nonlinear stochastic model form and exploits recursive algorithms to estimate the time variable parameters in this linearized representation.

There are two differences between the approaches of Young and Priestley. The first is conceptual and lies in the assumptions made about the time variability of the parameters. Based on the nature of the first order terms in the linearization expansion, Priestley notes that the parameters will be 'state dependent' and he uses this information to define the form of the stochastic model for the parameter variations. Young recognizes the possibility of this state dependency in a less formal manner but also allows for dependency on other variables that are not necessarily 'states' in the more limited definition of the state space employed by Priestley. The second difference is methodological. Priestley uses a recursive 'filtering' algorithm to estimate the time variable parameters in a similar manner to Young but then smooths these filtered estimates with a **separate** smoothing algorithm. Young, on the other hand, integrates the smoothing directly into the recursive estimation by employing the associated, fixed interval, smoothing recursions subsequent to the forward filtering pass through the data, thereby obtaining **lag-free** estimates of the time variable parameters and minimizing end effects.

In the present paper, we will explore further the concepts put forward by Young and Priestley and show how they can both be cast within a general recursive estimation and fixed interval smoothing context. The utility of the techniques will then be demonstrated by two illustrative examples: the first is based on simulated data from the famous Lorenz 'strange attractor' model, and the second is a practical one concerned with the modelling of nonlinear rainfall-flow processes.

26.2 THE NONLINEAR TIME SERIES MODEL AND LINEARIZATION

Following previous publications (e.g. Young and Runkle; 1989; Young, 1992) let us consider a scalar time series $y(k)$ which can be described by a nonlinear stochastic, dynamic equation of the form,

$$y(k) = f\{y(k-1), \ldots, y(k-n), \mathbf{u}(k), \ldots, \mathbf{u}(k-m), \ldots, \mathbf{U}(k), \ldots,$$
$$\mathbf{U}(k-q), e(k-1), \ldots, e(k-p)\} + e(k), \tag{26.1}$$

where $f\{\cdot\}$ is a reasonably behaved, nonlinear function dependent upon past values of $y(k)$, as well as present and past values of a deterministic input (or exogenous) variable vector $\mathbf{u}(k)$ with elements $u_i(k)$, $i = 1,2,\ldots,r$; the present and past values of a vector $\mathbf{U}(k)$ of other exogenous variables $U_j(k)$, $j = 1,2,\ldots,s$; and a white noise process $e(k)$. The vector $\mathbf{U}(k)$ represents any other associated variables which may affect the system nonlinearly but whose relevance in this regard is not clear prior to time series analysis. This model is very similar to that considered by Priestley (1980, 1988a,b) Priestley and Heravi (1985), and Haggan *et al.*, (1984) except for the inclusion here of the vector $\mathbf{U}(k)$ of 'other variables', the importance of which will become apparent as we proceed.

In this setting, $e(k)$ can be considered as an 'innovations' process, with the nonlinear function acting as a 'nonlinear predictor' or conditional expectation or the $y(k)$ given all information and data on the system up to the kth sample, i.e.

$$f\{\chi(k)\} = E\{y(k)|k\},$$

where $\chi(k)$ is, in general, a non-minimal state space (NMSS) vector (see Priestley, 1980; Young *et al.*, 1987) for the system with elements $y(k-i)$, $i = 1,2,\ldots,n$; $u_i(k-j)$, $i = 1,2,\ldots,r$; $j = 0,1,\ldots,m$; $\mathbf{U}_j(k-h)$, $j = 1,\ldots,s$; $h = 0,1,\ldots,q$; and $e(k-\ell)$, $\ell = 1,2,\ldots,p$.

Using the normal systems approach to linearization and, for simplicity, considering only a single exogenous variable $u(k)$, we can now expand the RHS of equation (26.1) in a Taylor series about $f\{\chi(k_0)\}$ at some sampling instant k_0, i.e.

$$\begin{aligned} y(k) = f\{\chi(k_0) &+ \sum_{i=1}^{n}\left[\frac{\partial f\{\chi(k)\}}{\partial y(k-i)}\right]_{k=k_0}\{y(k-i)-y(k_0-i)\} \\ &+ \sum_{j=0}^{m}\left[\frac{\partial f\{\chi(k)\}}{\partial u(k-j)}\right]_{k=k_0}\{u(k-j)-u(k_0-j)\} \\ &+ \sum_{\ell=1}^{p}\left[\frac{\partial f\{\chi(k)\}}{\partial e(k-\ell)}\right]_{k=k_0}\{e(k-\ell)-e(k_0-\ell)\} \\ &+ e(k) + \text{first order terms in } U(k-h),\ h = 1,2,\ldots,q \\ &+ \text{higher order terms}\ldots \end{aligned} \tag{26.2}$$

At this point, we assume that the first order sensitivity with respect to the $\mathbf{U}(k)$ variables is small enough for us to ignore them, in addition to the usual higher order terms in the other variables. Note that this does **not** mean that these variables are unimportant: clearly, the partial derivatives of $f\{\chi(k)\}$ with respect to the other variables may well be functions of the $\mathbf{U}(k)$ variables. In particular, we might expect these variables to influence the low frequency, wide ranging changes in these derivatives and, therefore, the resulting time variable parameters of the linearized model.

With some manipulation of equation (26.2), $y(k)$ can be represented in the form,

$$y(k) + \sum_{i=1}^{n} a_i[\chi(k)]y(k-i) = T[\chi(k)] + \sum_{j=0}^{m} b_j[\chi(k)]u(k-j)$$

$$+ \sum_{\ell=1}^{p} c_\ell[\chi(k)]e(k-\ell) + e(k). \tag{26.3}$$

In this equation, $a_i[\chi(k)]$, $b_j[\chi(k)]$, $c_\ell[\chi(k)]$ and $T[\chi(k)]$ are coefficients in the model which are functions of the NMSS vector and the sampling index k. Here, $T[\chi(k)]$ can be considered as a slowly varying 'trend' parameter which allows for long term changes in the mean of the series. All these parameters can be considered both as 'state dependent' (Priestley, 1980) or 'time variable' (Young, 1978) parameters, depending upon the perspective of the analyst.

As a specific and practical example of the TVP model, consider an aerospace vehicle designed to fly over an extended flight envelope. At any particular flight condition, the dynamic behaviour of the vehicle will be characterized by the perturbations of those variables which describe the motion relative to the local reference frame flight condition. Furthermore, local linearization of the nonlinear vehicle state equations at such a flight condition normally results in a linearized model such as (26.3), or its deterministic equivalent, with parameters that can be assumed sensibly constant at the chosen flight condition for purposes such as control system design. Over a complete fight mission, however, the coefficients of the linearized equations of motion (the 'stability and control derivatives') will also be functions of other 'flight condition' variables (playing the role of the $\mathbf{U}(k)$ variables in our formulation of the general model), such as dynamic pressure and altitude, which define the changing environment and significantly affect the dynamic characteristics of the vehicle. Consequently, the collection of all such linearized models over the whole flight envelope provides, in effect, a time varying parameter linear model for the vehicle which describes its dynamic behaviour at all flight conditions. Indeed, this is the motivation behind the self adaptive control system of Young (1969a,b; 1981) to which we shall refer later.

26.3 THE TIME VARIABLE PARAMETER (TVP) TIME SERIES MODEL

It is now convenient to write equation (26.3) in the following vector form,

$$y(k) = \mathbf{z}(k)^T \mathbf{a}(k) + e(k), \tag{26.4}$$

where,

$$\mathbf{z}(k)^T = [1, y(k-1), \ldots, y(k-n), u(k), \ldots, u(k-m), e(k-1), \ldots, e(k-p)]$$

$$\mathbf{a}(k)^T = [T(k), a_1(k), \ldots, a_n(k), b_0(k), \ldots, b_m(k), c_1(k), \ldots, c_p(k)],$$

and where the TVP nature of the model is denoted by the temporal dependence of the parameters in the **a** vector. This temporal dependence could, of course, be due to state dependence in the sense of Priestley and the later model identification and estimation procedures will acknowledge this possibility. For simplicity of exposition, however, we will drop the state dependent argument and proceed under the assumption that the parameters will, for various reasons, be dependent upon the time index k. Note that it is tempting, at this time, to compare this form of the model with the well known, constant parameter autoregressive moving average exogenous variable (ARMAX) model. In fact, the model (26.4) has much wider significance than the ARMAX model, as we shall see in later sections of the paper.

In order to complete the model description, it is now necessary to introduce some form of mathematical description for the temporal variation in the parameters of model (26.4). There are many different ways of approaching this problem, but here we will choose to 'model the parameter variations' (see Young, 1978, 1984) by the following Gauss—Markov (GM) process,

$$\mathbf{x}(k) = \mathbf{F}(k)\mathbf{x}(k-1) + \mathbf{G}(k)\boldsymbol{\eta}(k) \tag{26.5}$$

where $\mathbf{x}(k)$ is a 'state' vector representing the parameters in $\mathbf{a}(k)$ as well as any other elements required in the complete state description of their evolution through time. The dimension of $\mathbf{x}(k)$ will be equal to or greater than that of $\mathbf{a}$. The matrices $\mathbf{F}(k)$ and $\mathbf{G}(k)$ are, respectively, appropriately dimensioned transition and input matrices whose elements may also vary over time; while $\boldsymbol{\eta}(k)$ is a white noise vector with zero mean and (possibly time-variable) covariance matrix $\mathbf{Q}(k)$, i.e.

$$E\{\boldsymbol{\eta}(k)\boldsymbol{\eta}(j)^T\} = \mathbf{Q}(k)\delta_{k,j}; \quad \delta_{k,j} = \begin{cases} 1 \text{ for } k = j \\ 0 \text{ for } k \neq j \end{cases}.$$

The nature of the matrices $\mathbf{F}(k)$, $\mathbf{G}(k)$, $\mathbf{Q}(k)$ and the state vector $\mathbf{x}(k)$ (including various possible forms for their temporal dependence) will become clearer later in the paper, when we discuss special examples of the general model. For the moment, it will suffice to note that this model, in one form of another, has been employed on many occasions over the past 30 years as a device for modelling parameter variations. For example, with **F** and **G** both equal to the identity matrix, the model is simply the well known and used vector random walk (RW), as mentioned earlier.

The major estimation problem associated with the equations (26.4) and (26.5) arises from the presence of the unobservable stochastic terms $e(k-1)$ to $e(k-p)$ in $\mathbf{z}(k)$. However, the model can be simplified further to a linear TVP relationship if it is possible to assume that the stochastic influences in equation (26.4) reside completely in the additive white noise term $e(k)$, so that $\mathbf{z}(k)$ does not depend on the past values of this variable. The stochastic disturbance vector $\boldsymbol{\eta}(k)$ in the parameter variation equation (26.5) then constitutes the only other stochastic input to the system and, as we shall see, this can be associated directly with the constraints we choose to impose on the nature of the variable parameters in equation (26.4).

26.4. IDENTIFICATION AND ESTIMATION OF THE TVP MODEL

The model described by equations (26.4) and (26.5) can be represented in the following, well known, state space setting,

$$\mathbf{x}(k) = \mathbf{F}(k)\mathbf{x}(k-1) + \mathbf{G}(k)\boldsymbol{\eta}(k) \tag{26.6}$$

$$y(k) = \mathbf{H}(k)\mathbf{x}(k) + e(k) \tag{26.7}$$

where $\mathbf{H}(k)$ is an observation vector chosen so that the observation equation (26.7) represents the TVP model (26.4). The specific form of $\mathbf{H}(k)$ will, of course, depend upon the application but the specific examples discussed below will help to clarify the nature of this vector. Nominally, this model presents a quite formidable estimation problem since it involves the estimation of a combination of unknown, time variable parameters and states appearing in nonlinear ralation to each other. Let us consider first, therefore, the simpler, linear, TVP representation, where $\mathbf{z}(k)$ is assumed to be independent of the past values of $e(k)$.

26.4.1 The linear TVP model

The linear TVP form of the model (26.4) takes the special form,

$$y(k) = \mathbf{z}(k)^T \mathbf{a}(k) + e(k), \tag{26.8}$$

where,

$$\mathbf{z}(k)^T = [1, y(k-1), \ldots, y(k-n), u(k), \ldots, u(k-m)]$$

$$\mathbf{a}(k)^T = [T(k), a_1(k), \ldots, a_n(k), b_0(k), \ldots, b_m k)].$$

The recursive least squares (RLS) algorithm, suitably modified to allow for time variable parameters described by a Gauss—Markov (GM) model such as (26.6), can be applied directly to the model in this form. For this to be successful, however, the analyst must be able to specify the 'system' matrices $\mathbf{F}(k)$ and $\mathbf{G}(k)$, for all k, together with information on the statistical characteristics of the stochastic disturbances $e(k)$ and $\boldsymbol{\eta}(k)$ (see Young, 1984).

This latter requirement is eased somewhat by the scalar form of equation (26.8) which, depending upon the definition of $\mathbf{x}$ and the GM model (26.6), will define the 'observation' equation (26.7): for instance, if $\mathbf{x}(k) = \mathbf{a}(k)$ then $\mathbf{H}(k) = \mathbf{z}(k)^T$. It is easy to show that, for the purposes of estimation, it is not the absolute values of $\mathbf{Q}(k)$ and σ^2 that are important, but their relative values. As a result, without any loss of generality, we can define a 'noise variance ratio' (NVR) matrix $\mathbf{Q}_r(k)$, i.e.

$$\mathbf{Q}_r(k) = \mathbf{Q}(k)/\sigma^2 \tag{26.9}$$

which will replace $\mathbf{Q}(k)$ in the analysis and the recursive estimation algorithms. For simplicity, it is normally assumed that $\mathbf{Q}_r(k)$ is a diagonal matrix with elements (the NVR values) $q_{ii}(k)$, $i = 1, 2, \ldots, n + m + 2$, that are associated with the time variable nature of the parameters $a_i(k)$, $i = 1, 2, \ldots, n$; $b_j(k)$, $j = 0, 1, \ldots, m$; and $T(k)$.

The RLS filtering algorithm, with the TVP modification and the introduction of the NVR matrix, takes the following prediction-correction form (see e.g. Young, 1984),

Algorithm 26.1

Prediction

$$\begin{aligned}\hat{\mathbf{x}}(k/k-1) &= \mathbf{F}(k)\hat{\mathbf{x}}(k-1)\\ \mathbf{P}(k/k-1) &= \mathbf{F}(k)\mathbf{P}(k-1)\mathbf{F}(k)^T + \mathbf{G}(k)[\mathbf{Q}_r(k)]\mathbf{G}(k)^T.\end{aligned} \tag{26.10}$$

Correction

$$\begin{aligned}\hat{\mathbf{x}}(k) &= \hat{\mathbf{x}}(k/k-1) + \mathbf{P}(k/k-1)\mathbf{H}(k)^T[1 + \mathbf{H}(k)\mathbf{P}(k/k-1)\mathbf{H}(k)^T]^{-1}\\ &\quad \times \{y(k) - \mathbf{H}(k)\hat{\mathbf{x}}(k/k-1)\}\\ \mathbf{P}(k) &= \mathbf{P}(k/k-1) - \mathbf{P}(k/k-1)\mathbf{H}(k)^T[1 + \mathbf{H}(k)\mathbf{P}(k/k-1)\mathbf{H}(k)^T]^{-1}\\ &\quad \times \mathbf{H}(k)\mathbf{P}(k/k-1)\end{aligned} \tag{26.11}$$

Here, $\hat{\mathbf{x}}(k)$ denotes the recursive estimate of $\mathbf{x}(k)$ at the kth sampling instant, while $\hat{\mathbf{x}}(k/k-1)$ is the recursive estimate of $\mathbf{x}(k)$ at k conditional on data up to and including the $(k-1)$th sample. It can be shown that $\mathbf{P}(k)^* = \mathbf{P}(k)/\sigma^2$ provides an estimate of the covariance matrix for the estimate vector $\hat{\mathbf{x}}(k)$ and so, with this statistical interpretation, $\mathbf{P}(k/k-1)/\sigma^2$ is an estimate of the covariance at k conditional on the information processed up to the $(k-1)$th instant.

The algorithm 26.1 is, of course, identical in form to the Kalman filter algorithm. We choose to describe it within the RLS parameter estimation context because the vector $\mathbf{H}(k)$, which plays the role of the observation vector in conventional Kalman filter terms is, in part, composed here of stochastic variables measured in the presence of noise. Formally, the Kalman filter requires that the elements of this vector should be exactly known, deterministic variables. While this formal requirement is not critical to the success of the present algorithm in estimation terms, it is important that we recognize the differences between the present formulation and the more conventional Kalman filter. In this manner, it should be possible to ensure that these differences do not cause estimation problems (such as asymptotic bias on the estimates) or that we do not read more into the statistical properties of the estimates than is justified.

Bearing these caveats in mind, it is possible to proceed one step further in the estimation of $\mathbf{x}(k)$; namely the generation of a 'smoothed estimate' for the TVP vector. The algorithm 26.1 provides an estimate of $\mathbf{x}(k)$ at the kth sampling instant which is based on the data up to and including the kth sample, i.e. $\mathbf{x}(k)=\mathbf{x}(k/k)$. If we are pursuing off-line analysis and are confronted with a data set with $N>k$ samples, however, it is a distinct advantage in this TVP situation to obtain an estimate $\hat{\mathbf{x}}(k/N)$ at the kth instant conditional on all of the available data over the observation interval. This smoothed estimate will not then be affected by the phase lag which is inherent in the filtered estimate $\hat{\mathbf{x}}(k)$ and it will have lower estimation error variance. This argument suggests the generation of such a smoothed estimate by the use of a 'fixed interval smoothing' (FIS) algorithm (see e.g. Bryson and Ho, 1969; Gelb *et al.*, 1974).

There are a variety of FIS algorithms but the one we will consider here utilizes the following backwards recursive algorithm, subsequent to application of the above Kalman filtering forwards recursion (see e.g. Norton, 1975; Young, 1984).

Algorithm 26.2

$$\hat{\mathbf{x}}(k/N)=\mathbf{F}(k)^{-1}[\hat{\mathbf{x}}(k+1/N)+\mathbf{G}(k)\mathbf{Q}_r(k)\mathbf{G}(k)\mathbf{L}(k-1)], \qquad (26.12)$$

where

$$\mathbf{L}(N)=0,$$

N is the total number of observations (the 'fixed interval'), and

$$\mathbf{L}(k-1)=[\mathbf{I}-\mathbf{P}(k)\mathbf{H}(k)^T\mathbf{H}(k)][\mathbf{F}(k)^T\mathbf{L}(k)-\mathbf{H}(k)^T \times\{y(k)-\mathbf{H}(k)\mathbf{F}(k-1)\hat{\mathbf{x}}(k-1)\}] \qquad (26.13)$$

is an associated backwards recursion for the 'Lagrange Multiplier' vector $\mathbf{L}(k)$ required in the solution of this two point boundary value problem.

Finally, the covariance matrix $\mathbf{P}^*(k/N)=\sigma^2\mathbf{P}(k/N)$ for the smoothed estimate is obtained by reference to $\mathbf{P}(k/N)$ generated by the matrix recursion

$$\mathbf{P}(k/N)=\mathbf{P}(k)+\mathbf{P}(k)\mathbf{F}(k+1)^T[\mathbf{P}(k+1/k)]^{-1}\{\mathbf{P}(k+1/N)-\mathbf{P}(k+1/k)\}[\mathbf{P}(k+1/k)]^{-1}\mathbf{F}(k+1)\mathbf{P}(k), \qquad (26.14)$$

while the smoothed estimate of original series $y(k)$ is given simply by,

$$\hat{y}(k/N) = \mathbf{H}\hat{\mathbf{x}}(k/N), \tag{26.15}$$

i.e. the appropriate linear combination of the smoothed state variables. As in the forward filtering pass, the recursions (26.12)–(26.14) are only formally applicable if $\mathbf{z}(k)$ is a purely deterministic vector. Indeed, the problem here is rather more problematic than in the filtering case and the smoothing estimates obtained in this manner are sub-optimal in strict maximum likelihood or Bayesian sense. However, as we shall see, this sub-optimality is not of major practical significance in the present context.

When using the algorithms 26.1 and 26.2 it is often an advantage to obtain a TVP estimate of the variance of the white observation noise $e(k)$ in case it is heteroscedastic. Although no direct measurements $e(k)$ are available, it is possible to investigate any heteroscedasticity by computing an estimate of the variance from the FIS model residuals $\hat{e}(k/N)$, where,

$$\hat{e}(k/N) = y(k) - \mathbf{z}^T(k)\hat{\mathbf{a}}(k/N).$$

The same basic TVP recursive filtering-smoothing algorithm (26.10)–(26.14) is used for this secondary computation with the observation $y(k)$ replaced by $\hat{e}(k/N)^2$. But now there is only one unknown parameter to estimate, namely the white noise variance $\sigma^2(k)$, which is normally modelled by the simplest, scalar RW process (see section 26.5.1).

Finally, in practical time-series analysis and modelling, the exact nature of the parametric variation in TVP models is difficult to predict: while the changes in the behavioural characteristics of dynamic systems are often relatively slow and smooth, more rapid and violent changes do occur from time-to-time and lead to similarly rapid changes, or even discontinuities, in the nature of the related time series. One approach to this kind of problem is variance intervention (Young, 1989; Young and Ng, 1989; T.J. Young *et al.*, 1988), where instantaneous or short term increases in the diagonal elements of the NVR matrix $\mathbf{Q}_r(k)$ in equation (26.9) are introduced to allow for sudden changes in the corresponding FIS estimates of the parameters. For instance, if the IRW model (section 26.5.1) is applied to the modelling of trend behaviour in a time-series, then such 'variance interventions' can allow for discontinuities in the estimates of the trend $T(k)$ and/or its slope $d(k)$.

Before proceeding, it is important to note that Priestley (see e.g. 1988a, b) also utilizes the Kalman filter algorithm 26.1 for parameter estimation in a similar manner to that described above. However, while accepting the need for smoothing to offset the effects of noise, he does not exploit fixed interval smoothing of the kind discussed above. Rather, he accepts a less formal definition of smoothing and utilizes a two stage approach in which 'appropriate choice of the smoothing parameters' (as defined, in the present context, by the elements of NVR matrix $\mathbf{Q}_r$) is combined with a separate, multidimensional form of the nonparametric function fitting technique of Priestley and

Chao (1972). The advantages of the fixed interval smoothing algorithm 26.2 are three-fold: first, it is felt that the more formal definition of smoothing is preferable and provides a more 'natural' method for obtaining smoothed estimates when using an optimal state estimation approach; second, it yields lag-free estimates of the variable parameters, whereas there is an inevitable lag in the 'forward-pass' filtered estimates used by Priestley; thirdly, the covariance matrix $\mathbf{P}^*(k/N)$ associated with the smoothed estimates is provided automatically from equation (26.14); and, finally, the 'end effects' that are so prominent in Priestley's results (see e.g. Priestley, 1988a, p. 134 et seq.) are not so marked.

26.4.2 The pseudo-linear time series model

Strictly, the filtering and smoothing algorithms 26.1 and 26.2 are not directly applicable in the more general case of equation (26.4) where the $\mathbf{z}(k)$ vector is a function of past values of the unobserved $e(k)$ variable. Nevertheless, an approximate recursive solution can be evolved using a device first proposed by Young (1968) and Panuska (1969) where, at each recursion, the $e(k-i)$ elements in $\mathbf{z}(k)$ are replaced by their estimates $\hat{e}(k-i)$ obtained recursively from the equation

$$\hat{e}(k) = y(k) - \mathbf{H}(k)\hat{\mathbf{x}}(k) = y(k) - \mathbf{z}(k)^T\hat{\mathbf{a}}(k) \tag{26.16}$$

The resulting filtering algorithm has been termed either the approximate maximum likelihood (AML) or extended least squares (ELS) estimation procedure: it is an intuitively appealing approximation which allows us to develop a fairly simple recursive solution to the estimation problem posed by equation (26.4) using linear-like estimation procedures. It is not obvious, of course, that this 'pseudo-linear' RLS algorithm will converge under all conditions. However, it has been used successfully in many practical applications. Moreover, Solo (1980) has considered its convergence from a theoretical standpoint and shown that it possesses reasonable characteristics in this regard. The smoothing algorithm in this pseudo-linear case is less well known. As far as we are aware, only Norton (1975) has previously used the smoothing algorithm 26.2 in this context and, although we can confirm the generally good performance he reports, there is clearly need for further research on this topic.

26.4.3 The transfer function (TF) time-series model

In a similar manner to Priestley (1988, p. 98), we can consider the general time (or state) dependent 'transfer function' form of equation (26.3),

$$y(k) = t(k) + \frac{B(z^{-1})}{A(z^{-1})}u(k) + \frac{D(z^{-1})}{A(z^{-1})}e(k), \tag{26.17}$$

where $A(z^{-1}) = A(k, z^{-1})$, $B(z^{-1}) = B(k, z^{-1})$ and $D(z^{-1}) = D(k, z^{-1})$ are **time variable coefficient** polynomials in z^{-1}, each characterized, respectively, by the time variable parameters $a_i(k)$, $b_j(k)$ and $c_\ell(k)$; while $t(k)$ is a new trend variable defined as $T(k)/A(z^{-1})$. This representation, which is obtained simply by introducing the backward shift operator z^{-i}, i.e. $z^{-i}y(k) = y(k-i)$ into equation (26.3) and rearranging the equation, reveals the connection between the TVP model (26.17) and the equivalent constant parameter TF models which play such an important role in control and systems theory.

In (26.17), the system and noise transfer functions are both characterized by the same denominator polynomial $A(z^{-1})$. However, again following Priestley, if we choose to separate out the effect of $u(k)$ into a second nonlinear function $g\{.\}$ when formulating the original nonlinear model, i.e.

$$\begin{aligned} y(k) = f\{&y(k-1), \ldots, y(k-n), e(k-1), \ldots, e(k-p)\} \\ &+ g\{y(k-1), \ldots, y(k-n), u(k), \ldots, u(k-m)\} + e(k) \end{aligned}$$

then equation (26.17) would be transformed into the following alternative form,

$$y(k) = t(k) + \frac{B(z^{-1})}{A(z^{-1})}u(k) + \frac{D(z^{-1})}{C(z^{-1})}e(k) \qquad (26.18)$$

which will be recognized as the TVP version of the well known Box–Jenkins model (Box and Jenkins, 1970).

In the constant parameter recursive estimation situation, Jakeman and Young (1981, 1983) have shown that there are some advantages to considering this second Box–Jenkins model form rather than the common denominator 'ARMAX' form of equation (26.17). Note that, in the constant parameter version of this Box–Jenkins model, it is common to allow for the presence of a pure time delay δ between $u(k)$ and $y(k)$. For convenience, in the present context, we have not allowed explicitly for such a delay, but it can easily be accommodated by setting the δ leading coefficients of the $B(z^{-1})$ polynomial to zero. In the later practical examples, however, we will return to the more normal convention and introduce δ explicitly: the model is then denoted by the abbreviation $[n, m, \delta]$, with the leading coefficient of $B(z^{-1})$ always defined as b_0.

In this alternative setting, the recursive filtered estimates of the time variable parameters in (26.18) can be obtained by application of the recursive least squares (RLS), instrumental variable (IV), or prediction error method (PEM) algorithms (see e.g. Ljung and Soderstrom, 1983; Young, 1984); while the smoothed estimates can, under certain conditions, be obtained by application of the fixed interval smoothing algorithm 26.2.

26.5 SPECIAL EXAMPLES OF THE GM FOR THE PARAMETER VARIATIONS

Since the GM model (26.5) provides the main algorithmic device for specifying the nature of the expected parameter variations, its detailed specification is of considerable theoretical and practical importance. In this section, therefore, we discuss briefly those special model forms that have either proved particularly useful in practice, or are important in conceptual terms.

26.5.1 The generalized random walk (GRW)

The most popular GM model is the vector RW model mentioned in previous sections. This can be considered as the simplest member of the following, second order, family of generalized random walk (GRW) models,

$$\mathbf{x}_t(k) = \mathbf{F}_t\mathbf{x}_t(k-1) + \mathbf{G}_t\boldsymbol{\eta}_t(k) \tag{26.19}$$

where,

$$\mathbf{x}_t(k) = [t(k)d(k)^T \text{ and } \boldsymbol{\eta}_t(k) = [\eta_{t1}(k)\eta_{t2}(k)]^T$$

and,

$$\mathbf{F}_t = \begin{bmatrix} \alpha & \beta \\ 0 & \gamma \end{bmatrix}; \quad \mathbf{G}_t = \begin{bmatrix} 1 & 0 \\ 0 & 1 \end{bmatrix} \tag{26.20}$$

Here α, β, and γ are constant, scalar coefficients which need to be specified in some manner, while the subscript t is used merely to differentiate the matrices in this specific GM process from the general GM matrices **F** and **G**. The variables $\eta_{t1}(k)$ and $\eta_{t2}(k)$ represent zero mean, serially uncorrelated, discrete white noise inputs, with the vector $\boldsymbol{\eta}_t(k)$ normally characterized by a covariance matrix $\mathbf{Q}_t$. Unless there is evidence to the contrary, $\mathbf{Q}_t$ is assumed to be diagonal in form with unknown elements q_{t11} and q_{t22}, respectively, which are considered as noise variance ratios, as discussed above.

This GRW model subsumes, as special cases (see e.g. Young, 1984): the random walk itself (RW: $\alpha = 1$; $\beta = \gamma = 0$; $\eta_{t2}(k) = 0$); the smoothed random walk (SRW: $\beta = \gamma = 1$; $0 < \alpha < 1.0$; $\eta_{t1}(k) = 0$); and the integrated random walk (IRW: $\alpha = \beta = \gamma = 1$; $\eta_{t1}(k) = 0$). In the case of the IRW, we see that $t(k)$ and $d(k)$ can be interpreted as level and slope (time derivative) variables associated with the variations of the parameters in the model under consideration. If $\eta_{t1}(k) = 0$, the random disturbance $\eta_{t2}(k)$ only enters through the $d(k)$ equation. If $\eta_{t1}(k)$ is non-zero, however, then both the level and slope equations can have random fluctuations defined by $\eta_{t1}(k)$ and $\eta_{t2}(k)$, respectively. This variant has been termed the 'linear growth model' by Harrison and Stevens (1971, 1976).

The advantage of these random walk models is that they allow, in a very simple manner, for the introduction of non-stationarity into the time series models. By introducing a simple GM model of this type for each of the unknown parameters, we are assuming that they can be characterized by a variable mean value with stochastically variable level and/or slope. The nature of this variability will depend upon the specific form of the GRW chosen. For instance, the IRW model is particularly useful for describing large smooth changes in the parameters, while the RW model (in which the slope is not separately defined) provides for smaller scale, less smooth variations (Young, 1984). Clearly higher order RW-type models are possible, such as the double integrated random walk (DIRW) or the periodic random walk (PRW); see (Ng and Young, 1990).

26.5.2 The double integrated autoregressive (DIAR) model

The IRW model has been used for many years in the *micro*CAPTAIN program (Young and Benner, 1991), where the associated FIS algorithm (termed 'IRWSMOOTH') provides a simple but powerful approach to both trend estimation and numerical differentiation. The latter application exploits the fact that the second state variable in the IRW model is proportional to the derivative of the trend, so that the fixed interval smoothing estimate of this state provides a smoothed estimate of the trend time derivative. If the NVR value is chosen fairly high, so that the trend follows the data (but not any associated high frequency noise) then this smoothed derivative estimate has excellent properties, an illustration of which is given later in the example of section 26.7.1. It can be shown that the algorithm in this form is equivalent to non-recursive smoothing algorithms based on the minimization of 'regularization' integrals (see e.g. Young, 1991).

It is important to note that, in such applications, the IRW model is being employed merely as a device for TVP estimation and its use does not necessarily imply that the estimated variations will follow an IRW process. For example, if statistical tests indicate that the residual, as obtained by doubly differencing the FIS estimate of $t(k)$, is a serially correlated sequence describable by an AR(p) model, then the adequacy of the IRW model is thrown into doubt. In practice, the simple IRW model is rarely confirmed by such analysis. For example, the doubly differenced trend estimate obtained from the analysis of economic data often exhibits interesting long term spectral properties strongly redolent of trade or economic cycle effects (see Young, 1992). There are two alternative models in this situation. First, the following double integrated autoregressive (DIAR) process,

$$t(k) = t(k-1) + d(k-1)$$
$$d(k) = d(k-1) + \xi(k),$$

where $\xi(k)$ is an AR(p) process

$$\xi(k) = \frac{1}{1 + a_1 z^{-1} + a_2 z^{-2} + \cdots + a_p z^{-p}} \eta(k); \quad \eta(k) = N(0, \sigma_\eta^2)$$

or, equivalently,

$$\Delta^2 t(k) = \frac{1}{1 + a_1 z^{-1} + a_2 z^{-2} + \cdots + a_p z^{-p}} \eta(k-1).$$

This DIAR model can now be formulated straightforwardly in the following state space form,

$$\begin{bmatrix} t(k) \\ d(k) \\ \overline{\xi(k)} \\ \xi_2(k) \\ \xi_3(k) \\ \vdots \\ \xi_p(k) \end{bmatrix} = \begin{bmatrix} 1 & 1 & 0 & 0 & 0 & \cdots & 0 & 0 \\ 0 & 1 & 1 & 0 & 0 & \cdots & 0 & 0 \\ 0 & 0 & -a_1 & -a_2 & -a_3 & \cdots & -a_{p-1} & -a_p \\ 0 & 0 & 1 & 0 & 0 & \cdots & 0 & 0 \\ 0 & 0 & 0 & 1 & 0 & \cdots & 0 & 0 \\ \vdots & \vdots & \vdots & \vdots & \vdots & \cdots & \vdots & \vdots \\ 0 & 0 & 0 & 0 & 0 & \cdots & 1 & 0 \end{bmatrix} \begin{bmatrix} t(k-1) \\ d(k-1) \\ \xi(k-1) \\ \xi_2(k-1) \\ \xi_3(k-1) \\ \vdots \\ \xi_p(k-1) \end{bmatrix} + \begin{bmatrix} 0 \\ 0 \\ 1 \\ 0 \\ 0 \\ \vdots \\ 0 \end{bmatrix} \eta(k)$$

Alternatively, the doubly differenced trend series may be highly correlated with some other measured or estimated variable. For example, Young (1992) has shown how, in the case of quarterly $\log_e$ Unemployment of the USA, the series can be related to the doubly differenced trend in $\log_e$ GNP by a second order, constant parameter transfer function. In this situation, the AR(p) model can be replaced by the transfer function, with the $\log_e$ GNP input playing the role of a $\mathbf{U}(k)$ variable in equation (26.1).

26.5.3 The state dependent model of Young (SDM1)

The Taylor series linearization approach used in section 26.2 suggests that the variations in the linearized parameters will be time-dependent functions of the state $\chi(k)$. Probably the simplest general assumption which acknowledges this state dependency is that $\mathbf{a}(k)$ is linearly related to functions of $\chi(k)$, i.e.,

$$\mathbf{a}(k) = \mathbf{M}[\chi(k)]\boldsymbol{\alpha}(k) \tag{26.21}$$

or,

$$a_i(k) = \mathbf{m}_i(k)^T \alpha(k); \quad i = 1, 2, \ldots, n + m + p + 2 \tag{26.22}$$

where $\mathbf{M}[\chi(k)]$, which we will denote below simply as $\mathbf{M}(k)$, is a transformation matrix functionally dependent upon $\chi(k)$; $\mathbf{m}_i(k)^T$ is the ith row of $\mathbf{M}(k)$; $a_i(k)$ is the ith element of $\mathbf{a}(k)$; and $\boldsymbol{\alpha}(k)$ is a transformed parameter vector which, in certain, ideal circumstances, could have time-invariant elements. Given the generality of the model (26.4), however, it seems unlikely if such an ideal situation will apply in practice and it is necessary to add a statistical degree of freedom to the relationship by assuming that $\boldsymbol{\alpha}(k)$ can be modelled as a GM process. And, in the simplest case which certainly seems the most appropriate in general applications, we might assume that this GM process is a vector RW, e.g.

$$\boldsymbol{\alpha}(k) = \boldsymbol{\alpha}(k-1) + \eta_\alpha(k), \tag{26.23}$$

with the usual assumptions about the white noise vector $\eta_\alpha(k)$, which will be characterized by a NVR matrix $\mathbf{Q}_\alpha$.

If $\mathbf{M}$ is a square, non-singular matrix, then we can substitute from equation (26.21) into (26.23) and obtain a GM model for the variations of $\mathbf{a}(k)$ which is similar in form to equation (26.6), with

$$\mathbf{F}(k) = \mathbf{M}(k)\mathbf{M}(k-1)^{-1}; \quad \mathbf{G}(k) = \mathbf{M}(k) \tag{26.24}$$

and, in this case, $\mathbf{x}(k) = \mathbf{a}(k)$.

This particular approach to the modelling of parameter variations, which we will call the SDM1 model, was first used as a device for tracking the rapid variations in the coefficients of a linearized model of an airborne vehicle for the purposes of adaptive control (Young, 1969a, b; 1971a, 1981). In this example, $\mathbf{M}(k)$ was chosen to be diagonal in form with diagonal elements $m_{ii}(k)$ defined as physically motivated functions of certain 'air data' variables, such as dynamic pressure and altitude. These variables can be interpreted as 'extended' state variables and are associated with the elements of the $\mathbf{U}(k)$ vector in equation (26.1). In other words, the functional dependence is restricted to these other variables and a tighter state dependence in terms of the primary state variables of the system (i.e. $y(k)$, $u_i(k)$ and $e(k)$ in Priestley's definition of the state) was not found to be necessary in this particular application.

If $\mathbf{M}(k)$ is diagonal, then $\mathbf{F}(k)$ is also diagonal with elements $f_{ii}(k) = m_{ii}(k)/m_{ii}(k-1)$; in other words, this model has a particularly simple effect on the recursive estimation algorithm, with the ith parameter estimate $\hat{a}_i(k)$ being updated via a prediction equation (cf. Algorithm 26.1, equation (26.10),

$$\hat{a}_i(k/k-1) = \{m_{ii}(k)/m_{ii}(k-1)\}\,\hat{a}_i(k-1).$$

In this manner, a large increase (decrease) in $m_{ii}(k)$, in relation to its prior value at the previous sampling instant $m_{ii}(k-1)$, will lead to a similar proportionate increase (decrease) in the inter-sample predicted value of the parameter, which will then be updated on receipt of the next data sample by the correction equation (26.11).

26.5.4 The state dependent model of Priestley (SDM2)

In several important papers on nonlinear and non-stationary time series analysis and in a recent book, Priestley and his collaborators (Priestley 1980, 1988a, b; Haggan *et al.* 1984) have presented an SDM approach to nonlinear modelling which is similar to the procedures discussed in the present paper and which we will call the SDM2 model. This model uses the form of the resultant first order terms in the Taylor series expansion to define the parameter variation law directly in terms of the primary model variables $y(k)$, $e(k)$ and $u_i(k)$. In particular, he assumes that **each** unknown parameter evolves in time according to an equation of the general form,

$$a_i(k) = a_i(k-1) + \Delta\mathbf{z}(k)^T \boldsymbol{\alpha}_i(k-1); \quad i = 1, 2, \ldots, n+m+p+2, \tag{26.25}$$

where $\Delta\mathbf{z}(k) = \mathbf{z}(k) - \mathbf{z}(k-1)$ is the incremental change in the vector $\mathbf{z}$ over the sampling interval; while $\boldsymbol{\alpha}_i(k) = [\alpha_i(k), \ldots, \alpha_{n+m+p+2}(k)]_i^T$ is a vector of unknown 'gradient' parameters assumed to vary as a vector RW process, i.e.

$$\boldsymbol{\alpha}_i(k) = \boldsymbol{\alpha}_i(k-1) + \boldsymbol{\eta}_\alpha(k) \tag{26.26}$$

with $\boldsymbol{\eta}_\alpha(k)$ a white noise input vector defined in the usual manner. (Note that, for simplicity of presentation in the present context, these equations are close to but not identical to those of Priestley. (However, they represent the same general idea of state dependent modelling as proposed by Priestley.) This model can be put in the normal GM form of equation (26.5) with,

$$\mathbf{F}(k) = \left[\begin{array}{c|c} 1 & \Delta\mathbf{z}^T \\ \hline 0 & \mathbf{I}_{n+m+p+2} \end{array}\right]; \quad \mathbf{G}(k) = \left[\begin{array}{c} 0, \ldots, 0 \\ \hline \mathbf{I}_{n+m+p+2} \end{array}\right] \tag{26.27}$$

$$\mathbf{x}(k) = \mathbf{x}_i(k) = [a_i(k) \boldsymbol{\alpha}_i(k)^T]^T; \quad \boldsymbol{\eta}(k) = [0 \;\; \boldsymbol{\eta}_\alpha(k)^T]^T$$

where $\mathbf{I}_{n+m+p+2}$ is the $(n+m+p+2)$th order identity matrix. This GM can be compared directly with the IRW model: for example, in the case of a first order AR(1) model with $t(k) = 0$, we see that the identity matrix is reduced to a scalar of unity and so the only difference between the model (26.27) and the IRW is that the $f_{12}(k)$ element of $\mathbf{F}(k)$ is now defined as the change $y(k-1) - y(k-2)$, rather than unity.

Of course, for higher order equations, the GM model for each parameter is considerably more complex, and the complete GM model for the vector $\mathbf{x}(k)$, as obtained by combining the individual models (26.27) into a composite state space form, is of quite large dimension. (Note that, for clarity, we have concentrated here on the model at the individual parameter level. Priestley (1980, 1988a) presents the complete model in a block form with $\mathbf{F}(k)$ and $\mathbf{G}(k)$ defined accordingly.) As a result, the filtering and smoothing algorithms are relatively expensive in relation to the other GM models discussed previously. Also, the selection of this particular GM places quite heavy constraints on the nature of the parameter variations. This is, of course, an advantage if the linearization assumptions are appropriate to the nonlinear system under investigation. However, it could yield poor performance in prediction (forecasting) terms if the linearization assumptions are not appropriate.

Finally, two comments on the SDM2 approach are in order. First, we see from equation (26.25) that,

$$\Delta a_i(k) = \Delta\mathbf{z}(k)^T \boldsymbol{\alpha}_i(k-1); \quad i = 1, 2, \ldots, n+m+p+2.$$

If this is compared with equation (26.22) of the SDM1 approach, we see that the major difference in the assumptions are that here, in SDM2, the **changes** in the unknown parameters are related linearly to the **changes** in the model variables in **z**, while in SDM1 it is the **levels** that are related via $\mathbf{m}_i(k)$. Also, equations (26.21) and (26.22) permit nonlinear functions of $\chi(k)$ in the $\mathbf{M}(\chi(k))$ matrix. Secondly, we might question on practical grounds the insertion of differenced stochastic (measured) variables in the $\mathbf{F}(k)$ matrix, since it is well known that such differencing can cause high frequency noise amplification which, in turn, could lead to problems in the implementation of the filtering and smoothing algorithms.

This latter point certainly justifies the use of fixed interval smoothing, which should help to suppress some of the noise amplification effects. However, it may be better to look for other solutions such as replacing these differenced state elements in $\mathbf{F}(k)$ by their conditional expectations. For example, since only $\mathbf{F}(k)$ is required at the kth instant and this matrix depends only on $y(k-2)$ and $y(k-3)$, we could consider replacing these variables by their fixed lag smoothed estimates, i.e. $\hat{y}(k-2/k)$ and $\hat{y}(k-3/k)$, respectively.

26.6 CONTINUOUS TIME (LAPLACE AND DELTA OPERATOR) TF MODELS

The various techniques discussed in previous sections can also be applied to either continuous time, differential (s) or delta (δ) operator[4] models, as discussed in Young *et al.* (1991a). (The delta operator is the discrete differential operator defined as $\delta = (z-1)/\Delta t$, where Δt is the sampling interval; see also Middleton and Goodwin, 1990.) In the former case, for example, the general TF model equivalent to (26.18) takes the form,

$$y(t) = t_c(t) + \frac{B(s)}{A(s)}u(t) + \frac{D(s)}{C(s)}e(t)$$

where $s = \mathrm{d}/\mathrm{d}t$; $A(s)$, $B(s)$, $C(s)$ and $D(s)$ are appropriately defined polynomials in s, which may be characterized by time variable parameters; and $y(t)$, $u(t)$, $e(t)$ and $t_c(t)$ are appropriately defined continuous-time variables, This is equivalent to a simple TVP differential equation model of the form (with $t_c(t) = 0$ for simplicity),

$$\frac{\mathrm{d}^n y(t)}{\mathrm{d}t^n} + a_1(t)\frac{\mathrm{d}^{n-1} y(t)}{\mathrm{d}t^{n-1}} + \cdots + a_n(t)y(t) = b_0(t)u(t) + b_1(t)\frac{\mathrm{d}u(t)}{\mathrm{d}t} + \cdots$$
$$+ \frac{\mathrm{d}^m u(t)}{\mathrm{d}t^m} + \eta(t),$$

where $\eta(t)$ is appropriately defined coloured noise. We consider an example of TVP differential equation model estimation in section 26.7.1, but the reader is directed to Young *et al.* (1991a) for further information on this model and its estimation.

26.7 EXAMPLES

The general approach to non-stationary and nonlinear time series analysis and modelling outlined in previous sections of this paper has significance in many different areas where the adaptive extrapolation, interpolation, smoothing and modelling of non-stationary or nonlinear time series is important. These areas include: digital signal and image processing; forecasting and seasonal adjustment of socioeconomic, business, ecological and environmental data; geophysical, biological and medical data processing; and adaptive, learning, or self-tuning control. The results of such analysis in some of these areas are given in a number of recent papers by the author and his colleagues referred to earlier in this chapter. Because of space restrictions, therefore, we will consider here only two examples, one using the well known, Lorenz, nonlinear simulation model, and the order based on real data. The analytical results for both examples were obtained using Version 2.0 of the *micro*CAPTAIN package (Young and Benner, 1991).

26.7.1 A simulation example: SDM1 state dependent modelling of the Lorenz model

As an initial illustrative example, let us consider Lorenz's famous model of the interrelationships between temperature variation and convective motion in a fluid medium; the so-called Lorenz strange attractor. Although not very realistic in itself, the Lorenz model can be related to the more complex and justifiable Rayleigh–Bénard models of convection in a fluid between two horizontal, thermally conducting plates, with the lower one warmer than the top one. It can also be used to model the kinds of dynamic behaviour experienced by certain kinds of water-driven see-saws (Pippard, 1972; see also Young, 1988) or wheels (Gleick, 1987). The equations (Lorenz, 1963a, b, 1964) take the form of the following, three, coupled, nonlinear equations,

$$\frac{\mathrm{d}x(t)}{\mathrm{d}t} = -\sigma x(t) + \sigma y(t)$$

$$\frac{\mathrm{d}y(t)}{\mathrm{d}t} = x(t)z(t) + rx(t) - y(t)$$

$$\frac{\mathrm{d}z(t)}{\mathrm{d}t} = x(t)y(t) - az(t), \tag{26.28}$$

where $\sigma = 10$, $r = 28$ and $a = 8/3$. This is clearly a quite ordinary set of nonlinear dynamic equations: the first equation is linear in the variables, while the others each have a single, multiplicative nonlinear term.

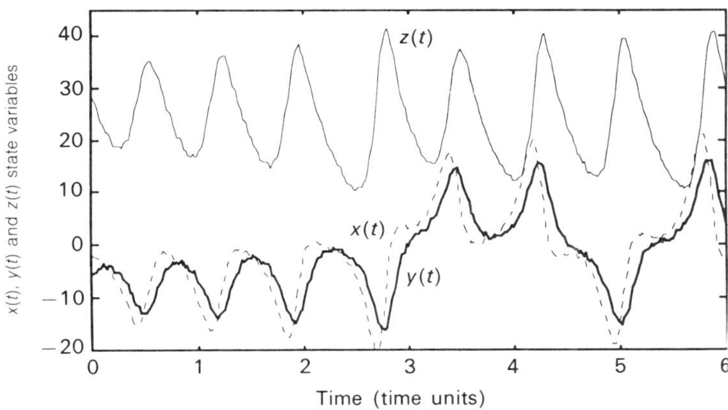

Figure 26.1. Lorenz example – plot of variations in the three state variables $x(t)$, $y(t)$ and $z(t)$ of the 'strange attractor' over a six time unit sample.

Figure 26.1 shows a typical simulation of the the equations (26.1) over a period of 6 time units: the trajectory loops around forever, never exactly repeating itself and switching, apparently (but not actually since it is a deterministic system) at random, between the two attraction points defined by the equilibrium solutions of (26.28) at $\{x = -8.4853, y = -8.4853, z = 27\}$ and $\{x = 8.4853, y = 8.453, z = 27\}$. To make the example more realistic, however, all the variables in Figure 26.1 have been corrupted by a small amount of measurement noise in the form of uniformity distributed white noise in the range $(-0.5, 0.5)$.

Here, for illustrative purposes, we will concentrate on the third equation in (26.28) which, following the nomenclature used in previous sections, can be considered in the following linear TVP form,

$$\frac{dz(t)}{dt} = -a_1(t)z(t) + b_{01}(t)x(t) + b_{02}(t)y(t) \tag{26.29}$$

In other words, this part of the model is assumed to be a linear TVP equation in $z(t)$ with three parameters that may be state dependent. From prior knowledge, of course, we know that $a_1(t) = a$ is actually a constant parameter and either $b_{01}(t) = y(t)$ or $b_{02}(t) = x(t)$, but we will assume that no such prior information is available and attempt to identify the most likely structure by repeated differential equation model parameter estimation (section 26.6) based on different model structures.

In these estimation studies, the analysis is based on 300 equi-spaced samples from the data shown in Figure 26.1, with a sampling interval of 0.02 time units. These rapidly sampled data are referred to subsequently as $x(k)$, $y(k)$ and $z(k)$, where the k in parentheses indicates the value of the continuous-time

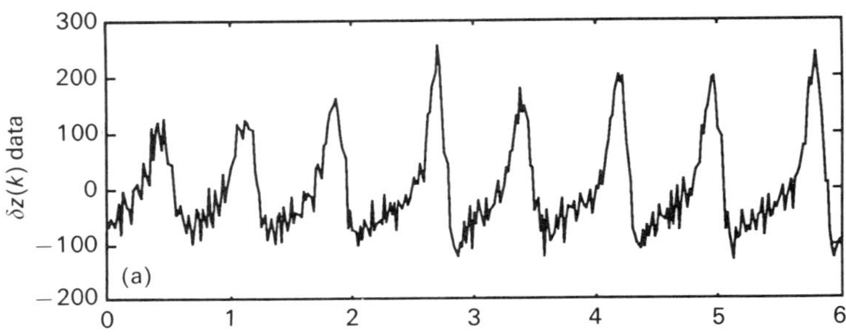

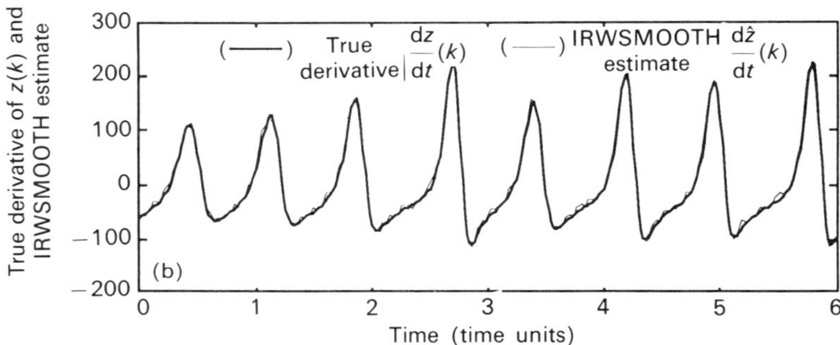

Figure 26.2. Lorenz example – numerical differentiation of the $z(k)$ variable: (a) numerical derivative based on crudely differenced data, $\delta z(k)$; (b) IRWSMOOTH estimate compared with true time derivative.

variable sampled at the kth sampling instant. Similarly, at the kth sampling instant, the time-derivative of $z(t)$ required for estimation is obtained using the IRWSMOOTH algorithm of *micro*CAPTAIN, as described in section 26.5.1, and is denoted by

$$\frac{\mathrm{d}z(k)}{\mathrm{d}t} = \left[\frac{\mathrm{d}z(t)}{\mathrm{d}t}\right]_{t=k\Delta t}.$$

Figure 26.2. compares the results obtained in this manner with those obtained by crude differencing of the data, as shown in Figure 26.2(a). The improvement in Figure 26.2(b) is obvious, with the IRWSMOOTH estimate very similar to the true time derivative.

One model that can be eliminated immediately is the constant parameter model, since it has a very low explanatory power, with a coefficient of determination $R_T^2 = 0.016$. (Here and subsequently, R_T^2 is defined in terms of the TF model fitting residuals and not the more usual one step ahead forecasting errors, which will normally produce a more inflated and misleading value.)

Next we try, in turn, various (non-exhaustive in this case) permutations and combinations of the model (26.29) with two parameters estimated as constants and the other allowed to vary each time. For these studies, it is assumed that the parameter variation can be modelled as an RW process with the NVR = 0.001 selected to ensure only relatively slow variation of the TVP estimate (otherwise all models would yield virtually perfect fits to the data). The preferred model is then the one which yields the best R_T^2 in these constrained circumstances. In this case, the model with $a_1(k)$ and $b_{01}(k)$ constant but $b_{02}(k)$ slowly variable appears best (this has an $R_T^2 = 0.89$ compared with $R_T^2 = 0.81$ and 0.67 for the other two models). Finally, in a similar manner, we investigate the single input models obtained with either

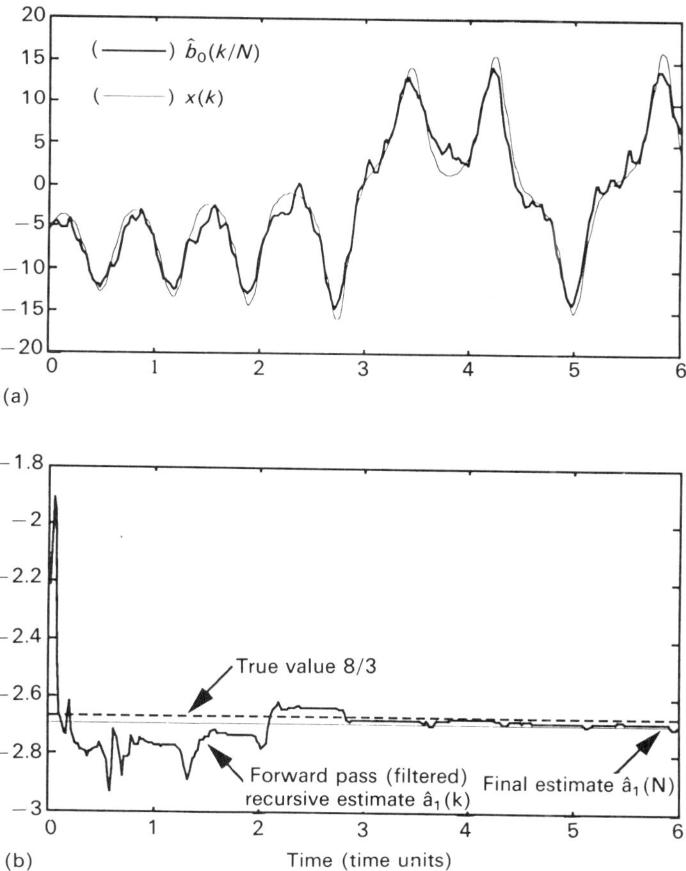

Figure 26.3. Lorenz example – TVP estimation results: (a) recursive smoothed TVP estimate $\hat{b}_0(k/N)$ (full line) compared with $x(k)$ state variable (fine line); (b) recursive filtered TVP estimate $\hat{a}_1(k)$ compared with true value.

$x(k)$ or $y(k)$ as 'input' (i.e. with either $b_{01}(k)$ or $b_{02}(k)$ set to zero). The following model (which corresponds to the true model if $b_0(k) = x(k)$ and $a_1 = a$) is then preferred,

$$\frac{dz(k)}{dt} = -a_1 z(k) + b_0(k) y(k), \tag{26.30}$$

with an $R_T^2 = 0.89$, compared with 0.81, 0.81 and 0.67 for the other model forms. Here and subsequently, since the model has only one 'input' $y(t)$, the $b_{02}(t)$ coefficient is denoted simply by $b_0(t)$.

Of course, this particular identification strategy is not exhaustive and is utilized here purely to illustrate the kind of approach that might be used. Certainly, an objective identification procedure for TVP models in a pure 'black box' situation is not currently available and will require much more research. Note, however, that in a practical setting, such as the rainfall-flow model in section 26.7.2, the identification problem will rarely be so difficult, since there is usually some prior knowledge about a physical system which can assist in identifying the most appropriate structure.

The next step in the TVP estimation procedure is to increase the NVR value in order to allow more variation in the $b_0(t)$ parameter estimate, while keeping the NVR value for the a_1 parameter set to zero under the assumption that it is constant. This naturally increases the quality of the model fit, but also increases the vulnerability of the parameter estimates to noise effects, so some compromise is necessary. In this case, however, it becomes immediately clear that the smoothed TVP estimate $\hat{b}_0(k/N)$ of $b_0(k)$ is highly correlated with the $x(t)$ variable, as we see in Figures 26.3 and 26.4 when the NVR chosen as 0.1. Figure 26.3(a) compares $\hat{b}_0(k/N)$ directly with $x(k)$;

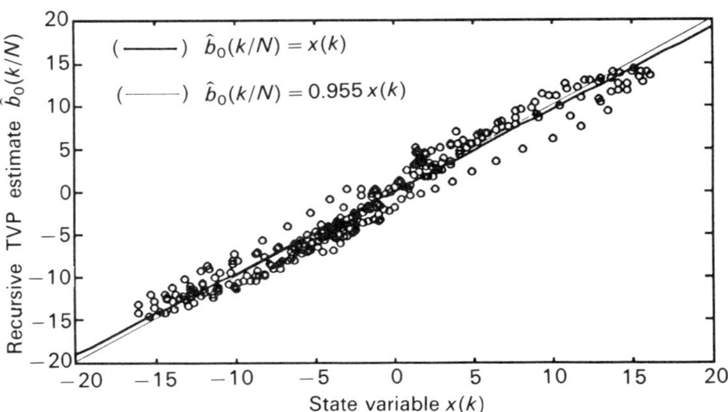

Figure 26.4. Lorenz example – scatter plot of recursive smoothed TVP estimate $\hat{b}_0(k/N)$ vs. the $x(k)$ state variable, showing true (fine line) and estimated (full line) linear relationships.

Figure 26.3(b) shows the associated filtered recursive estimate $\hat{a}_1(k)$ of a_1 (note that the smoothed estimate remains constant at the last filtered estimate since the estimate of a_1 has been constrained to be time invariant). In the scatter plot of the smoothed estimate $\hat{b}_0(k/N)$ vs. $x(k)$ shown in Figure 26.4, the predominantly linear relationship between the two variables is very clear indeed: the two lines on the graph represent the exact relationship $\hat{b}_0(k/N) = x(k)$ and the linear relationship estimated from the data using simple least squares.

As a final step having identified a plausible state dependent model, it is now possible to estimate the parameters in the following nonlinear (but linear in the parameters) model which is clearly equivalent to the third equation in the Lorenzian system, (26.28)

$$\frac{\mathrm{d}x(k)}{\mathrm{d}t} = \beta\{x(k)y(k)\} - \alpha z(k).$$

This yields least squares estimates (justified in this case by the low noise level) of of $\hat{\alpha} = -2.73$ (0.072) and $\hat{\beta} = 0.98(0.017)$, which correspond closely to the simulated equation.

The SDM1 form of this model is obvious. Since there is only one state dependent parameter, the general model in (26.22) and (26.23) takes the scalar form

$$b_0(k) = m(k)\alpha(k); \quad \alpha(k) = \alpha(k-1) + \eta_\alpha(k), \tag{26.31}$$

where $m(k) = x(k)$ and the white noise input η_α has variance $q_\alpha = 0$, because $\alpha(k) = \alpha = 1.0$ is a constant in this case. Of course, in more realistic circumstances it would be advisable to retain the statistical degree of freedom for possible variations in α and let q_α take on a small positive value. We will see the justification for such as assumption in the rainfall-flow example discussed in the next section.

26.7.2 A practical example: SDM1 state dependent modelling of rainfall-flow data

Figures 5(a) and (b) are plots of hourly rainfall and flow in a typical catchment. The modelling of such rainfall-flow processes is a major task of hydrologists who require such models for applications such as flow and flood forecasting (see e.g. Weyman, 1975; Kraijenhoff and Moll, 1986; Lees *et al.*, 1992). Visual inspection of the data in Figure 26.5 quickly reveals that the physical process involved is nonlinear, since 'antecedent' rainfall conditions clearly affect the subsequent flow behaviour. In particular, if the prior rainfall has been sufficient to thoroughly wet the soil in the catchment then river flow will be significantly higher than if the soil had dried out through lack of rainfall. Such 'soil moisture' nonlinearity is well known in hydrology and various models have been proposed, from the simple antecedent precipitation index' (API;

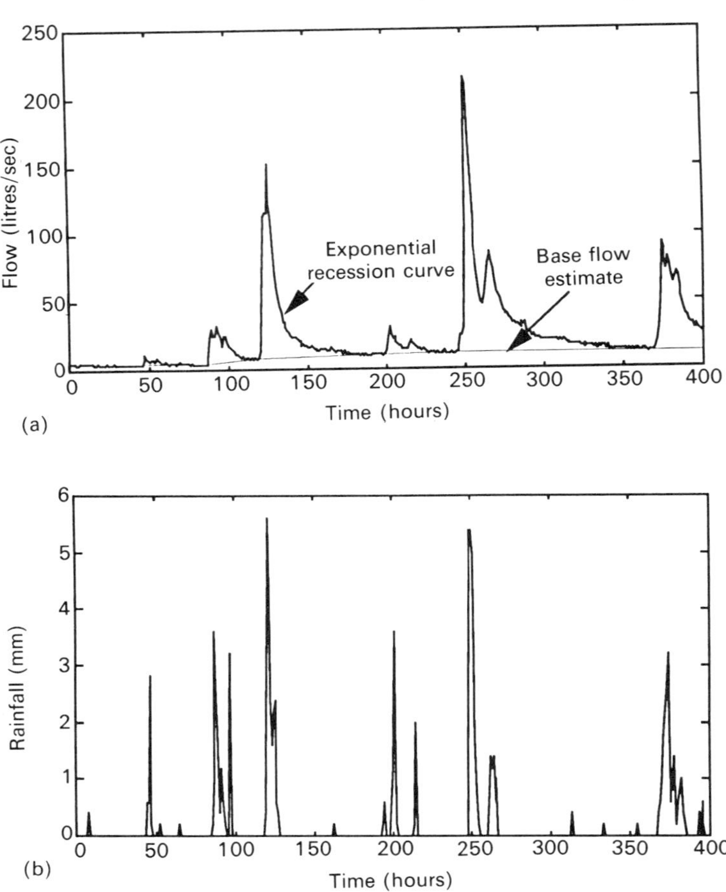

Figure 26.5. Rainfall-flow example – hourly rainfall-flow data from a catchment in Wales: (a) hourly flow, $y(k)$; (b) hourly rainfall $u(k)$ measured in same catchment as (a).

Weyman, 1975) approach, to the construction of large deterministic catchment simulation models such as the 'Stanford watershed model' (see e.g. Kraijenhoff and Moll, 1986, p. 30).

In the API approach, the 'effective rainfall' (i.e. the part of the rainfall which is effective in causing flow variations in the river) is obtained by multiplying the measured rainfall by an exponentially weighted index of past rainfall inputs which, it is argued, reflects the antecedent rainfall behaviour and the associated soil moisture conditions. A similar approach based on a TF model for soil moisture was first suggested by Young (1975b), and Whitehead *et al.* (1979). More recently, Jakeman *et al.* (1990) have used the same approach to model the data in Figure 26.5; in a comment on this latter analysis, Young and Beven (1991) have shown that a bilinear model, in which the effective rainfall input is obtained simply as the product of the rainfall and the flow, provides an alternative explanation of the data. Here, we will utilize the TVP and SDM1 estimation procedures advocated in this chapter to consider further the state dependent nature of the soil moisture nonlinearity.

Depending on the antecedent conditions, the immediate flow response to a rainfall event (or 'quick flow' as it is called) if fairly predictable and probably constitutes a first or second order dynamic process. However, the 'base level' of the flow, an estimate of which is shown as a fine line in Figure 26.5, indicates longer term or 'slow flow' processes, probably associated with the ground water. Hydrologists usually remove these base flow effects and consider the quick flow response separately. Here, however, we will analyse the data directly and initially fit a TF model to the data shown in Figure 26.5. Simple, constant parameter simplified refined instrumental variable (SRIV) identification (see e.g. Young, 1985; Young *et al.* 1991a) suggests a first order (i.e. $n = 1$, $m = 1$, $\delta = 0$, or a $[1, 1, 0]$ TF) model of the form

$$y(k) = \frac{b_0}{1 + a_1 z^{-1}} u(k) + \xi(k),$$

where $y(k)$ denotes the river flow, $u(k)$ the rainfall and $\xi(k)$ the residual noise process. Not surprisingly, this model does not fit the data too well, with a coefficient of determination $R_T^2 = 0.72$ and SRIV parameter estimates

$$\hat{a}_1 = -0.906(0.004) \hat{b}_0 = 8.09(0.31),$$

where here, and subsequently, the standard errors are shown in parentheses. The recursive estimates of these parameters indicate little variation in the lag parameter $\hat{a}_1$ but considerable changes in the $\hat{b}_0$ gain parameter. Also the residual series shows evidence of both the base flow variations and the amplitude variation caused by the soil moisture nonlinearity.

Bearing in mind these results, as well as the nature of the response and the effect of the soil moisture nonlinearity, it makes sense to constrain the $a_1(k)$ lag parameter to be constant and allow the $b_0(k)$ gain parameter to vary over the observation interval. In this manner, the relatively invariant shape of the quick flow response is preserved, since the time constant of the recession curve (see Figure 26.5(a)) is determined entirely by the lag parameter, while the $b_0(k)$ parameter variations allow for changes in the gain of the system, and the associated amplitude of the flow response, caused by the nonlinearity. Given the uncertainty about the nature of the parameter variation, it also seems reasonable to choose a RW model for the $b_0(k)$ parameter variations, with the NVR selected by trial-and error.

Figure 26.6(a) shows the output of the TVP model obtained in this manner, with the NVR = 0.1. This model explains the quickflow behaviour rather well but the base flow effects are not modelled at all. This is illustrated in

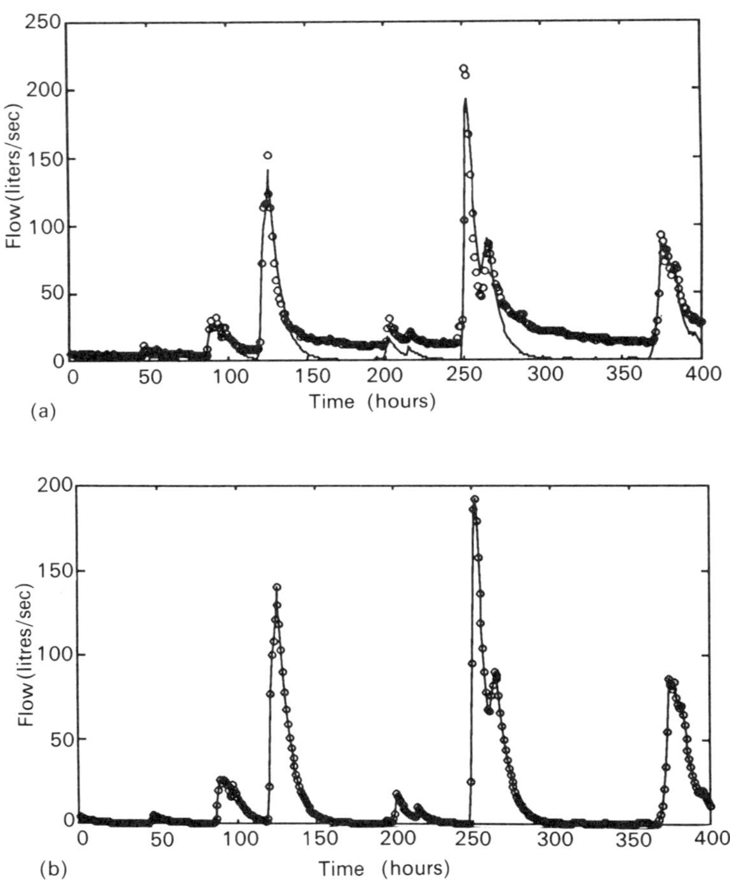

Figure 26.6. Rainfall-flow example – first order TVP model results for the rainfall-flow data: (a) comparison of the TVP model output (full time) with the original flow data $y(k)$ (circles); (b) comparison of the TVP model output (full line) with the flow data $y(k)$ (circles), after base-flow removal.

Figure 26.6(b), which compares the model output with the flow data adjusted by removal of the base flow in the normal hydrological manner. This is not essential to the analysis but, since the TVP model fits this ajusted data almost perfectly, it gives us confidence that the estimated parameter variations are allowing for the major nonlinearities in the quick flow characteristics.

Figures 26.7(a) and (b) show the TVP estimation results for the gain parameter $b_0(k)$. The forward pass recursive estimates (shown as a fine line) and the subsequent backward pass fixed interval smoothing estimates (full line) are compared in Figure 26.7(a), while the latter smoothed estimates are

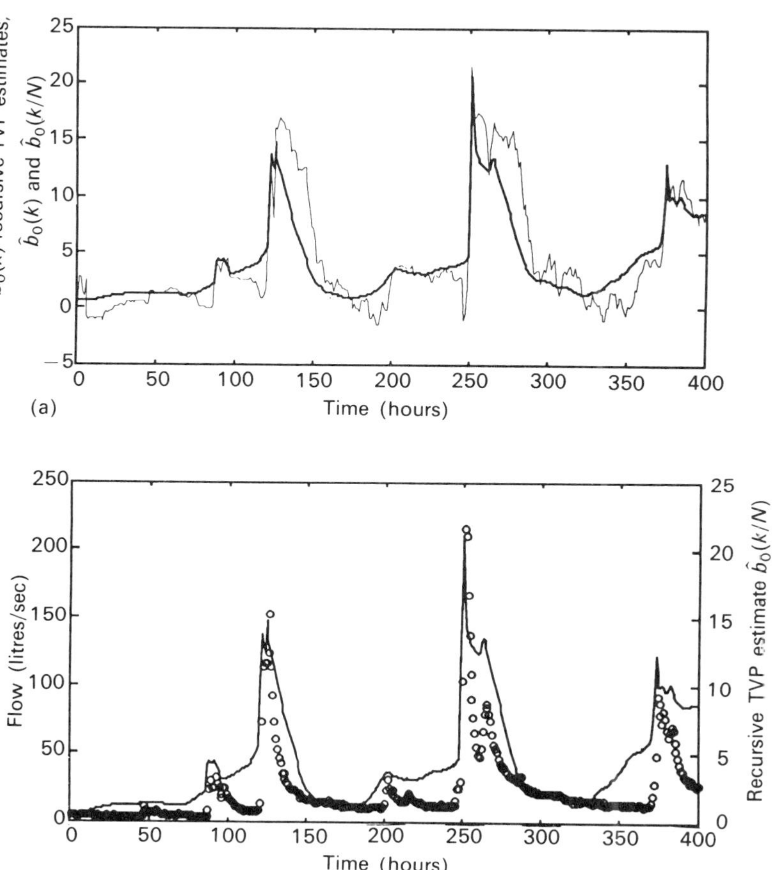

Figure 26.7. Rainfall-flow example – recursive smoothed TVP estimate $\hat{b}_0(k/N)$ of the gain parameter $b_0(k)$: (a) comparison of forward pass filtered TVP estimates $\hat{b}_c(k)$ (fine line) and backward pass smoothed TVP estimates $\hat{b}_0(k/N)$ (full line) of the gain parameter $b_0(k)$; (b) comparison of recursive smoothed estimate $\hat{b}_0(k/N)$ (full line) of the gain parameter $b_0(k)$ with the flow $y(k)$ (circles).

compared with the flow data $y(k)$ in Figure 26.7(b). The obvious high correlation between the fixed interval smoothing estimates $\hat{b}_0(k/N)$ and $y(k)$, and the consequent state dependency of the $b_0(k)$ parameter, is also illustrated in Figure 26.8, which shows a scatter plot of the two variables.

The relationship between $\hat{b}_0(k/N)$ and $y(k)$ in Figure 26.8 is clearly nonlinear but the exact nature of the nonlinear relationship is not clear because of the unavoidable uncertainty in the recursively estimated parameter variations. However, two possible nonlinear laws are shown by the lines in Figure 26.8. The first is a two-stage linear relationship of the form

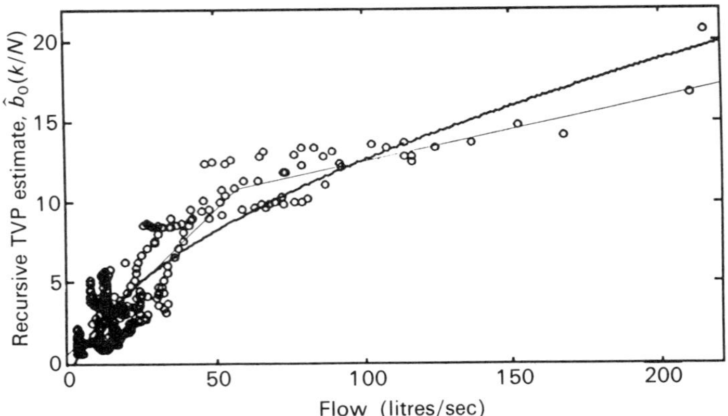

Figure 26.8. Rainfall-flow example – scatter plot of recursive smoothed TVP estimate $\hat{b}_0(k/N)$ of gain parameter vs flow, showing square root (full line) and two-stage linear (fine line) curve fits to the data.

$$\hat{b}_0(k/N) = 0.3038 + 0.1835\,y(k) \text{ for } y(k) < 60$$
$$\hat{b}_0(k/N) = 7.6709 + 0.0487\,y(k) \text{ for } y(k) > 60, \tag{26.32}$$

while the second law is a square root relationship of the form

$$\hat{b}_0(k/N) = -2.405 + 1.5014\,y(k)^{0.5} \tag{26.33}$$

Both lines were fitted by simple least squares, omitting the farthest point on the right of the graph, since it may well represent a high flow outlier (although it does not have a very large effect on the results). Clearly, other nonlinear laws could be evaluated but the results in Figure 26.8 are sufficient for the present illustrative purposes.

One way to proceed at this point is note that the relationships (26.32) and (26.33) are not exact and develop the SDM1 form of the model. This is particularly easy in this case since there is only one variable parameter and the general model in (26.22) and (26.23) takes the scalar form

$$b_0(k) = m(k)\alpha(k); \quad \alpha(k) = \alpha(k-1) + \eta_\alpha(k), \tag{26.34}$$

where $m(k)$ is defined by either (26.32) or (26.33); e.g. in the case of 26.33

$$m(k) = -2.405 + 1.5014\,y(k)^{0.5}.$$

The RW model for the 'linearized' parameter $\alpha(k)$ in (26.34) then introduces a statistical degree of freedom into the relationship, so allowing for some variation in the nonlinear law. This can be particularly useful in on-line adaptive applications of the model (see later).

A second approach is to assume that the nonlinearity is exactly described by (26.32) or (26.33), and to utilize it directly as a means of modifying the

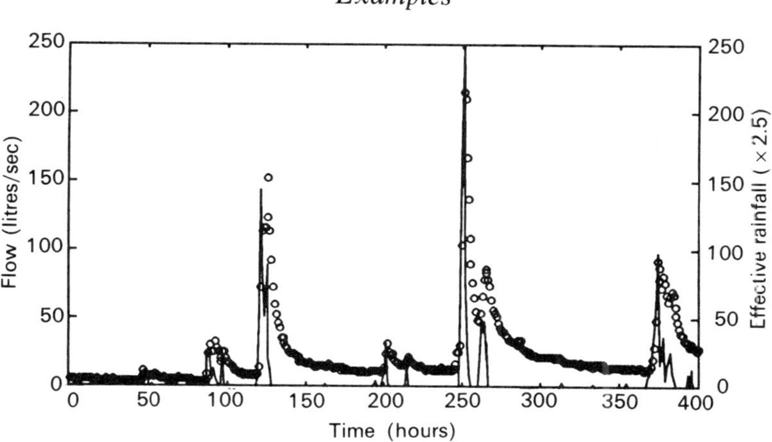

Figure 26.9. Rainfall-flow example – comparison of effective rainfall $u^*(k)$ based on the square root transformation law (full line) and flow $y(k)$ (circles).

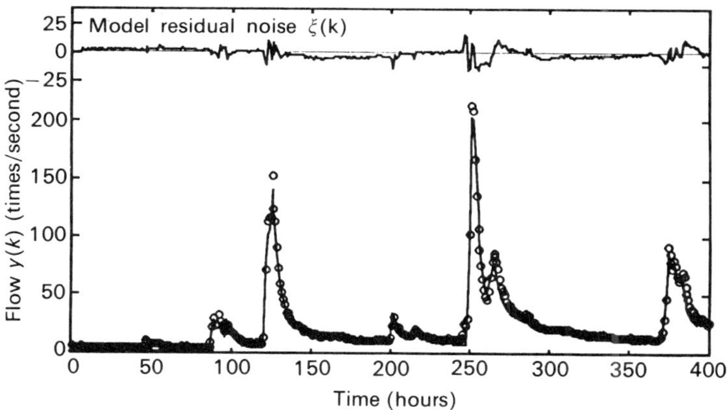

Figure 26.10. Rainfall-flow example – comparison of second order model output with measured flow, model residual error shown above.

rainfall series. The efficacy of the second relationship (26.33) in this regard is demonstrated in Figures 26.9 and 26.10. Figure 26.9 compares the modified or 'effective rainfall' $u_e(k)$, obtained from the equation

$$u_e(k) = \hat{b}_0(k/N)u(k) = \{-2.405 + 1.5014\, y(k)^{0.5}\}\, u(k)$$

with the flow $y(k)$. In contrast to the measured rainfall $u(k)$ in Figure 26.5(b), the amplitude variations of $u_e(k)$ show a visibly more linear relationship with the maximum amplitude variations of the flow $y(k)$. We can now assume, therefore, that any uncertainty in the nonlinear relationship (26.33) has been absorbed into the input–output TF and will appear in the residual error of this model.

For example, a first order [1, 1, 0] model of the form

$$y(k) = \frac{b_0}{1 + a_1 z^{-1}} u_e(k) + \xi(k) \tag{26.35}$$

yields SRIV estimates

$$\hat{a}_1 = -0.884(0.004), \quad \hat{b}_0 = 0.925(0.028),$$

and fits the data rather better than the previous first order model, with $R_T^2 = 0.81$. However, an examination of the model residuals $\xi(k)$, which are still fairly large, indicates clearly that the model error is due mainly to the presence of the base flow effects, which are not being explained by the first order dynamics, rather than any resudual error arising from the uncertainty in the nonlinear relations (26.33). Since the dominant input–output relationship between the effective rainfall $u_e(k)$ and flow $y(k)$ now appears linear, however, it makes sense to re-identify the model to see if a higher order model may not now be more appropriate than the first order model (26.35).

This is indeed the case: the SRIV identification analysis suggests a second order [2, 2, 0] model of the form

$$y(k) = \frac{b_0 + b_1 z^{-1}}{1 + a_1 z^{-1} + a_2 z^{-2}} u_e(k) + \xi(k), \tag{26.36}$$

and the associated SRIV estimates are obtained as

$$\hat{a}_1 = -1.7555(0.005) \quad \hat{a}_2 = 0.75839(0.005)$$
$$\hat{b}_0 = 1.1414(0.014) \quad \hat{b}_1 = -1.1031(0.013),$$

with an $R_T^2 = 0.978$ now quite close to unity.

Note that this high R_T^2 indicates a very good explanation of the data since it is based on the residual modelling errors and **not** the one step ahead forecasts. The excellent quality of the model is illustrated in Figure 26.10, which compares the model output $\hat{x}(k)$ and the flow data, with the model residual $\hat{\xi}(k)$ shown above. The second order dynamics have been able to successfully account for the base flow effects and the quite small residuals are now probably dominated by the remaining uncertainty in the nonlinearity (26.33). The model could be improved further, of course, either by fitting the complete nonlinear model by nonlinear optimization; by attempting to derive an improved nonlinear rainfall function; or by modelling the residuals as an autocorrelated process. But such exercises are only likely to introduce marginal improvement and the model is probably good enough, in its present form, for most hydrological purposes.

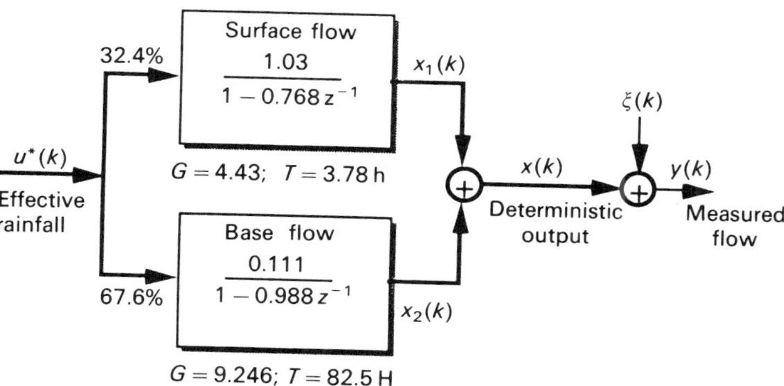

Figure 26.11. Rainfall-flow example – the [2, 2, 0] TF model considered as a parallel connection of two first order processes (G denotes steady state gain; and T the time constant or residence time).

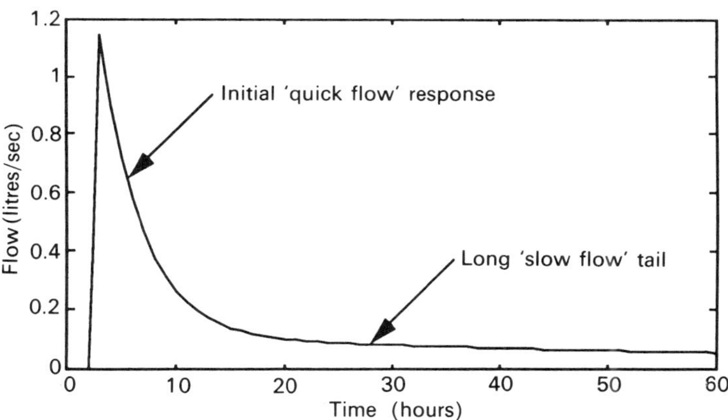

Figure 26.12. Rainfall-flow example – impulse response of the [2, 2, 0] TF model showing the long tail associated with the slow-flow parallel pathway.

In the latter regard, the second order model (26.36) not only explains the data very well, it is also of considerable physical interest. In particular, it is easily shown that the deterministic transfer function part of the model can be unambiguously decomposed, by partial fraction expansion, into the parallel structure shown in Figure 26.11, where we see that the model indicates that some 32.4% of the effective rainfall on the catchment reaches the river rapidly ('quick flow' with time constant or residence time TC = 3.78h), probably as surface flow; while the remaining 67.6% reaches the river much more slowly ('slow flow' with TC = 82.5h), probably via some complex process in the sub-surface and groundwater, as baseflow. This partitioning of rainfall is also illustrated in Figure 26.12 which shows the unit impulse response of the model: the effect of the quick flow pathway is obvious in the initial part of the response but, as this decays (with exponential time constant 3.78h), the slow-flow component dominates the flow in the form of the very long tail (with exponential time constant 82.5h).

The equivalent second order modelling results obtained with the two stage linear law (26.32) are quite similar to those for the square root law (26.33). The model explains the data marginally better with $R_T^2 = 0.980$ and suggests that 32.0% of the effective rainfall passes through the quick flow pathway with TC = 3.54 h, while 68.0% passes through the slow flow pathway with TC = 79.1 h. The physical interpretation of both second order models is in sympathy with the data-based mechanistic approach to modelling advocated for many years by the present author (see e.g. Young and Lees, 1992, for the most recent discussion on the subject) and it is also quite useful in practical terms. For example, the physical interpretation of the model as a parallel process has the additional desirable effect of allowing the modeller to objectively estimate the 'baseflow' element of the river flow. This is simply obtained as the output $x_2(k)$ of the lower TF in Figure 26.11 and is shown as the baseflow plotted in Figure 26.5(a). This can replace or supplement other methods of baseflow estimation used in conventional hydrological analysis.

Finally, it is interesting to note that on-line, adaptive versions of rainfall-flow models of this type can be useful in river management. A practical example of this is given in Figure 26.13, which compares the five-hour-ahead forecasts produced by a computer-based, adaptive flood warning system for the town of Dumfries in Scotland. This system, which is based on an adaptive SDM1 model similar to those discussed here, was developed recently by the

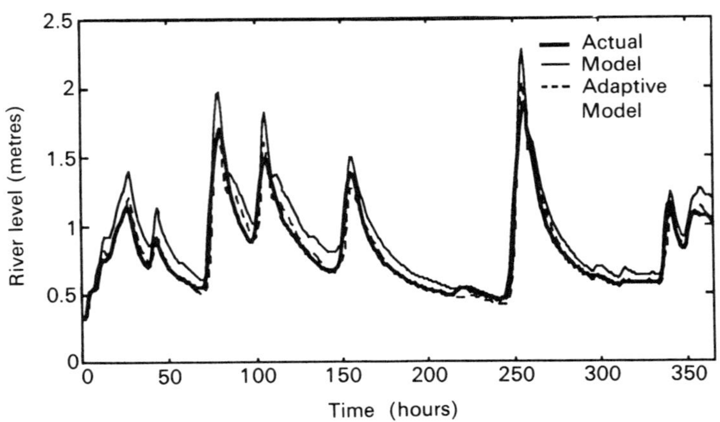

Figure 26.13. The Dumfries flood warning system – typical five-hour-ahead forecasting result.

author and his colleagues for the Solway River Purification Board (e.g. Lees *et al.*, 1992). The forecasts produced by an equivalent constant parameter model are also shown in Figure 26.13. This illustrates well the advantages of adaptive TVP models. While the constant gain linear model is able to predict the **shape** of the response, it frequently produces an incorrect forecast of the level, which is quickly and efficiently corrected by the adaptive gain element. The system was commissioned about one year ago and has so far provided correct forecasts of all inundations experienced by the town over this period.

26.8 CONCLUSIONS

This chapter has introduced an approach to nonlinear and non-stationary time series analysis for a fairly wide class of linear, time variable parameter (TVP) or nonlinear systems, an approach which is related to Maurice Priestley's own work in this area. The method theory presented here exploits recursive filtering and fixed interval smoothing algorithms to derive TVP linear model approximations to the nonlinear or non-stationary stochastic systems, on the basis of data obtained from the system during planned experiments or passive monitoring exercises. This TVP model includes the State Dependent type of Model (SDM) as a special case, and two particular SDM forms due to Priestley and the present author are discussed in detail.

The methodology presented here has wide application potential. In its SDM form, it is useful as an off-line method for the detection and identification of nonlinearities in time series. In this latter context, it can be considered as a pre-processing procedure, in which the recursive filtering and smoothing algorithms are used for identifying the nonlinear model structure, prior to more efficient parameter estimation based on these identification results. This complete identification–estimation approach is best illustrated by the rainfall-flow example, where the identified soil moisture nonlinearity is incorporated in the model, whose parameters are then estimated using the statistically more efficient SRIV algorithm applied to the resulting constant parameter, bilinear transfer function model.

However, the TVP estimation methodology is not limited to off-line analysis. Because of its fully recursive formulation, the off-line identification algorithms can be used to develop on-line procedures for the adaptive estimation, forecasting and control of nonlinear and non-stationary dynamic systems. Here, the recursive filtering algorithms used in the earlier off-line studies can form the basis for adaptive system design, so integrating the processes of systems analysis and synthesis in a rather useful and elegant manner. Recent examples of this approach are the adaptive control studies of Young *et al.* (e.g. 1991b) and the adaptive flood warning system of Lees *et al.* (1992).

Finally, the identification and estimation procedures discussed in this chapter are tools that have been developed specifically by the author, over a number of years, for the data-based mechanistic modelling dynamic systems, particularly in relation to the scientific investigation of environmental systems. The rainfall-flow example in this chapter, as well as other recent research (e.g. Young and Lees, 1992), has shown how these techniques can be exploited not only to characterize stochastic, dynamic systems in a parametrically efficient form, but also to allow the analyst to interpret the resulting model in physically meaningful terms. The theoretical and practical importance of such an interpretative step cannot be over-emphasized: while 'black box' time series models of a more conventional statistical type can prove very beneficial in the analysis, forecasting and control of dynamic systems, only models which are capable of adequate physical explanation are of real use in the scientific investigation and exploration of the physical world.

REFERENCES

Box, G.E.P., and Jenkins, G.M.(1970) *Time Series Analysis, Forecasting and Control.* Holden-Day; San Francisco.

Bryson, A.E., and Ho, Y.C. (1969) *Applied Optimal Control.,* Blaisdell; Mass.

Doan, T., Litterman, R., and Sims, C.A. (1984) Forecasting and conditional projection using realistic prior distributions. *Econometrics Reviews,* **3**, 1–100.

Gelb, A., Kasper, J.F., Nash, R.A., Price, C.F., and Sutherland, A.A. (1974) *Applied Optimal Estimation* MIT Press, Mass.

Engle, R.F., Brown, S.J., and Stern, G. (1988) A comparison of adaptive structural forecasting models for electricity sales. *J. of Forecasting*, **7**, 149–172

Gleick, J. (1987) *Chaos: Making a New Science,* Penguin Books; New York.

Haggan, V., Heravi, S.M., and Priestley, M.B. (1984) A study of the application of state-dependent models in nonlinear time-series analysis. *Journal of Time Series Analysis,* **5**, 69–102.

Harrison, P.J., and Stevens, C.F. (1971) A Bayesian approach to short-term forecasting. *Operational Res. Quarterly,* **22**, 341–362.

Harrison, P.J., and Stevens, C.F. (1976) Bayesian forecasting. *J. Royal Stat. Soc., Series B.,* **38**, 205–247.

Harvey, A.C. (1984) A unified view of statistical forecasting procedures. *J. of Forecasting,* **3**, 245–275.

Jakeman, A.J., and Young, P.C. (1979, 1984). Recursive filtering and the inversion of ill-posed causal problems. *Utilitas Mathematica*, 1984, **35**, 351–376 (appeared originally as *Report No. AS/R28/1979*, Centre for Resource and Environmental Studies, Australian National University, 1979).

Jakeman, A.J., and Young, P.C. (1981) On the decoupling of system and noise model parameter estimation in time-series analysis. *Int. J. of Control,* **34**, 423–431.

Jakeman, A.J., and Young, P.C. (1983) Advanced methods of recursive time-series analysis. *Int. J. of Control,* **37**, 1291–1310.

Jakeman, A.J., Littlewood, I.G., and Whitehead, P.G. (1990) Computation of the instantaneous unit hydrograph and identifiable component flows with application to two small upland catchments. *Journal of Hydrology,* **117**, 275–300.

Kalaba, R., and Tesfatsion, L. (1988). The flexible least squares approach to time-varying linear regression. *J. of Econ. Dynamics and Control* **12**, 43–48.

Kalman, R.E. (1960). A new approach to linear filtering and prediction problems. *ASME Trans., J. Basic Eng.*, **83-D**, 95–108.

Kalman, R.E., and Bucy, R.S. (1961) New results in linear filtering and prediction theory. *ASME Trans., J. Basic Eng.*, **83-D**, 95–108.

Kitagawa, G. (1981). A non-stationary time series model and its fitting by a recursive filter *J. of Time Series*, **2**, 103–116.

Kopp, R.E., and Orford, R.J. (1963) Linear regression applied to system identification for adaptive control systems, *AIAA Journal*, **1**, 2300–2306.

Kraijenhoff, D.A., Moll, J.R. (1986) *River Flow Modelling and Forecasting* (Water Science and Technology Library). D. Reidel, Dordrecht.

Lee, R.C.K. (1964) *Optimal Identification, Estimation and Control*, MIT Press; Cambridge Mass.

Lees, M., Young, P.C., and Ferguson, S. (1992) Adaptive flood warning, to appear in (ed.) P.C. Young, *Consise Encyclopedia of Environmental Systems*, Pergamon; Oxford.

Ljung, L., and Soderstrom, T. (1983) *Theory and Practice of Recursive Estimation.* MIT Press; Cambridge, Mass.

Lorenz, E. (1963a) Deterministic nonperiodic flow. *J. of Atmospheric Sciences*, **20**, 130–141.

Lorenz, E. (1963b) The mechanics of Vacillation. *J. of Atmospheric Sciences*, **20**, 448–464.

Lorenz, E. (1964) The problem of deducing climate from the governing equations, *Tellus*, **16**, 1–11.

Ng, C.N., and Young, P.C. (1990) Recursive estimation and forecasting of non-stationary time-series, *J. of Forecasting* (special issue on State Space Methods of Forecasting and Seasonal Adjustment), **9**, 173–204.

Ng, C.N., Young, P.C., and Wang, C.L. (1988) Recursive identification, estimation and forecasting of multivariate time-series, in (ed.) Chen, H.F. *Identification and System Parameter Estimation*, Pergamon Press; Oxford, pp. 860–865.

Norton, J.P. (1975) Optimal smoothing in the identification of linear time-varying systems. *Proc. IEE* **122**, 663–668.

Panuska, V. (1969) An adaptive recursive least squares identification algorithm. *Proc. 8th IEEE Symposium on Adaptive Processes.*

Pippard, A.B. (1972) *Reconciling Physics with Reality*, Cambridge University Press, Cambridge.

Priestley, M.B. (1981) *Spectral Analysis and Time Series*, Academic Press, London.

Priestley, M.B. (1980) State-Dependent Models: A general approach to nonlinear time-series analysis. *Journal of Time Series Analysis*, **1**, 47–71.

Priestley, M.B. (1988a) *Nonlinear and Nonstationary Time-Series Analysis*, Academic Press, London.

Priestley, M.B. (1988b) Current developments in time-series modelling. *J. of Econometrics*, **37**, 67–86.

Priestley, M.B., and Chao, M.T. (1972) Nonparametric function filtering. *J. Roy. Stat. Soc, Ser. B.*, **34**, 385–392.

Priestley, M.B., and Heravi, S.M. (1985) Identification of nonlinear systems using general state dependent models. *Journal App. Prob.*, **23A**, 257–274.

Rioul, O., and Vetterli, M. (1991) Wavelets and signal processing, *IEEE SP Magazine*, October, 14–38.

Solo, V. (1980) Some aspects of recursive parameter estimation. *Int. J. of Control*, **32**, 395–410.

Weyman, D.R. (1975) *Runoff Processes and Streamflow Modelling* Oxford University Press, Oxford.

Whitehead, P.G., Young, P.C., Hornberger, G.M. (1979) A systems model of stream flow and water quality in the Bedford-Ouse River; I stream flow modelling. *Water Research*, **13**, 1155–1169.

Widrow, B., and Hoff, M.E. (1960) Adaptive switching circuits, *IRE WESCON Convention Record*, Part 4, 96–104.

Willsky, A. (1969) *Digital Signal Processing and Control and Estimation Theory: Points of Tangency, Areas of Intersection and Parallel Directions*, MIT Press, Cambridge, Mass.

Young, P.C. (1968) The use of linear regression and related procedures for the identification of dynamic processes. *Proc. 7th IEEE Symposium on Adaptive Processes.*

Young, P.C. (1969a) *The Differential Equation Error Method of Process Parameter Estimation.* PhD. Thesis, University of Cambridge, England.

Young, P.C. (1969b) The use of *a priori* parameter variation information to enhance the performance of a recursive least squares estimator, *Tech. Note 404–90*, Naval Weapons Center, China Lake, California.

Young, P.C. (1970) An Instrumental Variable Method for Real-Time Identification of a Noisy Process, *Automatica*, **6**, 271–287.

Young, P.C. (1971a) A second generation adaptive pitch autostabilisation system for a missile or aircraft. *Tech. Note 404–109* Naval Weapons Centre, China Lake, California.

Young, P.C. (1971b) Comments on dynamic equations for economic forecasting. *J. Royal Stat. Soc., Series A*, **134**, 220–223.

Young, P.C. (1975a) Comments on techniques for testing the constancy of regression relationship over time. *J. Royal Stat. Soc.*, **37**, 149–192.

Young, P.C. (1975b) Recursive approaches to time-series analysis. *Bull. Inst. Math. and Applic.*, **10**, 209–224.

Young, P.C. (1978). A general theory of modelling for badly defined dynamic systems, in (ed.), G.C. Vansteenkiste, *Modeling, Identification and Control in Environmental Systems*, North-Holland, Amsterdam, pp. 103–135.

Young, P.C. (1981) A second generation adaptive autostabilisation system for airborne vehicles. *Automatica*, **17**, 459–469.

Young, C. (1984) *Recursive Estimation and Time-Series Analysis*, Springer–Verlag, Berlin.

Young, P.C. (1985) The instrumental variable method: a practical approach to identification and system parameter estimation, in (eds.) Barker, H.A. and Young, P.C., *Identification and System Parameter Estimation*, Pergamon Press; Oxford, 1–16.

Young, P.C. (1988) Recursive Extrapolation, Interpolation and Smoothing of Non-Stationary Time-Series, in (ed.) Chen, H.F. *Identification and System Parameter Estimation*, Pergamon Press, Oxford, pp. 33–44.

Young, P.C. (1989) Recursive estimation, forecasting and adaptive control, in (ed.) C.T. Leondes, *Control and Dynamic Systems: Advances in Theory and Applications*, Vol. 30, Academic Press; San Diego, pp. 119–166.

Young, P.C. (1991) Discussion on likelihood and cost as path integrals. *J. Royal Stat. Soc.*, **53**, 3, 529–531.

Young, P.C. (1993) Time variable parameter estimation methods for the analysis and modelling of nonstationary economic time-series. *J. of Forecasting*, to appear.

Young, P.C., and Benner, S. (1991) *microCAPTAIN Handbook: Version 2.0*, Centre for Research on Environmental Systems, University of Lancaster: Lancaster.

Young, P.C., Beven, K.J. (1991) Computation of the instantaneous unit hydrograph and identifiable component flows with application to two small upland catchments–comment. *Journal of Hydrology*, **129**, 389–396.

Young, P.C., and Ng, C.N. (1989) Variance intervention. *J of Forecasting*, **8**, 399–416.

Young, P.C., and Lees, M. (1992) The active mixing volume: a new concept in modelling environmentals systems, in (eds) V. Barnett and R. Turkman, *Statistics and the Environment*. J. Wiley, Chichester.

Young, P.C., and Runkle, D. (1989) Recursive estimation and modelling of non-stationary and nonlinear time-series, in *Adaptive System in Control and Signal Processing*, Vol. 1, Institute of Measurement and Control for IFAC: London, pp. 49–64.

Young, P.C., Behzadi, M.A., Wang, C., and Chotai, A. (1987) Direct digital and adaptive control by input-output, state variable feedback, *Int. J. of Control*, **46**, 1861–1881.

Young, P.C., Ng, C.N., and Armitage, P. (1989) A systems approach to recursive economic forecasting and seasonal adjustment, *Computers and Mathematics with Applications*, (special issue of System Theoretic Methods in Economic Modelling), **18**, 481–501.

Young, P.C., Chotai, A., and Tych, W. (1991a) Identification, estimation and control of continuous-time systems described by delta operator models, in (eds) N.K. Sinha and G.P. Rao, *Identification of Continuous-time Systems*, Kluwer Academic Publishers, Dordrecht, pp. 363–418.

Young, P.C., Chotai, A., and Tych, W. (1991b) True digital control: a unified design procedure for linear sampled data systems, in (ed.) K. Warwick, *Self-Tuning Control for Industrial Applications*, Springer-Verlag, Berlin, pp. 71–109.

Young, P.C., Ng, C.N., Lane, K., and Parker, D. (1991c) Recursive forecasting smoothing and seasonal adjustment of nonstationary environmental data. *J. of Forecasting*, **10**, 57–89.

Young, T.J., Ng, C.N., and Young, P.C. (1988) Seasonal Adjustment by Optimally Smoothed Time-Variable Parameter Estimation, in (ed.) H.F. Chen, *Identification and System Parameter Estimation*, Pergamon Press; Oxford, pp. 1349–1353.

27

Demodulation of phase modulated signals

T. Subba Rao and M. Yar

27.1 INTRODUCTION

Detection of the presence of a signal, when the observations are corrupted by noise, is a classic problem in communications engineering. Under the assumption that the noise is Gaussian and white, Fisher (1929) developed a statistical procedure widely known as Fisher's g test. Later, the method was extended by Bartlett (1966), Hannan (1961), Priestley (1962a) and Whittle (1952) to situations when the noise is a general stationary time series. A related problem is that of estimating the frequencies in a signal, this being a particularly important topic in signal processing. Indeed, the maximum entropy estimate of the spectral density function, and other high-resolution estimators such as Capon's and Pisarenko's, are aimed at estimating such frequencies. The object of this note is to discuss a related problem which arises in frequency modulated and phase modulated signals, and to consider an approach based on the techniques developed by Priestley (1962a, b) for analysing the 'mixed spectra' model.

It is well known that for efficient transmission of signals the base signals are shifted to higher frequencies. This can be done either by varying the phase or the frequency of a carrier wave. Consider a discrete parameter 'phase modulated' signal of the form

$$S_t = A\cos(\omega_c t + \beta_0 + \beta X_t),$$

where $\{X_t\}$ is a discrete parameter stationary process, called the 'base band signal' and A, ω_c, β_0, and β are constants. In the language of communications engineering we would call ω_c the 'carrier wave frequency', A the 'carrier wave amplitude', β_0 the 'carrier wave phase', and β the 'modulation index'. (The usual assumption that β_0 is uniformly distributed over $(-\pi, \pi)$ will be adopted.) Now suppose that S_t itself is not directly observable, but instead we observe the process

$$Y_t = S_t + Z_t,$$

where $\{Z_t\}$ is a stationary 'noise process'. The problem now is to 'estimate' the process $\{X_t\}$, given a finite number of observations on $\{Y_t\}$. The 'recovery' of $\{X_t\}$ from observations on $\{Y_t\}$ is known in the engineering literature as the problem of **demodulation**. Initially, we examime the case where $\{X_t\}$ is itself a harmonic process, and then consider the estimation of the relevant parameters. In the last section we briefly describe the method of 'phase locked loops' which is a widely used technique in communications engineering.

27.2 HARMONIC ANALYSIS

Here we assume that the base band signal can be written as a harmonic process, and then consider the statistical analysis. For simplicity, consider the case where $X_t = \sin \omega t$ $(t = 1,2,\ldots)$. Then we have

$$\begin{aligned} S_t &= A\cos(\omega_c t + \beta_0 + \beta \sin \omega t) \\ &= A\,\mathrm{Re}[\mathrm{e}^{i(\omega_c t + \beta_0)}\mathrm{e}^{i\beta \sin \omega t}] \\ &= A\,\mathrm{Re}\left[\mathrm{e}^{i(\omega_c t + \beta_0)} \sum_{m=-\infty}^{\infty} J_m(\beta)\mathrm{e}^{im\omega t}\right] \\ &= \sum_{-\infty}^{\infty} AJ_m(\beta)\cdot\cos(\omega_m t + \beta_0), \end{aligned} \tag{27.1}$$

where $\omega_m = \omega_c + m\omega$, and $J_m(\beta)$ denotes the Bessel function of the first kind of order m, i.e.

$$J_m(x) = \frac{1}{2\pi}\int_{-\pi}^{\pi} \cos(x \sin\theta - m\theta)\mathrm{d}\theta. \tag{27.2}$$

Note that $J_m(x)$ satisfies the difference equation (Abramowitz and Stegun 1970)

$$J_{m-1}(x) + J_{m+1}(x) = \frac{2m}{x} J_m(x). \tag{27.3}$$

For all real x, $J_m(x) \to 0$ as $m \to \infty$. In view of this we can, in fact, truncate the expression for S_t and write it in the form

$$S_t = \sum_{m=-m_1}^{m_1} A_m \cos(\omega_m t + \beta_0) \tag{27.4}$$

where $A_m = AJ_m(\beta)$. It is now easy to show that

$$\mathrm{E}(S_t) = 0$$

$$\mathrm{E}(S_t S_{t+s}) = R_S(s) = \frac{1}{2} \sum_{-m_1}^{m_1} A_m^2 \cos \omega_m s. \tag{27.5}$$

The problem we now consider is the detection of the presence of PM signals, and we then consider the estimation of the parameter $(\omega_c, \omega, \beta, A)$ when a sample $(Y_1, Y_2, \ldots, Y_n)$ is available. The advantage of writing the signal $S_t = A\cos(\omega_c t + \beta_0 + \beta \sin \omega t)$ in terms of expression (27.4), which is obtained by using the Bessel function expansion, is that this model (27.4) is very familiar in the time series literature and has been extensively discussed by Brillinger (1980), Hannan (1961), Priestley (1962a), Walker (1971), and several others. However, there is one important difference between this model and the usual multiple harmonic component model in that in (27.4) the frequencies ω_m and the amplitudes A_m satisfy certain linear relationships. For example, it is obvious that $\omega_m - \omega_{(m-1)} = \omega$, and in view of (27.3), we have the relation between amplitudes

$$xA_{m-1} + xA_{m+1} = 2mA_m. \tag{27.6}$$

In spite of this, we can still exploit the similarity between this PM model and the multiple harmonic model, and obtain various tests and also derive the maximum likelihood estimators of the parameters.

27.3 DETECTION IN THE PRESENCE OF GAUSSIAN WHITE NOISE

Let us assume that $\{Z_t\}$ is a Gaussian white noise process with mean zero and variance σ_Z^2, and is independent of the signal S_t. Then,

$$R_y(s) = \mathrm{E}(Y_t Y_{t+s}) = R_S(s) + \sigma_Z^2 \delta_{s,0} \tag{27.7}$$

where $\delta_{s,0}$ is a Kronecker delta function. If there is no base band signal present in S_t, i.e. if $\beta = 0$, then $R_S(s) = A^2/(2)\cos \omega_c s$, so that the covariance function consists of a single harmonic term corresponding to the carrier frequency ω_c; otherwise the covariance function consists of several harmonic terms by virtue of the relation (27.5).

To test for the presence of the phase modulation, we proceed as follows. Define the periodogram of the series $(Y_1, Y_2, \ldots, Y_n)$ at the frequency λ_p by

$$I_y(\lambda_p) = \frac{2}{n} \left| \sum_{t=1}^{n} (Y_t - \bar{Y}) e^{-i\lambda p t} \right|^2$$

$$= 2 \sum_{s=-(n-1)}^{n-1} \hat{R}_y(s) \cos \lambda_p s \tag{27.8}$$

where

$$\hat{R}_y(s) = \frac{1}{n}\sum_{t=1}^{n-s}(Y_t - \bar{Y})(Y_{t+s} - \bar{Y}), \quad (s = 0, 1, \ldots, n-1),$$

$$\bar{Y} = \frac{1}{n}\sum Y_t, \; \lambda_p = \frac{2\pi p}{n}, \; p = 0, 1, 2, \ldots \left[\frac{n}{2}\right].$$

Taking expectations both sides of (27.8) and then making use of the result (27.7), we obtain

$$\mathrm{E}(I_y(\lambda_p)) \simeq \frac{1}{2}\sum_{-m_1}^{m_1} A_m^2[2\pi F_n(\lambda_p + \omega_m) + 2\pi F_n(\lambda - \omega_p)] + 2\sigma_Z^2 \quad (27.9)$$

where

$$F_n(\lambda) = \frac{1}{2\pi n}\cdot\frac{\sin^2 n\lambda/2}{\sin^2 \lambda/2}.$$

From (27.9), if the phase modulation is present, we have

$$\mathrm{E}(I_y(\lambda_p)) = \begin{cases} O(n) & \text{if } \lambda_p = \omega_m, \quad (m = 1, \pm 1, \ldots \pm m_1) \\ O(1) & \text{otherwise.} \end{cases} \quad (27.10)$$

In other words, we observe sharp peaks in the periodogram at the frequencies λ_p at ω_c, $\omega_c \pm \omega$, $\omega_c \pm 2\omega, \ldots$ The peaks may be equidistant and symmetric about ω_c. (Note that the Bessel coefficients $J_m(\beta)$, can be zero for some values of m, and therefore some amplitudes A_m can be zero. At these values the peaks will not be very distinct.) We can now test for the significance of peaks using Fisher's test (for details see Priestley, 1981). Let us assume that the peaks at frequencies $\{\omega_{m_i}, (i = 1, 2, \ldots r)\}$ are shown to be significant. We must now consider the estimation of the associated parameters. We can show that maximizing the likelihood funçtion is the same as minimizing the function (Subba Rao and Yar, 1984a).

$$S_n = \sum_{t=1}^{n} Y_t^2 - \sum_{-m_1}^{m_1} I_y(\omega_m) \quad (27.11)$$

or, equivalently, maximizing the function

$$Q(\omega_c, \omega) = \sum_{m=-m_1}^{m_1} I_y(\omega_m). \quad (27.12)$$

The maximum likelihood estimates of ω_c, ω can then be shown to be given by,

$$\tilde{\omega}_c = \frac{\sum A_m^2 \omega_m}{A^2}, \quad \tilde{\omega} = \frac{\sum m A_m^2 \omega_m}{\sum m^2 A_m^2}. \quad (27.13)$$

The maximum likelihood estimate of A_m is then given by,

$$\hat{A}_{m_i} = \frac{2}{n}\sum Y_t \cos \omega_{m_i} t. \quad (i = 1,2,\ldots,r) \tag{27.14}$$

The estimate of the modulation index β can now be obtain by the iterative equation

$$\tilde{\beta}_{i+1} = \tilde{\beta}_i - \frac{\sum Y_t \sin(\hat{\omega}_c t + \tilde{\beta}_i \sin \tilde{\omega} t)\sin \tilde{\omega} t}{\sum Y_t \cos(\tilde{\omega}_c t + \tilde{\beta}_i \sin \tilde{\omega} t)\sin^2 \tilde{\omega} t}. \tag{27.15}$$

Note that from (27.6), we have, when $m \neq 0$,

$$\beta = \frac{2mA_m}{A_{m-1} + A_{m+1}}, \tag{27.16}$$

Here we chose minimum $m \neq 0$ such that $J_m(\beta) \neq 0$. This relation can be used to obtain an initial estimate of β, which could then be used to obtain the maximum likelihood estimate from equation (27.15). The distributional properties of these estimates were discussed by Subba Rao and Yar (1984a).

Note that in evaluating (27.13), we in fact use

$$\tilde{\omega}_c = \frac{\sum_i A_{m_i}^2 \omega_{m_i}}{A^2}, \quad \tilde{\omega} = \frac{\sum m_i A_{m_i}^2 \omega_{m_i}}{\sum m_i^2 A_{m_i}^2},$$

i.e. the frequencies $\{\omega_i\}$ at which the peak are found to be significant. This requires the knowledge of $\{m_i\}$, which in turn depends on ω_c and ω which we wish to estimate. Usually ω_c is known, and some idea of ω would help to get estimates of $\{m_i\}$. For example, if peaks at two successive frequencies are found to be significant, the difference will estimate the parameter ω, and this in turn could be used to estimate $\{m_i\}$.

27.4 DETECTION IN THE PRESENCE OF COLOURED GAUSSIAN NOISE

We now extend the discussion to the case where the noise process $\{Z_t\}$ is a zero-mean second-order stationary Gaussian process with a general spectral density function, $f_z(\omega)$ which is absolutely continuous. The signal S_t is as defined earlier in section 27.3. We now have,

$$\mathrm{E}(Iy(\lambda_p)) \simeq \frac{1}{2} \sum_{-m_1}^{m_1} A_m^2 [2\pi F_n(\lambda_p + \omega_m) + 2\pi F_n(\lambda_p - \omega_p)] + f_z(\omega). \tag{27.17}$$

As seen earlier, when $\lambda_p \to \omega_m$, $F_n(\lambda_p - \omega_m) = 0(n)$, therefore there will be sharp peaks in the periodogram at these frequencies. However, it is possible that there can be peaks in the spectral density function of $f_z(\omega)$ which will make it difficult to distinguish between the harmonic components of the signal S_t and the pseudo periods of the time series $\{Z_t\}$. In view of this features, Priestley (1962a) has suggested a modified form of the periodogram, known as Priestley's $P(\lambda)$ statistic, which can be used for the detection of these frequencies. We first note that since $\{Z_t\}$ has an absolutely continuous spectrum, the covariances $R_Z(s)$ will tend to zero as $|s| \to \infty$. But the covariance of S_t given by (27.6) will never die out. This suggests performing a Fourier analysis of the covariances $\{R_y(s), |s| \geqslant K_2\}$, where for some suitably large integer K_2 we will have $R_y(s) \simeq R_S(s)$.

Priestley (1962a) has suggested the statistic

$$P(\lambda) = \frac{1}{2\pi} \sum_{s=-(n-1)}^{n-1} \{\lambda^{(1)}(s) - \lambda^{(2)}(s)\} \hat{R}_y(s) \cos s\lambda \tag{27.18}$$

where $\lambda^{(1)}(s)$, $\lambda^{(2)}(s)$ are appropriately chosen lag windows.

If we choose the lag windows of the form

$$\lambda^{(1)}(s) = \begin{cases} 1 - \dfrac{|s|}{n}, & |s| \leqslant n \\ 0 & \text{otherwise} \end{cases}$$

$$\lambda^{(2)}(s) = \begin{cases} 1 - \dfrac{|s|}{K_2}, & |s| \leqslant K_2 \\ 0 & \text{otherwise} \end{cases} \tag{27.19}$$

then one can show that, when the signal is present, and $A_m \neq 0$

$$\mathrm{E}(P(\lambda)) \simeq \frac{A_m^2}{8\pi}(n - 2K_2) \quad \text{if } \lambda \simeq \omega_m. \tag{27.20}$$

In other words we observe sharp peaks in $P(\lambda)$ at frequencies $\lambda = \omega_c, \omega_c \pm \omega$, $\omega_c \pm 2\omega, \ldots$ if a PM signal is present. This is similar to the procedure suggested in section 27.3 for the case of Z_t for white noise processes, but using the $P(\lambda)$ statistic instead of the periodogram $I(\lambda)$.

Priestley (1962a) has shown that in the null case,

$$\operatorname{var}(P(\lambda)) = \frac{f_Z^2(\lambda)}{n} \sum_{s=-(n-1)}^{n-1} (\lambda^{(1)}(s) - \lambda^{(2)}(s))^2 + O\left(\frac{1}{n}\right),$$

and

$$\operatorname{cov}(P(\lambda), P(\mu)) = 0\left(\frac{1}{n}\right) \quad \text{if } |\lambda - \mu| \gg \frac{1}{K_2}.$$

For the choice of the windows given by (27.19), we have

$$\sum_s (\lambda^{(1)}(s) - \lambda^{(2)}(s))^2 = \left(\frac{2}{3}n - 2K_2 + \frac{2K_2^2}{n}\right) = \wedge\,(n, K_2), \text{ say.}$$

We can test for the significance of the peaks using the procedure suggested by Priestley (1962a). We observe that since $J_m(\beta) \to 0$ as $|m| \to \infty$, the amplitudes A_m tend to decrease and as such the heights of the peaks become smaller for large m. However, as pointed out earlier, for some values of m, A_m can be zero (because $J_m(\beta) = 0$ for some values of m) and this creates problems in the identification procedure which were discussed by Subba Rao and Yar (1984b). We now summarize Priestley's test procedure.

(a) Plot $P(\lambda_p)$ against $\lambda_p = 2\pi p/n$, $p = 0, 1, 2, \ldots [n/2]$. Select all the peaks to be tested for significance.
(b) Divide the frequency range $(0, \pi)$ into the intervals of length $2\pi/k_2$. Chose the first 'suspect' frequency (in order of frequency from 0 to π) and let δ be such that $2\pi_q/k_2 + \delta =$ first suspect frequency, where $0 \leqslant \delta \leqslant 2\pi/k_2$.
(c) Compute

$$J_q = \left(\frac{n}{K_2 \Lambda(n, K_2)}\right)^{1/2} \sum_{j=0}^{q} P\left(\frac{2\pi j}{K_2} + \delta\right) \Big/ \left(\frac{1}{2\pi} \hat{G}(\pi)\right)^{1/2}$$
$$\left(q = 0, 1, \ldots, \left[\frac{K_2}{2}\right]\right) \tag{27.21}$$

where

$$\hat{G}(\pi) = \frac{1}{4\pi}\left[2 \sum_{s=-K_2}^{K_2} \hat{R}_y^2(s) - \sum_{s=-2K_2}^{2K_2} \hat{R}_y^2(s)\right].$$

Test whether

$$\max_{0 < q \leqslant [K_2/2]} J_q \leqslant a,$$

where

$$\lim_{n \to \infty} P\left\{\max_{0 \leqslant q \leqslant [K_2/2]} J_q \leqslant a\right\} = 2\phi(a) - 1,$$

where $\phi(a)$ is the cumulative standard normal distribution function. Let us suppose the peak at $\hat{\omega}_{m_i}$ is significant. Then the amplitude A_{m_i}, can be estimated by

$$\hat{A}_{m_1}^2 = \frac{8\pi}{n - 2K_2} P(\hat{\omega}_{m_1}).$$

To test the significance of the next peak in $P(\lambda)$, we remove the contribution of the first peak from the covariance function $\hat{R}_y(s)$, and we form,

$$\hat{R}_{y,1}^{(s)} = \hat{R}_y(s) - \frac{\hat{A}_{m_1}^2}{2} \cdot \cos \hat{\omega}_{m_1} s,$$

estimate $P(\lambda)$ using $\hat{R}_{y,1}(s)$, and continue the test procedure until all significant peaks are detected. Let us suppose that the peaks at the frequencies $\tilde{\omega}_{m_1}, \hat{\omega}_{m_2}, \ldots, \hat{\omega}_{m_r}$ are found to be significant, and the estimates of the corresponding sequence of amplitudes are $\hat{A}_{m_1}, \hat{A}_{m_2}, \ldots, \hat{A}_{m_r}$. These estimates can then be used to obtain the maximum likelihood estimates of ω_c and ω. Note that the frequency ω_c corresponds to the frequency of the carrier wave and is usually known. In such a situation, if we find the peaks at frequencies $\hat{\omega}_{m_1}$ and $\hat{\omega}_{m_1-1}$ are significant, we can estimate the frequency ω from the relation $\omega_{m_1} - \omega_{(m_1-1)} = \omega$, where $\omega_m = \omega_c + m\omega$.

We can obtain the maximum likelihood estimates of the parameters following the procedure suggested by Hannan (1981). We now assume that the signal S_t is deterministic, so that $\mathrm{E}(Y_t) = S_t$. The problem then reduces to that of estimating the parameters of the mean of Y_t given a sample $(Y_1\ Y_2, \ldots, Y_n)$.

It can be shown that maximization the likelihood function of the Gaussian sample $(Z_1, Z_2, \ldots, Z_n)$ is equivalent to minimizing the criterion

$$Q = \sum_{p=0}^{[n/2]} \frac{I_Z(\lambda_p)}{f_Z(\lambda_p)}. \tag{27.22}$$

Following Hannan (1981) and Brillinger (1980), we can show that this leads to the maximization of

$$\tilde{Q} = \frac{n}{2\pi} \sum_i \left[\frac{A_{m_i}^2}{4} - \frac{n^2}{48} \cdot A_{m_i}^2 (\omega_{m_i} - \hat{\omega}_{m_i})^2 \right] \Big/ f_z(\omega_{m_i}), \tag{27.23}$$

where $\{\hat{\omega}_{m_i}\}$ are the frequencies at which the peaks in $\hat{P}(\lambda)$ have been found to be significant (see Subba Rao and Yar (1984b) for details).

Differentiating (27.23) with respect to ω_c and ω_m, we obtain the normal equations

$$\begin{bmatrix} \sum_i \frac{A_{m_i}^2}{f_z(\omega_{m_i})} & \sum m_i \frac{A_{m_i}^2}{f_z(\omega_{m_i})} \\ \sum_v m_i \frac{A_{m_i}^2}{f_z(\omega_{m_i})} & \sum m_i^2 \frac{A_{m_i}^2}{f_z(\omega_{m_i})} \end{bmatrix} \begin{bmatrix} \tilde{\omega}_c \\ \tilde{\omega} \end{bmatrix} = \begin{bmatrix} \sum A_{m_i}^2 \frac{\hat{\omega}_{m_i}}{f_z(\omega_{m_i})} \\ \sum m_i A_{m_i}^2 \frac{\hat{\omega}_{m_i}}{f_z(\omega_{m_i})} \end{bmatrix}, \tag{27.24}$$

where $\tilde{\omega}_c$ and $\tilde{\omega}$ are the maximum likelihood estimates of ω_c and ω respectively. To compute the above estimates, one needs to known the values of $\{m_i\}$, and this leads to the frequency identification problem described by Subba Rao and Yar (1984b). However, if one has previous knowledge of ω_c preliminary estimate of ω can be found, as above, and it is then possible to find the values of m_i from $(\hat{\omega}_{m_i}\}$. These values of $\{m_i\}$ can then be used to estimate ω_c and ω (see note at the end of section 27.3).

The maximum likelihood estimate of the modulation index β can be obtained by solving the equation

$$\frac{nA}{2}\sum_i \frac{J_m(\beta)J'_m(\beta)}{f_Z(\omega_{m_i})} - \sum_v \frac{J'_m(\beta)}{f_Z(\omega_{m_i})} \quad (\Sigma\, Y_t \cos\omega_{m_i} t) = 0$$

where $J'_m(\beta)$ is the derivative of $J_m(\beta)$ with respect to β. However, the above equation can be solved only iteratively.

As initial estimates of β, A_{m_i} and A, we can use (see section 27.3 for choice of m_i)

$$\hat{\beta} = \frac{2m\hat{A}_{m_i}}{\hat{A}_{m_i+1} + \hat{A}_{m_i-1}}, \quad m \neq 0,$$

$$\hat{A}_{m_i} = \frac{2}{n}\sum_{t=1}^{n} Y_t \cos\omega_{m_i} t,$$

$$\hat{A} = \frac{\hat{A}_{m_i}}{J_{m_i}(\hat{\beta})}, \quad J_{m_i}(\hat{\beta}) \neq 0. \tag{27.25}$$

Once the maximum likelihood estimates of the parameters β and ω have been obtained, the 'estimate' of the signal X_t can be constructed from $\hat{X}_t = \tilde{\beta}\sin\tilde{\omega}t$. In the computations described above it is assumed that $f_Z(\omega)$ is known. If it is unknown an estimate of $f_Z(\omega)$ can be obtained as described below.

Let $\hat{\omega}_{m_1}, \hat{\omega}_{m_2}, \ldots, \hat{\omega}_{m_r}$ be the frequencies at which the peaks in $\hat{P}(\lambda)$ are found to significant. Let $\hat{A}_{m_1}, \hat{A}_{m_2}, \ldots, \hat{A}_{m_r}$ be the corresponding estimates of $A_{m_1}, A_{m_2}, \ldots, A_{m_r}$ computed from (27.25). Obtain an estimate of the noise Z_t by

$$\hat{Z}_t = Y_t - \sum_{i=1}^{r} \hat{A}_{m_i} \cos\hat{\omega}_{m_i} t$$

and its autocovariance function by

$$\hat{R}_{\hat{Z}}(s) = \frac{1}{n}\sum_{t=1}^{n-s} (\hat{Z}_t - \bar{\hat{Z}})(\hat{Z}_{t+s} - \bar{\hat{Z}}), \quad (s = 0, 1, 2, \ldots, n-1).$$

The spectral density function $f_Z(\omega)$ can now be estimated from $\hat{R}_{\hat{Z}}(s)$ using the standard window technique (for example, see Priestley, 1981).

27.5 DEMODULATION BY PHASE LOCKED LOOP (PLL) METHOD

In earlier sections, we parameterized the signal X_t and then considered the problem of estimation of the parameters when the signal S_t is corrupted by noise. The problem of interest is to obtain an 'estimate' of X_t from the past values $(Y_t, Y_{t-1}, \ldots)$ when X_t is a second-order stationary process. A commonly used technique in communcations engineering is the method of

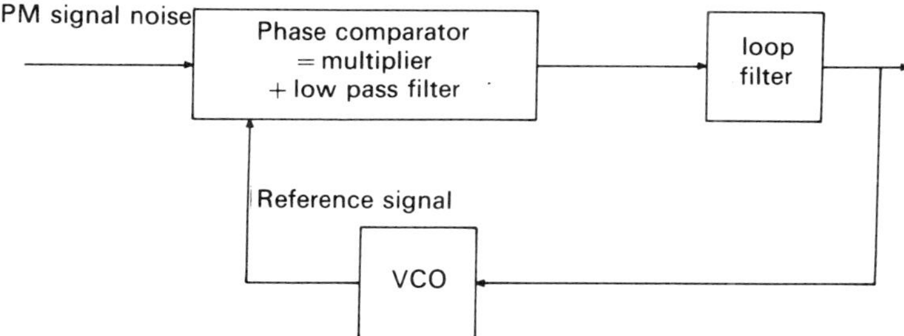

Figure 27.1. Phase locked loop system.

'phase locked loops' (for details see Viterbi, 1965). It is based upon a 'synchronization' or 'rythmic' principle. Here we briefly describe the PLL method. There are three basic components of phase locked loop, namely the phase comparator, the loop filter, and the voltage-controlled oscillator (VCO). A simple form of phase locked loop is shown schematically in Figure 27.1.

Let the signal received at the receiver be

$$\begin{aligned} Y_t &= S_t + Z_t \\ &= A_1 \cos(\omega_c t + \beta_0 + X_t) + Z_t \end{aligned} \tag{27.26}$$

where $\{Z_t\}$ is a sequence of independent, identically distributed normal variables with mean zero and variance σ_Z^2. The method consists of generating a reference signal r_t with some known parameters and an estimate of X_t, say $\hat{X}_t$. Let the reference signal r_t be

$$r_t = A_2 \sin(\omega_c t + \beta_0 + \hat{X}_t), \tag{27.27}$$

generated using the VCO.

Multiplying Y_t and r_t, we obtain $Y_t^* = Y_t r_t = S_t r_t + V_t^*$, where $V_t^* = Z_t r_t$. We then have

$$\begin{aligned} S_t r_t &= A_1 A_2 \cos(\omega_c t + \beta_0 + X_t) \sin(\omega_c t + \beta_0 + \hat{X}_t) \\ &= -\frac{A_1 A_2}{2}[\sin(2\omega_c t + 2\beta_0 + (X_t + \hat{X}_t)) - \sin(X_t - \hat{X}_t)]. \end{aligned}$$

When $S_t r_t$ is passed through a low-pass filter, the first component is effectively filtered out and we are left with essentially,

$$S_t r_t \approx b \sin(\hat{X}_t - X_t)$$

where $b = A_1 A_2/2$. Hence we have,

$$Y_t^* = b \sin(\hat{X}_t - X_t) + V_t^*. \tag{27.28}$$

The object is now to obtain an estimate of $(\hat{X}_t - X_t)$ from $\{Y_t^*, Y_{t-1}^*, \ldots\}$, and this is strictly a nonlinear filtering problem. However, as an approximation we may assume the estimate is linear. Let $e_t = \hat{X}_t - X_t$, and consider a linear estimate of e_t, say $\hat{e}_t = \sum_{u=0}^{\infty} a_u Y^*_{t-u}$. (The order of this summation is known as the order of the phase locked loop.) The sequence $\{a_u\}$ is constructed so that $\mathrm{E}[e_t - \sum_{u=0}^{\infty} a_u Y^*_{t-u}]^2$ is minimized. Suppose we decide on a first-order phase locked loop and find the optimum, a_0. Then the 'estimate' of X_t is obtained from the relation

$$\hat{\hat{X}}_t = \hat{X}_t - a_0 Y^*_t.$$

A new reference signal r_t is now generated from

$$r_t = A_2 \sin(\omega_c t + \beta_0 + \hat{\hat{X}}_t),$$

and the process continues until the difference between $\hat{X}_t$ and $\hat{\hat{X}}_t$ is negligible. The efficiency of these procedures have been widely studied, and an account of this work can be found in the book by Viterbi (1985).

REFERENCES

Abramowitz, M. and Stegun, I.A. (1970) *Handbook of Mathematical Functions*, Dover, New York.

Bartlett, M.S. (1966) *An Introduction to Stochastic Processes with Special Reference to Methods and Applications*, 2nd edn, Cambridge University Press.

Brillinger, D.R. (1980) The comparison of least-squares and third-order periodogram procedures in the analysis of bifrequency, *J. Time Ser. Anal.*, **1**, 95–102.

Fisher, R.A. (1929) Tests of significance in harmonic analysis, *Proc. Roy. Soc., Ser. A*, **125**, 54–59.

Hannan, E.J. (1961) Testing for a jump in the spectral function, *J. Roy. Statist. Soc. B*, **23**, 394–404.

Hannan, E.J. (1981) Nonlinear time series regression, *Jour. Appl. Prob.*, **8**, 767–780.

Priestley, M.B. (1962a) The analysis of stationary processes with mixed spectra—I, *J. Roy. Statist. Soc. B*, **24**, 215–233.

Priestley, M.B. (1962b) The analysis of stationary processes with mixed spectra—II, *J. Roy. Statist. Soc. B*, **24**, 511–529.

Priestley, M.B. (1981) Spectral Analysis and Time Series, vol I, Academic Press, New York.

Subba Rao, T. and Yar, M. (1984a) The demodulation of PM signals in the presence of white Gaussian noise, *IEEE Trans. Comm.*, **COM-32**, 288–297.

Subba Rao, T. and Yar, M. (1984b) Demodulation of PM signals in the presence of coloured Gaussian noise, *Technical Report* No. 167, Department of Mathematics, UMIST.

Viterbi, A.J. (1965) Principles of Coherent Communications, McGraw-Hill, New York.

Walker, A.M. (1971) On the estimation of harmonic components with stationary independent residuals, Biometrika, **58**, 21–36.

Whittle, P. (1952) The simultaneous estimation of a time series harmonic component and covariance structure, *Trab. Estad.*, **3**, 43–57.

Index